Tree Nuts

Composition, Phytochemicals, and Health Effects

NUTRACEUTICAL SCIENCE AND TECHNOLOGY

Series Editor

FEREIDOON SHAHIDI, PH.D., FACS, FAOCS, FCIC, FCIFST, FIAFoST, FIFT, FRSC
University Research Professor
Department of Biochemistry
Memorial University of Newfoundland
St. John's, Newfoundland, Canada

1. Phytosterols as Functional Food Components and Nutraceuticals, *edited by Paresh C. Dutta*
2. Bioprocesses and Biotechnology for Functional Foods and Nutraceuticals, *edited by Jean-Richard Neeser and Bruce J. German*
3. Asian Functional Foods, *John Shi, Chi-Tang Ho, and Fereidoon Shahidi*
4. Nutraceutical Proteins and Peptides in Health and Disease, *edited by Yoshinori Mine and Fereidoon Shahidi*
5. Nutraceutical and Specialty Lipids and their Co-Products, *edited by Fereidoon Shahidi*
6. Anti-Angiogenic Functional and Medicinal Foods, *edited by Jack N. Losso, Fereidoon Shahidi, and Debasis Bagchi*
7. Marine Nutraceuticals and Functional Foods, *edited by Colin Barrow and Fereidoon Shahidi*
8. Tea and Tea Products: Chemistry and Health-Promoting Properties, *edited by Chi-Tang Ho, Jen-Kun Lin, and Fereidoon Shahidi*
9. Tree Nuts: Composition, Phytochemicals, and Health Effects, *edited by Cesarettin Alasalvar and Fereidoon Shahidi*

Tree Nuts

Composition, Phytochemicals, and Health Effects

Edited by
Cesarettin Alasalvar
Fereidoon Shahidi

CRC Press
Taylor & Francis Group
Boca Raton London New York

CRC Press is an imprint of the
Taylor & Francis Group, an **informa** business

CRC Press
Taylor & Francis Group
6000 Broken Sound Parkway NW, Suite 300
Boca Raton, FL 33487-2742

Printed in the United States of America on acid-free paper
10 9 8 7 6 5 4 3 2 1

International Standard Book Number-13: 978-0-8493-3735-2 (Hardcover)

Library of Congress Cataloging-in-Publication Data

Tree nuts : composition, phytochemicals, and health effects / edited by Cesarettin Alasalvar and Fereidoon Shahidi.
p. ; cm. -- (Nutraceutical science and technology ; 8)
Includes bibliographical references and index.
ISBN 978-0-8493-3735-2 (hardcover : alk. paper)
1. Nuts--Composition. 2. Nuts--Health aspects. 3. Functional foods. 4. Food allergy. I. Alasalvar, Cesarettin. II. Shahidi, Fereidoon, 1951- III. Series.
[DNLM: 1. Nuts. 2. Phytosterols. 3. Phytotherapy. WB 431 T786 2008]

TX399.T74 2008
641.3'45--dc22 2008030867

Visit the Taylor & Francis Web site at
http://www.taylorandfrancis.com

and the CRC Press Web site at
http://www.crcpress.com

Contents

Preface

Tree nuts were long perceived as an unhealthy food due to their high fat content and caloric value. However, recent epidemiologic and numerous clinical studies have provided evidence that frequent nut consumption is associated with favorable plasma lipid profiles; reduced risk of coronary heart disease, certain types of cancer, stroke, atherosclerosis, type 2 diabetes, and inflammation; and several other chronic diseases. Despite the fact that tree nuts are high-fat and energy-dense foods, inclusion of the recommended amount (1.5 ounces/day = 42.5 g/day) of most tree nuts in the diet of free-living individuals does not lead to weight gain and renders health benefits. Recent recognition of nuts as "heart-healthy" foods by the Food and Drug Administration has provided a major boost to consumer knowledge of the beneficiary effects of tree nuts on human health.

Tree nuts are highly nutritious and provide macronutrients, micronutrients, and bioactive phytochemicals. Tree nuts and their by-products (skin or testa, hard shell, green leafy cover, hull, and leaf, among others) are rich sources of phytochemicals that possess multifunctional properties such as antioxidant and free radical scavenging activities and anticarcinogenic and antimutagenic effects as well as antiproliferative potential. Therefore, inclusion of bioactive phytochemicals from tree nut by-products into the diet is of great interest as these may provide inexpensive sources of natural antioxidants for use as functional food ingredients and nutraceuticals.

This book examines popular tree nuts (almond, Brazil nut, cashew, hazelnut, macadamia, pecan, pine nut, pistachio, and walnut) together with chestnut and heart nut, and describes each tree nut's compositional and lipid characteristics, phytochemicals, and health effects. Chemical composition of acorn nut, beech nut, coconut, and hickory are also briefly covered. In addition, the book provides a comprehensive assessment of allergens and antiaflatoxigenic activity of phytochemicals and sphingolipids, and health benefits of tree nuts as well as their flavor and volatile compounds. Where available, information on the bioactives and phytochemicals of tree nut by-products is included. Peanut, which is actually a legume, is not discussed in this book as a separate chapter, but where necessary it is used for comparison with tree nuts.

The book is of interest to biochemists, chemists, food scientists, nutritionists, and health professionals. It provides valuable information for senior undergraduate and graduate students as well as scientists from academia, government laboratories, and industry. Moreover, tree nut processors, exporters, and decision makers will obtain maximum benefit from this publication. We are indebted to the participating authors for their hard work and dedication in providing a state-of-the-art contribution and for their authoritative views, resulting from their latest investigations on different aspects of tree nut compositions, phytochemicals, and health effects.

Cesarettin Alasalvar and Fereidoon Shahidi

Editors

Cesarettin Alasalvar, PhD, PMIFT, MIFST, MACS, MISNFF, is the Chief Research Scientist at the Food Institute of TÜBİTAK Marmara Research Center and is also an Associate Professor of Food Science and Engineering. He received his PhD in Food Science and Technology from the University of Lincoln, U.K., in 1994. From 1995 to 1997, he was a postdoctoral fellow at the same university. Dr. Alasalvar was the recipient of a fellowship award from the Japanese Science and Technology Agency (1997–1998). He was then appointed as a senior research fellow/lecturer at Food Research Center and Department of Forensic and Biomedical Sciences at the University of Lincoln (1998–2005).

Dr. Alasalvar is the author of some 50 refereed research papers and book chapters, and editor of three books. He has given over 60 presentations at different scientific conferences. He has delivered invited lecturers, served as a session chairperson and poster-award chair for various international congresses, and organized international symposia. His research interests include different areas of nutraceuticals and functional foods as well as marine foods, tree nuts (particularly hazelnut), natural antioxidants, dietary phytochemicals, food bioactives as well as sensory, flavor, and lipid chemistry.

Dr. Alasalvar serves as the reviewer for several food journals including *Food Chemistry*; *Journal of Agricultural and Food Chemistry*; *Journal of Food Science*; *Molecular Nutrition and Food Research*; *Journal of Functional Foods*; *Journal of Food Lipids*; and *European Journal of Lipid Science and Technology*; among others. He is an editorial board member of *Food Chemistry* since 2007. Dr. Alasalvar served as the chair of the Nutraceuticals and Functional Foods division of the Institute of Food Technologists (IFT) and was a three-time recipient of the outstanding division volunteer award. Dr. Alasalvar organized several international congresses including International Congress on Functional Foods and Nutraceuticals, biannual European Congress on Fish Processing, and International Congress on Food and Nutrition.

Fereidoon Shahidi, PhD, FACS, FAOCS, FCIC, FCIFST, FIAFoST, FIFT, FRSC, is the University Research Professor in the Department of Biochemistry at Memorial University of Newfoundland (MUN). He is also cross-appointed additionally to the Department of Biology, Ocean Sciences Centre, and the aquaculture program at MUN.

Dr. Shahidi is the author of over 600 research papers and book chapters, has authored or edited 48 books, and has given over 400 presentations at scientific conferences. Dr. Shahidi's current research interests include different areas of nutraceuticals and functional foods as well as marine foods and natural antioxidants. Dr. Shahidi serves as the editor-in-chief of the *Journal of Food Lipids* and *Journal of Functional Foods*; an editor of *Food Chemistry* as well as an editorial board member of the *Journal of Food Science*; *Journal of Agricultural and Food Chemistry*; *Nutraceuticals and Food*; and the *International Journal of Food Properties*. He is also on the editorial advisory board of *Inform*.

Dr. Shahidi has received numerous awards, including the 1996 William J. Eva Award from the Canadian Institute of Food Science and Technology in recognition of his outstanding contributions to food science in Canada through research and service. He also received the Earl P. McFee Award from the Atlantic Fisheries Technological Society in 1998, the ADM Award from the American Oil Chemists' Society in 2002, and the Stephen Chang Award from the Institute of Food Technologists in 2005. In 2006, Dr. Shahidi was inducted as the Fellow of the International Academy of Food Science and Technology and was one of the most highly cited (seventh position) and most published (first position) individuals in the area of food, nutrition, and agricultural science for 1996–2006 as listed by ISI; the highly cited standing has now been revised to the fourth position. Dr. Shahidi was the recipient of the Advancement of Agricultural and Food Chemistry Award from the Agricultural and Food Chemistry Division of the American Chemical Society in 2007 and its Distinguished Service Award in 2008. He has served as an executive member of several societies and their divisions and organized many conferences and symposia. Dr. Shahidi served as a member of the Expert Advisory Panel of Health Canada on Standards of Evidence for Health Claims for Foods, the Standards Council of Canada on Fats and Oils, the Advisory Group of Agriculture and Agri-Food Canada on Plant Products, and the Nutraceutical Network of Canada. He was also a member of the Washington-based Council of Agricultural Science and Technology on Nutraceuticals. Dr. Shahidi is currently a member of the Expert Advisory Committee of the Natural Health Products Directorate of Health Canada.

Contributors

Cesarettin Alasalvar
Food Institute
TÜBİTAK Marmara Research Center
Gebze, Kocaeli, Turkey

Joana S. Amaral
Faculdade de Farmácia
Universidade do Porto
Porto, Portugal

Murat Ö. Balaban
Seafood Science Fishery Industrial Technology Center
University of Alaska Fairbanks
Kodiak, Alaska

Susan Bowerman
UCLA Center for Human Nutrition
David Geffen School of Medicine
University of California at Los Angeles
Los Angeles, California

Keith R. Cadwallader
Department of Food Science and Human Nutrition
University of Illinois
Urbana, Illinois

Bruce C. Campbell
Western Regional Research Center
Agricultural Research Service
U.S. Department of Agriculture
Albany, California

Ronald R. Eitenmiller
Department of Food Science and Technology
The University of Georgia
Athens, Georgia

Toshiyuki Fukuda
POLA Laboratories
POLA Chemical Industries, Inc.
Totsuka-ku, Yokohama, Japan

Manohar L. Garg
Nutraceuticals Research Group
School of Biomedical Sciences
The University of Newcastle
Callaghan, Australia

Constance J. Geiger
Division of Nutrition
University of Utah
Salt Lake City, Utah

and

Western Pistachio Association
Fresno, California

David Heber
UCLA Center for Human Nutrition
David Geffen School of Medicine
University of California at Los Angeles
Los Angeles, California

Chi-Tang Ho
Department of Food Science
Rutgers University
New Brunswick, New Jersey

Angela M. Hoffman
Department of Chemistry
The University of Portland
Portland, Oregon

Jong H. Kim
Western Regional Research Center
Agricultural Research Service
U.S. Department of Agriculture
Albany, California

Figen Korel
Department of Food Engineering
Izmir Institute of Technology
Urla, Izmir, Turkey

Harshal H. Kshirsagar
Department of Nutrition, Food and Exercise Sciences
Florida State University
Tallahassee, Florida

Karen Lapsley
Almond Board of California
Modesto, California

Li Li
Department of Chemistry
Changchun Normal College
Changchun, China

Noreen Mahoney
Western Regional Research Center
Agricultural Research Service
U.S. Department of Agriculture
Albany, California

Russell J. Molyneux
Western Regional Research Center
Agricultural Research Service
U.S. Department of Agriculture
Albany, California

Erin K. Monaghan
Department of Nutrition, Food and Exercise Sciences
Florida State University
Tallahassee, Florida

Irene A. Munro
Nutraceutical Research Group
School of Biomedical Sciences
The University of Newcastle
Callaghan, Australia

Beatriz P.P. Oliveira
Faculdade de Farmácia
Universidade do Porto
Porto, Portugal

Ronald B. Pegg
Department of Food Science and Technology
The University of Georgia
Athens, Georgia

Sirima Puangpraphant
Department of Food Science and Human Nutrition
University of Illinois
Urbana, Illinois

Kenneth H. Roux
Department of Biological Science
Florida State University
Tallahassee, Florida

Shridhar K. Sathe
Department of Nutrition, Food and Exercise Sciences
Florida State University
Tallahassee, Florida

Navindra P. Seeram
Department of Biomedical and Pharmaceutical Sciences
College of Pharmacy
University of Rhode Island
Kingston, Rhode Island

Fereidoon Shahidi
Department of Biochemistry
Memorial University of Newfoundland
St. John's, Newfoundland, Canada

Girdhari M. Sharma
Department of Nutrition, Food and Exercise Sciences
Florida State University
Tallahassee, Florida

Margaret Slavin
Department of Nutrition and Food Science
University of Maryland
College Park, Maryland

Di Tan
Department of Food Science
Rutgers University
New Brunswick, New Jersey

Zhuliang Tan
Department of Biochemistry
Memorial University of Newfoundland
St. John's, Newfoundland, Canada

Mauren E. Ternus
International Tree Nut Council Nutrition Research and Education Foundation
Davis, California

Rong Tsao
Food Research Program
Agriculture and Agri-Food Canada
Guelph, Ontario, Canada

Mahesh Venkatachalam
Department of Food Science
Purdue University
West Lafayette, Indiana

Yu Wang
Department of Food Science
Rutgers University
New Brunswick, New Jersey

Subhashinee S.K. Wijeratne
Department of Nutrition and Health Sciences
University of Nebraska–Lincoln
Lincoln, Nebraska

Liangli Yu
Department of Nutrition and Food Science
University of Maryland
College Park, Maryland

Yanjun Zhang
UCLA Center for Human Nutrition
David Geffen School of Medicine
University of California at Los Angeles
Los Angeles, California

Ying Zhong
Department of Biochemistry
Memorial University of Newfoundland
St. John's, Newfoundland, Canada

1 Tree Nuts: Composition, Phytochemicals, and Health Effects: An Overview

Cesarettin Alasalvar and Fereidoon Shahidi

CONTENTS

By definition, tree nuts are dry fruits with generally one seed in which the overall wall becomes hard at maturity. The most popular tree nuts include almond (*Prunus* spp.), Brazil nut (*Bertholletia excelsa*), cashew (*Anacardium occidentale*), hazelnut (*Corylus avellana*), macadamia (*Macadamia* spp.), pecan (*Carya illinoinensis*), pine nut (*Pinus* spp.), pistachio (*Pistacia vera*), and walnut (*Juglans regia*). In addition, acorn nut (*Quercus* spp.), beech nut (*Fagus* spp.), betel nut (*Areca catechu*), chestnut (*Castanea* spp.), coconut (*Cocos nucifera*), heartnut (*Juglan ailanthifolia* var. *cordiformis*), and hickory nut (*Carya* spp.) are also known as edible tree nuts. Peanut or groundnut (*Arachis hypogaea*), which is actually a legume, is not addressed in this book as a separate chapter, but where necessary, it is used to make comparison with tree nuts. Peanut shares a similar nutrient profile with tree nuts [1].

Considering the production of world's most popular tree nuts (Table 1.1), almond ranks first on a global basis with a production of 683,286 MT (shelled), followed by hazelnut (512,200 MT shelled), cashew (394,632 MT shelled), walnut (382,675 MT shelled), and pistachio (445,500 MT unshelled) in 2006–2007. The production of remaining four tree nuts (Brazil nut, macadamia, pecan, and pine nut) is around 132,918 MT (shelled) in the same year. Moreover, world's chestnut production is 1,164,959 MT (unshelled) in 2006 [2]. To the best of our knowledge, little information about the production of acorn nut, beech nut, betel nut, heartnut, and hickory nut is available.

With today's busy lifestyles, tree nuts are convenient, tasty, nutritious, and easy snack that contribute to a healthy lifestyle. They are typically consumed as whole nuts (either raw or roasted or salted) or used as ingredients in a variety of processed foods, especially in spreads, bakery, and confectionary products, among others. Tree nut oils, in particular hazelnut oil, are also used for several purposes such as cooking, salad dressings, and flavoring ingredients, among others [3–6]. In addition, tree nut oils (particularly hazelnut oil) are also components of some skin moisturizers and cosmetic products [7].

Tree nuts are highly nutritious and provide macronutrients (fat, protein, and carbohydrate)[1,8], micronutrients (minerals and vitamins)[1], fat-soluble bioactives (monounsaturated fatty acids [MUFA], polyunsaturated fatty acids [PUFA], monoacylglycerols [MAG], diacylglycerols [DAG], triacylglycerols [TAG], phospholipids, sterol esters, tocopherols, tocotrienols, phytosterols, phytostanols, squalene, terpenoids, sphingolipids, and essential oils, among others) [1,4–6,9–12] (Table 1.2), and phy-

TABLE 1.1
World Tree Nuts Production (Shelled) from 2000 to 2007 (in MT)

	2000/2001	2001/2002	2002/2003	2003/2004	2004/2005	2005/2006	2006/2007
Almond	483,000	403,907	465,335	491,627	504,000	562,500	683,286
Brazil nut	17,575	16,837	16,568	24,630	16,000	22,500	20,100
Cashew	925,000	225,666	242,676	297,108	351,540	525,158	394,632
Hazelnut	392,250	311,840	405,000	329,500	304,250	394,200	512,200
Macadamia	19,135	15,900	15,400	23,456	27,001	25,090	28,030
Pecan	129,500	140,712	118,997	84,069	67,011	88,078	69,908
Pine nut	na[b]	19,735	14,965	34,380	27,975	17,420	14,880
Pistachio[a]	487,179	247,710	444,870	373,600	420,100	438,800	445,500
Walnut	264,775	325,918	336,019	294,463	334,375	368,768	382,675

Source: From International Nut and Dried Fruit Council (INC), World Tree Nuts Production, Calle Boule 2, 3, 43201 Reus, Spain, 2008. With permission.

Note: Data are expressed as metric tons.

[a] Unshelled.

[b] na, not available.

tochemicals (phenolic acids, flavonoids [flavonols, flavones, flavanols or catechin, flavanones, anthocyanidins, isoflavanoids], stilbenes, lignan, hyrolyzable tannins, condensed tannins or proanthocyanidins, carotenoids, alkaloids, coumestan, phytates, and phytoestrogens, among others) [1,8,13–20] (Figure 1.1).

Phytochemicals are defined as nonnutritive, naturally occurring, biologically active, chemically derived compounds found in plant kingdoms. It is estimated that more than 5000 individual phytochemicals have been identified in plant-derived foods and their by-products, but a large percentage still remain unknown and need to be identified before we can fully understand the health benefits of phytochemicals in whole foods [21,22]. Phytochemicals consist of carotenoids, phenolics, organosulfur compounds, nitrogen-containing compounds, and alkoloids. Among them, tree nuts contain most phenolics and some carotenoids (Figure 1.1).

With respect to oxygen radical absorbance capacity (ORAC) and phenolic profiles of tree nuts and peanut, pecan has the highest total ORAC (L-ORAC and H-ORAC) [18], total phenolics [18], and

TABLE 1.2
Lipid Classes (g/100 g Oil) of Tree Nut Oils

Lipid Classes	Almond	Brazil Nut	Hazelnut	Pecan	Pine Nut	Pistachio	Walnut
Oil content (% w/w)	53.5 ± 0.2	68.9 ± 0.3	61.9 ± 0.2	73.4 ± 0.3	75.1 ± 0.2	54.1 ± 0.4	72.5 ± 0.3
Triacylglycerols	98.0 ± 0.1	96.6 ± 0.1	97.6 ± 0.1	96.3 ± 0.1	97.1 ± 0.1	95.8 ± 0.1	97.1 ± 0.1
Sterols	0.25 ± 0.03	0.19 ± 0.02	0.22 ± 0.02	0.28 ± 0.03	0.16 ± 0.01	0.21 ± 0.02	0.28 ± 0.02
Sterol esters	0.05 ± 0.01	0.05 ± 0.01	0.04 ± 0.01	0.07 ± 0.01	0.05 ± 0.01	0.03 ± 0.01	0.09 ± 0.01
Phosphotidylserine	0.32 ± 0.03	0.32 ± 0.02	0.36 ± 0.01	0.47 ± 0.01	0.33 ± 0.03	0.59 ± 0.04	0.46 ± 0.03
Phosphotidylinositol	0.17 ± 0.03	0.10 ± 0.02	0.08 ± 0.02	0.18 ± 0.03	0.19 ± 0.02	0.28 ± 0.03	0.31 ± 0.02
Phosphotidylcholine	0.56 ± 0.01	0.78 ± 0.06	0.48 ± 0.03	0.52 ± 0.04	0.37 ± 0.05	0.68 ± 0.04	0.52 ± 0.04
Phosphatidic acid	nd	nd	0.05 ± 0.01	nd	nd	nd	nd
Sphingolipids	0.63 ± 0.05	0.91 ± 0.02	0.32 ± 0.03	0.55 ± 0.04	0.57 ± 0.03	0.82 ± 0.01	0.68 ± 0.02

Source: From Miraliakbari, H. and Shahidi, F., *J. Food Lipids*, 15, 81, 2008. With permission.

Note: The oils of tree nuts were extracted using chloroform/methanol system; nd, not detected.

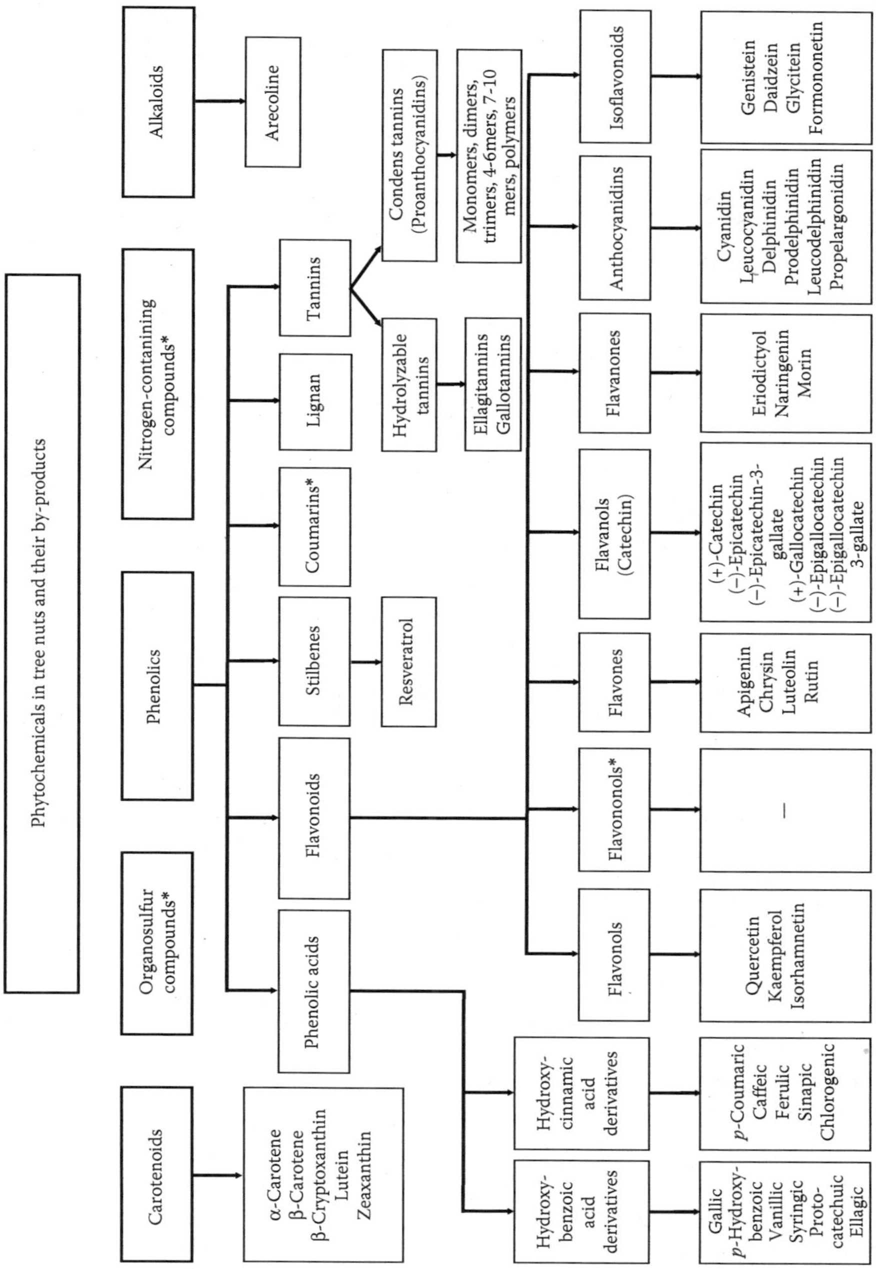

* Not reported in tree nuts and their by-products.

FIGURE 1.1 Phytochemicals in tree nuts and their by-products.

TABLE 1.3
Comparison of ORAC and Phenolics in Tree Nuts and Peanut

Tree Nuts and Peanut	Total ORAC (μmol of TE/g)[a] [18]	Total Phenolics (mg of GAE/g)[b] [18]	Total Flavonoids (mg/100g) [16]	Total Proanthocyanidins (mg/100g) [15,17]	Total Isoflavones (μg/100g) [19]	Total Lignans (μg/100g) [19]	Total Phytoestrogens (μg/100g) [19]
Almond	44.54	4.18	15.24	184.0	18.0	111.7	131.1
Brazil nut	14.19	3.10	nd[c]	nd	na	na	na
Cashew	19.97	2.74	1.98	8.7	22.1	99.4	121.9
Chestnut	na[d]	na	0.02	0.05	21.2	186.6	210.2
Hazelnut	96.45	8.35	11.96	**500.7**	30.2	77.1	107.5
Heartnut	na	na	na	na	na	na	na
Macadamia	16.95	1.56	nd	nd	na	na	na
Pecan	**179.40**	**20.16**	**34.01**	494.1	3.5	25.0	28.8
Pine nut	7.19	0.68	0.49	nd	na	na	na
Pistachio	79.83	16.57	14.37	237.3	**176.9**	**198.9**	**382.5**
Walnut[e]	135.41	15.56	2.71	67.3	53.3	85.7	139.5
Peanut	31.66	3.96	0.66	15.62	7.3	27.1	34.5

Note: Data are expressed as means of edible portion.
[a] Oxygen radical absorbance capacity (ORAC), expressed as micromoles of Trolox equivalents per gram (μmol of TE/g).
[b] Total phenolics, expressed as milligrams of gallic acid equivalents per gram (mg of GAE/g).
[c] nd, not detected.
[d] na, not available.
[e] English walnut.

total flavonoids[16], whereas pistachio has the highest total isoflavones, lignans, and phytoestrogens [19]. In addition, hazelnut contains the highest total proanthocyanidins[15,17] among tree nuts and peanut (Table 1.3). The values for heartnut and other tree nuts (acorn nut, beech nut, betel nut, and hickory) are not available in the literature.

Tree nuts and their by-products (skin or testa, hard shell, green leafy cover, hull, and leaf) are rich sources of phytochemicals that possess multifunctional properties, such as antioxidant and free-radical scavenging activities [14,18,23–31], and possess anticarcinogenic and antimutagenic effects [32] as well as antiproliferative potential [33]. These phytochemicals provide protection against harmful free radicals and are known to reduce the risk of certain types of cancer, coronary heart disease (CHD), stroke, atherosclerosis, osteoporosis, type-2 diabetes, inflammation, endothelial function, sudden death, and other neurodegenerative diseases associated with oxidative stress [14,32,34–43]. Therefore, the inclusion of fat-soluble bioactives and phytochemicals from tree nut by-products into human diet is of great interest as these may provide inexpensive sources of natural antioxidants for use as functional food ingredients and nutraceuticals.

Tree nuts and peanuts are now considered as important components of a healthy diet. Nuts are part of the U.S. Food Guide Pyramid and the Mediterranean Diet Pyramids [44]. Experts recommend eating a variety of foods from five food groups every day in order to obtain the required nutrients. Nuts fall into the "Meat, Poultry, Fish, Dry Beans, Eggs, and Nut Group" and can be eaten every day. The recommended number of servings for this group is two to three times per day [44,45]. Consumption of nuts is on the rise in the United States and elsewhere. A recent survey indicates that 72% of Americans are aware of the health benefits of nuts and 66% plan to increase their consumption of nuts [46]. Tree nut consumption is ~1.52 kg/person/year (shelled basis) in the United States and this amount increased in 2006 as compared to the previous year [47]. A recent European Prospective Investigation into Cancer and Nutrition (EPIC) cohorts from 10 European countries[48] indicated that three most popular tree nuts were walnut, almond, and hazelnut. In general, tree nuts (including almond, Brazil nut, cashew, hazelnut, macadamia, pecan, pine nut, pistachio, and walnut) were more widely consumed in Europe than peanuts or seeds. Based on this study, the data show that on the day of 24 h recall, 4.4% of all subjects consumed tree nuts, 2.3% consumed peanuts, 1.3% consumed nonspecific nuts, and 1.3% consumed seeds. The data show a clear northern (Sweden: mean intake = 0.15 g/d; average portion size = 15.1 g/d) to southern (Spain: mean intake = 2.99 g/d; average portion size = 34.7 g/d) increase in consumption.

The recent recognition of nuts as "heart-healthy" foods by the Food and Drug Administration (FDA) [49] has provided a major boost to the image of nuts. In 2003, the FDA authorized the health claim, which states that "scientific evidence suggests but does not prove that eating 1.5 ounces (~42.5 g) per day of most nuts as part of a diet low in saturated fat and cholesterol may reduce the risk of heart disease." These nuts include peanut and nine tree nuts (almond, Brazil nut, cashew, hazelnut, macadamia, pecan, pine nut, pistachio, and walnut). The FDA evaluated the data and determined that, although there is scientific evidence supporting this claim, the evidence is not conclusive.

Traditionally, tree nuts were perceived as being unhealthy due to their high fat content. However, recent epidemiologic and numerous clinical studies have provided evidence that frequent nut consumption is associated with favorable plasma lipid profiles [50–59] and reduced risk of CHD [42,60–74], various types of cancer [75–81], type-2 diabetes [38,40,82], gallstones [83], inflammation [43,84], endothelial function [43,85,86], and insulin resistance [87], among others. In addition, despite the fact that nuts are fat- and energy-dense foods, adding nuts to habitual diets of free-living individuals does not lead to any weight gain [87–94].

Although tree nuts have several beneficial effects, these together with peanuts are the culprits of IgE-mediated allergic reactions following their ingestion [95–104]. Nuts are a cause of food allergy, which affects approximately 1% of the general population in the United Kingdom and the United States [104]. While the majority of nut allergens are seed storage proteins, other nut allergens are profilins and pathogenesis-related protein homologues, considered as panallergens because of their widespread distribution in plants [104,105]. Allergic reactions to nuts appear to be particularly severe, sometimes even life-threatening, and fatal reactions following their ingestion have been documented [106–111]. Thus, individuals with nut allergy are advised to abstain from nut consumption altogether.

Apart from allergenicity of tree nuts, there is one more serious constraint in the marketing of tree nuts, namely the potential presence of aflatoxins. Contamination of human foods and animal feeds by these compounds has become an important international food safety and trade issue since aflatoxins are considered to be potent carcinogens and teratogens to humans and farm animals [112–116]. The aflatoxin action threshold level for tree nuts is set to 20 ng/g (ppb) by the U.S. FDA [117] for human consumption. The European Community (EC) has a higher concern over the issue of aflatoxin contamination and has set the threshold level for imported nuts intended for direct human consumption or use as an ingredient in foodstuffs at least five times lower, at 4 ng/g [116]. However, the low thresholds for aflatoxin level have significantly increased the probability for rejection of tree nut shipments by the major importing nations of the European countries and Japan. Therefore, research is ongoing by Codex Alimentarius to increase the threshold level. More recently, the Codex Committee on Contaminants in Foods (CCCF) [118] reached a consensus to increase the current maximum total aflatoxin levels to 10 ng/g for ready-to-eat almond, hazelnut, and pistachio, following International Nut and Dried Fruit Council (INC)'s recommendations. Several studies have been conducted to measure the aflatoxin level in tree nuts [113,119–122]. Due to strict safety regulations in the developed countries, aflatoxin contamination rarely occurs in tree nuts at levels above regulatory limits [119]. Research has also shown that aflatoxin production is markedly decreased by the presence of natural antioxidants that occur in tree nuts, including hydrolyzable tannins, flavonoids, and phenolic acids [122].

In conclusion, for all the healthy effects described, tree nuts are considered as natural functional foods and can be used to promote health by their easy incorporation into the usual diet of the population. Tree nuts, which are rich in several vitamins, minerals, unsaturated fatty acids, essential amino acids, soluble fiber, and fat-soluble bioactives, among others, contain numerous phytochemicals that contribute to promoting health and reducing the risk of cardiovascular diseases (CVDs), the greatest cause of morbidity and mortality in the world. However, as complete phytochemical profiles are lacking for most tree nuts, information is limited regarding their bioaccessibility, bioavailability, and metabolism, so further research on this topic is warranted. In addition, current knowledge suggests moderate nut consumption does not pose a threat for weight gain [94]. Allergens and aflatoxin in tree nuts should be considered despite their positive health effects.

REFERENCES

1. U.S. Department of Agriculture (USDA), USDA National Nutrition Database for Standard Reference, Release 20, National Technical Information Service, USDA, Springfield, VA, 2007.
2. Food and Agricultural Organization (FAO), Chestnut Production, Published online at: http://faostat.fao.org (accessed July 10, 2007).
3. Alasalvar, C., Amaral, J.S., and Shahidi, F., Functional lipid characteristics of Turkish Tombul hazelnut (*Corylus avellana* L.), *J. Agric. Food Chem.*, 54, 10177–10183, 2006.
4. Shahidi, F. and Miraliakbari, H., Tree nut oils, in *Bailey's Industrial Oil and Fat Products, Edible Oil and Fat Products: Specialty Oils and Oil Products*, Vol. 3, 6th ed., Shahidi F., Ed., Wiley Interscience, Hoboken, NJ, 2005, pp. 175–193.
5. Shahidi, F. and Miraliakbari, H., Tree nut oils and byproducts: Compositional characteristics and nutraceutical applications, in *Nutraceutical and Specialty Lipids and Their Co-Products*, Shahidi, F., Ed., CRC Press, Taylor & Francis Group, Boca Raton, FL, 2006, pp. 159–168.
6. Miraliakbari, H. and Shahidi, F., Lipid class compositions, tocopherols and sterols of tree nut oils extracted with different solvents, *J. Food Lipids*, 15, 81–96, 2008.
7. Madhaven, N., Final report on the safety assessment of *Corylus avellana* (Hazel) seed oil, *Corylus americana* (Hazel) seed oil, *Corylus avellana* (Hazel) seed extract, *Corylus americana* (Hazel) seed extract, *Corylus avellana* (Hazel) leaf extract, *Corylus americana* (Hazel) leaf extract, and *Corylus rostrata* (Hazel) leaf extract, *Int. J. Toxicol*, 20, 15–20, 2001.
8. Venkatachalam, M. and Sathe, S.K., Chemical composition of selected edible nut seeds, *J. Agric. Food Chem.*, 54, 4705–4714, 2006.
9. Maguire, L.S., O'Sullivan, S.M., Galvin, K., O'Connor, T.P., and O'Brien, N.M., Fatty acid profile, tocopherol, squalane and phytosterol content of walnuts, almonds, peanuts, hazelnuts and the macadamia nut, *Int. J. Food Sci. Nutr.*, 3, 171–178, 2004.
10. Phillips, K.M., Ruggio, D.M., and Ashraf-Khorassani, M., Phytosterol composition of nuts and seeds commonly consumed in the United States, *J. Agric. Food Chem.*, 53, 9436–9445, 2005.
11. Fang, F., Ho, C.-T., Sang, S., and Rosen, R.T., Determination of sphingolipids in nuts and seeds by a single quadrupole liquid chromatography–mass spectrometry method, *J. Food Lipids*, 12, 327–343, 2005.
12. Ryan, E., Galvin, K., O'Connor, T.P., Maguire, A.R., and O'Brien, N.M., Fatty acid profile, tocopherol, squalene and phytosterol content of brazil, pecan, pine, pistachio and cashew nuts, *Int. J. Food Sci. Nutr.*, 57, 219–228, 2006.
13. Ghelardini, C., Galeotti, N., Lelli, C., and Bartolini, A., M1 receptor activation is a requirement for arecoline analgesia, *IL Farmaco.*, 56, 383–385, 2001.
14. Shahidi, F. and Naczk, M., *Phenolics in Food and Nutraceuticals*, CRC Press, Boca Raton, FL, 2004.
15. U.S. Department of Agriculture (USDA), Database for the Proanthocyanidin Content of Selected Foods, August 2004, Published online at: http://www.nal.usda.gov/fnic/foodcomp (accessed September 28, 2006).
16. U.S. Department of Agriculture (USDA), Database for the Flavonoid Content of Selected Foods (Release 2), August 2006, Published online at: http://www.ars.usda.gov/nutrientdata (accessed September 28, 2006).
17. Gu, L., Kelm, M.A., Hammerstone, J.F., Beecher, G., Holden, J., Haytowitz, D., Gebhardt, S., and Prior, R.L., Concentrations of proanthocyanidins in common foods and estimations of normal consumption, *J. Nutr.*, 134, 613–617, 2004.
18. Wu, X., Beecher, G.R., Holden, J.M., Haytowitz, D.B., Gebhardt, S.E., and Prior, R.L., Lipophilic and hydrophilic antioxidant capacities of common foods in the United States, *J. Agric. Food Chem.*, 52, 4026–4037, 2004.
19. Thompson, L.U., Boucher, B.A., Liu, Z., Cotterchio, M., and Kreiger, N., Phytoestrogen content of foods consumption in Canada, including isoflavones, lignans, and coumestan, *Nutr. Cancer.*, 54, 184–201, 2006.
20. Oliver Chen, C.-Y. and Blumberg, J., Phytochemical composition of nuts, *Asia Pac. J. Clin. Nutr.*, 17, 329–332, 2008.
21. Lui, R.H.J., Health benefits of fruits and vegetables are from additive and synergestic combination of phytochemicals, *Am. J. Clin. Nutr.*, 78, 517S–520S, 2003.
22. Lui, R.H.J., Potential synergy of phytochemicals in cancer prevention: mechanism of action, *J. Nutr.*, 2004, 134, 3479S–3485S.

23. Siriwardhana, S.S.K.W. and Shahidi, F., Antiradical activity of extracts of almond and its by-products, *J. Am. Oil Chem. Soc.*, 79, 903–908, 2002.
24. Sang, S., Lapsley, K., Jeong, W-S., Lachance, P.A., Ho, C-T., and Rosen, R.T., Antioxidative phenolic compounds isolated from almond skins (*Prunus amygdalus* Batsch), *J. Agric. Food Chem.*, 50, 2459–2463, 2002.
25. Takeoka, G.R. and Dao, L.T., Antioxidant constituents of almond [*Prunus dulcis* (Mill.) D.A. Webb] hulls, *J. Agric. Food Chem.*, 51, 496–501, 2003.
26. Goli, A.H., Barzegar, M., and Sahari, M.A., Antioxidant activity and total phenolic compounds of pistachio (*Pistachia vera*) hull extracts, *Food Chem.*, 92, 521–525, 2005.
27. Wijeratne, S.S.K., Amarowicz, R., and Shahidi, F., Antioxidant activity of almonds and their by-products in food model systems, *J. Am. Oil Chem. Soc.*, 83, 223–230, 2006.
28. Wijeratne, S.S.K., Abou-Zaid, M.M., and Shahidi, F., Antioxidant polyphenols in almond and its coproducts, *J. Agric. Food Chem.*, 54, 312–318, 2006.
29. Alasalvar, C., Karamać, M., Amarowicz, R., and Shahidi, F., Antioxidant and antiradical activities in extracts of hazelnut kernel (*Corylus avellana* L.) and hazelnut green leafy cover, *J. Agric. Food Chem.*, 54, 4826–4832, 2006.
30. Fukuda, T., Polyphenols from walnuts: Structures and functions, presented at the Institute of Food Technologists Annual Meeting and Food Expo, Orlando, FL, June 24–28, 2006, Abstract 029-06.
31. Shahidi, F., Alasalvar, F., and Liyana-Pathirana, C.M., Antioxidant phytochemicals in hazelnut kernel (*Corylus avellana* L.) and hazelnut byproducts, *J. Agric. Food Chem.*, 55, 1212–1220, 2007.
32. Surh, Y.-J., Cancer chemoprevention with dietary phytochemicals, *Nat. Rev. Cancer*, 3, 768–780, 2003.
33. Yang, J., Halim, L., and Liu, R.H., Antioxidant and antiproliferative activities of common nuts, presented at the Institute of Food Technologists Annual Meeting and Food Expo, New Orleans, LA, July 16–20, 2005, Abstract 035-05.
34. Ness, A.R. and Powles, J.W., Fruit and vegetables, and cardiovascular disease: A review, *Int. J. Epidemiol.*, 26, 1–13, 1997.
35. Joseph, J.A., Shukitt-Hale, B., Denisova, N.A., Bielinski, D., Martin, A., McEwen, J.J., and Bickford, P.C., Reversals of age-related declines in neuronal signal transduction, cognitive, and motor behavioral deficits with blueberry, spinach, or strawberry dietary supplementation, *J. Neurosci.*, 19, 8114–8121, 1999.
36. Mazza, G., Fukumoto, L., Delaquis, P., Girard, B., and Ewert, B., Anthocyanins, phenolics, and color of Cabernet Franc, Merlot, and Pinot Noir wines from British Colombia, *J. Agric. Food Chem.*, 47, 4009–4017, 1999.
37. Hertog, M.G.L., Feskens, E.J.M., Hollman, P.C.H., Katan, M.B., and Kromhout, D., Dietary antioxidant flavonoids and risk of coronary heart disease: The Zutphen Elderly Study, *Lancet*, 342, 1007–1011, 1993.
38. Jiang, R., Manson, J.E., Stampfer, M.J., Liu, S., Willett, W.C., and Hu, F.B., Nut and peanut butter consumption and risk of type 2 diabetes in women, *JAMA*, 288, 2554–2560, 2002.
39. Watson, R.R., *Functional Foods and Nutraceuticals in Cancer Prevention*, Blackwell Publishing, Oxford, United Kingdom, 2003.
40. Tapsell, L.C., Gillen, L.J., Patch, C.S., Batterham, M., Owen, A., Bare, M., and Kennedy, M., Including walnuts in a low-fat/modified-fat diet improves HDL cholesterol-to-total cholesterol ratios in patients with type 2 diabetes, *Diabetes Care*, 27, 2777–2783, 2004.
41. Clifford, M.N. and Brown, J.E., Dietary flavonoids and health-broadening the perspective, in *Flavonoids: Chemistry, Biochemistry and Applications*, Andersen, Ø.M. and Markham, K.R., Eds., CRC Press, Taylor & Francis Group, Boca Raton, FL, 2005, pp. 319–370.
42. Kelly Jr, J.H. and Sabaté, J., Nuts and coronary heart disease: An epidemiological perspective, *Br. J. Nutr.*, 96 (Suppl. 2), 61S–67S, 2006.
43. Salas-Salvadó, J., Casas-Agustench, P., Murphy, M.M., López-Uriarte, P., and Bulló, M., The effect of nuts on inflammation, *Asia Pac. J. Clin.Nutr.*, 17 (Suppl.1), 333–336, 2008.
44. U.S. Department of Agriculture (USDA), Steps to a Healthier You, Published online at: http://www.mypyramid.gov (accessed April 17, 2006).
45. Brufau, G., Boatella, J., and Rafecas, M., Nuts: Source of energy and macronutrients, *Br. J. Nutr.*, 96 (Suppl. 2), 24S–28S, 2006.
46. *Hazelnut Consumer Attitude and Usage Tracking Wave IV*, The Hazelnut Council with TRD Frameworks, New York, 2005.
47. Pollack, S. and Perez A., Fruit and Tree Nuts Situation and Outlook Yearbook, Economic Research Service, 2007, Published online at: http://www.ers.usda.gov/publications/FTS/2007/Yearbook/FTS2007.pdf (accessed December 19, 2007).

48. Jenab, M., Sabaté, J., Slimani, N., Ferrari, P., Mazuir, M., Casagrande, C., Deharveng, G., Tjonneland, A., Olsen, A., Overvad, K., Boutron-Ruault, M.C., Clavel-Chapelon, F., Boeing, H., Weikert, C., Linseisen, J., Rohrmann, S., Trichopoulou, A., Naska, A., Palli, D., Sacerdote, C., Tumino, R., Mattiello, A., Pala, V., Bueno-de-Mesquita, H.B., Ocke, M.C., Peeters, P.H., Engeset, D., Skeie, G., Jakszyn, P., Ardanaz, E., Quiros, J.R., Chirlaque, M.D., Martinez, C., Amiano, P., Berglund, G., Palmqvist, R., Guelpen, B., Bingham, S., Key, T., and Riboli, E., Consumption and portion sizes of tree nuts, peanuts and seeds in the European Prospective Investigation into Cancer and Nutrition (EPIC) cohorts from 10 European countries, *Br. J. Nutr.*, 96 (Suppl. 2), 12S–23S, 2006.
49. Food and Drug Administration (FDA), Qualified Health Claims: Letter of Enforcement Discretion-Nuts and Coronary Heart Disease, Docket No 02P-0505, Food and Drug Administration, Washington, DC, 2003.
50. Kris-Etherton, P.M., You-Poth, S., Sabaté, J., Ratcliffe, H.E., Zhao, G., and Etherton, T.D., Nuts and their bioactive constituents: Effects on serum lipids and other factors that affect disease risk, *Am. J. Clin. Nutr.*, 70, 504S–511S, 1999.
51. Zambón, D., Sabaté, J., Muñoz, S., Campero, B., Casals, E., Merlos, M., Laguna, J.C., and Ros, E., Substituting walnuts for monounsaturated fat improves the serum lipid profile of hypercholesterolemic men and women: A randomized crossover trial, *Ann. Intern. Med.*, 132, 538–546, 2000.
52. Rajaram, S., Burke, K., Connell, B., Myint, T., and Sabaté, J., A monounsaturated fatty acid-rich pecan-enriched diet favorably alters the serum lipid profile of healthy men and women, *J. Nutr.*, 131, 2275–2279, 2001.
53. Hyson, D.A., Schneeman, B.O., and Davis, P.A., Almonds and almond oil have similar effects on plasma lipids and LDL oxidation in healthy men and women, *J. Nutr.*, 132, 703–707, 2002.
54. Garg, M.L., Blake, R.J., and Wills, R.B.H., Macadamia nut consumption lowers plasma total and LDL cholesterol levels in hypercholesterolemic men, *J. Nutr.*, 133, 1060–1063, 2003.
55. Mukuddem-Petersen, J., Oosthuizen, W., and Jerling, J.C., A systematic review of the effects of nuts on blood lipid profiles in humans, *J. Nutr.*, 135, 2082–2089, 2005.
56. Griel, A.E. and Kris-Etherton, P.M., Tree nuts and the lipid profile: A review of clinical studies, *Br. J. Nutr.*, 96 (Suppl. 2), 68S–78S, 2006.
57. Kocyigit, A., Koylu, A.A., and Keles, H., Effects of pistachio nuts consumption on plasma lipid profile and oxidative status in healthy volunteers, *Nutr. Metab. Cardiovasc. Dis.*, 16, 202–209, 2006.
58. Mercanlıgil, S.M., Arslan, P., Alasalvar, C., Okut, E., Akgül, E., Pınar, A., Geyik, P.Ö., Tokgözoğlu, L., and Shahidi, F., Effects of hazelnut-enriched diet on plasma cholesterol and lipoprotein profiles in hypercholesterolemic adult men, *Eur. J. Clin. Nutr.*, 61, 212–220, 2007.
59. Griel, A.E., Cao, Y., Bagshaw, D.D., Cifelli, A.M., Holub, B., and Kris-Etherton, P.M., A macadamia nut-rich diet reduces total and LDL-cholesterol in mildly hypercholesterolemic men and women, *J. Nutr.*, 138, 761–767, 2008.
60. Fraser, G.E., Sabaté, J., Beeson, W.L., and Strahan, T.M., A possible protective effect of nut consumption on risk of coronary heart disease. The adventist health study, *Arch. Intern. Med.*, 152, 1416–1424, 1992.
61. Dreher, M.L., Maher, C.V., and Kearney, P., The traditional and emerging role of nuts in healthful diets, *Nutr. Rev.*, 54, 241–245, 1996.
62. Hu, F.B., Stampfer, M.J., Manson, J.E., Rimm, E.B., Colditz, G.A., Rosner, B.A., Speizer, F.E., Hennekens, C.H., and Willett, W.C., Frequent nut consumption and risk of coronary heart disease in women: Prospective cohort study, *Br. Med. J.*, 317, 1341–1345, 1998.
63. Hu, F.B. and Stampfer, M.J., Nut consumption and risk of coronary heart disease: A review of epidemiologic evidence, *Curr. Ather. Rep.*, 1, 204–209, 1999.
64. Colquhoun, D.M., Food for prevention of coronary heart disease: Beyond the low fat, low cholesterol diet, *Asia Pac. J. Clin. Nutr.*, 9 (Suppl.1), 86S–90S, 2000.
65. de Lorgeril, M., Salen, P., Laporte, F., Boucher, F., and De Leiris, J., Potential use of nuts for the prevention and treatment of coronary heart disease: From natural to functional foods, *Nutr. Metab. Cardiovasc. Dis.*, 11, 362–371, 2001.
66. Ellsworth, J.L., Kushi, L.H., and Folsom, A.R., Frequent nut intake and risk of death from coronary heart disease and all causes in postmenopausal women: The Iowa women's health study, *Nutr. Metab. Cardiovasc. Dis.*, 11, 372–377, 2001.
67. Sabaté, J., Radak, T., and Brown Jr, J., The role of nuts in cardiovascular disease prevention, in *Handbook of Nutraceuticals and Functional Foods*, Wildman, R.E.C., Ed., CRC Press, Boca Raton, FL., 2001, pp. 477–495.
68. Albert, C.M., Gaziano, J.M., Willett, W.C., and Manson, J.E., Nut consumption and decreased risk of sudden cardiac death in the Physicians' Health Study, *Arch. Intern. Med.*, 162, 1382–1387, 2002.

69. Hu, F.B., Plant-based foods and prevention of cardiovascular disease: An overview, *Am. J. Clin. Nutr.*, 78, 544S–551S, 2003.
70. Fraser, G.E., Nut consumption, lipids, and risk of a coronary event, *Asia Pac. J. Clin. Nutr.*, 9 (Suppl.1), 28S–32S, 2000.
71. Kris-Etherton, P.M., Zhao, G., Binkoski, A.E., Coval, S.M., and Etherton, T.D., The effects of nuts on coronary heart disease risk, *Nutr. Rev.*, 59, 103–111, 2001.
72. Nash, S.D. and Westpfal, M., Cardiovascular benefits of nuts, *Am. J. Cardiol.*, 95, 963–965, 2005.
73. Ros, E. and Mataix, J., Fatty acid composition of nuts—Implications for cardiovascular health, *Br. J. Nutr.*, 96 (Suppl. 2), 29S–35S, 2006.
74. Salas-Salvadó, J., Bulló, M., Pérez-Heras, A., Ros, E., Dietary fibre, nuts and cardiovascular diseases, *Br. J. Nutr.*, 96 (Suppl. 2), 45S–51S, 2006.
75. Kune, S., Kune, G.A., and Watson, L.F., Case-control study of dietary etiological factors: The Melbourne colorectal cancer study, *Nutr. Cancer.*, 9, 21–42, 1987.
76. Mills, P.K., Beeson, W.L., Phillips, R.L., and Fraser, G.E., Cohort study of diet, lifestyle, and prostate cancer in Adventist men, *Cancer*, 64, 598–604, 1989.
77. Hebert, J.R., Hurley, T.G., Olendzki, B.C., Teas, J., Ma, Y., and Hampl, J.S., Nutritional and socioeconomic factors in relation to prostate cancer mortality: A cross-national study, *J. Natl. Cancer Inst.*, 90, 1637–1647, 1998.
78. Singh, P.N. and Fraser, G.E., Dietary risk factors for colon cancer in a low-risk population, *Am. J. Epidemiol.*, 148, 761–774, 1998.
79. Jain, M.G., Hislop, G.T., Howe, G.R., and Ghadirian, P., Plant foods, antioxidants, and prostate cancer risk: Findings from case–control studies in Canada, *Nutr. Cancer,* 34, 173–184, 1999.
80. Jenab, M., Ferrari, P., Slimani, N., et al., Association of nut and seed intake with colorectal cancer risk in the European prospective investigation into cancer and nutrition, *Cancer Epidemiol. Biomarkers Prev.*, 13, 1595–1603, 2004.
81. González, C.A. and Salas-Salvadó, J., The potential of nuts in the prevention of cancer, *Br. J. Nutr.*, 96 (Suppl. 2), 87S–94S, 2006.
82. Garg, A., High monounsaturated-fat diets for patients with diabetes mellitus: A meta-analysis, *Am. J. Clin. Nutr.*, 67, 577S–582S, 1998.
83. Tsai, C.-J., Leitzmann, M.F., Hu, F.B., Willett, W.C., and Giovannucci, E.L., Frequent nut consumption and decreased risk of cholecystectomy in women, *Am. J. Clin. Nutr.*, 80, 76–81, 2004.
84. Jiang, R., Jacobs Jr, D.R., Mayer-Davis, E., Szklo, M., Herrington, D., Jenney, N.S., Kronmal, R., and Barr, R.G., Nut and seed consumption and inflammatory markers in the multi-ethnic study of atherosclerosis, *Am. J. Epidemiol.*, 163, 222–231, 2006.
85. Ros, E., Núñez, I., Pérez-Heras, A., Serra, M., Gilabert, R., Casals, E., and Deulofeu, R., A walnut diet improves endothelial function in hypercholesterolemic subjects: A randomized crossover trial, *Circulation*, 109, 1609–1614, 2004.
86. Cortés, B., Núñez, I., Cofán, M., Gilabert, R., Pérez-Heras, A., Casals, E., Deulofeu, R., and Ros, E., Acute effects of high-fat meals enriched with walnuts or olive oil on postprandial endothelial function. *J. Am. Coll. Cardiol.*, 48, 1666–1671, 2006.
87. Wien, M.A., Sabaté, J.M., Iklé, D.N., Coles, S.E., Kandeel, F.R., Almonds vs complex carbohydrates in a weight reduction program, *Int. J. Obes. Relat. Metab. Disord.*, 27, 1365–1372, 2003.
88. García-Lorda, P., Rangil, I.M., and Salas-Salvadó, J., Nut consumption, body weight and insulin resistance, *Eur. J. Clin. Nutr.*, 57 (Suppl. 1), 8S–11S, 2003.
89. Piers, L.S., Walker, K.Z., Stoney, R.M., Soares, M.J., and O'Dea, K., Substitution of saturated with monounsaturated fat in a 4-week diet affects body weight and composition of overweight and obese men, *Br. J. Nutr.*, 90, 717–728, 2003.
90. Sabaté, J., Nut consumption and body weight, *Am. J. Clin. Nutr.*, 78, 647S–650S, 2003.
91. St-Onge, M.P., Dietary fats, teas, dairy, and nuts: Potential functional foods for weight control? *Am. J. Clin. Nutr.*, 81, 7–15, 2005.
92. Rajaram, S. and Sabaté, J., Nuts, body weight and insulin resistance, *Br. J. Nutr.*, 96 (Suppl. 2), 79S–86S, 2006.
93. Sabaté, J., Nut consumption and change in weight: The weight of the evidence, *Br. J. Nutr.*, 98, 456–457, 2007.
94. Mattes, R.D., The energetics of nut consumption, *Asia Pacific J. Nutr.*, 17, 337–339, 2008.
95. Ortolani, C., Ballmer-Weber, B.K., Hansen, K.S., Ispano, M., Wüthrich, B., Bindslev-Jensen, C., Ansaloni, R., Vannucci, L., Pravettoni, V., Scibilia, J., Poulsen, L.K., and Pastorello, E.A., Hazelnut allergy: A double-blind, placebo-controlled food challenge multicenter study, *J. Allergy Clin. Immunol.*, 105, 577–581, 2000.

96. Burks, W., Helm, R., Stanley, S., and Bannon, G.A., Food allergens, *Curr. Opin. Allergy Clin. Immunol.*, 1, 243–248, 2001.
97. Holzhauser, T., Stephan, O., and Vieths, S., Detection of potentially allergenic hazelnut (*Corylus avellana*) residues in food: A comparative study with DNA PCR-ELISA and protein sandwich-ELISA, *J. Agric. Food Chem.*, 50, 5808–5815, 2002.
98. Pastorello, E.A., Vieths, S., Pravettoni, V., Farioli, L., Trambaioli, C., Fortunato, D., Lüttkopf, D., Calamari, M., Ansaloni, R., Scibilia, J., Ballmer-Weber, B.K., Poulsen, L.K., Wüthrich, B., Hansen, K.S., Robino, A.M., Ortolani, C., and Conti, A., Identification of hazelnut major allergens in sensitive patients with positive double-blind, placebo-controlled food challenge results, *J. Allergy Clin. Immunol.*, 109, 563–570, 2002.
99. Rance, F., Bidat, E., Bourrier, T., and Sabourand, D., Cashew allergy: Observations of 42 children without associated peanut allergy, *Allergy*, 58, 1311–1314, 2003.
100. Roux, K.H., Teuber, S.S., Robotham, J.M., and Sathe, S.K., Detection and stability of the major almond allergens in foods, *J. Agric. Food Chem.*, 49, 2131–2136, 2001.
101. Roux, K.H., Teuber, S.S., and Sathe, S.K., Tree nut allergens, *Int. Arch. Allergy Immuol.*, 131, 234–244, 2003.
102. Teuber, S.S., Sathe, S.K., Peterson, W.R., and Roux, K.H., Characterization of the soluble allergenic proteins of cashew nut (*Anacardium occidentale* L.), *J. Agric. Food Chem.*, 50, 6543–6549, 2002.
103. Teuber, S.S., Comstock, S.S., Sathe, S.K., and Roux, K.H., Tree nut allergy, *Curr. Allergy Asthma Rep.*, 3, 54–61, 2003.
104. Crespo, J.F., James, J.M., Fernandez-Rodriguez, C., and Rodriguez, J., Food allergy: Nuts and tree nuts, *Br. J. Nutr.*, 96 (Suppl. 2), 95S–102S, 2006.
105. Midoro-Horiuti, T., Brooks, E.G., and Goldblum, R.M., Pathogenesis-related proteins of plants as allergens, *Ann. Allergy Asthma Immunol.*, 87, 261–271, 2001.
106. Bock, S.A. and Atkins, F.M., The natural history of peanut allergy, *J. Allergy Clin. Immunol.*, 83, 900–904, 1989.
107. Ewan, P.W., Clinical study of peanut and nut allergy in 62 consecutive patients: New features and associations, *Br. Med. J.*, 312, 1074–1078, 1996.
108. Kay, A.B., Allergy and allergic diseases: First of two parts, *N. Eng. J. Med.*, 344, 30–37, 2001.
109. Sicherer, S.H., Burks, A.W., and Sampson, H.A., Clinical features of acute allergic reactions to peanut and tree nuts in children, *Pediatrics*, 102, 1–6, 1998.
110. Eigenmann, P.A., Future therapeutic options in food allergy, *Allergy*, 58, 1217–1223, 2003.
111. Tariq, S.M., Stevens, M., Matthews, S., Ridout, S., Twiselton, R., and Hide, D.W., Cohort study of peanut and tree nut sensitisation by age of 4 years, *Br. Med. J.*, 313, 514–517, 1996.
112. Henry, S.H., Bosch, F.X., Troxell, T.C., and Bolger, P.M., Reducing liver cancer: Global control of aflatoxin, *Science*, 286, 2453–2454, 1999.
113. Campbell, B.C., Molyneux, R.J., and Schatzki, T.F., Current research on reducing pre- and post-harvest aflatoxin contamination of U.S. almond, pistachio, and walnut, *Toxin Rev.*, 22, 225–266, 2003.
114. Kuiper-Goodman, T., Risk assessment and risk management of mycotoxins in food, in *Mycotoxins in Food-Detection and Control*, Magan, N. and Olsen, M., Eds., Woodhead Publishing Limited, Cambridge, United Kingdom, 2004, pp. 3–31.
115. Wang, J.-S. and Tang, L., Epidemiology of aflatoxin exposure and human liver cancer, *J. Toxicol. Toxin Rev.*, 23, 249–271, 2004.
116. European Commission (EC), Commission Regulation No. 1881/2006 of 19 December 2006—Setting maximum levels for certain contaminants in foodstuffs, *Off. J. Eur. Commun.* L 364/5–24, 2006.
117. Food and Drug Administration (FDA), Compliance Policy Guides Manual, Sec. 555.400, 268; Sec. 570.500, 299, 1996.
118. Codex Committee on Contaminants in Foods (CCCF), Outcomes of the 2nd Session of the Codex Committee on Contaminants in Foods, The Hague, Netherlands, March 31–April 4, 2008.
119. Food Standard Agency (FSA), Aflatoxins in Nuts Survey, June 2004, Published online at: http://www.food.gov.uk/news/newsarchive/2004/jun/aflatoxinsurvey (accessed December 5, 2007).
120. Saleemullah, Iqbal, A., Khalil, I.A., and Shah, H., Aflatoxin contents of stored and artificially inoculated cereals and nuts, *Food Chem.*, 98, 699–703, 2006.
121. Chun, H.S., Kim, H.J., Ok, H.E., Hwang, J.-B., and Chung, D.-H., Determination of aflatoxin levels in nuts and their products consumed in South Korea, *Food Chem.*, 102, 385–391, 2007.
122. Molyneux, R.J., Mahoney, N., Kim, J.H., and Campbell, B.C., Mycotoxin in edible nuts, *Int. J. Food Microbiol.*, 119, 72–78, 2007.
123. International Nut and Dried Fruit Council (INC), World Tree Nuts Production, Calle Boule 2, 3, 43201 Reus, Spain, 2008.

2 Chemical Composition of Edible Nut Seeds and Its Implications in Human Health

Shridhar K. Sathe, Erin K. Monaghan, Harshal H. Kshirsagar, and Mahesh Venkatachalam

CONTENTS

2.1 INTRODUCTION

Historically, edible nut seeds have enjoyed universal acceptance as an important component of the human food supply. Among the edible nut seeds, peanut (a legume that is also used as an oilseed) consumption in the United States has steadily increased from 2.45 kg (1967) to 2.86 kg (2003) per capita/annum. During the same period, tree nut (excluding coconut) consumption also increased from 0.77 kg to 1.45 kg, with almonds (0.51 kg) leading the way [1]. Almond, Brazil nut, cashew, chestnut, coconut, hazelnut, macadamia, pecan, pine nut, pistachio, and walnut are some of the commercially important cultivated tree nuts. Tree nuts such as hickory that include walnuts and pecans growing in the wild, and several other minor nut seeds (such as nangai, tiger, beech, and acorn) may be mainly grown and consumed in some regions. Globally, tree nuts are valued for their nutrient content and sensory properties. Nutty, sweet, and mellow flavors and soft yet crunchy textural attributes unique to each nut seed are two of the most important sensory properties that are valued by the consumers and therefore typically determine the acceptance of a certain nut type.

Edible nut seeds are a rich source of lipids and therefore they are calorie-dense (~500–700 kcal/100 g edible portion). In addition to lipids, nut seeds are a good source of quality proteins, certain minerals (such as selenium in Brazil nut) and vitamins (such as vitamin E in almond and hazelnut), and depending

TABLE 2.1
Edible Nut Production and Consumption (Shelled Basis)

Nut Seed	World Production (× 1000 MT)	Major Producer (%)	United States	
			Production (%)	Consumption[a]
Peanut (*Arachis hypogea*)	34,856	China (40)	6	6.7
Total tree nuts	10,179	China (16)	16	3.36
Acorn nut (*Quercus macrocarpa*)	na	na	na	na
Almond (*Prunus dulcis*)	2,065	United States (41)	41	1.01
Beechnut (*Fagus grandifolia*)	na	na	na	na
Brazil nut (*Bertholletia excelsa*)	5,285	Côte d'Iuoire (89)	na	T[b]
Cashew (*Anacardium occidentale*)	3,186	India (20)	na	T[b]
Chestnut (*Castanea sativa*)	1,223	China (73)	na	na
Coconut (*Cocos nucifera*)	54,716	Indonesia (30)	na	0.54
Hazelnut (*Corylus avellana* L.)	777	Turkey (67)	3	0.08
Hickory nut (*Carya sect. Carya*)	na	na	na	na
Macadamia (*Macadamia integrifolia*)	108*	Australia (31)	19	0.13
Pecan (*Carya illinoinensis*)	191*	United States (69)	69	0.49
Pine nut (*Pinus pinea*)	17*	Spain, Italy (30)	na	T[b]
Pistachio (*Pistacia vera*)	501	Iran (38)	28	0.16
Walnut (*Juglans regia*)	1,695	China (28)	21	0.53

Source: From U.S. Department of Agriculture (USDA), *Fruit and Nut Situation and Outlook Yearbook*, USDA, Springfield, VA, October 2007; FAO, *FAO Statistical Yearbooks 2005/2006*, Published online at: http://www.fao.org/statistics/ (accessed October 2008); The Craker, *Global Statistics*, International Tree Nut Council Foundation, Modesto, CA, September 2006.

Note: Production statistics are for in-shell nuts in 2005 (FAO, FAO Statistical Yearbooks 2005/2006, Published online at: http://www.fao.org/statistics/ (accessed October 2008); *The Craker, Global Statistics*, International Tree Nut Council Foundation, Modesto, CA, September 2006); na, not available.

[a] Consumption data (U.S. Department of Agriculture [USDA], *Fruit and Nut Situation and Outlook Yearbook*, USDA, Springfield, VA, October 2007) represent per capita estimated consumption of nuts (lb) for 2006/2007.

[b] T = 1.23 is the total consumption of all other edible nuts including the ones incorporated in the table.

on the nut seed type, source of dietary fiber (beechnut). Many tree nuts are ecologically important. For example, the cashew nut tree from Brazil was apparently introduced by Portuguese to western India to help reduce and prevent soil erosion caused by rapid water runoffs from hilly regions during monsoon season. Due to favorable climatic conditions, the cashew nut crop flourished and is now one of the major export commodities for Indian agriculture. Brazil nut tree is another example. The tree may grow in the wild to great heights (30–45 m tall) and its large canopy is known to support a small, independent, and self-sufficient habitat to multiple species. Almonds were originally introduced in the United States (state of Texas) by the early Spanish settlers, as almonds were already popular in Spain long before the Spanish traveled and settled in new areas. Almonds were subsequently introduced and systematically cultivated in California and are one of the major export crops of economical importance to the U.S. agriculture (exports ~$900 million in 2005) [2].

This chapter provides an overview of chemical composition of tree nuts and attempts to discuss some of the potential health benefits that may result from moderate tree nut consumption. The botanical names, overall estimated world major producers as well as consumption data for several edible nut seeds are summarized in Table 2.1.

2.2 CHEMICAL COMPOSITION

2.2.1 Proximate Composition

The proximate composition (Table 2.2) varies considerably depending on the seed type and growing conditions.

2.2.1.1 Moisture

Typically, moisture content of harvested nuts is <10% and fluctuates significantly depending on the time of harvest, the climatic conditions (wet/dry conditions) during growth, and storage [3,4]. For example, kernel moisture in western Schley pecans grown in Australia decreased from 47.69 g/100 g dry-weight basis (DWB) harvested in March to 4.21 g/100 g harvested in June 1999, whereas in 2000, the crop moisture content decreased from 31.87 g/100 g (in March) to 3.32 g/100 g (in June) [5]. Low moisture content is important for an extended shelf life and sensory quality of nuts, as low moisture helps reduce microbial growth and various undesirable biochemical changes that often accompany it.

2.2.1.2 Lipids

Most edible nuts are rich in lipids, ranging from 26.1% in coconut to 75.8% in macadamia, with few exceptions such as chestnut (Table 2.2), which contribute to the major portion of the energy obtained from these seeds (total calories range from 557 kcal for pistachio to 713 kcal for macadamia per 100 g edible portion) [6]. Typically, Brazil nut, hazelnut, macadamia, pecan, pine nut, and walnut seeds contain higher lipid (>60%) as compared to almond, cashew nut, and pistachio (44%–51%). Again, cultivar type, geographical locations, and growing conditions influence the lipid content of the mature kernels. For example, García-Lórda et al. [7] observed a range in lipid content (53.1%–61.7%) of 19 almond cultivars grown in different countries (four United States, seven Spanish, three Italian, and one Australian). Ruggeri et al. [8] reported a range of 52.5%–57% in lipid content of four almond varieties grown in Italy. Sathe [9] reported a range of 53.6%–56.1% for five major U.S. almond varieties (Mission, Nonpareil, Carmel, Neplus, and Peerless). Similarly, recent data from our laboratory indicate a range in the lipid content (67%–78.1%) for 27 pecan cultivars grown in different regions of the United States [10]. Parcerisa et al. [11] investigated the lipid content of four Spanish hazelnut varieties cultivated in two different geographical origins (Reus and Falset) in Spain over three crop years. Significant differences in the lipid content of hazelnut samples were noted as a function of collection period (1990–1992; 59.75 ± 1.71 g/100 g edible portion, $P = 0.0002$) and geographical origin (59.75 ± 1.39 g/100 g edible portion, $P = 0.0247$), but not variety.

TABLE 2.2
Proximate Composition (g/100 g) of Edible Tree Nuts

		Acorn Nut	Almond	Beechnut	Brazil Nut	Cashew	Chestnut[a]	Coconut	Hazelnut	Hickory Nut	Macadamia	Pecan	Pine Nut	Pistachio	Walnut
Moisture	Low		3.1		3.1	4.4	45.3	47.0	4.0		1.4	3.5	1.5	4.0	2.7
	High		9.5		3.5	8.0	52.0	51.9	5.3		2.1	7.4	2.3	5.7	4.7
	Mean	27.9	6.3	6.6	3.3	6.2	48.7	49.4	4.6	2.7	1.7	5.5	1.9	4.9	3.7
Lipid	Low		43.3		66.4	42.8	1.3	26.1	59.8		66.2	66.2	61.7	44.4	64.5
	High		50.6		66.7	43.7	4.0	33.5	61.5		75.8	72.0	68.4	45.1	65.2
	Mean	23.9	47.0	50.0	66.6	43.2	2.6	29.8	60.6	64.4	71.0	69.1	65.1	44.8	64.9
Protein	Low		19.5		13.9	18.2	1.6	3.3	14.1		7.9	7.5	13.1	19.8	13.5
	High		23.3		14.3	20.9	7.4	3.9	20.6		8.4	9.2	13.7	20.6	15.2
	Mean	6.2	21.4	6.2	14.2	19.6	4.5	3.6	17.3	12.7	8.2	8.3	13.4	20.2	14.3
Ash	Low		2.5		3.3	2.5	1.0	0.9	2.0		1.1	1.5	2.5	3.0	1.8
	High		4.6		3.5	2.8	2.9	1.0	2.3		1.2	1.9	2.6	3.2	1.8
	Mean	1.4	3.5	3.7	3.4	2.7	1.9	1.0	2.2	2.0	1.2	1.7	2.5	3.1	1.8
Carbohydrate	Low		19.7			24.1	44.2	15.2	10.0						
(by difference)	High		27.0			30.2	62.3	17.2	16.7						
	Mean	40.8	23.4	na	12.3	27.2	53.2	16.2	13.4	18.25	13.8	13.9	13.1	28.0	13.7
Sugars	Low		2.1		0.7	4.0	9.5		1.4		1.4	1.6	1.8	1.5	2.1
	High		7.5		2.3	5.9	17.1		4.3		4.6	4.0	3.6	7.6	2.6
	Mean	na	4.8	na	1.5	4.9	13.3	6.2	2.9	na	3.0	2.8	2.7	4.6	2.3
Dietary fiber	Low		11.8			1.4	2.3	8.7	3.4						
	High		13.0			3.3	3.7	9.0	9.7						
	Mean	na	12.4	33.5	7.5	2.4	3.0	8.9	6.5	6.4	8.6	9.6	3.7	10.3	6.7

Source: From U.S. Department of Agriculture (USDA), *National Nutrient Database for Standard Reference, Release 19,* Published online at: http://www.nal.usda.gov/fnic/foodcomp/search/ (accessed February 2007); Ruggeri, S., Cappelloni, M., Gambelli, L., and Carnovale, E., *Ital. J. Food Sci.*, 10, 243, 1998; Venkatachalam, M. and Sathe, S.K., *J. Agric. Food Chem.*, 54, 4705, 2006; Çağlarırmak, N., *Nahrung,* 47, 28, 2003; Çağlarırmak, N. and Batkan A.C., *J. Food Proc. Preserv.*, 29, 407, 2005; de Leon S.Y. and Delores M.I., in *Processing Fruits: Science and Technology,* 2nd ed., Barrett, D.M., Somogyi, L., and Ramaswamy, H.S., Eds., CRC Press, Boca Raton, FL, 2004, 707–727; Pereira-Lorenzo, S., Ramos-Cabrera, A.M, Díaz-Hernándeza, M.B, Ciordia-Arab, M., and Ríos-Mesac, D., *Sci. Hortic.*, 107, 306, 2006.

Note: The numbers are rounded to the first digit after decimal point; na, not available.

[a] European chestnut.

In general, tree nuts are rich in monounsaturated fatty acids (MUFA) and polyunsaturated fatty acids (PUFA) (Table 2.3). MUFA content in edible portions ranges from 0.4 g/100 g in chestnut to 58.9 g/100 g in macadamia, while PUFA vary from 0.4 g/100 g in coconut to 47.2 g/100 g in walnut. With the exception of coconut, total saturated fatty acids (SFA) content is lower than that compared to the total unsaturated fatty acids, ranging from 0.2 g/100 g in chestnut to 15.1 g/100 g in Brazil nut. Fatty acid composition of coconut meat is mostly saturated in nature, with lauric acid (12:0) being the predominant SFA (14.9 g/100 g edible portion). MUFA and PUFA levels in chestnut are comparable, whereas walnut has over five times as much PUFA as MUFA. The PUFA in walnut are the essential fatty acids linoleic (18:2ω6) and α-linolenic (18:3ω3) acids. MUFA predominate over PUFA in acorn nut, almond, Brazil nut, beechnut, cashew, hazelnut, hickory nut, macadamia, pecan, and pistachio, with oleic acid (18:1ω9) being the most predominant fatty acid.

Most tree nuts contain phytosterols and sphingolipids (Table 2.4). Among tree nuts, phytosterol concentrations range from 0.05 mg/100 g (coconut) to 214 mg/100 g (pistachio) edible portion with pine nut and pistachio being the richest sources, 141 and 214 mg/100 g edible portion, respectively. β-Sitosterol is the major phytosterol (70–90%) in tree nuts. The U.S. Department of Agriculture (USDA) database values [6] are largely in agreement with those reported in the literature [12–14].

2.2.1.3 Proteins

The total nitrogen content varies from 1.7 g/100 g (macadamia) to 4.06 g/100 g (almond and pistachio), of which <0.4 g/100 g account for nonprotein nitrogen [15]. Almond, pistachio, and cashew appear to have the highest protein content (18.2–23.3 g/100 g) followed by walnut, hazelnut, Brazil nut, and pine nut (13.7–15.2 g/100 g) while coconut (3.6 g/100 g) and chestnut (4.5 g/100 g) seem to be the lowest in protein content.

2.2.1.4 Amino Acids

Acidic amino acids (aspartic acid + glutamic acid) predominate in tree nuts (Table 2.5). Similar to other plant proteins, tree nut proteins are incomplete proteins. When compared to the Food and Agricultural Organization (FAO) and World Health Organization (WHO)-recommended pattern for essential amino acids for a 2-5 year old, tryptophan is the first limiting amino acid in all tree nuts except macadamia, where lysine is the first. However, compared to the FAO- and WHO-recommended essential amino acid pattern for an adult, only almond is deficient in sulfur amino acids (methionine + cysteine), whereas all others contain adequate amounts of all of the essential amino acids.

Tree nuts are a rich source of arginine (ranging from 9.15 g/100 g of protein in pistachio to 15.41 g/100 g of protein in pine nut) equivalent to 1812 and 2016 mg of arginine/100 g of edible nuts and comparable to 2140 mg/110 g sirloin steak, 2150 mg/182.55 g whiting fish, and 2140 mg arginine/68.39 g in peanut, estimated to be supplied by some of the high-arginine foods in the U.S. food supply [16]. In general, tree nuts appear to have essential to total amino acid ratio between 31.2% for almond and 53.1% for European chestnut, with chestnut and Brazil nut (in that order) having the highest proportion of essential amino acids (Table 2.5).

2.2.1.5 Carbohydrates

The total carbohydrate content of tree nuts (calculated by difference) ranges from 12.3% (Brazil nut) to 62.3% (chestnut). Tree nuts, unlike other plant foods such as cereals and tubers, do not contain large amounts of starch (typically nut seeds have <1.5%) as a storage polysaccharide. Carbohydrates are essentially composed of fibers (3.3% for cashew to 13.0% for almond) and simple sugars (2.3% for Brazil nut to 17.1% for chestnut). Sucrose (2.3%–7.0%) is the major simple sugar that accounts for >95% of the total sugars. Minor saccharides include glucose (0%-0.27%), fructose (0%-0.17%), maltose (0%-0.2%), and others in trace amounts. The sugar content

TABLE 2.3
Fatty Acid Composition (%) of Edible Tree Nuts

Fatty Acid	Acorn Nut	Almond	Beechnut	Brazil Nut	Cashew	Chestnut[a]	Coconut	Hazelnut	Hickory Nut	Macadamia	Pecan	Pine Nut	Pistachio	Walnut
6:0	0.0	0.0	0.0	0.0	0.0	0.0	0.2	0.0	0.0	0.0	0.0	0.0	0.0	0.0
8:0	0.0	0.0	0.0	0.0	0.0	0.0	2.3	0.0	0.0	0.0	0.0	0.0	0.0	0.0
10:0	0.0	0.0	0.0	0.0	0.0	0.0	1.9	0.0	0.0	0.0	0.0	0.0	0.0	0.0
12:0	0.0	0.0	0.0	0.0	0.0	0.0	14.9	0.0	0.0	0.1	0.0	0.0	0.0	0.0
14:0	0.0	0.0	0.1	0.05*	0.0	0.0	5.9	0.0	0.0	0.7	0.0	0.0	0.0	0.0
16:0	2.9	3.2	3.6	9.1	4.2	0.2	2.8	3.1	5.4	6.0	4.4	3.7	4.9	4.4
17:0	0.0	0.0	0.0	nd	0.1	0.0	0.0	0.0	0.0	0.1	0.0	nd	0.0	0.0
18:0	0.3	0.7	1.2	5.8	3.4	0.0	1.7	1.3	1.4	2.3	1.7	1.7	0.5	1.7
20:0	0.0	0.0	0.0	0.2	0.3	0.0	0.0	0.1	0.0	1.9	0.1	nd	0.0	0.1
22:0	0.0	0.0	0.0	0.0	0.2	0.0	0.0	0.0	0.0	0.6	0.0	nd	0.0	0.0
24:0	0.0	0.0	0.0	nd	0.1	0.0	0.0	0.0	0.0	0.3	0.0	nd	0.0	0.0
Total SFA	3.1	3.9	5.7	15.1	8.3	0.2	29.7	4.5	7.0	12.1	6.2	7.8	5.4	6.1
16:1	0.0	0.2	0.3	0.2	0.1	0.0	0.0	0.1	0.3	13.0	0.0	0.2	0.5	0.0
17:1	nd	nd	nd	0.04*	nd	nd	nd	nd	nd	nd	nd	nd	nd	nd
18:1ω9	15.1	31.9	18.8	24.2	25.2	0.4	1.4	45.4	32.0	43.8	40.6	17.9	22.7	8.8
20:1ω9	0.0	0.0	2.8	0.05*	0.1	0.0	0.0	0.1	0.0	1.9	0.2	1.0	0.2	0.1
22:1ω9	0.0	0.0	0.0	0.0	0.0	0.0	0.0	0.0	0.0	0.2	0.0	0.0	0.0	0.0
24:1ω9	0.0	0.0	0.0	nd	nd	0.0	0.0	0.0	0.0	0.0	0.0	nd	0.0	0.0
Total MUFA	15.0	32.2	21.9	24.6	25.5	0.4	1.4	45.7	32.6	58.9	40.8	19.1	23.3	8.9
18:2ω6	4.6	12.2	18.4	20.5	8.3	0.4	0.4	7.8	20.6	1.3	20.6	20.7	13.2	38.1
18:3	0.0	0.0	1.7	0.04*	0.1	0.1	0.0	0.1	1.0	0.2	1.0	0.7	0.3	9.1
18:3ω3	nd	nd	nd	0.02*	nd	nd	nd	nd	nd	nd	nd	nd	nd	nd
18:3ω6	nd	nd	nd	0.02*	nd	nd	nd	nd	nd	nd	nd	nd	nd	nd
20:4ω6	0.0	0.0	0.0	0.0	0.0	0.0	0.0	0.0	0.0	0.0	0.0	0.0	0.0	0.0
Total PUFA	4.6	12.2	20.1	20.6	8.4	0.5	0.4	7.9	21.9	1.5	21.6	21.3	13.5	47.2

Source: From U.S. Department of Agriculture (USDA), *National Nutrient Database for Standard Reference, Release 19,* Published online at: http://www.nal.usda.gov/fnic/foodcomp/search/ (accessed February 2007).

Note: The numbers are rounded to the first digit after decimal point, except when indicated by *, where the numbers are rounded to second digit after the decimal point; nd, not detected.

[a] European chestnut.

TABLE 2.4
Cholesterol, Phytosterol Composition, and Sphingolipids (mg/100 g) in Edible Tree Nuts

	Acorn Nut	Almond	Beechnut	Brazil Nut	Cashew	Chestnut[a]	Coconut	Hazelnut	Hickory Nut	Macadamia	Pecan	Pine Nut	Pistachio	Walnut
Cholesterol	0.0	0.0	0.0	0.0	0.0	0.0	0.0	0.0	0.0	0.0	0.0	0.0	0.0	0.0
Phytosterols	0.0	120.0	0.0	nd	nd	0.0	0.05	96.0	0.0	116.0	102.0	141.0	214.0	72.0
Stigmasterol	0.0	4.0	0.0	nd	nd	0.0	nd	1.0	0.0	0.0	3.0	nd	5.0	1.0
Campesterol	0.0	5.0	0.0	nd	nd	0.0	nd	6.0	0.0	8.0	5.0	nd	10.0	7.0
β-sitosterol	0.0	111.0	0.0	nd	nd	0.0	nd	89.0	0.0	108.0	89.0	nd	198.0	64.0
Cerebroside														
D18:2-C16:0h-Glucose	0.0	6.8	0.0	nd	3.9	0.0	nd	2.1	0.0	nd	nd	4.2	nd	2.5

Source: From U.S. Department of Agriculture (USDA), *National Nutrient Database for Standard Reference, Release 19,* Published online at: http://www.nal.usda.gov/fnic/foodcomp/search/ (accessed February 2007).

Note: The numbers are rounded to the first digit after decimal point; nd, not detected.

[a] European chestnut.

TABLE 2.5
Amino Acid Content (g/100 g) of Edible Tree Nuts

Amino Acid	FAO/ WHO[a]	Acorn Nut	Almond	Beechnut	Brazil Nut	Cashew	Chestnut[b]	Coconut	Hazelnut	Hickory Nut	Macadamia	Pecan	Pine Nut	Pistachio	Walnut
Tryptophan	**1.1**	**0.1**	**0.2**	**0.1**	**0.1**	**0.3**	**0.0**	**0.0**	**0.2**	**0.1**	0.1	**0.1**	**0.3**	**0.3**	**0.2**
Threonine	3.4	0.2	0.7	0.2	0.4	0.7	0.1	0.1	0.5	0.4	0.4	0.3	0.8	0.7	0.6
Isoleucine	2.8	0.3	0.7	0.2	0.5	0.8	0.6	0.1	0.5	0.6	0.3	0.3	0.9	0.9	0.6
Leucine	6.6	0.5	1.5	0.4	1.2	1.5	0.1	0.2	1.1	1.0	0.6	0.6	1.7	1.6	1.2
Lysine	5.8	0.4	0.6	0.4	0.5	0.9	0.1	0.1	0.4	0.5	**0.0**	0.3	0.9	1.2	0.4
Methionine	2.5	**0.1**	**0.2**	0.1	1.0	0.4	0.0	0.1	0.2	0.3	0.0	0.2	0.4	0.3	0.2
Cysteine		**0.1**	0.3	0.2	0.4	0.4	0.1	0.1	0.3	0.3	0.0	0.2	0.4	0.4	0.2
Phenylalanine	6.3	0.3	1.1	0.3	0.6	1.0	0.1	0.2	0.7	0.7	0.7	0.4	0.9	1.1	0.7
Valine	3.5	0.3	0.8	0.3	0.8	1.1	0.1	0.2	0.7	0.7	0.4	0.4	1.2	1.2	0.8
Histidine	1.9	0.2	0.6	0.0	0.4	0.5	0.0	0.1	0.4	0.4	0.2	0.3	0.6	0.5	0.4
Tyrosine		0.2	0.5	0.2	0.4	0.5	0.0	0.1	0.4	0.5	0.5	0.2	0.9	0.4	0.4
Arginine		0.5	2.5	0.4	2.2	2.1	0.2	0.5	2.2	2.1	1.4	1.2	4.7	2.0	2.3
Alanine		0.4	1.0	0.4	0.6	0.8	0.1	0.2	0.7	0.7	0.4	0.4	1.3	0.9	0.7
Aspartic acid		0.6	2.7	1.1	1.4	1.8	0.3	0.3	1.7	1.4	1.1	0.9	2.2	1.8	1.8
Glutamic acid		1.0	5.2	0.8	3.2	4.5	0.2	0.8	3.7	2.9	2.3	1.8	4.1	3.8	2.8
Glycine		0.3	1.5	0.3	0.7	0.9	0.1	0.2	0.7	0.7	0.5	0.5	1.2	1.0	0.8
Proline		0.2	1.0	0.3	0.8	0.8	0.1	0.1	0.6	0.6	0.5	0.4	1.3	0.8	0.7
Serine		0.3	1.0	0.3	0.7	1.1	0.1	0.2	0.7	0.8	0.4	0.5	1.0	1.2	0.9
TAA		5.9	21.3	6.1	14.3	18.2	2.3	3.6	15.0	14.6	7.9	9.2	24.0	20.6	15.2
EAA/TAA (%)[c]		41.9	31.2	36.7	45.1	40.7	53.1	34.7	33.5	34.7	33.2	33.3	34.3	39.1	34.7

Source: From U.S. Department of Agriculture (USDA), *National Nutrient Database for Standard Reference, Release 19,* Published online at: http://www.nal.usda.gov/fnic/foodcomp/search/ (accessed February 2007).

Note: The numbers are rounded to the first digit after decimal point.

[a] The values presented in bold represent the first limiting essential amino acid as per FAO/WHO (1990) recommendation for children (2–5 years).

[b] European chestnut.

[c] Percentage ratio of essential amino acids (EAA)/total amino acids (TAA).

of tree nuts also varies considerably, depending on growing conditions, seed maturity, cultivar, and growth location [3,4,17].

2.2.2 Vitamins

Chestnuts (Table 2.6) have the highest amounts of vitamin C (40.2 mg/100 g edible portion) among tree nuts. Macadamia nuts have highest amounts of thiamin. Almonds are a rich source of riboflavin (B_2), niacin (B_3), folate, and α-tocopherol (vitamin E) (Table 2.6). Li et al. [18] estimated relative amounts of four tocopherol homologues (α, β, δ, and γ) for walnut and heartnut and found γ-tocopherol to be the main homologue in both followed by δ- and α-tocopherols. Heartnut and hickory nuts have the highest content of pantothenic acid (B_5) while pistachio nuts have the highest content of vitamin A and pyridoxine (B_6) [6]. Acorn nuts, hazelnuts, and beechnuts contain comparable levels of folate (113–114 μg/100 g edible portion) and are the highest among tree nuts. Pine nuts and cashews are both rich in vitamin K [6].

Tree nuts can be "good" dietary sources (>10% of recommended dietary allowances [RDA] or adequate intake [AI]) of most vitamins except vitamins A and cobalamin (B_{12}). At suggested consumption level (1.5 ounces or ~42.5 g/day), acorn and pine nuts are excellent sources of folate and vitamin K (Table 2.7). In case of thiamin (B_1), niacin (B_3), and pyridoxine (B_6), tree nuts offer alternative sources of these vitamins when more abundant animal sources are not available or are not consumed.

2.2.3 Minerals

Ash, which represents mineral content of the edible nuts, ranges from 0.9% coconut to 3.7% Beechnut, suggesting significant differences in the mineral composition of tree nuts (Table 2.2). In general, tree nuts are rich in Mn, Mg, P, and K (Table 2.8). Mineral content of walnuts, as reported by the USDA database [6], is in agreement with that reported by Çağlarırmak [19] except for K (268 mg/100 g). Çağlarırmak and Batkan [20] reported twice as much Ca and K, about 8 times as much Fe, and 10 times as much Na in pistachio as that reported in the USDA database.

Some tree nuts are comparable to other abundant dietary sources of Fe, Mg, Mn, P, Se, and Zn (Table 2.9). Brazil nuts are an excellent source of Se, P, and Mg. Pine nuts are rich in Zn and Mn, almonds are a good nondairy source of Ca, and cashew nuts are a rich source of Fe.

2.2.4 Phenolics and Antioxidants

Depending on sample freshness, particle size, extraction solvent (ethanol, methanol, acidified methanol, and acetone), reference phenolic compound (catechin vs. gallic acid), and detection method (vanillin vs. Folin Ciocalteu reagent) used for analysis, one may obtain widely varying results for phenolics of the same seed sample [15,21–25]. However, the *Juglandaceae* family pecan and walnut typically contain the highest amount of phenolics per unit weight (Table 2.10), followed by pistachio nuts.

Fukuda et al. [26] identified 16 hydrolysable tannins in walnuts, including a new class of ellagitannin derivatives, glansrins, which exhibit a strong free radical scavenging activity comparable to ascorbic acid and gallic acid. In addition to the kernel, walnut liquor is another edible walnut product, which is derived from husks of immature walnut fruit. Walnut liquor has been shown to be a rich source of 13 phenolic compounds (1968 mg total phenolics/L of nocino liquor) [27], such as gallic acid (~68 mg/L), (–)-epicatechin (25 mg/L), and syringic acid (~23 mg/L) [28]. Pistachio nuts also have been reported to contain the polyphenolic phytoalexin reservatrol, which is associated with reduced cardiovascular disease (CVD) and reduced cancer risk, in the range of 0.09–1.67 μg/g [29]. A recent investigation found that certain tree nuts (almond, cashew, and walnut) contain the

TABLE 2.6
Vitamin Content of Edible Tree Nuts

Vitamin	Unit	Acorn Nut	Almond	Beechnut	Brazil Nut	Cashew	Chestnut[a]	Coconut	Hazelnut	Hickory Nut	Macadamia	Pecan	Pine Nut	Pistachio	Walnut
Vitamin A (international unit)	IU/100 g	51.0	5.0	0.0	0.0	0.0	26.0	0.0	20.0	**131.0**	0.0	56.0	29.0	***553.0***	20.0
Vitamin A (retinol activity equivalents)	μg/100 g	3.0	0.0	0.0	0.0	0.0	1.0	0.0	1.0	7.0	0.0	0.0	0.0	***28.0***	1.0
Retinol	μg/100 g	0.0	0.0	0.0	0.0	0.0	0.0	0.0	0.0	0.0	0.0	0.0	0.0	0.0	0.0
Vitamin C (total ascorbic acid)	mg/100 g	0.0	0.0	**15.5**	0.7	0.5	***40.2***	3.3	6.3	2.0	1.2	1.1	0.8	5.0	1.3
Vitamin E (α-tocopherol)	mg/100 g	0.0	***25.9***	0.0	5.7	0.9	0.0	0.2	15.0	0.0	0.5	1.4	9.3	2.3	**13.0**
β-Tocopherol	mg/100 g	0.0	***0.4***	0.0	0.0	0.0	0.0	0.0	0.3	0.0	0.0	***0.4***	0.0	0.0	0.0
γ-Tocopherol	mg/100 g	0.0	0.9	0.0	7.9	5.3	0.0	0.5	0.0	0.0	0.0	***24.4***	11.2	22.6	**20.8**
δ-Tocopherol	mg/100 g	0.0	0.3	0.0	0.8	0.4	0.0	0.0	0.0	0.0	0.0	0.5	0.0	0.8	**13.0**
Vitamin K (phylloquinone)	μg/100 g	0.0	0.0	0.0	0.0	**34.1**	0.0	0.2	14.2	0.0	0.0	3.5	***53.9***	0.0	0.3
Thiamin (B_1)	mg/100 g	0.1	0.2	0.3	0.1	0.4	0.1	0.1	0.6	0.9	***1.2***	0.7	0.4	0.9	0.3
Riboflavin (B_2)	mg/100 g	0.2	***0.8***	0.4	0.04	0.1	0.0	0.0	0.1	0.1	0.2	0.1	0.2	0.2	0.2
Niacin (B_3)	mg/100 g	2.4	3.9	0.9	0.3	1.1	1.1	0.5	1.8	0.9	2.5	1.2	***4.4***	1.3	1.1
Pantothenic acid (B_5)	mg/100 g	0.9	0.3	0.9	0.2	0.9	0.5	0.3	0.9	***1.7***	0.8	0.9	0.3	0.5	0.6
Pyridoxine (B_6)	mg/100 g	0.7	0.1	0.7	0.1	0.4	0.4	0.1	0.6	0.2	0.3	0.2	0.1	***1.7***	0.5
Cobalamin (B_{12})	μg/100 g	0.0	0.0	0.0	0.0	0.0	0.0	0.0	0.0	0.0	0.0	0.0	0.0	0.0	0.0
Betaine	mg/100 g	0.0	0.5	0.0	0.4	0.0	0.0	0.0	0.4	0.0	0.0	***0.7***	0.4	0.0	0.3
Folate	μg/100 g	***87.0***	**50.0**	**113.0**	**22.0**	25.0	**58.0**	**26.0**	**113.0**	40.0	**11.0**	22.0	34.0	51.0	**98.0**

Source: From U.S. Department of Agriculture (USDA), *National Nutrient Database for Standard Reference, Release 19,* Published online at: http://www.nal.usda.gov/fnic/foodcomp/search/ (accessed February 2007).

Note: The numbers are rounded to the first digit after decimal point; Most abundant tree nut source for each vitamin is indicated in bold italic, while the most abundant vitamin in a single tree nut is presented in bold.

[a] European chestnut.

TABLE 2.7
Vitamin Content of Edible Tree Nuts in Relation to Dietary Requirements and in the Context of Other Foods Abundant in the Same Vitamins

Vitamin	Unit	RDA or AI for Adult Males (Per Day)[a]	RDA or AI for Adult Females (Per Day)[a]	Most Abundant Food Source		Normal Food Intake (g)	Effective Vitamin Intake	Percentage of RDA or AI for Adult Males	Percentage of RDA or AI for Adult females
Vitamin A (retinol activity equivalents)	μg	900	700	Tree nuts	Hickory nut	28.8	2.0	0.2	0.3
				Other	Beef liver	30.0	1014.0	112.7	144.9
Vitamin C (total ascorbic acid)	mg	90	75	Tree nuts	Chestnut	28.8	11.6	12.9	15.4
				Other	Parsley	100.0	130.0	144.4	173.3
Vitamin E (α-tocopherol equivalents)	mg	15	15	Tree nuts	Almond	28.8	7.5	49.7	49.7
				Other	Sunflower seed	30.0	46.7	311.1	311.1
Vitamin K (phylloquinone)	μg	120*	90*	Tree nuts	Pine nut	28.8	15523.2	12936.0*	17248.0*
				Other	Kale	100.0	650.0	541.7*	722.2*
Thiamin (B_1)	mg	1.2	1.1	Tree nuts	Macadamia	28.8	0.3	28.7	31.3
				Other	Pork chop	120.0	7.5	625.0	681.8
Riboflavin (B_2)	mg	1.3	1.1	Tree nuts	Almond	28.8	0.2	18.0	21.2
				Other	Beef liver	30.0	4.0	307.7	363.6
Niacin (B_3)	mg	16	14	Tree nuts	Almond	28.8	1.1	7.1	8.1
				Other	Steak	120.0	4.7	29.2	33.3
Pantothenic acid (B_5)	mg	5*	5*	Tree nuts	Hickory nut	28.8	0.5	10.1*	10.1*
				Other	Sunflower seed	50.0	4.6	92.0*	92.0*
Pyridoxine (B_6)	mg	1.3	1.3	Tree nuts	Pistachio	28.8	0.5	37.7	37.7
				Other	Salmon	90.0	888.9	68376.1	68376.1
Cobalamin (B_{12})	μg	2.4	2.4	Tree nuts	None	28.8	0.0	0.0	0.0
				Other	Beef liver	30.0	4.0	166666.7	166666.7
Betaine	mg	na	na	Tree nuts	Pecan	28.8	0.7	na	na
				Other	Wheat bran	100.0	1.6	na	na
Folate	μg	400	400	Tree nuts	Acorn nut	28.8	32832.0	8208.0	8208.0
				Other	Lentil	100.0	179000.0	44750.0	44750.0

Source: From Institute of Medicine of the National Academics. *Dietary Reference Intakes,* Published Online at http://www.iom.edu/object. File/Master/7/296/webtablevitamins.pdf (accessed January 2007).

Note: The numbers are rounded to the first digit after decimal point. na, not available.

[a] Recommended dietary allowances (RDA) or adequate intake (AI) for adults (aged 19–50 years).

* Values are expressed as AI.

TABLE 2.8
Mineral Content of Edible Tree Nuts

Mineral	Unit	Acorn Nut	Almond	Beechnut	Brazil Nut	Cashew	Chestnut[a]	Coconut	Hazelnut	Hickory Nut	Macadamia	Pecan	Pine Nut	Pistachio	Walnut
Aluminum (Al)	mg/100 g	nd	nd	nd	nd	nd	nd	nd	***5.02***	nd	nd	nd	nd	nd	nd
Boron (B)	mg/100 g	nd	nd	nd	nd	nd	nd	nd	nd	nd	nd	nd	nd	nd	nd
Cadmium (Cd)	mg/100 g	nd	nd	nd	nd	nd	nd	nd	***0.01***	nd	nd	nd	nd	nd	nd
Calcium (Ca)	mg/100 g	43.0	***248.0***	1.0	160.0	37.0	19.0	14.0	102.5	61.0	85.0	70.0	16.0	107.0	98.0
Cobalt (Co)	mg/100 g	nd	nd	nd	nd	nd	nd	nd	***0.22***	nd	nd	nd	nd	nd	nd
Copper (Cu)	mg/100 g	0.6	1.1	0.7	1.7	***2.2***	0.4	0.4	1.0	0.02*	0.8	1.2	1.3	0.7	1.6
Iron (Fe)	mg/100 g	1.2	4.3	2.5	2.4	6.7	0.9	2.4	0.55	0.04*	3.7	2.5	***5.5***	0.5	2.9
Magnesium (Mg)	mg/100 g	110.0	275.0	nd	***376.0***	292.0	30.0	32.0	141.5	7.8	130.0	121.0	251.0	116.0	158.0
Manganese (Mn)	mg/100 g	1.7	2.5	1.3	1.2	1.7	0.3	1.5	3.3	4.6	4.1	4.5	***8.8***	1.2	3.4
Phosphorus (P)	mg/100 g	103.0	474.0	nd	***725.0***	**502.3**	38.0	113.0	355.7	336.0	188.0	277.0	575.0	490.0	346.0
Potassium (K)	mg/100 g	**712.0**	**728.0**	***1017.0***	**659.0**	251.1	**484.0**	**356.0**	**502.9**	**436.0**	**368.0**	**410.0**	**597.0**	**633.1**	**441.0**
Selenium (Se)	μg/100 g	nd	2.8	nd	***1917.0***	19.9	nd	10.1	0.06	8.1	3.6	3.8	0.7	7.0	4.9
Sodium (Na)	mg/100 g	nd	1.0	***38.0***	3.0	12.0	2.0	20.0	0.9	1.0	5.0	nd	2.0	**1.0**	2.0
Vanadium (V)	mg/100 g	nd	nd	nd	nd	nd	nd	nd	0.08*	nd	nd	nd	nd	***2.7***	nd
Zinc (Zn)	mg/100 g	0.6	3.4	0.4	4.1	**5.8**	0.5	1.1	1.9	4.3	1.3	4.5	***6.5***	2.2	3.1

Source: From U.S. Department of Agriculture (USDA), *National Nutrient Database for Standard Reference, Release 19,* Published online at: http://www.nal.usda.gov/fnic/foodcomp/search/ (accessed February 2007).

Note: The numbers are rounded to the first digit after decimal point, except when indicated by *, where the numbers are rounded to second digit after the decimal point; Most abundant tree nut source for each vitamin is indicated in bold italic, while the most abundant vitamin in a single tree nut is given in bold; nd, not detected.

[a] European chestnut.

TABLE 2.9
Mineral Content of Edible Tree Nuts in Relation to Dietary Requirements and in the Context of Other Foods Abundant in the Same Minerals

Mineral	Unit	RDA or AI for Adult Males (Per Day)[a]	RDA or AI for Adult Females (Per Day)[a]	Most Abundant Food Source		Normal Food Intake (g)	Effective Mineral Intake	Percentage of RDA or AI for Adult Males	Percentage of RDA or AI for Adult Females
Calcium	g	1*	1*	Tree nuts	Almond	28.8	0.1	7.1*	7.1*
				Other	Parmesan cheese	60.0	1.1	111.7*	111.7*
Copper	µg	900	900	Tree nuts	Cashew	28.8	0.6	100.0	100.0
				Other	Beef liver	30.0	4.3	500.0	500.0
Iron	mg	8	18	Tree nuts	Cashew	28.8	1.9	24.0	10.7
				Other	Sirloin steak	150.0	3.2	40.0	17.8
Magnesium	mg	400–420	310–320	Tree nuts	Brazil nut	28.8	108.3	25.8	34.4
				Other	Wheat germ	50.0	180.0	42.9	57.1
Manganese	mg	2.3*	1.8*	Tree nuts	Pine nut	28.8	2.5	110.2*	140.8*
				Other	Whole wheat	100.0	3.8	165.2*	211.1*
Phosphorous	mg	700	700	Tree nuts	Brazil nut	28.8	208.8	29.8	29.8
				Other	Sardines	90.0	466.7	66.7	66.7
Selenium	µg	55	55	Tree nuts	Brazil nut	28.8	552.1	1003.8	1003.8
				Other	Sirloin steak	150.0	31.7	57.7	57.7
Zinc	mg	11	8	Tree nuts	Pine nut	28.8	1.9	16.9	23.2
				Other	Sirloin steak	120.0	6.2	56.1	77.1

Source: From Institute of Medicine of the National Academies, *Dietary Reference Intakes*, Published online at: http://www.iom.edu/Object.File/Master/7/296/webtablevitamins.pdf (accessed January 2007).

Note: The numbers are rounded to the first digit after decimal point.

[a] Recommended dietary allowances (RDA) or adequate intake (AI) for adults (aged 19–50 years).

* Values are expressed as AI.

TABLE 2.10
Comparison of Total Antioxidant Capacity of Major Edible Nuts, Tomato, and Cranberry

Nut	Total Phenolics[a] (mg of GAE/g)	L-$ORAC_{FL}$[b] (μmol of TE/g)	H-$ORAC_{FL}$ (μmol of TE/g)	Total Antioxidant Capacity[c] (μmol of TE/g)	Total Antioxidant Capacity[d] (Per Serving)
Almond	4.18±0.84	1.72±0.50	42.82±8.71	44.54	1265
Brazil nut	3.10±0.96	5.57±2.17	8.62±2.06	14.19	403
Cashew	2.74±0.39	4.74±1.38	15.23±2.04	19.97	567
Hazelnut	8.35±2.16	3.70±2.66	92.75±17.78	96.45	2739
Macadamia	1.56±0.29	2.52±0.57	14.43±2.31	16.95	481
Peanut	3.96±0.54	2.73±1.04	28.93±2.36	31.66	899
Pecan	20.16±1.03	4.16±0.98	175.24±10.36	179.40	5095
Pine nut	0.68±0.25	2.76±0.60	4.43±1.11	7.19	204
Pistachio	16.57±1.21	4.25±1.46	75.57±10.50	79.83	2267
Walnut	15.56±4.06	4.84±1.25	130.57±35.20	135.41	3846
Tomato	0.80±0.12	0.24±0.07	3.13±0.69	3.37	415
Cranberry	7.09±0.07	2.00±0.38	92.56±1.38	94.56	8983

Source: From Wu, X., Beecher, G.R., Holden, J.M., Haytowitz, D.B., Gebhardt, S.E., and Prior, R.L., *J. Agric. Food Chem.*, 52, 4026, 2004. With permission.

Note: Data are expressed as means ± SD on the "as is" weight basis.

[a] Total phenolics, expressed in milligrams of gallic acid equivalents per gram (mg of GAE/g).

[b] $ORAC_{FL}$, expressed in micromoles of Trolox equivalents per gram (μmol of TE/g).

[c] Total antioxidant capacity (activity) = L-$ORAC_{FL}$ + H-$ORAC_{FL}$.

[d] One serving of nut seeds = 28.8 g.

polyphenolics, lignans, in amounts comparable to certain rice varieties (346–486 μg of lignans/100 g edible portion), but lower than other cereals such as rye (10377 μg of lignan/100 g) and wheat (7548 μg of lignan/100 g) [30]. Almonds, cashews, and walnuts contain between 344 and 912 μg of lignan/100 g edible portion, with cashew nut being the most abundant tree nut source (912 μg of lignan/100 g) [30].

The estimated total antioxidant capacity of the major tree nuts, as measured by the oxygen radical scavenging ability, is mostly exhibited by the hydrophilic portion of the nut seeds [31]. The two exceptions are Brazil nut and pine nut, which have about 40% of their antioxidant activity concentrated in the lipophilic portion of the seeds (Table 2.10). Anthocyanins partially contribute to antioxidant activity of tree nuts. Gu et al. [32] found the proanthocyanidin (PA) content of certain tree nuts (almond, cashew, hazelnut, pecan, walnut, and pistachio) to be comparable (8.7–500 mg/100 g edible portion) to that of berries (8.2–663 mg/100 g edible portion) and beans (8.1–456 mg/100 g edible portion) in over 40 naturally occurring foods tested. Among tree nuts tested (hazelnut, pecan, pistachio, almond, walnut, and cashew), hazelnut and pecan had the highest PA concentration (500.7 and 494.1 mg/100 g, respectively). Over 50% of total PA in tree nuts, however, were heptamers or larger polymers, a form that is not absorbed intact through the small intestine and is slowly (over 4 h) broken down to monomeric and dimeric flavonoids. Though low in concentration (from 6.7 mg/100 g in cashew to 17.2 mg/100 g in pecan), monomeric flavonoids have been identified in almond kernel [22], almond skin [33], pistachio [34], and pistachio skin [35]. A recent study concluded that consumption of ~1 g PA may be essential to observe beneficial effects such as antithrombotic activity in humans [36].

2.3 HEALTH IMPLICATIONS

Although rich in lipids, tree nuts possess a desirable lipid profile (>75% are MUFA and PUFA, Table 2.3). In addition, tree nuts are also a good source of certain vitamins, minerals, and several bioactive constituents. Together, these constituents are believed to be beneficial in improving overall human health while reducing the risk of certain chronic diseases such as heart disease [37–41], cancers [41,42], diabetes [41,43], and oxidative stress-related inflammation [44]. Based on the correlation of tree nut consumption to reduced risk of certain diseases, the U.S. Food and Drug Administration (FDA) in 2003 approved the following qualified health claim for use in advertising and package labels for nut and nut-containing products: "Scientific evidence suggests but does not prove that eating 1.5 ounces (~42.5 g) per day of most nuts as part of a diet low in saturated fat and cholesterol may reduce the risk of heart disease" [45].

2.3.1 Lipids

2.3.1.1 Unsaturated Fats

The favorable fatty acid profile of tree nuts has been suggested to play a cardioprotective role [38,46,47]. Incorporation of almonds [48–52], hazelnuts [53], macadamias [54,55], pecans [56,57], pistachios [58,59], or a combination of several tree nuts [43,60,61] in the diet has been demonstrated to lower plasma cholesterol. Walnut consumption has been shown to not only lower plasma total cholesterol [48,62–67], but also plasma low-density lipoprotein (LDL) cholesterol with a concurrent improved vasodilation [64]. Clinical studies indicate that tree nut lipids improve [68] or have no negative effect on insulin sensitivity [69,70] as well as improve glycemic control in diabetic patients [71,75]. In one study [68], almond consumption lowered blood glucose and insulin levels, and increased serum protein thiol concentrations (a marker for protein oxidative damage), indicating less oxidative protein damage in healthy individuals compared to the control group. Thus, despite being high in fat, nut seed consumption does not seem to affect glucose homeostasis and may help reduce the risk of diabetes and oxidative stress. Nut consumption (as low as once a week to more than five times a week) has been reported to significantly lower levels of several molecules (C-reactive protein [CRP], interleukin-6 and fibrinogen) that are often used as markers of inflammation [44]. Several tree nut seed constituents, PUFA, antioxidants, dietary fiber, and arginine are known to play an important role in modulating inflammation indicated by CRP, interleukin 6, and soluble tumor necrosis factor receptors 1 and 2 (sTNF-R1 and sTNF-R2) [16,72–75], which may partially explain the inverse association of nut consumption with CVD and diabetes risk.

2.3.1.2 Essential Fatty Acids

Tree nuts, particularly walnuts, contain α-linolenic acid (18:3ω3), (Table 2.3) that has been shown to lower CVD risk by reducing vascular inflammation, blood clot formation, and blood cholesterol [73,75–77]. α-Linolenic acid reduces the risk of CVD by various mechanisms including (1) interference with platelet aggregation by inhibiting thromboxane (TXA2), a type of eicosanoid, which assists blood clotting and increases stickiness of platelets, (2) reduction in plaque formation by reducing the release of proinflammatory substances (e.g., adhesion molecules and chemokines) from endothelial cells lining of the artery walls, and (3) reduction in serum triacylglycerols (TAG) levels by decreasing hepatic secretion of TAG-rich very low-density lipoprotein (VLDL) [78–81].

2.3.1.3 Sphingolipids

Sphingolipids in tree nuts are given in Chapter 5. They are a class of lipids that are hydrolyzed throughout the gastrointestinal tract into metabolites that include ceramides and sphingoid bases that are used by cells to regulate growth, differentiation, apoptosis, and other cellular functions. Sphingolipids also play a structural role by shielding the cell surface from harmful environmental factors by forming a mechanically stable and chemically resistant outer leaflet of the plasma membrane [82]. Although present in most foods, dairy products (12 mg/100 g edible portion in milk and more than 98 mg/100 g in cheese), meat products (30–38 mg/100 g), eggs (181 mg/100 g), and soybeans (>180 mg/100 g) typically contain much higher amounts of sphingolipids than fruits and vegetables (usually less than 8 mg/100 g) [83,84]. The sphingolipid contents mentioned were converted from μmol/kg to mg/100 g using the average molecular weights of 751 g/mol and 747 g/mol for sphingomylein and glycosylceramide, respectively [84]. The major sphingolipids found in plants are cerebrosides and free ceramides (Table 2.4). The cerebroside content of almonds, cashews, hazelnuts, pine nuts, and walnuts range from 2.1 to 6.8 mg/100 g edible portion [85]. Studies have shown that mice treated with a colon carcinogen (1,2-dimethylhydrazine) had 70% reduced formation of aberrant crypt foci (ACF), a precursor for colon carcinogenesis, after feeding a diet containing sphingolipids compared to the control group that did not contain any sphingolipids [86]. Another study found up to 31% of the 1,2-dimethylhydrazine-induced tumors of sphingolipid-fed mice were adenomas, whereas all the tumors of the control group were malignant adenocarcinomas [87]. In a more recent study, a diet of whole almonds decreased azoxymethane induced ACF by 30% compared to wheat bran diet and by 40% compared to cellulose diet in a rat model of colon carcinogenesis [88].

2.3.1.4 Phytosterols

Tree nuts a good source of phytosterols, mainly contain β-sitosterol (Table 2.4). Phytosterols interfere with cholesterol absorption, which results in reduction of serum LDL cholesterol levels [89]. The mechanism of action for phytosterols still remains unclear. One of the suggested mechanisms is that phytosterols being more hydrophobic than cholesterol have a higher affinity for micelles and may compete with cholesterol for incorporation into mixed micelles in the intestinal tract, thus resulting in reduced cholesterol absorption and higher fecal excretion of cholesterol [90,91]. Another mechanism is that phytosterols increase cholesterol efflux out of the intestinal enterocytes back into the lumen by activating higher expression of adenosine triphosphate-binding cassette (ABC) transporter A1, G5, and G8, transporter proteins that actively transport cholesterol across cell membranes. Therefore, less cholesterol is incorporated into chylomicrons for entry into circulation. A lower level of intestinal-derived cholesterol prompts cells to restore cellular cholesterol homeostasis by other mechanisms. These alternative mechanisms include (1) increasing the expression of total LDL receptors that in turn decreases LDL formation along the apolipoprotein B cascade [90] and (2) increase in cholesterol synthesis [90,91]. The resulting effect of reduced serum LDL cholesterol has been suggested as a reason for the role of phytosterols in decreasing atherosclerosis through decreased plaque formation [91,92].

2.3.2 Amino Acids

Tree nuts have a high arginine to lysine ratio (Table 2.5), typically 1.75–7.69 [93]. High lysine levels can negatively influence *in vivo* arginine uptake by cells because lysine shares the same transport system with arginine [94], and therefore the low lysine levels in tree nuts may not necessarily be a negative attribute. Arginine is the precursor for nitric oxide (NO), which is oxidized by NO synthases (NOS), (endothelial NOS ([eNOS] or inducible NOS [iNOS]) to produce NO. NO has a wide range of biological properties that maintain vascular homeostasis, including vasodilation, antioxidative, and antiplatelet effects [95–97]. NO increases the concentration of cyclic guanosyl monophosphate (cGMP), which produces relaxation in vascular smooth muscle and decreases platelet activation and adhesion on the endothelial surface [95]. Increased incidence of CVD is associated with reduced biological activity of NO, also known as endothelial dysfunction [95]. The risk of developing hypercholesterolemia and atherosclerosis (both markers of CVD) is increased when foods consumed have a high lysine to arginine ratio [93]. Ros et al. [64] have shown that incorporating 8–13 whole walnuts (~40–65 g) per day increased dietary L-arginine by 0.9 to 1.4 g/day. Other studies have shown that oral supplementation of L-arginine has improved endothelial function in subjects with impaired NO synthesis [98–100]. Wells et al. [16] reported that diets containing foods with high levels of arginine, including pecans, pistachios, and cashews, resulted in lower levels of CRP (a marker for inflammation).

2.3.3 Antioxidants and Phenolic Compounds

Available evidence suggests that free radical-mediated oxidative stress increases the risk of many diseases, including CVD, diabetes, and certain cancers. Oxidative stress is generally mediated by reactive oxygen species (ROS), which in excess can cause oxidative damage to nucleic acids, proteins, and lipids, thus impairing their normal biological function. Antioxidants such as vitamin C, vitamin E, selenium, and certain phenolic compounds counter the negative effects of ROS. Chestnuts are a good source of vitamin C (12.2 mg vitamin C per 28.35 g [1 ounce] serving) [6], which is required for collagen synthesis, carnitine synthesis, neurotransmitter synthesis (e.g., norepinephrine and serotonin), and as a reducing agent or electron donor (antioxidant activity). The RDA for vitamin C is 90 mg/day for men (19–50 years of age) and 75 mg/day for women (19–50 years of age) [101]. Almond is an excellent source of vitamin E (7.4 mg α-tocopherol per 28.35 g serving) [6], which may inhibit lipid peroxidation and platelet aggregation and act as an anti-inflammatory agent [102]. The RDA for vitamin E is 15 mg/day for both men and women (19–50 years of age) [101]. Jambazian et al. [49] found that incorporating almonds into the diet significantly increased blood α-tocopherol levels. Brazil nuts contain significant amounts of selenium. It is an essential cofactor for glutathione peroxidase (GPX), which acts to neutralize or eliminate ROS and thus prevents lipid peroxidation and cellular damage. Selenium availability has been shown to affect mRNA concentrations of GPX, which leads to decreased GPX synthesis and may result in higher risk of cellular damage from ROS. Several studies have investigated the role of selenium as a chemopreventive agent for a variety of cancers, including breast, colorectal, esophageal, lung, prostate, and stomach [103–106]. Selenium and vitamin E work synergistically. Selenium prevents free radical production by reducing peroxide concentrations in the cell and vitamin E neutralizes the free radicals once they are produced. Several studies suggest that both selenium and vitamin E are effective in reducing the risk of developing prostate cancer in humans [107–109]. One study found Brazil nuts administered to rats lead to increased selenium retention in the liver, kidney, mammary gland, and plasma, which was associated with increased prevention of mammary cancer [110]. The RDA for selenium in both men and women aged 19–50 years is 55 µg/day [101]. The selenium content in Brazil nuts is highly dependent on the soil selenium concentration as well as growth location [111]. On an average, Brazil nuts provide 543.5 µg of selenium per 28.35 g serving [6], which exceeds the U.S. RDA for selenium (Table 2.8).

Tree nuts contain many phenolic compounds, including ellagic acid, resveratrol, and flavonoids (e.g. quercetin and kaempferol). Phenolic compounds have been identified in seeds of almond [21–23], hazelnut [112], macadamia [113], walnut [114], cashew [115,116], and pistachio [35]. Several studies have suggested that phenolic compounds have anticarcinogenic properties [72,117,118] and the ability to reduce the risk of CVD by inhibiting LDL oxidation, platelet aggregation, and lowering blood pressure [117,119]. Polyphenolics in almonds and walnuts have been shown to be effective inhibitors of LDL oxidation in animal [120,121] and human [114,120,122] *in vitro* studies. Ellagic acid, a phenolic compound present in walnuts and pecans, has been shown to be antimutagenic, anticarcinogenic, antiviral, antioxidative, and anti-inflammatory in animal studies [117,123]. One study also demonstrated that ellagic acid was effective at reducing serum TAG and total cholesterol and exhibited free-radical scavenging activities and inhibited LDL oxidation in hyperlipidemic rabbits. The study suggests that ellagic acid may prevent CVD via suppression of oxidative stress and apoptosis [123]. In another study, ellagic acid fed to mice was shown to significantly reduce

N-nitrosodiethylamine-induced lung tumor incidence to 20% from the control value of 72.2% [124]. The study also determined that ellagic acid caused a significant increase in reduced glutathione (GSH), an antioxidant responsible for reduced lipid peroxidation. Other phenolic phytochemicals present in tree nuts, resveratrol and flavonoids, have been shown to increase NO synthesis by endothelial cells and platelets [94].

2.3.4 Folate

Tree nuts, especially acorn nut, beechnut, and hazelnut, are a good source of folate (ranging from 11 to 114 μg/100 g edible portion, Table 2.6), an essential micronutrient that must be obtained from the diet [125]. The current RDA for adults aged 19–50 years (excluding pregnant and lactating women) is 400 μg daily [101]. Folate plays an important role in nucleotide synthesis, methylation, gene expression, and protein synthesis. Folate deficiency may lead to megaloblastic anemia, congenital defects (e.g., neural tube defects like spina bifida, conotruncal heart defects [also known as outflow tract defects] and craniofacial malformations), increased risk for CVD by elevating levels of homocysteine (hyperhomocyteinemia), and increased risk for certain cancers [125,126]. Consuming certain tree nuts as part of a healthy diet may contribute to an adequate folate status.

2.3.5 Fiber

Tree nuts are a good source of fiber, ranging from 2.4 to 33.5 g/100 g edible portion (Table 2.2) when consumed with seed skins. Dietary fiber is perceived to have a wide range of health benefits, including protective roles in both gastrointestinal and cardiovascular health. Among the various effects of dietary fiber on the gastrointestinal system, a decrease in the length of intestinal transit time and an increase in fecal volume appear to be beneficial as the former leads to quick removal of carcinogens while the latter helps to dilute carcinogens, possibly reducing the risk of colorectal cancer [42,127]. Fiber is also beneficial to cardiovascular health as it enhances satiety, which is associated with lower body mass index (BMI), reduced blood cholesterol, and postprandial glucose response [74]. Current fiber intake recommendations are 38 g/day for men and 25 g/day for women [101].

2.3.6 Weight Control

Tree nuts, due to their high fat content, have been perceived to be associated with weight gain. However, several studies suggest that nut consumption leads to no change [46–48,50,52,53,56,58,59,64,66,67,128,129] or a decrease [55,130] in body weight. One study found small but a significant (0.9 kg for men and 0.3 kg for women) weight gain associated with nut consumption over 4 weeks [51]. The reasons for lack of weight gain remain unclear although several suggested causes include incomplete fat absorption, changes in fat metabolism and storage, and satiating properties of nut seed fat and fiber [43,93,131]. Increased excretion of dietary fat in feces has been observed with consumption of pecans and almonds [132,133]. One possible explanation is chewing only damaged the first layer of cells on the nut, resulting in an incomplete release of the fatty acids [132]. The unsaturated fat content in tree nuts may also have a positive effect on fat metabolism and storage by increasing fat oxidation through enhancing diet-induced thermogenesis, resulting in less body fat accumulation [132].

Nut seeds are also high in protein and low in carbohydrates. Diets high in protein and low carbohydrate have been shown to decrease hunger, reduce body fat, lower blood pressure, and improve blood lipid levels by decreasing TAG and LDL cholesterol levels [43,93,134–137]. Thus, inclusion of moderate amounts of nut seeds as a part of a well-balanced food intake, ideally when substituted for other fat or protein sources in the diet, may be useful in designing adequate diets intended for weight loss and weight control.

2.3.7 Allergy

Detailed information on tree nut allergens is given in Chapter 4. While tree nuts have been shown to play a protective role is health and disease, it is important to note that they can also induce adverse reactions in susceptible individuals. Tree nuts are one of the "Big 8" allergy-causing foods with symptoms varying substantially depending on the individual, ranging from hives, itching, and swelling to life-threatening anaphylaxis. It is estimated that 12 million Americans have food allergies with 3.3 million allergic to peanuts or tree nuts (~1.1% of the population) [138]. Every year in the United States, food allergies result in over 30,000 emergency room visits and between 150 and 200 fatalities [138]. Of the Americans with peanut or tree nut allergies, 50% were reactive to peanuts, 30% to walnuts, 10% to almonds, and 4% to both peanuts and tree nuts [139] and 10% of allergic individuals were reactive to two or more nuts [140].

2.4 CONCLUSION

Although calorie-rich, tree nuts are a good source of several nutrients including proteins, certain vitamins and minerals, and several bioactive components such as sterols. The high calories in tree nuts are mainly due to their high lipid content. However, tree nut lipids, with few exceptions, are rich in MUFA and PUFA. Literature reports suggest environment to influence chemical composition of nut seeds, indicating the need for further research to carefully evaluate the possibilities of producing nut seeds with desirable quality traits. When consumed in moderation (<1.5 ounces or ~42.5 g/day) as a part of a balanced and varied food intake, tree nuts may offer several health benefits that may include possible reduction in CVD risks, help manage satiety, management of obesity, and obesity associated risks such as type-2 diabetes. The role of nut seed chemical constituents needs to be carefully evaluated. Of particular interest would be to learn the possible protective effects of nut seed components in prevention of chronic diseases in humans.

ACKNOWLEDGEMENT

Partial financial support from the Department of Nutrition, Food and Exercise Sciences and USDA NRICGP (#9901530; #2004-35503-14117; #2006-35503-17229) is gratefully acknowledged.

REFERENCES

1. U.S. Department of Agriculture (USDA), *Fruit and Nut Situation and Outlook Yearbook*, USDA, Springfield, VA, October 2007.
2. The Almond Board of California, *Almond Almanac*, Almond Board of California, Modesto, CA, 2006.
3. Maskan, M. and Karatas, S., Storage stability of whole-split pistachio nuts (*Pistacia vera* L.) at various conditions, *Food Chem.*, 66, 227–233, 1999.
4. Nanos, G.D., Kazantzis, I., Kefalas, P., Petrakis, C., and Stavroulakis, G.G., Irrigation and harvest time affect almond kernel quality and composition, *Sci. Hortic.*, 96, 249–256, 2002.
5. Singanusong, R., Mason, R.L., D'Arcy, B.R., and Nottingham, S.M., Compositional changes of Australia-grown Western Schley pecans [*Carya illinoinensis* (Wangenh.) K. Koch] during maturation, *J. Agric. Food Chem.*, 51, 406–412, 2003.
6. U.S. Department of Agriculture (USDA), *National Nutrient Database for Standard Reference, Release 19,* Published online at: http://www.nal.usda.gov/fnic/foodcomp/search/ (accessed February 2007).
7. García-Lórda, C., Grané-Teruel, N., Berenguer-Navarro, V., García-García, J.E., and Martín Carratal, M.L., Major fatty acid composition of 19 almond cultivars of different origins. A chemometric approach, *J. Agric. Food Chem.*, 44, 1751–1755, 1996.
8. Ruggeri, S., Cappelloni, M., Gambelli, L., and Carnovale, E., Chemical composition and nutritive value of nuts grown in Italy, *Ital. J. Food Sci.*, 10, 243–252, 1998.
9. Sathe, S.K., Solubilization, electrophoretic characterization and in vitro digestibility of almond (*Prunus amygdalus*) proteins, *J. Food Biochem.*, 16, 249–264, 1993.

10. Venkatachalam, M., Kshirsagar, H.H., Seeram, N.P., Heber, D., Thompson, T.E., Roux, K.H. and Sathe, S.K., Biochemical composition and immunological comparison of select pecan [*Carya illinoinensis* (Wangenh.) K. Koch] cultivars, *J. Agric. Food Chem.*, 55, 9899-9907, 2007.
11. Parcerisa, J., Boatella, J., Codony, R., Farra, A., Garcia, J., Lopez, A., Rafecas, M., and Romero, A., Influence of variety and geographical origin on the lipid fraction of hazelnuts (*Corylus avellana* L.) from Spain: I. Fatty acid composition, *Food Chem.*, 48, 411–414, 1993.
12. Ryan, E., Galvin, K., O'Connor, T.P., Maguire, A.R., and O'Brien, N.M., Fatty acid profile, tocopherol, squalene and phytosterol content of Brazil, pecan, pine, pistachio and cashew nuts, *Int. J. Food Sci. Nutr.*, 57, 219–228, 2006.
13. Phillips, K.M., Ruggio, D.M., and Shraf-Khorassani, M. Phytosterol composition of nuts and seeds commonly consumed in the United States, *J. Agric. Food Chem.*, 53, 9436–9445, 2005.
14. Maguire, L.S., O'Sullivan, S.M., Galvin, K., O'Connor, T.P., and O'Brien, N.M., Fatty acid profile, tocopherol, squalene and phytosterol content of walnuts, almonds, peanuts, hazelnuts and the macadamia nut, *Int. J. Food Sci. Nutr.*, 55, 171–178, 2004.
15. Venkatachalam, M. and Sathe, S.K., Chemical composition of selected edible nut seeds, *J. Agric. Food Chem.*, 54, 4705–4714, 2006.
16. Wells, B.J., Mainous III, A.G., and Everett, C.J., Association between dietary arginine and C-reactive protein, *Nutrition*, 21, 125–130, 2005.
17. Alasalvar, C., Shahidi, F., Ohshima, T., Wanasundara, U., Yurttas, H.C., Liyanapathirana, C.M., and Rodrigues, F.B., Turkish Tombul hazelnut (*Corylus avellana* L.). 1. Compositional characteristics, *J. Agric. Food Chem.*, 51, 3790–3805, 2003.
18. Li, L., Tsao, R., Yang, R., Kramer, J.K., and Hernandez, M., Fatty acid profiles, tocopherol contents, and antioxidant activities of heartnut (*Juglans ailanthifolia* var. *cordiformis*) and Persian walnut (*Juglans regia* L.), *J. Agric. Food Chem.*, 55, 1164–1169, 2007.
19. Çağlarırmak, N., Biochemical and physical properties of some walnut genotypes (*Juglans regia* L.), *Nahrung,* 47, 28–32, 2003.
20. Çağlarırmak, N. and Batkan A.C., Nutrients and biochemistry of nuts in different consumption types in Turkey, *J. Food Proc. Preserv.*, 29, 407–423, 2005.
21. Amarowicz, R., Troszynska, A., and Shahidi, F., Antioxidant activity of almond seed extract and its fractions, *J. Food Lipids*, 12, 344–358, 2005.
22. Milbury, P.E., Chen, C.Y., Dolnikowski, G.G., and Blumberg, J.B., Determination of flavonoids and phenolics and their distribution in almonds, *J. Food Proc. Preserv.*, 54, 5027–5033, 2006.
23. Kornsteiner, M., Wagner, K.H., and Elmadfa, I., Tocopherols and total phenolics in 10 different nut types, *Food Chem.*, 98, 381–387, 2006.
24. Siriwardhana, S.S.K.W. and Shahidi, F., Antiradical activity of extracts of almond and its by-products, *J. Am. Oil Chem. Soc.*, 79, 903–908, 2002.
25. Shahidi, F., Alasalvar, C., and Liyana-Pathirana, C.M., Antioxidant phytochemicals in hazelnut kernel (*Corylus avellana* L.) and hazelnut byproducts, *J. Agric. Food Chem.*, 55, 1212–1220, 2007.
26. Fukuda, T., Ito, H., and Yoshida, T., Antioxidative polyphenols from walnuts (*Juglans regia* L.), *Phytochemistry*, 63, 795–801, 2003.
27. Alamprese, C., Pompei, C., and Scaramuzzi, F., Characterization and antioxidant activity of nocino liqueur, *Food Chem.*, 90, 495–502, 2005.
28. Stampar, F., Solar, A., Hudina, M., Veberic, R., and Colaric, M., Traditional walnut liqueur—cocktail of phenolics, *Food Chem.*, 95, 627–631, 2006.
29. Tokusoglu, O., Unal, M.K., and Yemis, F., Determination of the phytoalexin resveratrol (3,5,4′-trihydroxystilbene) in peanuts and pistachios by high-performance liquid chromatographic diode array (HPLC-DAD) and gas chromatography–mass spectrometry (GC–MS), *J. Agric. Food Chem.*, 53, 5003–5009, 2005.
30. Smeds, A.I., Eklund, P.C., Sjöholm, R.E., Willför, S.M., Nishibe, S., Deyama, T., and Holmbom, B.R., Quantification of a broad spectrum of lignans in cereals, oilseeds, and nuts, *J. Agric. Food Chem.*, 55, 1337–1346, 2007.
31. Wu, X., Beecher, G.R., Holden, J.M., Haytowitz, D.B., Gebhardt, S.E., and Prior, R.L., Lipophilic and hydrophilic antioxidant capacities of common foods in the United States, *J. Agric. Food Chem.*, 52, 4026–4037, 2004.
32. Gu, L., Kelm, M.A., Hammerstone, J.F., Beecher, G., Holden, J., Haytowitz, D., Gebhardt, S., and Prior, R.L., Concentrations of proanthocyanidins in common foods and estimations of normal consumption, *J. Nutr.*, 134, 613–617, 2004.

33. Sang, S., Lapsley, K., Jeong, W.S., Lachance, P.A., Ho, C.-T., and Rosen, R.T., Antioxidative phenolic compounds isolated from almond skins (*Prunus amygdalus* Batsch), *J. Agric. Food Chem.*, 50, 2459–2463, 2002.
34. Wu, X. and Prior, R.L., Identification and characterization of anthocyanins by high-performance liquid chromatography–electrospray ionization–tandem mass spectrometry in common foods in the United States: Vegetables, nuts, and grains, *J. Agric. Food Chem.*, 53, 3101–3113, 2005.
35. Seeram, N., Zhang, Y, Henning, S.M., Lee, R., Niu, Y., Lin, G., and Heber, D., Pistachio skin phenolics are destroyed by bleaching resulting in reduced antioxidative capacities, *J. Agric. Food Chem.*, 54, 7036–7040, 2006.
36. Sano, T, Oda, E., Yamashita, T., Naemura, A., Ijiri, Y., Yamakoshi, J., and Yamamoto, J., Anti-thrombotic effect of proanthocyanidin, a purified ingredient of grape seed, *Thromb. Res.*, 115, 115–121, 2005.
37. Griel, A.E. and Kris-Etherton, P.M., Tree nuts and the lipid profile: A review of clinical studies, *Br. J. Nutr.*, 96 (Suppl. 2), 68S–78S, 2006.
38. Hu, F.B., Plant-based foods and prevention of cardiovascular disease: An overview, *Am. J. Clin. Nutr.*, 78, 544S–551S, 2003.
39. Kelly, J.H. and Sabaté, J., Nuts and coronary heart disease: An epidemiological perspective, *Br. J. Nutr.*, 96, 61S–67S, 2006.
40. Ros, E. and Mataix, J., Fatty acid composition of nuts—Implications for cardiovascular health, *Br. J. Nutr.*, 96 (Suppl. 2), 29S–35S, 2006.
41. Sabaté, J, Ros, E., and Salas-Salvado J., Nuts: Nutrition and health outcomes, *Br. J. Nutr.*, 96 (Suppl. 2), 1S–2S, 2006.
42. Gonzalez, C.A. and Salas-Salvado J., The potential of nuts in the prevention of cancer, *Br. J. Nutr.*, 96 (Suppl. 2), 87S–94S, 2006.
43. García-Lórda, P., Megias R.I., and Salas-Salvado J., Nut consumption, body weight and insulin resistance, *Eur. J. Clin. Nutr.*, 57 (Suppl.1), 8S–11S, 2003.
44. Jiang, R., Jacobs, D.R. Jr., Mayer-Davis, E., Szklo, M., Herrington, D., Jenny, N.S., Kronmal, R., and Barr, R.G., Nut and seed consumption and inflammatory markers in the multi-ethnic study of atherosclerosis, *Am. J. Epidemiol.*, 163, 222–231, 2006.
45. Food and Drug Administration (FDA), Qualified Health Claims: Letter of Enforcement Discretion—Nuts and Coronary Heart Disease, Docket No 02P-0505, Food and Drug Administration, Washington, DC, 2003.
46. Albert, C.M., Gaziano, J.M., Willett, W.C., and Manson, J.E., Nut consumption and decreased risk of sudden cardiac death in the physicians' health study, *Arch. Intern. Med.,* 162, 1382–1387, 2002.
47. Fraser, G.E., Sabaté, J., Beeson, W.L., and Strahan, T.M., A possible protective effect of nut consumption on risk of coronary heart disease. The adventist health study, *Arch. Intern. Med.*, 152, 1416–1424, 1992.
48. Abbey, M., Noakes, M., Belling, G.B., and Nestel, P.J., Partial replacement of saturated fatty acids with almonds or walnuts lowers total plasma cholesterol and low-density-lipoprotein cholesterol, *Am. J. Clin. Nutr.*, 59, 995–999, 1994.
49. Jambazian, P.R., Haddad, E., Rajaram, S., Tanzman, J., and Sabaté, J., Almonds in the diet simultaneously improve plasma [α]-tocopherol concentrations and reduce plasma lipids, *J. Am. Diet. Assoc.*, 105, 449–454, 2005.
50. Jenkins, D.J.A., Kendall, C.W.C., Marchie, A., Parker, T.L., Connelly, P.W., Qian, W., Haight, J.S., Faulkner, D., Vidgen, E., Lapsley, K.G., and Spiller, G.A., Dose response of almonds on coronary heart disease risk factors: Blood lipids, oxidized low-density lipoproteins, lipoprotein (a), homocysteine, and pulmonary nitric oxide: A randomized, controlled, crossover trial, *Circulation*, 106, 1327–1332, 2002.
51. Lovejoy, J.C., Most, M.M., Lefevre, M., Greenway, F.L., and Rood, J.C., Effect of diets enriched in almonds on insulin action and serum lipids in adults with normal glucose tolerance or type 2 diabetes, *Am. J. Clin. Nutr.*, 76, 1000–1006, 2002.
52. Spiller, G.A., Jenkins, D.A.J., Bosello, O., Gates, J.E., Cragen, L.N., and Bruce, B., Nuts and plasma lipids: An almond-based diet lowers LDL-C while preserving HDL-C, *J. Am. Coll. Nutr.*, 17, 285–290, 1998.
53. Mercanlıgil, S.M., Arslan, P., Alasalvar, C., Okut, E., Pınar, A., Geyik, P.Ö., Tokgözoğlu, L., and Shahidi, F., Effects of hazelnut-enriched diet on plasma cholesterol and lipoprotein profiles in hypercholesterolemic adult men, *Eur. J. Clin. Nutr.*, 61, 212–220, 2007.
54. Curb, J.D., Wergowske, G., Dobbs, J.C., Abbott, R.D., and Huang, B., Serum lipid effects of a high-monounsaturated fat diet based on macadamia nuts, *Arch. Intern. Med.*, 160, 1154–1158, 2000.
55. Garg, M.L., Blake, R.J., and Wills, R.B.H., Macadamia nut consumption lowers plasma total and LDL cholesterol levels in hypercholesterolemic men, *J. Nutr.*, 133, 1060–1063, 2003.

56. Morgan, W.A. and Clayshulte, B.J., Pecans lower low density lipoprotein cholesterol in people with normal lipid levels, *J. Am. Diet. Assoc.*, 100, 312–318, 2000.
57. Rajaram, S., Burke, K., Connell, B., Myint, T., and Sabaté, J., A monounsaturated fatty acid-rich pecan-enriched diet favorably alters the serum lipid profile of healthy men and women, *J. Nutr.*, 131, 2275–2279, 2001.
58. Edwards, K., Kwaw, I., Matud, J., and Kurtz, I. Effect of pistachio nuts on serum lipid levels in patients with moderate hypercholesterolemia, *J. Am. Coll. Nutr.*, 18, 229–232, 1999.
59. Kocyigit, A., Koylu, A.A., and Keles, H., Effects of pistachio nuts consumption on plasma lipid profile and oxidative status in healthy volunteers, *Nutr. Metab. Card. Dis.*, 16, 202–209, 2006.
60. Chisholm, A., Mc Auley, K., Mann, J., Williams, S., and Skeaff, M., Cholesterol lowering effects of nuts compared with a canola oil enriched cereal of similar fat composition, *Nutr. Metab. Card. Dis.*, 15, 284–292, 2005.
61. Mukuddem-Petersen, J., Oosthuizen, W., and Jerling, J.C., A systematic review of the effects of nuts on blood lipid profiles in humans, *J. Nutr.*, 135, 2082–2089, 2005.
62. Iwamoto, M., Imaizumi, K., Sato, M., Hirooka, Y., Sakai, K., Takeshita, A., and Kono, M., Serum lipid profiles in Japanese women and men during consumption of walnuts, *Eur. J. Clin. Nutr.*, 56, 629–637, 2002.
63. Munoz, S., Merlos, M., Zambón, D., Rodríguez, C., Sabaté, J., Ros, E., and Laguna, J.C., Walnut-enriched diet increases the association of LDL from hypercholesterolemic men with human HepG2 cells, *J. Lipid Res.*, 42, 2069–2076, 2001.
64. Ros, E., Núnez, I., Pérez-Heras, A., Serra, M., Gilabert, R., Casals, E., and Deulofeu, R., A walnut diet improves endothelial function in hypercholesterolemic subjects: A randomized crossover trial, *Circulation*, 109, 1609–1614, 2004.
65. Sabaté, J., Fraser, G.E., Burke, K., Knutsen, S.F., Bennett, H., and Lindsted, K.D., Effects of walnuts on serum lipid levels and blood pressure in normal men, *N. Engl. J. Med.,* 328, 603–607, 1993.
66. Tapsell, L.C., Gillen, L.J., Patch, C.S., Batterham, M., Owen, A., Baré, M., and Kennedy, M., Including walnuts in a low-fat/modified-fat diet improves HDL cholesterol-to-total cholesterol ratios in patients with type 2 diabetes, *Diabetes Care*, 27, 2777–2783, 2004.
67. Zambón, D., Sabaté, J., Munoz, S., Campero, B., Casals, E., Merlos, M., Laguna, J.C., and Ros, E., Substituting walnuts for monounsaturated fat improves the serum lipid profile of hypercholesterolemic men and women: A randomized crossover trial, *Ann. Intern. Med.*, 132, 538–546, 2000.
68. Jenkins, D.J.A., Kendall, C.W.C., Josse, A.R., Salvatore, S., Brighenti, F., Augustin, L.S.A., Ellis, P.E., Vidgen, E., and Rao, A.V., Almonds decrease postprandial glycemia, insulinemia, and oxidative damage in healthy individuals, *J. Nutr.*, 136, 2987–2992, 2006.
69. Jiang, R., Manson, J.E., Stampfer, M.J., Liu, S., Willett, W.C., and Hu, F.B., Nut and peanut butter consumption and risk of type 2 diabetes in women, *J. Am. Med. Assoc.*, 288, 2554–2560 2002.
70. Parillo, M., Riviellesse, A.A., Ciardullo, A.V., Capaldo, B., Giacco, A., Genovese, S., and Riccardi, G., A high-monounsaturated-fat/low-carbohydrate diet improves peripheral insulin sensitivity in non-insulin-dependent diabetic patients, *Metabolism*, 41, 1373–1378, 1992.
71. Garg, A., Bonanome, A., Grundy, S.M., Zhang, Z.J., and Unger, R.H., Comparison of a high-carbohydrate diet with a high-monounsaturated-fat diet in patients with non-insulin-dependent diabetes mellitus, *N. Engl. J. Med.,* 316, 829–834, 1988.
72. Blomhoff, R., Carlsen, M.H., Anderson, L.F., and Jacobs, D.R. Jr., Health benefits of nuts: Potential role of antioxidants, *Br. J. Nutr.*, 96 (Suppl. 2), 52S–60S, 2006.
73. Pischon, T., Hankinson, S.E., Hotamisligil, G.S., Rifai, N., Willett, W.C., and Rimm, E.B., Habitual dietary intake of *n*–3 and *n*–6 fatty acids in relation to inflammatory markers among US men and women, *Circulation*, 108, 155–160, 2003.
74. Salas-Salvadó J., Bulló, M., Pérez-Heras, A., and Ros, E., Dietary fibre, nuts and cardiovascular diseases, *Br. J. Nutr.*, 96 (Suppl. 2), 45S–51S, 2006.
75. Zhao, G., Etherton, T.D., Martin, K.R., West, S.G., Gillies, P.J., and Kris-Etherton, P.M., Dietary α-linolenic acid reduces inflammatory and lipid cardiovascular risk factors in hypercholesterolemic men and women, *J. Nutr.*, 134, 2991–2997, 2004.
76. Mutanen, M. and Freese, R., Fats, lipids and blood coagulation, *Curr. Opin. Lipidol.*, 12, 25–29, 2001.
77. Rallidis, L.S., Paschos, G., Liakos, G.K., Velissaridou, A.H., Anastasiadis, G., and Zampelas, A., Dietary α-linolenic acid decreases C-reactive protein, serum amyloid A and interleukin-6 in dyslipidaemic patients, *Atherosclerosis*, 167, 237–242, 2003.
78. Calder, P.C., Polyunsaturated fatty acids and inflammation, *Biochem. Soc. Trans.*, 33, 423–427, 2005.

79. Harris, W.S. and Bulchandani, D., Why do omega-3 fatty acids lower serum triglycerides? *Curr. Opin. Lipidol.*, 17, 387–393, 2006.
80. Sacks, F.M. and Campos, H., Polyunsaturated fatty acids, inflammation, and cardiovascular disease: Time to widen our view of the mechanisms, *J. Clin. Endocrinol. Metab.*, 91, 398–400, 2006.
81. Yaqoob, P., Fatty acids and the immune system: From basic science to clinical applications, *Proc. Nutr. Soc.*, 63, 89–104, 2004.
82. Perrotta, C., Palma, C.D., Falcone, S., Sciorati, C., and Clementi, E., Nitric oxide, ceramide and sphingomyelinase-coupled receptors: A tale of enzymes and messengers coordinating cell death, survival and differentiation, *Life Sci.*, 77, 1732–1739, 2005.
83. Schmelz, E.M., Dietary sphingomyelin and other sphingolipids in health and disease, *Nutr. Bullet.*, 25, 135–139, 2007.
84. Vesper, H., Schmelz, E.M., Nikolova-Karakashian, M.N., Dillehay, D.L., Lynch, D.V., and Merrill, A.H. Jr., Sphingolipids in food and the emerging importance of sphingolipids to nutrition, *J. Nutr.*, 129, 1239–1250, 1999.
85. Fang, F., Ho, C.-T., Sang, S., and Rosen, R.T., Determination of sphingolipids in nuts and seeds by a single quadrupole liquid chromatography–mass spectrometry method, *J. Food Lipids*, 12, 327–343, 2005.
86. Dillehay, D.L., Webb, S.K., Schmelz, E.M., and Merrill, A.H. Jr., Dietary sphingomyelin inhibits 1,2-dimethylhydrazine-induced colon cancer in CF1 mice, *J. Nutr.*, 124, 615–620, 1994.
87. Schmelz, E.M., Dillehay, D.L., Webb, S.K., Reiter, A., Adams, J., and Merrill, A.H. Jr., Sphingomyelin consumption suppresses aberrant colonic crypt foci and increases the proportion of adenomas versus adenocarcinomas in CF1 mice treated with 1,2-dimethylhydrazine: Implications for dietary sphingolipids and colon carcinogenesis, *Cancer Res.*, 56, 4936–4941, 1996.
88. Davis, P.A. and Iwahashi, C.K., Whole almonds and almond fractions reduce aberrant crypt foci in a rat model of colon carcinogenesis, *Cancer Lett.*, 165, 27–33, 2001.
89. Thompson, G.R. and Grundy, S.M., History and development of plant sterol and stanol esters for cholesterol-lowering purposes, *Am. J. Cardiol.*, 96 (Suppl. 1), 3S–9S, 2005.
90. Plat, J. and Mensink, R.P., Plant stanol and sterol esters in the control of blood cholesterol levels: Mechanism and safety aspects, *Am. J. Cardiol.*, 96 (Suppl. 1), 15S–22S, 2005.
91. de Jong, A., Plat, J., and Mensink, R.P., Metabolic effects of plant sterols and stanols (review), *J. Nutr. Biochem.*, 14, 362–369, 2003.
92. Moghadasian, M.H., McManus, B.M., Pritchard, P.H., and Frohlich, J.J., Tall oil-derived phytosterols reduce atherosclerosis in ApoE-deficient mice, *Arterioscler. Thromb. Vasc. Biol.*, 17, 119–126, 1997.
93. Brufau, G., Boatella, J., and Rafecas, M., Nuts: Source of energy and macronutrients, *Br. J. Nutr.*, 96 (Suppl. 2), 24S–28S, 2006.
94. Wu, G. and Meininger, C.J., Regulation of nitric oxide synthesis by dietary factors, *Annu. Rev. Nutr.*, 22, 61–86, 2002.
95. Cannon, R.O., III. Role of nitric oxide in cardiovascular disease: Focus on the endothelium, *Clin. Chem.*, 44, 1809–1819, 1998.
96. de Lorgeril, M., Dietary arginine and the prevention of cardiovascular diseases, *Cardiovasc. Res.*, 37, 560–563, 1998.
97. Yetik-Anacak, G. and Catravas, J.D., Nitric oxide and the endothelium: History and impact on cardiovascular disease, *Vascul. Pharmacol.*, 45, 268–276, 2006.
98. Brown, A.A. and Hu, F.B., Dietary modulation of endothelial function: Implications for cardiovascular disease, *Am. J. Clin. Nutr.*, 73, 673–686, 2001.
99. Lerman, A., Burnett, J.C. Jr., Higano, S.T., McKinley, L.J., and Holes, D.R. Jr., Long-term L-arginine supplementation improves small-vessel coronary endothelial function in humans, *Circulation*, 97, 2123–2128, 1998.
100. Preli, R.B., Klein, K.P., and Herrington, D.M., Vascular effects of dietary-arginine supplementation, *Atherosclerosis*, 162, 1–15, 2002.
101. Institute of Medicine of the National Academies, *Dietary Reference Intakes*, Published online at: http://www.iom.edu/Object.File/Master/7/296/webtablevitamins.pdf (accessed January 2007).
102. Jialal, I. and Devaraj, S., Scientific evidence to support a vitamin E and heart disease health claim: Research needs, *J. Nutr.*, 135, 348–353, 2005.
103. Connelly-Frost, A., Poole, C., Satia, J.A., Kupper, L.L., Millikan, R.C., and Sandler, R.S., Selenium, apoptosis, and colorectal adenomas, *Cancer Epidemiol. Biomarkers Prev.*, 15, 486–493, 2006.

104. Lee, S.O., Nadiminty, N., Wu, X.X., Lou, W., Dong, Y., Ip, C., Onate, S.A., and Gao, A.C., Selenium disrupts estrogen signaling by altering estrogen receptor expression and ligand binding in human breast cancer cells, *Cancer Res.*, 65, 3487–3492, 2005.
105. Mark, S.D., Qiao, Y.L., Dawsey, S.M., Wu, Y.P., Katki, H., Gunter, E.W., Fraumeni, J.F. Jr., Blot, W.J., Dong, Z.W., and Taylor, P.R., Prospective study of serum selenium levels and incident esophageal and gastric cancers, *J. Natl. Cancer Inst.*, 92, 1753–1763, 2000.
106. Patrick, L., Selenium biochemistry and cancer: A review of the literature, *Altern. Med. Rev.*, 9, 239–258, 2004.
107. Brenneisen, P., Steinbrenner, H., and Sies, H., Selenium, oxidative stress, and health aspects, *Mol. Asp. Med.*, 26, 256–267, 2005.
108. Patel, R., Garg, R., Erande, S., and Maru, G.B., Chemopreventive herbal anti-oxidants: Current status and future perspectives, *J. Clin. Bio. Nutr.*, 40, 82–91, 2007.
109. Klein, E.A., Lippman, S.M., Thompson, I.M., Goodman, P.J., Albanes, D., Taylor, P.R., and Coltman, C., The selenium and vitamin E cancer prevention trial, *World J. Urol.*, 21, 21–27, 2003.
110. Ip, C. and Lisk, D.J., Bioactivity of selenium from Brazil nut for cancer prevention and selenoenzyme maintenance, *Nutr. Cancer*, 21, 203–212, 1994.
111. Dumont, E., Vanhaecke, F., and Cornelis, R., Selenium speciation from food source to metabolites: A critical review, *Anal. Bioanal. Chem.*, 385, 1304–1323, 2006.
112. Alasalvar, C., Karamać, M., Amarowicz, R., and Shahidi, F., Antioxidant and antiradical activities in extracts of hazelnut kernel (*Corylus avellana* L.) and hazelnut green leafy cover, *J. Agric. Food Chem.*, 54, 4826–4832, 2006.
113. Quinn, L.A. and Tang, H.H., Antioxidant properties of phenolic compounds in macadamia nuts, *J. Am. Oil Chem. Soc.*, 73, 1585–1588, 1996.
114. Anderson, K.J., Teuber, S.S., Gobeille, A., Cremin, P., Waterhouse, A.L., and Steinberg, F.M., Walnut polyphenolics inhibit in vitro human plasma and LDL oxidation, *J. Nutr.*, 131, 2837–2842, 2001.
115. Paramashivappa, R., Kumar, P.P., Vithayathil, P.J., and Rao, A.S., Novel method for isolation of major phenolic constituents from cashew (*Anacardium occidentale* L.) nut shell liquid, *J. Agric. Food Chem.*, 49, 2548–2551, 2001.
116. Trevisan, M.T.S., Pfundstein, B., Haubner, R., Würtele, G., Spiegelhalder, B., Bartsch, H., and Owen, R.W., Characterization of alkyl phenols in cashew (*Anacardium occidentale*) products and assay of their antioxidant capacity, *Food Chem. Toxicol.*, 44, 188–197, 2006.
117. Vattem, D.A. and Shetty, K., Biological functionality of ellagic acid: A review, *J. Food Biochem.*, 29, 234–266, 2005.
118. Yang, C.S., Landau, J.M., Huang, M.T., and Newmark, H.L., Inhibition of carcinogenesis by dietary polyphenolic compounds, *Annu. Rev. Nutr.*, 21, 381–406, 2001.
119. Kris-Etherton, P.M., Hecker, K.D., Bonanome, A., Coval, S.M., Binkoski, A.E., Hilpert, K.F., Griel, A.E., and Etherton, T.D., Bioactive compounds in foods: Their role in the prevention of cardiovascular disease and cancer, *Am. J. Med.*, 113 (Suppl. 2), 71S–88S, 2002.
120. Chen, C.Y., Milbury, P.E., Lapsley, K., and Blumberg, J.B., Flavonoids from almond skins are bioavailable and act synergistically with vitamins C and E to enhance hamster and human LDL resistance to oxidation, *J. Nutr.*, 135, 1366–1373, 2005.
121. Milbury, P., Chen, C.Y., Kwak, H.K., and Blumberg, J., Almond skins polyphenolics act synergistically with alpha-tocopherol to increase the resistance of low-density lipoproteins to oxidation, *Free Radical Res.*, 36, 78–80, 2002.
122. Wijeratne, S.S.K., Abou-Zaid, M.M., and Shahidi, F., Antioxidant polyphenols in almond and its coproducts, *J. Agric. Food. Chem.*, 54, 312–318, 2006.
123. Yu, Y.M., Chang, W.C., Wu, C.H., and Chiang, S.Y., Reduction of oxidative stress and apoptosis in hyperlipidemic rabbits by ellagic acid, *J. Nutr. Biochem.*, 16, 675–681, 2005.
124. Khanduja, K.L., Gandhi, R.K., Pathania, V., and Syal, N., Prevention of *N*-nitrosodiethylamine-induced lung tumorigenesis by ellagic acid and quercetin in mice, *Food Chem. Toxicol.*, 37, 313–318, 1999.
125. Segura, R., Javierre, C., Lizarraga, M.A., and Ros, E., Other relevant components of nuts: Phytosterols, folate and minerals, *Br. J. Nutr.*, 96 (Suppl. 2), 36S–44S, 2006.
126. Finnell, R.H., Shaw, G.M., Lammer, E.J., Brandl, K.L., Carmichael, S.L., and Rosenquist, T.H., Gene-nutrient interactions: Importance of folates and retinoids during early embryogenesis, *Toxicol. Appl. Pharmacol.*, 198, 75–85, 2004.
127. Park, Y., Hunter, D.J., Spiegelman, D., Bergkvist, L., Berrino, F., van den Brandt, P.A., Buring, J.E., Colditz, G.A., Freudenheim, J.L., Fuchs, C.S., Giovannucci, E., Goldbohm, R.A., Graham, S., Harnack, L., Hartman, A.M., Jacobs, D.R. Jr., Kato, I., Krogh, V., Leitzmann, M.F., McCullough, M.L., Miller,

A.B., Pietinen, P., Rohan, T.E., Schatzkin, A., Willett, W.C., Wolk, A., Zeleniuch-Jacquotte, A., Zhang, S.M., and Smith-Warner, S.A., Dietary fiber intake and risk of colorectal cancer: A pooled analysis of prospective cohort studies, *J. Am. Med. Assoc.*, 294, 2849–2857, 2005.

128. Almario, R.U., Vonghavaravat, V., Wong, R., and Kasim-Karakas, S.E., Effects of walnut consumption on plasma fatty acids and lipoproteins in combined hyperlipidemia, *Am. J. Clin. Nutr.*, 74, 72–79, 2001.
129. Hu, F.B., Stampfer, M.J., Manson, J.E., Rimm, E.B., Colditz, G.A., Rosner, B.A., Speizer, F.E., Hennekens, C.H., and Willett, W.C., Frequent nut consumption and risk of coronary heart disease in women: Prospective cohort study, *Br. Med. J.*, 317, 1341–1345, 1998.
130. Wien, M.A., Sabaté, J., Iklé, D.N., Cole, S.E., and Kandeel, F.R., Almonds vs complex carbohydrates in a weight reduction program, *Int. J. Obes.*, 27, 1365–1372, 2003.
131. Sabaté, J., Nut consumption and body weight, *Am. J. Clin. Nutr.*, 78, 647S–650S, 2003.
132. Rajaram, S. and Sabaté, J., Nuts, body weight and insulin resistance, *Br. J. Nutr.*, 96 (Suppl. 2), 79S–86S, 2006.
133. Zemaitis, J. and Sabaté, J., Effect of almond consumption on stool weight and stool fat, FASEBJ., 15 (Abstract), A602, 2001.
134. Chen, C.Y., Lapsley, K., and Blumberg, J., A nutrition and health perspective on almonds, *J. Sci. Food Agri.*, 86, 2245–2250, 2006.
135. Hu, F.B., Protein, body weight, and cardiovascular health, *Am. J. Clin. Nutr.*, 82, 242S–247S, 2005.
136. St-Onge, M.P., Dietary fats, teas, dairy, and nuts: Potential functional foods for weight control? *Am. J. Clin. Nutr.*, 81, 7–15, 2005.
137. Westerterp-Plantenga, M.S., Luscombe-Marsh, N., Lejeune, M.P.G.M., Diepvens, K., Nieuwenhuizen, A., Engelen, M.P.K.J., Deutz, N.E.P., Azzout-Marniche, D., Tome, D., and Westerterp, K.R., Dietary protein, metabolism, and body-weight regulation: Dose–response effects, *Int. J. Obes.*, 30 (Suppl. 3), 16S–23S, 2006.
138. Food Allergy and Anaphylaxis Network, *Food Allergy Facts and Statistics*, Patients and Consumers, Fairfax, VA, 2006.
139. Sicherer, S.H., Muñoz-Furlong, A., and Sampson, H.A., Prevalence of peanut and tree nut allergy in the United States determined by means of a random digit dial telephone survey: A 5-year follow-up study, *J. Allergy Clin. Immunol.*, 112, 1203–1207, 2003.
140. Poltronieri, P., Cappello, M.S., Dohmae, N., Conti, A., Fortunato, D., Pastorello, E.A., Ortolani, C., and Zacheo, G., Identification and characterization of the IgE-binding proteins 2S albumin and conglutin gamma in almond (*Prunus dulcis*) seeds, *Int. Arch. Allergy Immunol.*, 128, 97–104, 2002.
141. de Leon, S.Y. and Delores, M.I., Coconut, in *Processing Fruits: Science and Technology,* 2nd ed., Barrett, D.M., Somogyi, L., and Ramaswamy, H.S., Eds., CRC Press, Boca Raton, FL, 2004, pp. 707–727.
142. Pereira-Lorenzo, S., Ramos-Cabrera, A.M., Díaz-Hernándeza, M.B., Ciordia-Arab, M., and Ríos-Mesac, D., Chemical composition of chestnut cultivars from Spain, *Sci. Hortic.*, 107, 306–314, 2006.
143. FAO, *FAO Statistical Yearbooks 2005/2006*, Published online at: http://www.fao.org/statistics/ (accessed October 2008).
144. The Craker, *Global Statistics*, International Tree Nut Council Foundation, Modesto, CA, September, 2006.

3 Health Benefits of Tree Nuts

Mauren E. Ternus, Karen Lapsley, and Constance J. Geiger

CONTENTS

3.1 INTRODUCTION

Tree nuts are produced in more than 40 developed and less-developed countries on millions of hectares of land, providing economic livelihood for hundreds of thousands of producers and small family operations [1]. The International Nut and Dried Fruit Council Foundation (INC, formerly known as the International Tree Nut Council) was formed in 1983 and represents all sectors of the dried fruit and tree nut production and trade of almonds, Brazil nuts, cashews, hazelnuts, macadamias, peanuts, pecans, pine nuts, pistachios, and walnuts [1]. The world tree nuts production from 2000 to 2007 is given in Chapter 1.

In 1995, researchers concluded that tree nuts, as one of the integral plant foods in the traditional Mediterranean diet, had been overlooked by the nutrition research community [2]. In 1997, the World Cancer Research Fund (WCRF) acknowledged that nuts and seeds, which have been common in human diets since preagricultural times, are nutrient dense and a good source of unsaturated fats, protein, dietary fiber, phytochemicals, and micronutrients [3]. Interestingly, the WCRF referred to Brazil nuts, macadamias, and cashews as seeds [3]. According to the WCRF, nuts needed to be identified separately in human studies to better evaluate their protective effects on human health [3].

This chapter will review the nutritional composition of tree nuts as well as the growing body of research on nuts and health. Research over the last decade has focused primarily on nuts and heart disease, but other research areas are emerging as experts examine the effects of nuts on diabetes, satiety, maintenance of healthy body weight, and cancer prevention.

3.1.1 Worldwide Tree Nut Availability

Tree nuts are widely consumed in both raw and processed forms. Unlike groundnuts and seeds, which are used predominantly for oil, tree nuts are consumed primarily as whole foods, as ingredients in foods, or in medicinal preparations. For example, in several Asian cultures, almonds play a significant role in Ayurvedic preparations, a philosophy that for thousands of years has promoted the interrelationship of nutrition and diet with healing, prevention, and longevity.

Tree nut availability varies both among and within the regions of tree nut production (Table 3.1) [4]. For instance, the European region has a tree nut supply of 4.9 kg per person/year overall. However, Greece, Spain, and Italy have all increased nut supply since 1960. Greece has maintained the highest levels and increased nut supply from 8.3 kg/person in 1960 to 11.9 kg/person in 2001. The availability in Spain has increased from 6.2 kg/person in 1960 to 7.3 kg/person in 2001 [4,5]. Over the last 40 years, nut supply in Africa and South America has remained steady at about 1 and 0.3 kg per person/year, respectively. Nut supply increased in North and Central America and in Asia during the same time period (from 1.3 to 1.9 kg per person/year in North and Central America and 0.4 to 1.0 kg per person/year in Asia) [4]. In North America, supply increased from 1.3 kg/person in 1961 to 1.9 kg/person in 2001. U.S. Department of Agriculture (USDA) analysis of U.S. consumption patterns of tree nuts found in 1994–96 that for the 12% of males and 14% of females who ate tree nuts (a range of 1–10 g daily), over 51% of tree nuts were consumed as snacks [6].

Despite a growing visibility of and increasing consumer interest in tree nuts, as well as a greater availability of information on the health effects of higher nut consumption, there is very little objective and reliable information on nut intake profiles and qualitative and quantitative differences in nut consumption patterns globally. Most descriptive information on nut consumption has been based on estimates from food disappearance or market data rather than on individual dietary intakes. In many dietary surveys, questions on nut intake have been asked in insufficient detail or not at all.

Within the European Prospective Investigation into Cancer and Nutrition (EPIC), a cohort study of 520,000 subjects from 23 centers in 10 countries of western Europe provided a unique opportunity to evaluate how tree nuts, peanuts, and seeds are eaten-whole, in hidden sources and in spreads.

TABLE 3.1
Global per Capita Tree Nut Supply

	Per Capita Supply Changes between 1960 and 2001	
Location	**Kg Supply/Year (1960)**	**Kg Supply/Year (2001)**
Africa	1.0	1.0
North and Central America	1.3	1.9
South America	0.2	0.4
Asia	0.4	1.0
Europe	2.7	4.9
Oceania	0.4	3.2

Source: From American Institute for Cancer Research and World Cancer Research Fund, in *Food, Nutrition and the Prevention of Cancer: A Global Perspective,* BANTA Book Group, Menasha, WI, 1997.

From an 8% subset (~36,000 subjects), standardized information on intakes of nuts and seeds was collected via computerized 24 h recalls. The standardized data include;

- Population mean intakes across 10 countries for various nuts and nut forms,
- Comparison of average portion size by country,
- How nuts are consumed across Europe and diversity of dietary and cultural patterns,
- Intake for a particular survey day (habitual versus occasional consumers not identified).

Overall, whole tree nuts were consumed by 4.4% of the population versus 2.3% for peanuts and 1.3% for seeds. Population mean intakes did go as high as 8.3% for Spain and down to 1% for Sweden, and there was a consumption gradient from north (2%) to central (4.3%) to south (6.3%) [7]. In a British health conscious group studied, 15.5% consumed tree nuts versus 2.7% for the general U.K. population cohort. The most common tree nuts consumed were walnuts, almonds, and hazelnuts, with consumption ratios varying by country and regions (north to south) [7].

Tree nut consumption in North America has traditionally been lower than that in European and Mediterranean countries, but consumption has been rising. USDA recently analyzed dietary intake data of 17,306 subjects from 2001 to 2004 to determine daily nut (tree nuts and peanuts) consumption and the contribution to nutrient intakes. USDA found that 34% of all individuals reported consuming nuts in some form with 25% consuming them as ingredients in foods, 28% as peanut butter, and 6% "out of hand" [8]. A consumer obsession to lower fat took its toll on the consumption of tree nuts in North America in the 1980s and the 1990s. As more recent research emerged on unsaturated fats and health, dietary recommendations have moved toward moderate rather than low fat, with specific focus on the need for unsaturated fatty acids in the diet. In 2000, USDA economists concluded that nut eaters had a slightly better quality diet than nonnut eaters and that "although nut consumption is low compared with other protein sources, nuts provide many of the same nutrients to the diet and have potential health benefits [9].

3.1.2 Implications and Directions for Increased Trade and Consumption

Exports of tree nuts have increased over the last decade and continue to rise [1]. Growth has occurred across several key producing origins in northern and southern hemispheres, in response to rising production, as well as expanding consumer awareness of the positive aspects of tree nuts, particularly their health and nutritional benefits. With further production increases projected for several tree nuts such as almonds and cashews, there is an opportunity to maintain this positive momentum. Expanded consumption of tree nuts is certainly a favorable prospect in increasingly prosperous emerging markets such as India and China, but greater market penetration is no less an opportunity in a number of countries such as in western Europe and Japan where tree nut usage is more established. In addition, bilateral and multilateral negotiations are resulting in agreements that reduce tariffs and duties, thereby improving market access. However, sanitary and phytosanitary requirements continue to present potential barriers to trade.

The World Health Organization (WHO) suggested a minimum of 30 g/day of a combination of nuts, seeds, and pulses as part of a recommendation focusing on prevention of some types of cancer and coronary heart disease (CHD) [3]. Nutrition research has demonstrated that as little as 30 g of nutrient-dense nuts per day can have a positive effect on health. Moreover, tree nuts are a nutrient-dense, shelf-stable, nonperishable, whole food source of valuable micro- and macronutrients, which are crucial considerations in countries with limited controlled storage or processing facilities.

For Canada and the United States, the National Academy of Sciences' (NAS) Institute of Medicine issued an updated report "Dietary Reference Intakes for energy, carbohydrates, fiber, fat, protein, and amino acids" or the "Macronutrient Report" establishing dietary fat goals at a wider range, from 20% up to 35% of calories, or from low to moderate amounts of fat for a healthy diet [10]. Although there were no specific values set for monounsaturated fatty acids (MUFA), the NAS report does recommend switching from saturated fatty acids (SFA) to MUFA and polyunsaturated fatty acids (PUFA). The report also set values for the essential fatty acids such as linoleic acid (18:2ω6). Tree nuts contain anywhere from 0.5 g of 18:2ω6 in macadamias to 11 g in walnuts per 1 ounce serving. The average tree nut content of 18:2ω6 (4.9 g) provides 29% of the 17 g/day for men and 41% of the 12 g/day for women.

The 2005 Dietary Guidelines for Americans recommended "Keeping total fat intake between 20% and 35% of calories, with most fats coming from sources of MUFA and PUFA, such as fish, nuts, and vegetable oil" [11]. Similarly, the Third Report of the Expert Panel on Detection, Evaluation and Treatment of High Blood Cholesterol in Adults, released by the National Cholesterol Education Program (NCEP), recommends keeping total fat in the diet between 25% and 35% of calories and MUFA up to 20% of calories [12]. The American Heart Association (AHA) emphasizes decreasing SFA to less than 7% of calories and *trans* fatty acids to less than 1% of energy [13]. The AHA also supports the Dietary Approach to Stop Hypertension (DASH) program and the Therapeutic Lifestyle Changes (TLC) from NCEP, both of which include nuts in their dietary patterns. The DASH diet recommends four to five servings of nuts per week and the TLC diet includes nuts in the vegetables servings of five per day [14].

3.2 TREE NUT COMPOSITION

In addition to this chapter, detailed information about chemical composition of tree nuts is given in Chapter 2. According to Sabaté et al. [15], nuts are nutrient-dense foods (high in protein and lipid with a variety of micronutrients) that have constituted a part of humankind's diet since preagricultural times. Venkatachalam and Sathe [16] found that regardless of the nut, lipid is mainly composed of MUFA and PUFA (>75% of the total lipid), with a low level of SFA. Walnuts, which have a distinctive fatty acid profile, are richer in 18:2ω6 and 18:3ω3 than other tree nuts and peanuts [17]. In addition, tree nuts are a source of valuable micronutrients such as folic acid; niacin; vitamins B_6 and E; and minerals such as calcium, magnesium, copper, zinc, and phosphorus [18]. They also contain important phytochemicals, which have been shown to have beneficial effects on overall health. Each tree nut has its own special attributes (see Chapter 2). For example, almonds and hazelnuts are excellent sources of α-tocopherol with 1 ounce (~28.35 g) providing 7.3 mg α-tocopherol [19], and 1 ounce of Brazil nuts provides 544 μg of selenium (780% of the selenium needed per day for adults) [10,19].

Tree nuts are energy dense, with 45%–75% of their calories from lipid. The absorption of energy from nuts is incomplete, probably due to the fibrous cellular structure of nuts, which results in incomplete release of fatty acids during digestion [20]. This is an active area of research showing incomplete digestion of nuts may be linked to an increase in satiety [21–23]. Coates and Howe [24] conclude that appetite suppression may result from the satiating effects of high fiber, protein, and energy content of nuts.

The protein content of most nuts is about 10%–25% of calories, making them a good source of plant protein [25]. However, the biological value of nut protein is not as high as animal protein as nuts are missing some essential amino acids [18]. A study comparing the amino acid composition of walnuts and hazelnuts to whole eggs [18] found that nuts contained 25–40 mg/g protein of threonine (the limiting amino acid in nuts) and whole eggs contained 44 mg/g protein. Nuts tend to be lower than cereals in lysine as well, but in contrast to other vegetable proteins, nuts are high in arginine. This results in a low lysine to arginine ratio for tree nuts from 0.13 to 0.57, compared to a higher ratio in soy protein (0.5–1.0) and whole milk (2.4). Interestingly, lower ratios are linked with lower cardiovascular risk [26].

Nuts are a good source of dietary fiber. After cereals, nuts are the plant food highest in fiber, followed by legumes, whole grain bread, fruits, and vegetables [27]. Yet, this positive attribute of nuts is not well-known or publicized despite the fact that most populations globally do not consume the WHO recommended level of 25 g of fiber/d [28]. A standard serving of tree nuts provides 5%–10% of daily fiber requirements [24].

In addition to the macronutrients, nuts contain significant amounts of essential micronutrients such as vitamin E, calcium, magnesium, potassium, and folate [29]. Nuts are also one of the richest whole foods sources of vitamin E and were specifically recommended in the 2005 Dietary

Guidelines for Americans [11], because over 50% of the U.S. population does not meet their recommended dietary allowances (RDA) of 15 mg vitamin E [30]. Kornsteiner et al. [31] examined tocopherol and carotenoid content of oil extracted from all nine tree nuts and peanuts. Almonds and hazelnuts had the highest concentrations of α-tocopherol, with 1 ounce providing 7.5 mg or half the RDA. The USDA [11] recommended that the serving size of 1 ounce of almonds can daily be isocalorically substituted for white bread, crackers, chips, and similar refined products to provide 7.4 mg of vitamin E. Brazil nuts, pecans, pistachios, and walnuts contain vitamin E in the forms of β- and γ-tocopherols. Among tree nuts, pistachio is the only tree nut with detectable levels of carotenoids.

Polyphenolic compounds are found in small quantities in numerous plant foods, including nuts, and they are divided into two main classes: flavonoids and phenolic acids. Recently the hypothesis that dietary flavonoids play a significant role as antioxidants *in vivo*, thereby reducing chronic disease risk, has been challenged by studies on the bioavailability of flavonoids, which indicate they reach only very low concentration in human plasma after consumption of flavonoid-rich foods [32]. In this evolving research area, one must first quantify the levels of these varied compounds in plant-based foods and then show that they themselves, or their metabolites, are actually absorbed in the human gastrointestinal tract to impart a health benefit. Among the fatty fraction of nuts, there are noncholesterol sterols belonging to a heterogeneous group of compounds known as plant sterols or phytosterols. These compounds are known to interfere with cholesterol absorption, thereby imparting a cardiovascular health benefit [29]. The range of flavonoids (see Chapter 13) and phytosterols (see Chapter 2) found in tree nuts is explained in detail.

Flavonoids are a broad class of polyphenolic compounds that can be water-soluble, include anthocyanidins and represent ~20% of the total phenolic compounds in foods. Among tree nuts, hazelnuts and pecans have the highest total flavonoid content with over 500 mg/100 g [19]. These values are higher than many of the fruits and vegetables most commonly cited as flavonoid-rich foods [32]. Proanthocyanidins, better known as condensed tannins, are flavanol forms of flavonoids of varying sizes and degrees of polymerization. They are known for contributing astringent flavor to foods and may also help reduce the risk of blood clotting and urinary tract infections. Gu et al. [33] found that total proanthocyanidins are in the highest concentrations in hazelnuts, pecans, pistachios, and almonds, and with 185–500 mg/100 g compare very well with dark chocolate (250 mg/100 g), berries (150–660 mg/100 g), red wine (300 mg/L), and grape juice (525 mg/L). Wu and Prior [34] determined the anthocyanin levels in vegetables, nuts, and grains and found that pistachio was the only tree nut to contain anthocyanins at low levels.

Some 80% of polyphenolic compounds in nuts are phenolic acids, with walnuts and pecans having the highest levels for all nuts [35]. Kornsteiner et al. [31] in Austria determined the mean content of total phenolics ranged from 32 mg of gallic acid equivalents (GAE)/100 g of pine nuts to 1625 mg of GAE/100 g in walnuts, which compare well with total phenolic values in nuts as determined by Wu et al. [36] in the United States. The total phenolic values in nuts also compare favorably with a wide range of fruits (e.g., watermelon [590 GAE/100 g] and grapes [1450 GAE/100 g]) [36], but nuts do not appear in very many charts ranking high food sources. Kornsteiner et al. [31] concluded that due to the heterogenic amounts and range of phenolic compounds in nuts, it is best to consume a variety of mixed nuts.

In a series of papers on the complete characterization and quantification of almond polyphenolics, Frison-Norrie and Sporns [37], Sang et al. [38,39], and Milbury et al. [40] were the first to quantify the total flavonoids in eight varieties of California almonds. They found the total phenolics ranged from 127 to 241 mg of GAE/100 g of fresh weight and for the aglycone form of flavonoids, as reported to the USDA flavonoid database, whole almonds contain 17.31 mg isorhamnetin, 12.21 mg kaempferol, and 1.93 mg catechin per 100 g of fresh weight, respectively. This approach allows for comparisons of flavonoid intake from different foods on a weight basis, showing that almonds provide an amount of flavonols similar to that of red onions, but 9-fold more isorhamnetin than white onions. Kampferol and quercetin contents of almonds are comparable to those of broccoli and its concentration of catechin is between that of brewed black and green tea [40].

Plant sterols or phytosterols are important structural components of plant membranes and seem to play an important role in cholesterol lowering in humans, by blocking cholesterol absorption [29].

Phytosterols are found in all tree nuts, especially pistachios, pine nuts, almonds, and cashews, in a range of 72–214 mg/100g, with β-sitosterol the most abundant form (see Chapter 2). Segura et al. [29] contrasted the total phytosterol values cited in the USDA nutrient database with the values obtained by Phillips et al. [41] who reported values ranging from 95 to 279 mg/100 g and are considered the most accurate and higher, as they include steryl glycoside forms. These forms represent a significant portion of the sterols in nuts, but they are not usually measured [41].

In conclusion, the high contents of healthy unsaturated fatty acids (MUFA and PUFA), protein with a low lysine to arginine ratio, carbohydrates mainly present as dietary fiber, and a range of bioactives make the addition of tree nuts to healthy diets beneficial. These compositional benefits for nuts have only recently been researched and recognized more fully.

3.3 SUMMARY OF TREE NUT RESEARCH

The epidemiological and clinical studies reviewed in this chapter offer compelling evidence that nuts, eaten as part of a balanced diet, contribute to reduction of CHD and related risk factors. Common foods such as nuts, fish, fruits, and vegetables are increasingly recognized as important sources of bioactive nutrients. Collectively, they offer an alternative approach for prevention and management of CHD [42]. The relationship between nut consumption and decreased incidence of other lifestyle diseases is emerging.

3.3.1 Role of Tree Nuts in Mediterranean Diet

Epidemiological studies have shown a positive association between the Mediterranean diet and a longer life span and reduced mortality and morbidity for cardiovascular disease (CVD). As a primarily plant food-based diet, the Mediterranean diet is characterized by a large variety of foods high in β-carotene, vitamins C and E, minerals, and phytochemicals. Trichopoulou and Vasilopoulou [43] concluded that a diet that adheres to the principles of the traditional Mediterranean eating plan is associated with longer survival. This traditional diet also has a moderate to high fat content (e.g., Italy 30% and Greece 40% of RDA), compared to widely promoted lower fat diets [44]. The traditional diet is typically rich in beneficial unsaturated fatty acids and is low in SFA.

The beneficial effects of the Mediterranean diet in the prognosis of CHD were demonstrated in two large randomized, secondary prevention trials, the Lyon Heart Study [45] and the Indo-Mediterranean Diet Study [46]. The Lyon Heart Study was a randomized, secondary prevention trial that tested whether a Mediterranean-type diet containing large amounts of fiber, antioxidants, minerals, vegetable proteins, and vitamins compared with a prudent western-type diet would reduce recurrence after a first myocardial infarction. After 27 months, overall mortality was reduced by 70% and cardiac mortality was reduced by 81% in the intervention group. de Lorgeril et al. [47] concluded that a Mediterranean diet rich in fruits, vegetables, nuts, and a high intake of 18:3ω3 was more efficient in the secondary prevention of cardiac deaths than diets routinely recommended by hospital dietitians or physicians. de Lorgeril et al. [47] make a compelling argument that scientists and physicians should give a higher priority to studying natural foods, such as nuts, to better understand the nature of their cardioprotective properties since they do better than aspirin or prevastatin for primary prevention of CHD. The Mediterranean diet as a secondary prevention measure is also inexpensive compared to other diets or drug treatments [48].

In the Indo-Mediterranean diet [46], 1000 patients with CHD in northern India were randomized into two groups, one receiving a diet rich in whole grains, fruits, vegetables, and nuts and the other group which received a diet similar to the NCEP prudent diet. The intervention group consumed 573 g fruits, vegetables, legumes, walnuts, and almonds versus a 231 g intake of the low-cholesterol-prudent diet. After 2 years of follow-up, the Indo-Mediterranean diet was associated with a 37% reduction in total mortality and a 49% reduction in cardiovascular mortality. Singh et al. [46] concluded that a diet enriched with fruit, vegetables, nuts, whole grains, and mustard or soy bean oil was associated with a pronounced decline in CHD morbidity and mortality without an increase in noncardiac deaths and in the presence of improved metabolic profiles.

The results of the Lyon Heart Study and the Indo-Mediterranean Diet Study are compatible with a large observational study, the Greece EPIC cohort, which demonstrates that the reduction in overall mortality is associated with increased adherence to the Mediterranean diet [49].

Serra-Majem et al. [50] reviewed 35 experimental studies on the Mediterranean diet and concluded they did show favorable results for lipoprotein levels, endothelium vasodilation, myocardial, and cardiovascular mortality. The Spanish have conducted a significant amount of research examining the effects of tree nuts on health promotion and disease prevention and have the distinction of conducting the first primary prevention, large-scale, long-term clinical trial that incorporates tree nuts. In the PREDIMED study, in late 2003, 9000 high-risk people 55–80 years of age with no history of heart disease were randomized into a control group or intervention group with 1 L of olive oil weekly or 30 g nuts daily (walnuts, hazelnuts, and almonds). Participants were assigned to a low-fat diet or to one of two Mediterranean diets. After 3 months, compared to the low-fat diet, the two Mediterranean diets produced beneficial changes in most outcomes including blood glucose, systolic blood pressure, and total cholesterol to high-density lipoprotein (HDL) ratio [51].

3.3.2 Nuts and Cardiovascular Disease

To date, more than 20 clinical and observational mixed-nut studies have been conducted on the effects of tree nuts on reducing the risk of CVD. The following is an overview of the research to date.

3.3.2.1 Observational Studies and CVD

Fraser et al. [52] conducted the first observational study among 31,208 non-Hispanic, white, California Seventh-Day Adventists with detailed information on how the nut-eating cohorts were healthier and less obese. Subjects who consumed nuts frequently (more than four times per week) had less CHD, despite all covariant adjustments. The apparent protective effect of nuts was independent of established coronary risk factors such as age, sex, smoking habits, history of hypertension, relative weight, and physical exercise. Fraser et al. [52] also found that consumers of nuts were less obese.

Hu and Stampfer [53] summarized the relationship between nut consumption and risk of CHD. Five large prospective cohort studies (Adventist Health Study, Iowa Women Health Study, Nurses' Health Study, Physicians' Health Study, and the CARE Study) examined the relationship between nut consumption and the risk of CHD. All the studies found an inverse association. Based on the data from the Nurses' Health Study, Hu and Stampfer [53] estimated that substitution of the fat from 1 ounce of nuts for equivalent energy from carbohydrate in an average diet was associated with a 30% reduction in CHD risk and the substitution of nut fat for saturated fat was associated with 45% reduction in risk. They concluded that regular nut consumption can be recommended in the context of a healthy and balanced diet. Hu et al. [54] found that 5% of the nurses in the health study, or 4423 of the 86,016 people, reported eating 1 ounce of nuts more than five times per week. The women who consumed nuts were leaner (Body mass index, BMI = 23.4) than the nonnut eaters (BMI = 24.8). The nut eaters also ate less meat and had lower intakes of *trans* fatty acids and higher intakes of unsaturated fatty acids and dietary fiber. Hu et al. [54] concluded that the regular nut eaters tended to weigh less, indicating that in practice, the energy contained in nuts can be readily balanced by reductions in other energy sources or by increased physical activity.

Hu and Stampfer [53] also estimated that substituting 1 ounce (~28.35 g) of nuts for the equivalent energy from carbohydrate in an average diet was associated with a 30% reduction in CHD risk. The substitution of nut fat for saturated fat was associated with a 45% reduction in risk. Hu and Stampfer [53] concluded regular nut consumption can be recommended in the context of a healthy and balanced diet.

Ellsworth et al. [55] examined the specific relationship between the frequency of nut intake and risk of death in the 34,111 postmenopausal women in the 12-year Iowa Women's Health Study and concluded that frequent nut consumption may offer these women modest protection against the risk of death from all causes and CHD. They recommended that future studies should consider using a more precise way to measure nut intake.

In order to further understand the mechanism underlying the apparent protective effect of nut consumption, Albert et al. [56] examined the association between nut consumption and the risk of sudden cardiac death in the Physicians' Health Study. Investigations revealed that dietary nut intake was associated with a significantly reduced risk of sudden cardiac death. Those who consumed nuts two or more times per week had a 47% lower risk of sudden cardiac death and a 30% lower risk of total CHD death, and nonfatal myocardial infarction.

In another study, Mantzoros et al. [57] examined the effect of the Mediterranean diet on adiponectin concentrations. Adiponectin is an adipose tissue-secreted cytokine that has been shown to regulate glucose and lipid metabolism, improve insulin sensitivity, and possess pronounced antiatherosclerotic effects. Mantzoros et al. [57] evaluated dietary data and plasma adiponectin in 987 diabetic women, with no history of CVD, from the Nurses' Health Study. Women with the greatest adherence to a Mediterranean-type diet had higher plasma adiponectin concentrations than women with the lowest adherence. The data suggested that several components of the Mediterranean diet (alcohol, nuts, and whole grains) showed the strongest association with higher concentrations of adiponectin.

Jiang et al. [58] examined nut and seed consumption and inflammatory markers (C-reactive protein, interleukin-6, and fibrinogen) in the Multi-Ethnic Study of Atherosclerosis. This cross-sectional analysis included 6080 U.S. participants aged 45–84 years. Nut and seed consumption was categorized as never/rare, less than once/week, one to four times per week, and five or more times/week. Associations of nut and seed consumption with these biomarkers were not modified by BMI, waist to hip ratio, or race/ethnicity. Frequent nut and seed consumption was associated with lower levels of inflammatory markers, which may partially explain the inverse association of nut consumption with levels of C-reactive protein, interleukin-6, and fibrinogen.

The association of nut intake and CHD mortality was examined in 399,633 subjects, including 1158 cases, enrolled in the EPIC study [59]. When comparing the highest (>13 g/day) and the lowest (<1 g/day) nut intake categories, and after adjusting for coronary risk factors and dietary variables, the hazard ratio (HR) was 0.71 [95% confidence interval (CI), 0.51–0.98]. An HR of 0.74 (0.57–0.96) was estimated for every 8 g/day of increased nut intake, and after adjustments, the HR was 0.89 (0.74–1.08). Nut intake was associated with a decreased risk of CHD mortality. In approximately half of a population that rarely consumes nuts, an intake of only two servings of nuts per week (8 g/day) may reduce CHD mortality by 11%.

3.3.2.2 Clinical Trials and CVD

In July 2003, the U.S. Food and Drug Administration (FDA) approved a qualified health claim stating, "Scientific evidence suggests but does not prove that eating 1.5 ounces (~42.5 g) per day of most nuts as part of a diet low in saturated fat and cholesterol may reduce the risk of heart disease" [60]. This heart disease risk reduction health claim is the result of a petition filed by the International Tree Nut Council Nutrition Research and Education Foundation in August 2002. Both observational and clinical studies were considered during the approval process.

Clinical trials involving mixed nuts have been conducted in six countries: Australia, Canada, Israel, India, New Zealand, and the United States [61–67]. In the studies that reported blood lipid values, the "nut diets" significantly reduced total cholesterol from 7% to 25% and low-density lipoprotein (LDL) cholesterol by 10%–33%. No studies found any significant effect on HDL cholesterol, and two found a significant decrease in triacylglycerols (TAG) [63,66]. Furthermore, both the Lyon Heart Study and the Indo-Mediterranean Diet Study showed a decrease in overall mortality and cardiovascular mortality [45,46].

The Jerusalem Nutrition Study [62,68] was a randomized, controlled crossover design with 18 young men that examined the effects of a high-MUFA diet (including almonds, olive oil, and avocado) versus a high-PUFA diet (including walnuts, safflower, and soy oils) during two 12-week dietary periods. The MUFA diet lowered total cholesterol by ~10% and LDL cholesterol by 14% compared to baselines, and the PUFA diet lowered total cholesterol by 16% and LDL cholesterol by 21%. There were no significant effects on HDL cholesterol or TAG.

The first Australian study involved 16 normolipidemic men in a consecutive, supplemental field study that lasted for three 3-week dietary periods [61]. During the first 3 weeks, subjects consumed a reference diet that included a background diet supplemented with 50 g/day of peanuts, 40 g/day of coconut, and 50 g/day of a coconut confectionary bar. During the second 3 weeks, subjects consumed the background diet supplemented with almonds (84 g/day) and during weeks 7–9, subjects consumed the background diet supplemented with walnuts (68 g/day). Compared to the reference diet, the almond diet lowered total cholesterol by 7% and LDL cholesterol by 10%, and the walnut diet lowered total cholesterol by 5% and LDL cholesterol by 9%, with no significant effect on HDL cholesterol or TAG.

Another Australian intervention study involved 15 adults who ate a plant-based diet that included whole grains, sun-dried raisins, and mixed nuts (almonds, hazelnuts, and walnuts) and nut/seed butters (almond and sesame). After 4 weeks, both total and LDL cholesterols decreased by 8% and 15%, respectively, and HDL cholesterol was not significantly affected [63]. Another study involved 12 hyperlipidemic women in a crossover design that lasted for two 4-week periods [64]. Subjects first consumed a refined-food diet and then switched to a phytochemical-rich diet primarily consisting of whole grains, legumes, fruits, vegetables, seeds, and two tablespoons of almonds, hazelnuts, or pecans per day. Compared to the refined-food diet, the phytochemical-rich diet lowered total cholesterol by 13% and LDL cholesterol by 16%, with no significant changes in HDL cholesterol or TAG.

A recent New Zealand randomized, controlled trial of 28 hyperlipidemic men and women [65] showed that a 30 g serving of nuts consumed daily over 6 weeks was found to be as effective as one serving of a canola oil enriched cereal at reducing total and LDL cholesterols. They concluded that since the fatty acid profiles of the nut- and canola-enriched cereals were similar, and the results were similar, the fatty acid profile is one of the major factors responsible for the beneficial effects of nuts.

A Canadian study involved 10 adults in a randomized, crossover design that lasted for two 2-week periods [66]. The control diet was the subjects' habitual diet. The study diet consisted of mainly vegetables, fruit, avocados, and nuts (limited to 60–120 g/day, average consumption was 100 g/day). Compared to the control diet, total cholesterol, LDL cholesterol, and TAG were reduced by ~25%, 33%, and 20%, respectively, with no significant change in HDL cholesterol.

An American study by Kris-Etherton et al. [69] examined the beneficial effects of diets high in MUFA and concluded that such diets are effective for reducing total and LDL cholesterols and may have more beneficial changes in CVD risk than low-fat and high-carbohydrate diets. They estimated a 16–25% reduction in CVD risk in subjects on high-MUFA diets compared with a 12% reduction in risk in subjects on low-fat and high-carbohydrate diets.

An Indian study [67] involved 406 patients who were recruited 24 to 48 h after having an acute myocardial infarction (MI). The randomized, single-blind intervention study lasted 6 weeks. Subjects were divided into two groups. One group consumed Diet A in which meat and eggs were replaced by fish, vegetarian meat substitutes and nuts (almonds and walnuts). Those following Diet B ate a low-calorie, typical hospital diet, followed by a diet prescribed by their doctors. Those consuming Diet A had ~9%, 10%, and 9% decrease in total cholesterol, LDL cholesterol and TAG, respectively compared with the initial levels in Diet group A. Those following Diet A also had a 36% decrease in cardiovascular events compared to those consuming Diet B.

3.3.2.3 Clinical Single-Nut Trials and CVD

To date, 25 randomized clinical trials in seven countries (Australia, Canada, the United States, Japan, New Zealand, Spain, and Turkey) have examined the effects of single nuts on risk factors for CHD (Table 3.2) [70–94]. In most of the studies, the "nut diets" significantly reduced total cholesterol by

TABLE 3.2
Single-Nut Studies

						Nut Diet End vs. Control Diet End (%)			
Nut/Reference	Population, Sample Size, Length of Study, and Study Design	Comparison Made	Mean Amount of Nuts/Day	Total Fat and SFA from Nut Diet (%)	Total Fat and SFA from Control Diet (%)	Total Cholesterol (% Change)	LDL Cholesterol (% Change)	HDL Cholesterol (% Change)	TAG (% Change)
Almond									
Spiller et al. [91]	26 hypercholesterolemic subjects (13 M, 13 W) 9 week; single intervention	Baseline diet	100 g/day	36.9 5.9	28.5 4.9	−8.9 ($P < 0.05$)	−12.4 ($P < 0.01$)	NS	−4.4 NS
Spiller et al. [90]	45 hypercholesterolemic subjects (12 M, 33 W) 4 week; randomized, controlled, parallel design	Control diet (dairy based)	100 g/day	39 5	35 17	−15.6 ($P < 0.001$)	−19.0 ($P < 0.001$)	NS	NS
		Olive oil-based diet	100 g/day	39 5	35 6	−8.7 NS	−9.8 NS	NS	NS
Hyson et al. [78]	22 normocholesterolemic subjects (10 W, 12 W) 6 week; randomized, crossover design	Baseline diet	66 ± 5 g/day WA	30 8 WA and AO combined	29 10 WA and AO combined	−4.3 ($P < 0.05$)	−6.0 ($P < 0.05$)	+4.3 ($P < 0.05$)	−14.5 ($P < 0.05$)
		Baseline diet	35 ± 2 g/day AO	30 8 WA and AO combined	29 10 WA and AO combined	−4.5 ($P < 0.05$)	−6.6 ($P < 0.05$)	+6.9 ($P < 0.05$)	−15.3 ($P < 0.05$)
Jenkins et al. [80]	27 subjects (15 hyperlipidemic M, 12 postmenopausal W) 4 week; randomized, crossover design	Muffins-control diet	25–50 g/day (half almond dose)	32.1 7.5	26.3 7	−2.0 ($P < 0.05$)		+2.2 NS	−3.2 NS
		Muffins-control diet	50–100 g/day (full almond dose)	36 7.2	26.3 7	−4.4 ($P \leq 0.01$)	−7.1 ($P \leq 0.01$)	+2.2 ($P < 0.05$)	−5.3 NS

TABLE 3.2 (continued)
Single-Nut Studies

Nut/Reference	Population, Sample Size, Length of Study, and Study Design	Comparison Made	Mean Amount of Nuts/Day	Total Fat and SFA from Nut Diet (%)	Total Fat and SFA from Control Diet (%)	Nut Diet End vs. Control Diet End (%): Total Cholesterol (% Change)	LDL Cholesterol (% Change)	HDL Cholesterol (% Change)	TAG (% Change)
Lovejoy et al. [82]	30 type 2 diabetics (13 M, 17 W)	High-fat control diet	57–113 g/day	39 6.8	36.8 7.4	−1.3 NS	−2.7 NS	−3.4 ($P = 0.002$)	+5.4 NS
	4 week; randomized, double-blind, crossover design	Low-fat control diet	57–113 g/day	27.2 4.8	26 5.3	NS	NS	−2.6 ($P = 0.002$)	+5.0 NS
Sabaté et al. [89]	25 healthy subjects (14 M, 11 W) 4 week; randomized crossover design	Step I diet	±27 g/day (10% of total energy, low almond diet)	35 8	30 8	−0.9 NS	−1.1 NS	−0.9 NS	NS
		Step I diet	±54 g/day (20% of total energy, high almond diet)	39 8	30 8	−4.4 ($P < 0.05$)	−7.0 ($P < 0.05$)	+1.7 NS	−3.3 NS
Spiller et al. [92]	38 hypercholesterolemic subjects (12 M, 26 W)	Baseline diet	100 g/day raw almond	45 8	36 12	−6.9 ($P < 0.01$)	−11.9 ($P < 0.002$)	NS	+10.5 NS
	4 week; randomized, controlled, parallel design	Baseline diet	100 g/day roasted almond	44 7	31 11	−4.9 ($P < 0.034$)	−7.3 ($P < 0.012$)	NS	NS
		Baseline diet	100 g/day almond butter	44 9	30 9	−4.5 NS	−7.0 ($P < 0.034$)	+8.3 NS	−17.5 NS
Hazelnut									
Durak et al. [74]	17 healthy subjects (18 M, 12 W) 4 week; single intervention	Baseline diet	1 g/kg body weight/day	NR	NR	−6.1 ($P < 0.005$)	−18.7 ($P < 0.001$)	+7.2 ($P < 0.05$)	+24.8 ($P < 0.001$)
Mercanligil et al. [83]	15 M (hypercholesterolemic) 8 week; controlled crossover design	Baseline diet	40 g/day	37 8	31 9	−5.2 NS	−3.3 NS	+12.5 ($P < 0.05$)	−31.8 ($P < 0.05$)

(continued)

TABLE 3.2 (continued)
Single-Nut Studies

Nut/Reference	Population, Sample Size, Length of Study, and Study Design	Comparison Made	Mean Amount of Nuts/Day	Total Fat and SFA from Nut Diet (%)	Total Fat and SFA from Control Diet (%)	Nut Diet End vs. Control Diet End (%) Total Cholesterol (% Change)	LDL Cholesterol (% Change)	HDL Cholesterol (% Change)	TAG (% Change)
Macadamia									
Colquhoun et al. [72]	14 hypercholesterolemic subjects (7M, 7W) 4 week; dietary advice, randomized crossover design	Low-fat, high-complex carbohydrate diet	50–100 g/day (20% of total energy)	42.4 11.3	21.3 8.7	+0.2 NS	−0.3 NS	+9.1 NS	−13.3 NS
Curb et al. [73]	30 normo- and hyperlipidemic subjects (15M, 15W)	Step I diet	NR	35 9	30 9	−0.8 NS	+0.3 NS	+2.2 NS	−16 NS
	4 week; randomized, controlled, crossover design	AAD diet	NR	35 9	35 14	−4.8 ($P < 0.01$)	−4.5 ($P < 0.05$)	−4.2 ($P < 0.01$)	−9.2 ($P < 0.05$)
Garg et al. [76]	17M (hypercholesterolemic) 4 week; single intervention	Baseline diet	40–90 g/day (15% of total energy)	37.6 13.4	31.2 12.7	−3.2 ($P < 0.05$)	−6.0 ($P < 0.05$)	+6.7 ($P < 0.05$)	−2.8 NS
Hiraoka-Yamamoto et al. [77]	71W (healthy) 3 week; intervention	Baseline diet	20 g/day	NR	NR	−6.1 ($P < 0.05$)	−7.2 ($P < 0.05$)	−8.0 NS	−5.2 NS
Pecan									
Morgan and Clayshulte [84]	19 normolipidemic subjects (4M, 15W) 8 week; randomized, controlled, parallel design	Control diet	68 g/day	20 n/a	17 n/a	−10.7 ($P < 0.05$)	−16.3 ($P < 0.05$)	+1.2 ($P < 0.05$)	−9.3 NS

TABLE 3.2 (continued)
Single-Nut Studies

Nut/Reference	Population, Sample Size, Length of Study, and Study Design	Comparison Made	Mean Amount of Nuts/Day	Total Fat and SFA from Nut Diet (%)	Total Fat and SFA from Control Diet (%)	Nut Diet End vs. Control Diet End (%): Total Cholesterol (% Change)	LDL Cholesterol (% Change)	HDL Cholesterol (% Change)	TAG (% Change)
Rajaram et al. [86]	23 normal-high cholesterol subjects (14M, 9W) 4 week; single-blind, randomized, controlled, crossover design	Step I diet	72 g/day	39.6 8.1	28.3 8.2	−6.7 ($P \leq 0.01$)	−10.4 ($P \leq 0.01$)	+5.6 ($P \leq 0.01$)	−11.1 ($P \leq 0.01$)
Pistachio									
Edwards et al. [75]	10 hypercholesterolemic subjects (4M, 6W) 3 week; controlled, randomized, crossover design	Habitual diet	20% of total energy	39 7.5	37 11	−3.7 ($P < 0.04$)	−6.1 NS	+8.0 ($P < 0.09$)	−5.3 NS
Kocyigit et al. [81]	44 healthy subjects (24M, 20W) 3 week; randomized controlled design	Control diet	65–75 g/day (20% of total energy)	40 7	38 11	−11.0 ($P < 0.05$)	−1.0 NS	+29.3 ($P < 0.001$)	−10.1 NS
Walnut									
Sabaté et al. [88]	18M (healthy) 4 week; randomized, controlled, single-blind crossover design	Step I diet	84 g/day (20% of total energy)	31.3 6	29.3 9	−12.4 ($P < 0.001$)	−16.3 ($P < 0.001$)	−4.9 ($P = 0.009$)	−8.3 NS
Chisholm et al. [71]	21M (hyperlipidemic) 4 week; randomized, crossover design	Low-fat diet	78 g/day	38 10	30 12	−2.0 NS	−3.9 NS	+2.5 NS	+7.5 NS

(continued)

TABLE 3.2 (continued)
Single-Nut Studies

Nut/Reference	Population, Sample Size, Length of Study, and Study Design	Comparison Made	Mean Amount of Nuts /Day	Total Fat and SFA from Nut Diet (%)	Total Fat and SFA from Control Diet (%)	Nut Diet End vs. Control Diet End (%) Total Cholesterol (% Change)	LDL Cholesterol (% Change)	HDL Cholesterol (% Change)	TAG (% Change)
Zambon et al. [93]	49 hypercholesterolemic subjects (26M, 23W) 6 week; randomized, crossover design	Mediterranean diet	41–56 g/day (18% of total energy	33.2 6	31.2 6.9	−4.1 ($P < 0.001$)	−5.9 ($P < 0.001$)	+3.2 NS	−6.1 NS
Almario et al. [70]	18 subjects: 5M (hyperlipidemic), 13W (hyperlipidemic, post-menopausal)	Habitual diet (4 week)	48 g/day	37.2 9.8	31.4 11	−3.0 NS	−1.7 NS	−10.2 ($P < 0.01$)	−10.1 NS
	6 week; sequential intervention periods	Low-fat diet	48 g/day	33.7 8.2	19.7 7.5	−7.7 ($P < 0.01$)	−12.3 ($P < 0.01$)	+1.8 NS	−1.3 NS
Munoz et al. [85]	10M (polygenic hypercholesterolemic; subset of the Zambon study) 6 week; randomized, crossover design	Mediterranean diet	41–56 g/day	31.8 5.5	30.9 6	−4.2 NS	−6.0 NS	0.0 NS	−5.1 NS
Iwamoto et al. [79]	40 healthy subjects (20M, 20W) 4 week; randomized, crossover design	Japanese diet	44–58 g/day (12.5% of total energy)	26 4.8	24 6.9	−4.5 ($P = 0.001$)	−9.8 ($P = 0.001$)	−1.3 NS	0.0 NS

TABLE 3.2 (continued)
Single-Nut Studies

Nut/Reference	Population, Sample Size, Length of Study, and Study Design	Comparison Made	Mean Amount of Nuts /Day	Total Fat and SFA from Nut Diet (%)	Total Fat and SFA from Control Diet (%)	Nut Diet End vs. Control Diet End (%) Total Cholesterol (% Change)	LDL Cholesterol (% Change)	HDL Cholesterol (% Change)	TAG (% Change)
Ros et al. [87]	20 hypercholesterolemic subjects (8 M, 12 W) 4 week; randomized, crossover design	Mediterranean diet	40–65 g/day (18% of total energy)	33 4.8	33.2 5.5	−4.3 ($P = 0.02$)	−6.7 ($P = 0.01$)	−1.3 NS	8.3 NS
Zhao et al. [94]	23 hypercholesterolemic subjects (20 M, 3 W) 6 week, randomized, controlled, 3-diet, 3-period, crossover design	AAD-control diet	High PUFA (18:3ω3)	37.6 8.2	34.5 12.7	−10.7 ($P < 0.05$)	−10.9 ($P < 0.05$)	−5.9 ($P < 0.05$)	−18.4 ($P < 0.05$)
			High PUFA (18:2ω6)	37.1 8.5	34.5 12.7	−10.9 ($P < 0.05$)	−12.3 ($P < 0.05$)	−2.54 NS	−18.4 ($P < 0.05$)

Source: From Mukuddem-Petersen, J., Oosthuizen, W., and Jerling, J.C., *J. Nutr.*, 135, 2082, 2005. With permission.

Abbreviations: M, men; W, women; LDL, low-density lipoprotein; HDL, high-density lipoprotein; TAG, triacylglycerols; NR, not reported; NS, not significant; PUFA, polyunsaturated fatty acids; WA, whole almond; AO, almond oil; AAD, average American diet.

2–15% and LDL cholesterol by 5–19%. About a third of the studies showed a significant increase in HDL cholesterol and only a couple found a significant decrease in TAG. Mukuddem-Petersen et al. [95] reviewed 23 studies that showed decreases in total cholesterol between 2% and 16% and LDL cholesterol between 2% and 19% compared with subjects consuming control diets. Recently, Griel and Kris-Etherton [96] reviewed 24 studies and found that tree nuts reduce LDL cholesterol by 3%–19% compared with western and lower fat diets. These results were demonstrated consistently in all population groups studied, including men and women of all ages, Caucasians, African American, and persons with diabetes.

3.3.3 Nuts and Satiety and Maintenance of Healthy Body Weight

Addressing the increasing problem of obesity globally is becoming a major focal point of dietary recommendations. This shift in thinking ideally suits the recommendation for individual or whole foods to help build healthy eating patterns. However, caloric intake then assumes greater importance. Misperceptions continue that although nuts are high in unsaturated fat and calories, they are a food to be avoided, despite emerging research showing nuts may play a role in weight maintenance via increased satiety levels, increased resting energy expenditure, or energy malabsorption [20,23,97,98].

Sabaté et al. reported in 2003 [89] that there was no epidemiological data that related long-term nut consumption with obesity, despite the fact that nuts are an energy-dense food. He pointed out that per capita nut consumption in the Mediterranean populations is double that in the United States but the Mediterranean obesity rate has been much lower. Further evidence supporting an inverse association between frequency of nut consumption and BMI for Americans came from the data of the 1994–1996 Continuing Survey of Food Intakes by Individuals conducted by the U.S. Department of Agriculture. They compared BMI and total energy intake of nut eaters with that of nonnut eaters [89]. Dietary intake data from a nationally representative sample were collected on two nonconsecutive 24 h recalls. Those reporting consumption of tree nuts, peanuts, or seeds on either of the 2 days were included in the nut eater group. Data results indicate that young and adult nut eaters had a lower BMI compared with nonnut eaters. Notably, energy intake was higher among nut consumers. This analysis indicates an inverse relationship or no relationship between intake of nuts and BMI in the U.S. population. As well, within the large cohort studies that reported a decrease in the risk of CHD with frequent nut consumption, all show an inverse or no relationship between frequency of nut consumption and BMI [52,54,56,99].

In a recent review by Rajaram and Sabaté [100], the authors noted "nuts have a tendency to lower body weight and fat mass. In the context of calorie-restricted diets, adding nuts produces a more lasting and greater magnitude of weight loss among obese subjects while improving insulin sensitivity." Garcia-Lorda et al. [101] examined American and European epidemiological and clinical data on the effect of nut consumption on body weight and insulin resistance and concluded that although more research is needed, evidence indicated that nuts can help regulate bodyweight and protect against type-2 diabetes.

3.3.3.1 Observational Trials and Satiety/Body Weight

Bes-Rastrollo et al. [102] have been examining the association between frequency of nut consumption and the risk of weight gain in a Mediterranean cohort of Spanish university graduates over the past decade. The ongoing study collects data on 16,000 participants using biannual surveys on diet and food consumption, behavior, and physical activity, and on a variety of health conditions, including weight gain and obesity. After accounting for factors such as age, sex, activity level, and smoking, researchers found that among study participants, those who ate nuts (two or more times per week) had a reduced incidence of weight gain, whereas those who abstained from nuts gained more weight than their nut-eating counterparts. Among nut eaters, higher frequency consumption (up to more than five times per week) was associated with even further reductions in chances of obesity or weight gain. The results regarding tree-nut consumption include data from 8800 individuals with a median time in the study of 28 months. Bes-Rastrollo et al. [102] noted that since this is an observational study, not a controlled feeding experiment, results indicate associations but do not prove cause and effect. The authors also highlighted some possible explanations for encouraging nut-consumption results, including that since nuts are high in fiber and PUFA, this may help trigger satiety, reducing the impulse to continue eating, or nuts may be substituting for other high-calorie snacks with lower beneficial characteristics.

Sabaté and Blix [103] summarized the frequency of nut consumption in relation to all-cause mortality in several California Seventh-Day Adventist subpopulations and found reduced hazard ratios for nut consumers, regardless of age, sex, race, or fitness status. In three dietary subgroups, the prevalence of obesity at baseline between vegetarian and nonvegetarian groups was also strikingly different. For both men and women, BMI increased as the frequency of meat consumption increased. Vegetarian men and women had a two-point lower BMI value than nonvegetarians. Although these results were for middle aged (45–60 years) subjects, similar results were also observed for other ages. Summarizing BMI data from the four large published studies of adult vegetarians, Key et al. [104] compared nonvegetarian counterparts of the same cohort. Vegetarians in each study on average had one to two points lower BMI values than meat eaters within the same group. This same difference between vegetarians and meat eaters was observed in both men and women. There was a substantial variation in BMI values, with the U.S. population having greater BMI than European cohorts. This can be attributed to methodological differences in data collection, geographic location, secular trends, and ethnic or genetic differences. Overall, these epidemiological data clearly suggest that meatless diets are associated with lower overall BMI scores and lower prevalence of obesity in adults. It is well known that vegetarians consume more nuts than nonvegetarians (3.7 servings per week versus 2.1 servings per week, respectively) [105].

3.3.3.2 Clinical Mixed-Nut Trials and Satiety/Body Weight

Sabaté et al. [89] reported that various dietary intervention studies on different nuts provided evidence that short-term consumption of moderate to large amounts of nuts did not increase body weight. Well-controlled metabolic-type feeding studies did not show significant changes in body weight for nut diets versus nonnut control diets. Furthermore, in their well-controlled feeding studies at Loma Linda University with walnuts, pecans, and almonds, subjects on nut diets tended to be hungrier and require more energy intake to maintain body weight [89]. In a recent review of potential foods for weight control, St-Onge [106] concluded that nuts may play a role in weight maintenance, possibly due to increased satiety and food substitutions.

The results from the McManus et al. [107] ($n = 101$) study give a long-term perspective for the dietary fat and obesity debate. Subjects in the moderate fat group who followed a regime reflecting the dietary pattern of the Mediterranean maintained a 4.8 kg weight loss over 2 years, whereas the low-fat group could not adhere to the diet over the long term, and ended up with weight gain. The increase in olive oil enhanced the palatability of the regime and facilitated a larger vegetable intake. The moderate fat dietary pattern was higher in vegetables, peanuts, and tree nuts, which improved the fiber intake and may have contributed to satiety. This study demonstrated that nuts are able to enhance long-term dietary compliance and may have a role to play in the context of a supervised weight-reduction program.

In a small ($n = 8$), short-term Australian study on overweight and obese men where dietary SFA were replaced by MUFA from olive oil, nuts, and avocadoes, there was a small but significant loss of body weight and fat mass [108]. Nash and Westpfal [109] recently questioned the value of low-fat diets (with a high concentration of simple sugars) or low-carbohydrate diets (with a high concentration of animal fats) and suggested that substituting nuts for simple sugars as a snack or for meats or dairy as a source of fats and vegetable protein may have significant health benefits.

3.3.3.3 Clinical Single-Nut Trials and Satiety/Body Weight

A southern California study showed that adding a modest quantity of almonds (65 g) to the diet for 6 months resulted in increased unsaturated fat intake with no significant changes in body weight for 81 subjects [110]. The authors reported that food displacement occurred after almond supplementation and over 54%–78% (from food diary and 24 h recalls, respectively) of the extra calories from almonds were displaced by a decrease in intake of other less healthy foods in the habitual diet [111]. In another almond study, Wien et al. [112] discovered that a moderate-fat diet with almonds resulted in more weight loss than a low-fat diet in a 6-month study of 65 overweight and obese individuals, even though the total number of calories was the same for both groups. In addition, the almond group had a 50% greater reduction in waist circumference and a 62% greater reduction in fat mass than the low-fat diet group.

Sabaté et al. [98] examined the effect of daily walnut consumption (~12% energy intake) on body weight in 90 free-living subjects in a 12-month randomized crossover trial. Subjects consumed a walnut-supplemented diet (1–2 ounce of walnuts/day) for 6 months and a control diet with no supplemented walnuts for the other 6 months. The walnut-supplemented diet resulted in an increase in weight, BMI, fat mass, and lean mass, but after adjusting for energy differences between the control and walnut supplemented diets, no significant differences were observed. The weight gain from adding walnuts to the diet (control→walnut diet) was less than the weight loss from withdrawing walnuts from the diet (walnut diet→control). The results show that regular walnut intake resulted in weight gain much lower than expected. The gain became insignificant after controlling for differences in energy intake.

Several mechanisms can potentially explain why nuts do not bring about a weight gain. Nuts are energy-dense foods with high fiber, protein, and low glycemic index, all of which are dietary factors that have been shown to increase satiety and suppress appetite [24,113]. Kirkmeyer and Mattes [22] found that hunger ratings following consumption of 500 kcal preloads of peanuts, peanut butter, or almonds were significantly lower than with low calorie preloads. A number of researchers have reported that subjects in their nut feeding trials compensated within their dietary regime for nut intake [23,61,110]. Additionally, Iyer et al. [114] found that in a three-country (Brazil, Ghana, and the United States), placebo-controlled 8-week study among 129 healthy adults who consumed peanut and peanut oil, that there was a greater compensatory effect on food intake with whole nuts versus just the lipid fraction. Burton-Freeman et al. [21] found that the satiety response to dietary fat as almonds was influenced by sex and appeared to be dependant on the availability of the fat to stimulate hormone release in women but not in men. These data support the need to better understand sex-specific differences in order to design diets that provide optimal food intake and satiety.

Another explanation for the lack of weight gain might be due to the emerging research that a portion of energy from peanuts, pecans, and almonds is unavailable because the cell walls act as a physical barrier to gastrointestinal digestion of nutrients [20,97,115–117]. As a result, the excretion of fat in the stools, coupled with the displacement of foods from habitual diets, may account for the lack of weight gain observed in nut eaters [100]. Animal and human studies have shown that unsaturated fat increases fat oxidation because of higher diet-induced thermogenesis. Peanut supplementation for 19 weeks resulted in a 11% increase in resting energy expenditure [23], but daily almond supplementation for 6 months did not change resting energy expenditure [110].

3.3.4 Nuts and Diabetes

The American Diabetes Association (ADA) currently recommends an individualized dietary pattern that is based on the nutritional assessment and desired outcome of each patient. This approach takes into consideration patient preferences, control of high blood sugar, high blood pressure, and high blood lipids [118]. To achieve these nutritional goals, a SFA diet is currently advised. A meta-analysis of various studies comparing two approaches (either a high-MUFA + PUFA or a SFA and high carbohydrate diet) to diet therapy in patients with type-2 diabetes revealed that high-MUFA diets improve lipoprotein profiles as well as glycemic control [119]. Furthermore, there is no evidence that high-MUFA + PUFA diets induce weight gain in patients with diabetes mellitus, provided that energy intake is controlled. Coates and Howe [24] have recently speculated that the MUFA + PUFA from nuts may have an insulin sensitizing effect.

Several studies have shown that the risk of type-2 diabetes is lowered with higher intakes of dietary fiber and lower glycemic loads [120–124]. Jenkins [125], who coined the phrase glycemic index, has evaluated the ability of either mixed nuts or almonds to prevent blood glucose spikes that occur after consuming carbohydrate-rich foods that commonly raise blood sugar levels. They found that in healthy men and women eating mixed nuts with a carbohydrate-rich meal blunted the glycemic and insulin responses of the body to a significant degree. Jenkins et al. [126] and Josse et al. [127] also assessed the effect of decreasing postprandial glucose excursions on measures of oxidative damage. Fifteen healthy subjects ate two bread control meals and three test meals: almonds and bread; parboiled rice; and instant mashed potatoes, balanced in carbohydrate, fat, and protein, using butter and cheese. Blood samples were obtained at baseline and for 4h postprandially. Glycemic indices (GI) for the rice and almond meals were less than for the potato meal, as were the postprandial areas under the insulin concentration time curve. No postmeal treatment differences were seen in total antioxidant capacity. However, the serum protein thiol concentration increased following the almond meal, indicating less oxidative protein damage, and decreased after the control bread, rice, and potato meals, when data from these three meals were pooled. The change in protein thiols was also negatively related to the postprandial incremental peak glucose and peak insulin responses observations. Therefore, lowering postprandial glucose excursions may decrease the risk of oxidative damage to proteins.

3.3.4.1 Observational Studies and Diabetes

The Nurses' Health Study demonstrated that the consumption of nuts and peanut butter was inversely associated with the risk of type-2 diabetes, independent of known risk factors for type-2 diabetes [128]. In a 10-year follow-up, among 32,826 women in the Nurses' Health Study, Lopez-Garcia et al. [129] demonstrated that elevated plasma levels of inflammatory markers, especially C-reactive protein (CRP), were independent predictors of type-2 diabetes. Dietary intervention concurrent with healthy lifestyle strategies to increase exercise and encourage weight loss was effective in lowering CRP and other inflammatory markers.

3.3.4.2 Clinical Single-Nut Trials and Diabetes

The role that nuts may play in the diet of persons with diabetes is an emerging research area that is currently being pursued by INC. The MUFA, PUFA, vegetable protein, dietary fiber, and polyphenols found in nuts have, in a limited number of studies, independently been shown to have a number of effects including blunting the postprandial glucose rise, improving carbohydrate tolerance, and reducing risk factors for diabetic complications [24,120,130,131].

There have been a limited number of clinical trials on almonds and walnuts and the effects of consumption for diabetic health. In a randomized, double blind, crossover study, 34 men and women with type-2 diabetes were assigned to four high- and low-MUFA diets for 4 weeks, with the almond-enriched diets using 10% energy from almonds [82]. Results showed that there was no significant effect of fat amount or source on plasma glucose or insulin levels during a 2h oral glucose tolerance test. Lovejoy et al. [82] found that almonds had beneficial effects on serum lipids in these patients. It is likely that nut intervention longer than 4 weeks and greater than 10% of total energy intake is required to modify insulin sensitivity and glycemic control [100].

Similarly, 30g daily of walnuts added to the low-fat diet of type-2 diabetic patients in Australia improved their blood lipid profiles without affecting glycosylated hemoglobin levels [132]. Gillen et al. [133] conducted a parallel-design, walnut trial with 55 free-living men and women with established type-2 diabetes mellitus. Participants were randomly assigned to three diet groups: low-fat, modified low-fat, and a modified low-fat including 30 g of walnuts per day. Dietary intakes and clinical outcomes were measured at baseline, 3, and 6 months. The results showed that at baseline dietary intakes were not significantly different among the groups. Only a few individuals (10%) and no groups were consuming enough PUFA. At 3 and 6 months, calories and macronutrient intakes were similar among the groups. However, the walnut group was the only group to achieve all fatty acid targets. This group had the greatest proportion of subjects achieving targets. Walnuts were the primary source of dietary fat (31%) and omega-3 fatty acids (50%), and 350 g of oily fish per day provided an additional 17% of omega-3 fatty acids consumed by this group. Gillen et al. [133] concluded that recommending patients with type-2 diabetes to include walnuts regularly in their diet can help them achieve optimal fat intake without any adverse effects on total fat or calorie intakes.

In the Wien et al. [112] 6-month weight loss study, 70% of the participants had type-2 diabetes and the remainder were insulin-resistant. There was a 54% reduction in fasting insulin levels in the 84 g/day almond supplemented group compared to the carbohydrate group. Insulin resistance was decreased significantly in both diet groups, but improved beta cell function was observed only on the almond diet. Among subjects with type-2 diabetes, diabetes medication reduction was either sustained or reduced further in 96% on the almond diet compared to only 50% among the carbohydrate group. More studies are needed to support these observations.

3.3.5 Nuts as Antioxidants and Cancer Prevention

To date, little research has been done on the potential of nuts in cancer prevention. A report by the WCRF and the American Institute for Cancer Research (AICR) concluded, "While there are theoretical reasons to believe that diets high in nuts and seeds might protect against some cancers, the evidence is currently lacking" [3]. However, nuts contain several vitamins, micro, and phytonutrients that have possible biological mechanisms of action that may decrease the risk of cancer. Gonzalez and Salas-Salvado [134] concluded that the most effective strategy for decreasing cancer risk is to encourage a dietary pattern rich in a variety of fruits, vegetables, legumes, whole cereals, and nuts, because it is more likely that nut components add to and interact with the nutritional components of fruits and vegetables.

In respect to prevention of a broad range of degenerative diseases in addition to CVD and cancer, the importance of antioxidant-rich foods with daily nutrition is increasingly recognized. Ginter [135] reported that the CVD epidemic in central and eastern Europe seems to be only partially associated with a high prevalence of traditional risk factors (hypercholesterolemia, hypertension, and smoking, among others). Hence, the effect of traditional risk factors may be intensified by additional unidentified factors, such as environmental and psychosocial problems, and specific nutritional antioxidant deficiencies (antioxidant vitamins, flavonoids, and folic acid, among others). The intake of antioxidants from fruits, vegetables, nuts, and vegetable oils was reported to be substantially lower in most East European countries than in the West [135]. Thus, in Eastern Europe factors increasing free radical production may not be sufficiently counterbalanced by protective nutritional antioxidants, and oxidative stress plays a crucial role in CVD pathogenesis.

The question of how important dietary antioxidants and oxidative stress are in overall development of chronic disease is a "hot topic" in nutrition research globally. Nuts have been identified as being especially rich in antioxidants, but not enough recognition is given to them [35,36,136]. Nuts constitute one of the most nutritionally concentrated kinds of food available and most have a remarkably long shelf-life [137]. There is also a very active debate internationally about how the antioxidant capacity of plant foods should be assessed and whether *in vitro* tests have any validity *in vivo* for humans. Italian researchers have done some of the most comprehensive screening of a wide range of foods, including nuts, for both water and fat soluble components "total antioxidant capacity" [35]. Pellegrini et al. [35] assessed the contribution of bound antioxidant compounds to the total antioxidant capacity value in fiber-rich foods, such as cereals, legumes, and nuts, where phenolics are present in both free and bound forms. Pellegrini et al. [35] found that walnuts and pistachios had the highest values and that there was a high contribution of bound antioxidants. The fate of these phytochemicals is to reach the colon in an undigested form, where they can be modified by microflora, yielding other compounds that may be absorbed. Unabsorbed phytochemicals may also play a role in protection of the gastrointestinal tract from certain cancers. This is also the general conclusion of a recent extensive review on flavonoid-rich foods, where bioavailability has been shown to be low in human plasma, that health benefits may be much broader, but more complex to determine. Nichenametla et al. [138] concluded that due to mechanistic interactions, much more work is needed to determine the effects of plant foods or combinations of different polyphenolics on the prevention of cancer.

3.3.5.1 Observational Studies and Antioxidants/Cancer

Gonzalez and Salas-Salvado [134] examined 15 epidemiological studies with results on the consumption of nuts and cancer. Several of the studies considered nuts, seeds, and legumes together. Gonzalez and Salas-Salvado [134] also elaborated on the potential mechanisms of action by which compounds in nuts may prevent cancer. They identified eight studies, mostly from the United States, on the risk of colon and rectum cancers, before the EPIC study was published. In a 6-year prospective study, Singh and Fraser [139] found that those who ate nuts more than four times per week had a lower incidence of colon cancer than those who never ate nuts or ate them one to four times per week. Kune et al. [140] compared 715 cases with 727 age- and sex-matched community controls in Australia. The combination of a high-fiber and high-vegetable intake (including nuts) was found to be protective against large bowel, colon, and rectal cancers.

In 2004, Jenab et al. [141] investigated the effect of nut and seed intake on colorectal cancer in men and women participating in the EPIC study across Europe. Total nut intake was determined from country-specific dietary questionnaires for 855 (327 men and 528 women) colon cancer cases, 474 (215 men and 259 women) rectal cancer cases, and 478,040 (141,988 men and 336,052 women) total subjects. Division of the data into colon and rectal cancers showed a significant protective effect of nut intake on colon cancer in females at the highest quintile of intake, which was greater than 6 g daily. The results showed a significant protective effect of increased nut intake on colon cancer in women, with no effects on rectal cancer for either gender.

In 1998, a study was published on prostate cancer, the most prevalent cancer among men. Prostate cancer represents a large and growing health problem in the United States and other western countries [142]. In this large, 59-country study, Hebert et al. [142] tried to identify predictive measures for prostate cancer mortality. They concluded by stating that "the results from this study are consistent with previous information and support the hypothesis that grains, cereals, and nuts are protective against prostate cancer" [142].

3.3.5.2 Clinical Trials and Antioxidants/Cancer

The only clinical research on nuts and cancer has been conducted with rats. Davis and Iwahashi [143] showed that almond consumption may provide a measure of protection against developing colon cancer. The findings suggest a need to reassess the current view that intake of high-fat foods invariably has deleterious health effects.

The effects of almond consumption on DNA damage and oxidative stress among cigarette smokers were studied. Results from this pilot study indicate that almond consumption has preventive effects on oxidative stress and DNA damage caused by smoking [144].

Finally, a limited number of studies with almonds and walnuts have evaluated their antioxidant effect in human health. Chen et al. [145] extracted almond skin flavonoids, which possess antioxidant capacity *in vitro*; they are bioavailable and act in synergy with vitamins C and E to protect LDL cholesterol against oxidation in hamsters and humans. Anderson et al. [146] had shown that walnut polyphenolics inhibit *in vitro* oxidation. Further, Jambazian et al. [147] demonstrated the dose–response effect of almond intake on plasma and red blood cell tocopherol concentrations in healthy adults while enrolled in a randomized, crossover feeding trial. Participants were 16 men and women, aged 41 ± 13 years. Incorporating almonds into the diet helped the participants meet the revised RDA of 15 mg/day vitamin E.

3.4 CONCLUSION

Over the last decade, the body of research on the effects of tree nuts on various chronic diseases has grown dramatically. The data have repeatedly shown the positive effect of nuts in reducing the risk of heart disease, and new research on nuts and satiety, body weight, and diabetes looks very promising. While each individual nut contains its own special attributes, tree nuts as a group contain high levels of MUFA, PUFA, protein with a low lysine to arginine ratio, fiber, and a variety of important vitamins and minerals such as folic acid, vitamin E, and selenium. In addition, tree nuts contain a wide range of bioactives and phytochemicals that have been shown to have beneficial effects on overall health. While more research is needed, the data presented in this chapter support the recommendation to include a handful of nuts in the diet every day.

REFERENCES

1. International Nut and Dried Fruit Council Foundation (INC), Published online at: http://www.nutfruit.org.html (accessed April 10, 2007).
2. Oldways, *Tree Nuts, Health and the Mediterranean Diet. A Summary Report*, Oldways Preservation & Exchange Trust, Boston, MA, 1995.
3. American Institute for Cancer Research and World Cancer Research Fund, *Food, Nutrition and the Prevention of Cancer: A Global Perspective,* BANTA Book Group, Menasha, WI, 1997.
4. Aranceta, J., Perez Rodrigo, C., Naska, A., Vadillo, V.R., and Trichopoulou, A., Nut consumption in Spain and other countries, *Br. J. Nutr.*, 96 (Suppl. 2), 3S–11S, 2006.
5. Food and Agricultural Organization (FAO), Published online at: http://faostat.fao.org (accessed July 10, 2007).
6. Lin, B.H., Frazao, E., and Allhouse, J., U.S. consumption patterns of tree nuts, *Food Rev.*, 24, 54–58, 2001.
7. Jenab, M., Sabaté, J., Slimani, N., Ferrari, P., Mazuir, M., Casagrande, C., Deharveng, G., Tjonneland, A., Olsen, A., Overvad, K., Boutron-Ruault, M.C., Clavel-Chapelon, F., Boeing, H., Weikert, C., Linseisen, J., Rohrmann, S., Trichopoulou, A., Naska, A., Palli, D., Sacerdote, C., Tumino, R., Mattiello, A., Pala, V., Bueno-de-Mesquita, H.B., Ocke, M.C., Peeters, P.H., Engeset, D., Skeie, G., Jakszyn, P., Ardanaz, E., Quiros, J.R., Chirlaque, M.D., Martinez, C., Amiano, P., Berglund, G., Palmqvist, R., Guelpen, B., Bingham, S., Key, T., and Riboli, E., Consumption and portion sizes of tree nuts, peanuts and seeds in the European Prospective Investigation into Cancer and Nutrition (EPIC) cohorts from 10 European countries, *Br. J. Nutr.*, 96 (Suppl. 2), 12S–23S, 2006.
8. Moshfegh, A., Ingwersen, I., and Goldman, J., *Nut Consumption on the U.S. and the Contribution to Nutrient Intakes*, USDA, ARS, Beltsville, MD, 2007.
9. Lino, M., Marcoe, K., Dinkins, J.M., Hiza, H., and Anand, R., The role of nuts in a healthy diet, *Nutr. Insight*, 23, 1–2, 2000.
10. *Dietary Reference Intakes for Vitamin C, Vitamin E, Selenium, and Carotenoids*, The National Academy Press, Washington, DC, 2000.
11. United States Department of Health and Human Services and United States Department of Agriculture, *Dietary Guidelines for Americans*, 6th ed., Washington, DC, 2005.
12. Expert Panel on Detection, Evaluation, and Treatment of High Blood Cholesterol in Adults, Executive Summary of the Third Report of The National Cholesterol Education Program (NCEP) Expert Panel on Detection, Evaluation, and Treatment of High Blood Cholesterol in Adults (Adult Treatment Panel III), *J. Am. Med. Assoc.*, 285, 2486–2497, 2001.
13. Lichtenstein, A.H., Appel, L.J., Brands, M., Carnethon, M., Daniels, S., Franch, H.A., Franklin, B., Kris-Etherton, P., Harris, W.S., Howard, B., Karanja, N., Lefevre, M., Rudel, L., Sacks, F., Van Horn, L., Winston, M., and Wylie-Rosett, J., Diet and lifestyle recommendations revision 2006: A scientific statement from the American Heart Association Nutrition Committee, *Circulation*, 114, 82–96, 2006.

14. Sacks, F.M., Appel, L.J., Moore, T.J., Obarzanek, E., Vollmer, W.M., Svetkey, L.P., Bray, G.A., Vogt, T.M., Cutler, J.A., Windhauser, M.M., Lin, P.H., and Karanja, N., A dietary approach to prevent hypertension: A review of the Dietary Approaches to Stop Hypertension (DASH) Study, *Clin. Cardiol.*, 22 (Suppl. 7), III6–10, 1999.
15. Sabaté, J., Ros, E., and Salas-Salvado, J., Nuts: Nutrition and health outcomes, *Br. J. Nutr.*, 96 (Suppl. 2), 1S–2S, 2006.
16. Venkatachalam, M. and Sathe, S.K., Chemical composition of selected edible nut seeds, *J. Agric. Food Chem.*, 54, 4705–4714, 2006.
17. Ros, E. and Mataix, J., Fatty acid composition of nuts—Implications for cardiovascular health, *Br. J. Nutr.*, 96 (Suppl. 2), 29S–35S, 2006.
18. Brufau, G., Boatella, J., and Rafecas, M., Nuts: Source of energy and macronutrients, *Br. J. Nutr.*, 96 (Suppl. 2), 24S–28S, 2006.
19. U.S. Department of Agriculture (USDA). *National Nutrition Database for Standard References Release No. 19*, National Technical Information Service, USDA, Springfield, VA, 2006.
20. Ellis, P.R., Kendall, C.W., Ren, Y., Parker, C., Pacy, J.F., Waldron, K.W., and Jenkins, D.J., Role of cell walls in the bioaccessibility of lipids in almond seeds, *Am. J. Clin. Nutr.*, 80, 604–613, 2004.
21. Burton-Freeman, B., Davis, P.A., and Schneeman, B.O., Interaction of fat availability and sex on post-prandial satiety and cholecystokinin after mixed-food meals, *Am. J. Clin. Nutr.*, 80, 1207–1214, 2004.
22. Kirkmeyer, S.V. and Mattes, R.D., Effects of food attributes on hunger and food intake, *Int. J. Obes. Relat. Metab. Disord.*, 24, 1167–1175, 2000.
23. Alper, C.M. and Mattes, R.D., Effects of chronic peanut consumption on energy balance and hedonics, *Int. J. Obes. Relat. Metab. Disord.*, 26, 1129–1137, 2002.
24. Coates, A.M. and Howe, P.R., Edible nuts and metabolic health, *Curr. Opin. Lipidol.*, 18, 25–30, 2007.
25. Sathe, S.K., Kshirsagar, H.H., and Roux, K.H., Advances in seed protein research: A perspective on seed allergens, *J. Food Sci.*, 70, 93–120, 2005.
26. Kritchevsky, D., Protein and atherosclerosis, *J. Nutr. Sci. Vitaminol. (Tokyo)*, 36 (Suppl. 2), 81–86, 1990.
27. Salas-Salvadó, J., Bulló, M., Pérez-Heras, A., and Ros, E., Dietary fibre, nuts and cardiovascular diseases, *Br. J. Nutr.*, 96 (Suppl. 2), 45S–51S, 2006.
28. Gray, J., *Dietary Fibre—Definition, Analysis, Physiology & Health*, International Life Sciences Institute Europe, Brussels, Belgium, 2006.
29. Segura, R., Javierre, C., Lizarraga, M.A., and Ros, E., Other relevant components of nuts: Phytosterols, folate and minerals, *Br. J. Nutr.*, 96 (Suppl. 2), 36S–44S, 2006.
30. Gao, X., Wilde, P.E., Lichtenstein, A.H., and Tucker, K.L., The 2005 USDA Food Guide Pyramid is associated with more adequate nutrient intakes within energy constraints than the 1992 Pyramid, *J. Nutr.*, 136, 1341–1346, 2006.
31. Kornsteiner, M., Wagner, K.-H., and Elmadfa, I., Tocopherols and total phenolics in 10 different nut types, *Food Chem.*, 98, 381–387, 2006.
32. Lotito, S.B. and Frei, B., Consumption of flavonoid-rich foods and increased plasma antioxidant capacity in humans: Cause, consequence, or epiphenomenon? *Free Radic. Biol. Med.*, 41, 1727–1746, 2006.
33. Gu, L., Kelm, M.A., Hammerstone, J.F., Beecher, G., Holden, J., Haytowitz, D., Gebhardt, S., and Prior, R.L., Concentrations of proanthocyanidins in common foods and estimations of normal consumption, *J. Nutr.*, 134, 613–617, 2004.
34. Wu, X. and Prior, R.L., Identification and characterization of anthocyanins by high-performance liquid chromatography–electrospray ionization–tandem mass spectrometry in common foods in the United States: Vegetables, nuts, and grains, *J. Agric. Food Chem.*, 53, 3101–3113, 2005.
35. Pellegrini, N., Serafini, M., Salvatore, S., Del Rio, D., Bianchi, M., and Brighenti, F., Total antioxidant capacity of spices, dried fruits, nuts, pulses, cereals and sweets consumed in Italy assessed by three different in vitro assays, *Mol. Nutr. Food Res.*, 50, 1030–1038, 2006.
36. Wu, X., Beecher, G.R., Holden, J.M., Haytowitz, D.B., Gebhardt, S.E., and Prior, R.L., Lipophilic and hydrophilic antioxidant capacities of common foods in the United States, *J. Agric. Food Chem.*, 52, 4026–4037, 2004.
37. Frison-Norrie, S. and Sporns, P., Identification and quantification of flavonol glycosides in almond seedcoats using MALDI–TOF MS, *J. Agric. Food Chem.*, 50, 2782–2787, 2002.
38. Sang, S., Kikuzaki, H., Lapsley, K., Rosen, R.T., Nakatani, N., and Ho, C.-T., Sphingolipid and other constituents from almond nuts (*Prunus amygdalus* Batsch), *J. Agric. Food Chem.*, 50, 4709–4712, 2002.
39. Sang, S., Lapsley, K., Jeong, W.S., Lachance, P.A., Ho, C.-T., and Rosen, R.T., Antioxidative phenolic compounds isolated from almond skins (*Prunus amygdalus* Batsch), *J. Agric. Food Chem.*, 50, 2459–2463, 2002.

40. Milbury, P.E., Chen, C.Y., Dolnikowski, G.G., and Blumberg, J.B., Determination of flavonoids and phenolics and their distribution in almonds, *J. Agric. Food Chem.*, 54, 5027–5033, 2006.
41. Phillips, K.M., Ruggio, D.M., and Ashraf-Khorassani, M., Phytosterol composition of nuts and seeds commonly consumed in the United States, *J. Agric. Food Chem.*, 53, 9436–9445, 2005.
42. Franco, O.H., Bonneux, L., de Laet, C., Peeters, A., Steyerberg, E.W., and Mackenbach, J.P., The Polymeal: A more natural, safer, and probably tastier (than the Polypill) strategy to reduce cardiovascular disease by more than 75%, *Br. Med. J.*, 329, 1447–1450, 2004.
43. Trichopoulou, A. and Vasilopoulou, E., Mediterranean diet and longevity, *Br. J. Nutr.*, 84 (Suppl. 2), 205S–209S, 2000.
44. Trichopoulou, A. and Lagiou, P., Healthy traditional Mediterranean diet: An expression of culture, history, and lifestyle, *Nutr. Rev.*, 55, 383–389, 1997.
45. de Lorgeril, M., Renaud, S., Mamelle, N., Salen, P., Martin, J.L., Monjaud, I., Guidollet, J., Touboul, P., and Delaye, J., Mediterranean alpha-linolenic acid-rich diet in secondary prevention of coronary heart disease, *Lancet*, 343, 1454–1459, 1994.
46. Singh, R.B., Dubnov, G., Niaz, M.A., Ghosh, S., Singh, R., Rastogi, S.S., Manor, O., Pella, D., and Berry, E.M., Effect of an Indo-Mediterranean diet on progression of coronary artery disease in high risk patients (Indo-Mediterranean Diet Heart Study): A randomised single-blind trial, *Lancet*, 360, 1455–1461, 2002.
47. de Lorgeril, M., Salen, P., Laporte, F., Boucher, F., and de Leiris, J., Potential use of nuts for the prevention and treatment of coronary heart disease: From natural to functional foods, *Nutr. Metab. Cardiovasc. Dis.*, 11, 362–371, 2001.
48. Colquhoun, D.M., Food for prevention of coronary heart disease: Beyond the low fat, low cholesterol diet, *Asia Pac. J. Clin. Nutr.*, 9 (Suppl. 1), 86S–90S, 2000.
49. Trichopoulou, A. and Critselis, E., Mediterranean diet and longevity, *Eur. J. Cancer Prev.*, 13, 453–456, 2004.
50. Serra-Majem, L., Roman, B., and Estruch, R., Scientific evidence of interventions using the Mediterranean diet: A systematic review, *Nutr. Rev.*, 64 (Suppl. 1), 27S–47S, 2006.
51. Estruch, R., Martinez-Gonzalez, M.A., Corella, D., Salas-Salvado, J., Ruiz-Gutierrez, V., Covas, M.I., Fiol, M., Gomez-Gracia, E., Lopez-Sabater, M.C., Vinyoles, E., Aros, F., Conde, M., Lahoz, C., Lapetra, J., Saez, G., and Ros, E., Effects of a Mediterranean-style diet on cardiovascular risk factors: A randomized trial, *Ann. Intern. Med.*, 145, 1–11, 2006.
52. Fraser, G.E., Sabaté, J., Beeson, W.L., and Strahan, T.M., A possible protective effect of nut consumption on risk of coronary heart disease. The Adventist Health Study, *Arch. Intern. Med.*, 152, 1416–1424, 1992.
53. Hu, F.B. and Stampfer, M.J., Nut consumption and risk of coronary heart disease: A review of epidemiologic evidence, *Curr. Atheroscler. Rep.*, 1, 204–209, 1999.
54. Hu, F.B., Stampfer, M.J., Manson, J.E., Rimm, E.B., Colditz, G.A., Rosner, B.A., Speizer, F.E., Hennekens, C.H., and Willett, W.C., Frequent nut consumption and risk of coronary heart disease in women: Prospective cohort study, *Br. Med. J.*, 317, 1341–1345, 1998.
55. Ellsworth, J.L., Kushi, L.H., and Folsom, A.R., Frequent nut intake and risk of death from coronary heart disease and all causes in postmenopausal women: The Iowa Women's Health Study, *Nutr. Metab. Cardiovasc. Dis.*, 11, 372–377, 2001.
56. Albert, C.M., Gaziano, J.M., Willett, W.C., and Manson, J.E., Nut consumption and decreased risk of sudden cardiac death in the Physicians' Health Study, *Arch. Intern. Med.*, 162, 1382–1387, 2002.
57. Mantzoros, C.S., Williams, C.J., Manson, J.E., Meigs, J.B., and Hu, F.B., Adherence to the Mediterranean dietary pattern is positively associated with plasma adiponectin concentrations in diabetic women, *Am. J. Clin. Nutr.*, 84, 328–335, 2006.
58. Jiang, R., Jacobs, D.R., Jr., Mayer-Davis, E., Szklo, M., Herrington, D., Jenny, N.S., Kronmal, R., and Barr, R.G., Nut and seed consumption and inflammatory markers in the multi-ethnic study of atherosclerosis, *Am. J. Epidemiol.*, 163, 222–231, 2006.
59. Sabaté, J., Jenab, M., Ferrari, P., Norat, T., Slimani, N., and Riboli, E., Nut consumption and protection against coronary heart disease death in the European Prospective Investigation into Cancer and Nutrition (EPIC), presented at World Congress of Cardiology, Barcelona, Spain, September 2–6, 2006, Abstract 270.
60. Food and Drug Administration (FDA), Qualified Health Claims: Letter of Enforcement Discretion—Nuts and Coronary Heart Disease, Docket No 02P-0505, Food and Drug Administration, Washington, DC, 2003.

61. Abbey, M., Noakes, M., Belling, G.B., and Nestel, P.J., Partial replacement of saturated fatty acids with almonds or walnuts lowers total plasma cholesterol and low-density-lipoprotein cholesterol, *Am. J. Clin. Nutr.*, 59, 995–999, 1994.
62. Berry, E.M., Eisenberg, S., Haratz, D., Friedlander, Y., Norman, Y., Kaufmann, N.A., and Stein, Y., Effects of diets rich in monounsaturated fatty acids on plasma lipoproteins—the Jerusalem Nutrition Study: High MUFAs vs high PUFAs, *Am. J. Clin. Nutr.*, 53, 899–907, 1991.
63. Bruce, B., Spiller, G.A., and Farquhar, J.W., Effects of a plant-based diet rich in whole grains, sun-dried raisins and nuts on serum lipoproteins, *Veg. Nutr.*, 1, 58–63, 1997.
64. Bruce, B., Spiller, G.A., Klevay, L.M., and Gallagher, S.K., A diet high in whole and unrefined foods favorably alters lipids, antioxidant defenses, and colon function, *J. Am. Coll. Nutr.*, 19, 61–67, 2000.
65. Chisholm, A., McAuley, K., Mann, J., Williams, S., and Skeaff, M., Cholesterol lowering effects of nuts compared to with a Canola enriched cereal of similar fat content, *Nutr. Metab. Cardiovasc. Dis.*, 15, 284–292, 2005.
66. Jenkins, D.J., Popovich, D.G., Kendall, C.W., Vidgen, E., Tariq, N., Ransom, T.P., Wolever, T.M., Vuksan, V., Mehling, C.C., Boctor, D.L., Bolognesi, C., Huang, J., and Patten, R., Effect of a diet high in vegetables, fruit, and nuts on serum lipids, *Metabolism*, 46, 530–537, 1997.
67. Singh, R.B., Rastogi, S.S., Verma, R., Bolaki, L., and Singh, R., An Indian experiment with nutritional modulation in acute myocardial infarction, *Am. J. Cardiol.*, 69, 879–885, 1992.
68. Berry, E.M., Eisenberg, S., Friedlander, Y., Harats, D., Kaufmann, N.A., Norman, Y., and Stein, Y., Effects of diets rich in monounsaturated fatty acids on plasma lipoproteins—the Jerusalem Nutrition Study. II. Monounsaturated fatty acids vs carbohydrates, *Am. J. Clin. Nutr.*, 56, 394–403, 1992.
69. Kris-Etherton, P.M., Pearson, T.A., Wan, Y., Hargrove, R.L., Moriarty, K., Fishell, V., and Etherton, T.D., High-monounsaturated fatty acid diets lower both plasma cholesterol and triacylglycerol concentrations, *Am. J. Clin. Nutr.*, 70, 1009–1015, 1999.
70. Almario, R.U., Vonghavaravat, V., Wong, R., and Kasim-Karakas, S.E., Effects of walnut consumption on plasma fatty acids and lipoproteins in combined hyperlipidemia, *Am. J. Clin. Nutr.*, 74, 72–79, 2001.
71. Chisholm, A., Mann, J., Skeaff, M., Frampton, C., Sutherland, W., Duncan, A., and Tiszavari, S., A diet rich in walnuts favourably influences plasma fatty acid profile in moderately hyperlipidaemic subjects, *Eur. J. Clin. Nutr.*, 52, 12–16, 1998.
72. Colquhoun, D.M., Effects of a macadamia nut enriched diet on serum lipids and lipoproteins compared to a low fat diet, *Food Australia*, 48, 216–222, 1996.
73. Curb, J.D., Wergowske, G., Dobbs, J.C., Abbott, R.D., and Huang, B., Serum lipid effects of a high-monounsaturated fat diet based on macadamia nuts, *Arch. Intern. Med.*, 160, 1154–1158, 2000.
74. Durak, I., Köksal, I., Kaçmaz, M., Büyükkoçak, S., Çimen, B.M.Y., and Öztürk, H.S., Hazelnut supplementation enhances plasma antioxidant potential and lowers plasma cholesterol levels, *Clin. Chim. Acta*, 284, 113–115, 1999.
75. Edwards, K., Kwaw, I., Matud, J., and Kurtz, I., Effect of pistachio nuts on serum lipid levels in patients with moderate hypercholesterolemia, *J. Am. Coll. Nutr.*, 18, 229–232, 1999.
76. Garg, M.L., Blake, R.J., and Wills, R.B., Macadamia nut consumption lowers plasma total and LDL cholesterol levels in hypercholesterolemic men, *J. Nutr.*, 133, 1060–1063, 2003.
77. Hiraoka-Yamamoto, J., Ikeda, K., Negishi, H., Mori, M., Hirose, A., Sawada, S., Onobayashi, Y., Kitamori, K., Kitano, S., Tashiro, M., Miki, T., and Yamori, Y., Serum lipid effects of a monounsaturated (palmitoleic) fatty acid-rich diet based on macadamia nuts in healthy, young Japanese women, *Clin. Exp. Pharmacol. Physiol.*, 31 (Suppl. 2), 37S–38S, 2004.
78. Hyson, D.A., Schneeman, B.O., and Davis, P.A., Almonds and almond oil have similar effects on plasma lipids and LDL oxidation in healthy men and women, *J. Nutr.*, 132, 703–707, 2002.
79. Iwamoto, M., Imaizumi, K., Sato, M., Hirooka, Y., Sakai, K., Takeshita, A., and Kono, M., Serum lipid profiles in Japanese women and men during consumption of walnuts, *Eur. J. Clin. Nutr.*, 56, 629–637, 2002.
80. Jenkins, D.J., Kendall, C.W., Marchie, A., Parker, T.L., Connelly, P.W., Qian, W., Haight, J.S., Faulkner, D., Vidgen, E., Lapsley, K.G., and Spiller, G.A., Dose response of almonds on coronary heart disease risk factors: Blood lipids, oxidized low-density lipoproteins, lipoprotein(a), homocysteine, and pulmonary nitric oxide: A randomized, controlled, crossover trial, *Circulation*, 106, 1327–1332, 2002.
81. Kocyigit, A., Koylu, A.A., and Keles, H., Effects of pistachio nuts consumption on plasma lipid profile and oxidative status in healthy volunteers, *Nutr. Metab. Cardiovasc. Dis.*, 16, 202–209, 2006.

82. Lovejoy, J.C., Most, M.M., Lefevre, M., Greenway, F.L., and Rood, J.C., Effect of diets enriched in almonds on insulin action and serum lipids in adults with normal glucose tolerance or type 2 diabetes, *Am. J. Clin. Nutr.*, 76, 1000–1006, 2002.
83. Mercanlıgil, S.M., Arslan, P., Alasalvar, C., Okut, E., Akgül, E., Pınar, A., Geyik, P.Ö., Tokgözoğlu, L., and Shahidi, F., Effects of hazelnut-enriched diet on plasma cholesterol and lipoprotein profiles in hypercholesterolemic adult men, *Eur. J. Clin. Nutr.*, 61, 212–220, 2007.
84. Morgan, W.A. and Clayshulte, B.J., Pecans lower low-density lipoprotein cholesterol in people with normal lipid levels, *J. Am. Diet. Assoc.*, 100, 312–318, 2000.
85. Munoz, S., Merlos, M., Zambon, D., Rodriguez, C., Sabaté, J., Ros, E., and Laguna, J.C., Walnut-enriched diet increases the association of LDL from hypercholesterolemic men with human HepG2 cells, *J. Lipid Res.*, 42, 2069–2076, 2001.
86. Rajaram, S., Burke, K., Connell, B., Myint, T., and Sabaté, J., A monounsaturated fatty acid-rich pecan-enriched diet favorably alters the serum lipid profile of healthy men and women, *J. Nutr.*, 131, 2275–2279, 2001.
87. Ros, E., Nunez, I., Perez-Heras, A., Serra, M., Gilabert, R., Casals, E., and Deulofeu, R., A walnut diet improves endothelial function in hypercholesterolemic subjects: A randomized crossover trial, *Circulation*, 109, 1609–1614, 2004.
88. Sabaté, J., Fraser, G.E., Burke, K., Knutsen, S.F., Bennett, H., and Lindsted, K.D., Effects of walnuts on serum lipid levels and blood pressure in normal men, *N. Engl. J. Med.*, 328, 603–607, 1993.
89. Sabaté, J., Haddad, E., Tanzman, J.S., Jambazian, P., and Rajaram, S., Serum lipid response to the graduated enrichment of a Step I diet with almonds: A randomized feeding trial, *Am. J. Clin. Nutr.*, 77, 1379–1384, 2003.
90. Spiller, G.A., Jenkins, D.A., Bosello, O., Gates, J.E., Cragen, L.N., and Bruce, B., Nuts and plasma lipids: An almond-based diet lowers LDL-C while preserving HDL-C, *J. Am. Coll. Nutr.*, 17, 285–290, 1998.
91. Spiller, G.A., Jenkins, D.J., Cragen, L.N., Gates, J.E., Bosello, O., Berra, K., Rudd, C., Stevenson, M., and Superko, R., Effect of a diet high in monounsaturated fat from almonds on plasma cholesterol and lipoproteins, *J. Am. Coll. Nutr.*, 11, 126–130, 1992.
92. Spiller, G.A., Miller, A., Olivera, K., Reynolds, J., Miller, B., Morse, S.J., Dewell, A., and Farquhar, J.W., Effects of plant-based diets high in raw or roasted almonds, or roasted almond butter on serum lipoproteins in humans, *J. Am. Coll. Nutr.*, 22, 195–200, 2003.
93. Zambón, D., Sabaté, J., Muñoz, S., Campero, B., Casals, E., Merlos, M., Laguna, J.C., and Ros, E., Substituting walnuts for monounsaturated fat improves the serum lipid profile of hypercholesterolemic men and women. A randomized crossover trial, *Ann. Intern. Med.*, 132, 538–546, 2000.
94. Zhao, G., Etherton, T.D., Martin, K.R., West, S.G., Gillies, P.J., and Kris-Etherton, P.M., Dietary alpha-linolenic acid reduces inflammatory and lipid cardiovascular risk factors in hypercholesterolemic men and women, *J. Nutr.*, 134, 2991–2997, 2004.
95. Mukuddem-Petersen, J., Oosthuizen, W., and Jerling, J.C., A systematic review of the effects of nuts on blood lipid profiles in humans, *J. Nutr.*, 135, 2082–2089, 2005.
96. Griel, A.E. and Kris-Etherton, P.M., Tree nuts and the lipid profile: A review of clinical studies, *Br. J. Nutr.*, 96 (Suppl. 2), 68S–78S, 2006.
97. Kendall, C.W.C., Jenkins, D.J.A., Marchie, A., Ren, Y., Ellis, P.R., and Lapsley, K.G., Energy availability from almonds: implications for weight loss and cardiovascular health. A randomized controlled dose–response trial, *FASEB J.*, 17 (Abstract), A339, 2003.
98. Sabaté, J., Cordero-Macintyre, Z., Siapco, G., Torabian, S., and Haddad, E., Does regular walnut consumption lead to weight gain? *Br. J. Nutr.*, 94, 859–864, 2005.
99. Kushi, L.H., Folsom, A.R., Prineas, R.J., Mink, P.J., Wu, Y., and Bostick, R.M., Dietary antioxidant vitamins and death from coronary heart disease in postmenopausal women, *N. Engl. J. Med.*, 334, 1156–1162, 1996.
100. Rajaram, S. and Sabaté, J., Nuts, body weight and insulin resistance, *Br. J. Nutr.*, 96 (Suppl. 2), 79S–86S, 2006.
101. Garcia-Lorda, P., Megias Rangil, I., and Salas-Salvado, J., Nut consumption, body weight and insulin resistance, *Eur. J. Clin. Nutr.*, 57 (Suppl. 1), 8S–11S, 2003.
102. Bes-Rastrollo, M., Sabaté, J., Gomez-Gracia, E., Alonso, A., Martinez, J.A., and Martinez-Gonzalez, M.A., Nut consumption and weight gain in a Mediterranean cohort: The SUN study, *Obesity (Silver Spring)*, 15, 107–116, 2007.
103. Sabaté, J. and Blix, G., Vegetarian diets and obesity prevention, in *Vegetarian Nutrition*, Sabaté, J., Ed., CRC Press, Boca Raton, FL, 2001, pp. 91–107.

104. Key, T.J., Fraser, G.E., and Thorogood, M., Mortality in vegetarians and non-vegetarians: A collaborative analysis of 8300 deaths among 76,000 men and women in five prospective studies, *Pub. Health Nutr.*, 1, 33–41, 1998.
105. Rajaram, S. and Wien, M., Vegetarian diets in the prevention of osteoporosis, diabetes, and neurological disorders, in *Vegetarian Nutrition*, Sabaté, J., Ed., CRC Press, Boca Raton, FL, 2001, pp. 109–134.
106. St-Onge, M.P., Dietary fats, teas, dairy, and nuts: Potential functional foods for weight control? *Am. J. Clin. Nutr.*, 81, 7–15, 2005.
107. McManus, K., Antinoro, L., and Sacks, F., A randomized controlled trial of a moderate-fat, low-energy diet compared with a low fat, low-energy diet for weight loss in overweight adults, *Int. J. Obes. Relat. Metab. Disord.*, 25, 1503–1511, 2001.
108. Piers, L.S., Walker, K.Z., Stoney, R.M., Soares, M.J., and O'Dea, K., Substitution of saturated with monounsaturated fat in a 4-week diet affects body weight and composition of overweight and obese men, *Br. J. Nutr.*, 90, 717–727, 2003.
109. Nash, S.D. and Westpfal, M., Cardiovascular benefits of nuts, *Am. J. Cardiol.*, 95, 963–965, 2005.
110. Fraser, G.E., Bennett, H.W., Jaceldo, K.B., and Sabaté, J., Effect on body weight of a free 76 Kilojoule (320 calorie) daily supplement of almonds for six months, *J. Am. Coll. Nutr.*, 21, 275–283, 2002.
111. Jaceldo-Siegl, K., Sabaté, J., Rajaram, S., and Fraser, G.E., Long-term almond supplementation without advice on food replacement induces favourable nutrient modifications to the habitual diets of free-living individuals, *Br. J. Nutr.*, 92, 533–540, 2004.
112. Wien, M.A., Sabaté, J.M., Ikle, D.N., Cole, S.E., and Kandeel, F.R., Almonds vs complex carbohydrates in a weight reduction program, *Int. J. Obes. Relat. Metab. Disord.*, 27, 1365–1372, 2003.
113. Holt, S.H., Miller, J.C., Petocz, P., and Farmakalidis, E., A satiety index of common foods, *Eur. J. Clin. Nutr.*, 49, 675–690, 1995.
114. Iyer, S.S., Boateng, L.A., Sales, R.L., Coelho, S.B., Lokko, P., Monteiro, J.B., Costa, N.M., and Mattes, R.D., Effects of peanut oil consumption on appetite and food choice, *Int. J. Obes.*, 30, 704–710, 2006.
115. Levine, A.S. and Silvis, S.E., Absorption of whole peanuts, peanut oil, and peanut butter, *N. Engl. J. Med.*, 303, 917–918, 1980.
116. Haddad, E.H. and Sabaté, J., Effect of pecan consumption on stool fat, *FASEB J.*, 14 (Abstract), A294, 2000.
117. Zemaitis, J. and Sabaté, J., Effect of almond consumption on stool weight and stool fat, *FASEB J.*, 15 (Abstract), A602, 2001.
118. ADA, Nutrition recommendations and interventions for diabetes: A position statement of the American Diabetes Association, *Diabetes Care*, 30 (Suppl. 1), 48S–65S, 2007.
119. Garg, A., High-monounsaturated-fat diets for patients with diabetes mellitus: A meta-analysis, *Am. J. Clin. Nutr.*, 67, 577S–582S, 1998.
120. Chandalia, M., Garg, A., Lutjohann, D., von Bergmann, K., Grundy, S.M., and Brinkley, L.J., Beneficial effects of high dietary fiber intake in patients with type 2 diabetes mellitus, *N. Engl. J. Med.*, 342, 1392–1398, 2000.
121. Jarvi, A.E., Karlstrom, B.E., Granfeldt, Y.E., Bjorck, I.E., Asp, N.G., and Vessby, B.O., Improved glycemic control and lipid profile and normalized fibrinolytic activity on a low-glycemic index diet in type 2 diabetic patients, *Diabetes Care*, 22, 10–18, 1999.
122. Salmeron, J., Ascherio, A., Rimm, E.B., Colditz, G.A., Spiegelman, D., Jenkins, D.J., Stampfer, M.J., Wing, A.L., and Willett, W.C., Dietary fiber, glycemic load, and risk of NIDDM in men, *Diabetes Care*, 20, 545–550, 1997.
123. Salmeron, J., Manson, J.E., Stampfer, M.J., Colditz, G.A., Wing, A.L., and Willett, W.C., Dietary fiber, glycemic load, and risk of non-insulin-dependent diabetes mellitus in women, *J. Am. Med. Assoc.*, 277, 472–477, 1997.
124. Luscombe, N.D., Noakes, M., and Clifton, P.M., Diets high and low in glycemic index versus high monounsaturated fat diets: Effects on glucose and lipid metabolism in NIDDM, *Eur. J. Clin. Nutr.*, 53, 473–478, 1999.
125. Kendall, C.W., Marchie, A., Parker, T.L., Augustin, L.S., Ellis, P.R., Lapsley, K.G., Ternus, M., and Jenkins, D.J., Effect of nut consumption on postprandial starch digestion—a dose response study, *Ann. Nutr. Metab.*, 47 (Abstract PS S5), A636, 2003.
126. Jenkins, D.J., Kendall, C.W., Josse, A.R., Salvatore, S., Brighenti, F., Augustin, L.S., Ellis, P.R., Vidgen, E., and Rao, A.V., Almonds decrease postprandial glycemia, insulinemia, and oxidative damage in healthy individuals, *J. Nutr.*, 136, 2987–2992, 2006.
127. Josse, A.R., Kendall, C.W., Augustin, L.S., Ellis, P.R., and Jenkins, D.J., Almonds and postprandial glycemia—a dose-response study, *Metabolism*, 56, 400–404, 2007.

128. Jiang, R., Manson, J.E., Stampfer, M.J., Liu, S., Willett, W.C., and Hu, F.B., Nut and peanut butter consumption and risk of type 2 diabetes in women, *J. Am. Med. Assoc.*, 288, 2554–2560, 2002.
129. Lopez-Garcia, E., Schulze, M.B., Fung, T.T., Meigs, J.B., Rifai, N., Manson, J.E., and Hu, F.B., Major dietary patterns are related to plasma concentrations of markers of inflammation and endothelial dysfunction, *Am. J. Clin. Nutr.*, 80, 1029–1035, 2004.
130. Miller, J.C., Importance of glycemic index in diabetes, *Am. J. Clin. Nutr.*, 59, 747S–752S, 1994.
131. Sarkkinen, E., Schwab, U., Niskanen, L., Hannuksela, M., Savolainen, M., Kervinen, K., Kesaniemi, A., and Uusitupa, M.I., The effects of monounsaturated-fat enriched diet and polyunsaturated-fat enriched diet on lipid and glucose metabolism in subjects with impaired glucose tolerance, *Eur. J. Clin. Nutr.*, 50, 592–598, 1996.
132. Tapsell, L.C., Gillen, L.J., Patch, C.S., Batterham, M., Owen, A., Bare, M., and Kennedy, M., Including walnuts in a low-fat/modified-fat diet improves HDL cholesterol-to-total cholesterol ratios in patients with type 2 diabetes, *Diabetes Care*, 27, 2777–2783, 2004.
133. Gillen, L.J., Tapsell, L.C., Patch, C.S., Owen, A., and Batterham, M., Structured dietary advice incorporating walnuts achieves optimal fat and energy balance in patients with type 2 diabetes mellitus, *J. Am. Diet. Assoc.*, 105, 1087–1096, 2005.
134. Gonzalez, C.A. and Salas-Salvado, J., The potential of nuts in the prevention of cancer, *Br. J. Nutr.*, 96 (Suppl. 2), 87S–94S, 2006.
135. Ginter, E., Cardiovascular disease prevention in eastern Europe, *Nutrition*, 14, 452–457, 1998.
136. Halvorsen, B.L., Holte, K., Myhrstad, M.C., Barikmo, I., Hvattum, E., Remberg, S.F., Wold, A.B., Haffner, K., Baugerod, H., Andersen, L.F., Moskaug, O., Jacobs, D.R., Jr., and Blomhoff, R., A systematic screening of total antioxidants in dietary plants, *J. Nutr.*, 132, 461–471, 2002.
137. Blomhoff, R., Carlsen, M.H., Andersen, L.F., and Jacobs, D.R., Jr., Health benefits of nuts: Potential role of antioxidants, *Br. J. Nutr.*, 96 (Suppl. 2), 52S–60S, 2006.
138. Nichenametla, S.N., Taruscio, T.G., Barney, D.L., and Exon, J.H., A review of the effects and mechanisms of polyphenolics in cancer, *Crit. Rev. Food Sci. Nutr.*, 46, 161–183, 2006.
139. Singh, P.N. and Fraser, G.E., Dietary risk factors for colon cancer in a low-risk population, *Am. J. Epidemiol.*, 148, 761–774, 1998.
140. Kune, S., Kune, G.A., and Watson, L.F., Case–control study of dietary etiological factors: The Melbourne Colorectal Cancer Study, *Nutr. Cancer*, 9, 21–42, 1987.
141. Jenab, M., Ferrari, P., Slimani, N., Norat, T., Casagrande, C., Overad, K., Olsen, A., Stripp, C., Tjonneland, A., Boutron-Ruault, M.C., Clavel-Chapelon, F., Kesse, E., Nieters, A., Bergmann, M., Boeing, H., Naska, A., Trichopoulou, A., Palli, D., Krogh, V., Celentano, E., Tumino, R., Sacerdote, C., Bueno-de-Mesquita, H.B., Ocke, M.C., Peeters, P.H., Engeset, D., Quiros, J.R., Gonzalez, C.A., Martinez, C., Chirlaque, M.D., Ardanaz, E., Dorronsoro, M., Wallstrom, P., Palmqvist, R., Van Guelpen, B., Bingham, S., San Joaquin, M.A., Saracci, R., Kaaks, R., and Riboli, E., Association of nut and seed intake with colorectal cancer risk in the European Prospective Investigation into Cancer and Nutrition, *Cancer Epidemiol. Biomarkers Prev.*, 13, 1595–1603, 2004.
142. Hebert, J.R., Hurley, T.G., Olendzki, B.C., Teas, J., Ma, Y., and Hampl, J.S., Nutritional and socioeconomic factors in relation to prostate cancer mortality: A cross-national study, *J. Natl. Cancer Inst.*, 90, 1637–1647, 1998.
143. Davis, P.A. and Iwahashi, C.K., Whole almonds and almond fractions reduce aberrant crypt foci in a rat model of colon carcinogenesis, *Cancer Lett.*, 165, 27–33, 2001.
144. Jia, X., Li, N., Zhang, W., Zhang, X., Lapsley, K., Huang, G., Blumberg, J., Ma, G., and Chen, J., A pilot study on the effects of almond consumption on DNA damage and oxidative stress in smokers, *Nutr. Cancer.*, 54, 179–183, 2006.
145. Chen, C.Y., Milbury, P.E., Lapsley, K., and Blumberg, J.B., Flavonoids from almond skins are bioavailable and act synergistically with vitamins C and E to enhance hamster and human LDL resistance to oxidation, *J. Nutr.*, 135, 1366–1373, 2005.
146. Anderson, K.J., Teuber, S.S., Gobeille, A., Cremin, P., Waterhouse, A.L., and Steinberg, F.M., Walnut polyphenolics inhibit in vitro human plasma and LDL oxidation, *J. Nutr.*, 131, 2837–2842, 2001.
147. Jambazian, P.R., Haddad, E., Rajaram, S., Tanzman, J., and Sabaté, J., Almonds in the diet simultaneously improve plasma alpha-tocopherol concentrations and reduce plasma lipids, *J. Am. Diet. Assoc.*, 105, 449–454, 2005.

4 Tree Nut Allergens

Shridhar K. Sathe, Girdhari M. Sharma, and Kenneth H. Roux

CONTENTS

4.1 INTRODUCTION

Tree nuts are consumed by a majority of individuals in one form or another, depending on the type of tree nut available in the geographical area. Tree nuts are liked by many due to their pleasing taste and potential health benefits. Although rich in calories, with a range of 570–720 kcal/100 g edible portion, they are a good source of mono- and polyunsaturated fatty acids (MUFA and PUFA). The most common and widely consumed tree nuts include almond, Brazil nut, cashew, hazelnut, macadamia, pecan, pine nut, pistachio, and walnut. Among them, almond, walnut, and hazelnut are the important tree nuts in terms of global production and trade.

4.1.1 Tree Nut Crops and Their Commercial Importance

Tree nuts are of economic importance in animal and human food supply and nutrition. Consumed in raw and processed forms, they are readily accepted by many. When used judiciously, tree nuts offer a valuable ingredient in the development of food products with pleasing sensory characteristics. Such products, often described as value added, typically provide good economic returns for the processor.

For some unexplained reasons, food allergies seem to be on the rise in the Western countries. A recent publication provides a good overview on food allergy and anaphylaxis [1]. Although safely enjoyed by many, tree nuts pose health problems to sensitive individuals. These individuals, for reasons not yet fully understood, encounter negative physiological outcome upon exposure to tree nuts. The reactions may range from temporary mild irritation to, in severe cases, anaphylaxis. Such reactions to tree nut consumption may include simple local chemical reactions to complicated systemic (e.g., IgE-mediated) ones. In the literature, IgE-mediated reaction in response to an offending food protein is referred to as food allergy [2]. Tree nuts are among the eight major food groups widely recognized as being responsible for human allergies. While many food allergies may be outgrown (e.g., infancy/childhood milk allergies), tree nut allergies are often considered to be permanent. Recent research findings suggest, at least in certain individuals, tolerance may be developed and in certain instances, tree nut allergies may be reversible [3]. Often, sensitive individuals exhibit allergies to multiple nut seeds [4].

In recent years, edible nut seed allergies are receiving increased attention. Reasons for increased attention include increase in food allergies, increased awareness of food allergies, improved documentation and reporting of food allergy incidences, and passage of food allergen labeling laws in the United States [5] and European Union [6]. Major population centers (Asia and Africa) around the globe are also becoming increasingly aware of food allergies and such recognition is expected to enhance global documentation of food allergies. Why otherwise well-tolerated food proteins adversely affect sensitive individuals remains unclear. However, IgE cross-linking on the surface of mast cells and basophils by the offending food protein appears to be the key step in developing food allergy. Lack of proven medical treatments for food allergies dictates that avoidance of the offending food to be the best choice for sensitive individuals.

4.1.2 Food Allergy

An allergy refers to the immune response to foreign substances when they come in contact with the body via inhalation (plant pollens and mold spores), injection (insect venoms and vaccines), ingestion (foods and drugs), or dermal contact (plant leaves, metals, and synthetic chemicals). Most often, these substances are harmless and not identified as a threat by the immune system of a normal person. However, sometimes the system fails to distinguish between the harmful and harmless substances in allergic patients, and launches an attack against the otherwise innocuous foreign material and produces cutaneous (urticaria, angioedema, and rash), respiratory (coughing, wheezing, and sneezing), and/or gastrointestinal (nausea, cramping, vomiting, and diarrhea) symptoms. The symptoms of such adverse reactions may range from minor skin rash to life-threatening anaphylaxis. Systemic anaphylaxis is an acute allergic reaction that affects multiple systems in the body, where the allergen is disseminated by the bloodstream and causes widespread activation of the connective tissue mast cells associated with blood vessels and mucosa. Food-induced anaphylactic reactions account for more than one-third of the anaphylactic reactions and are most often attributable to peanut, tree nuts, or shell fish [7]. In the United States, there are about 30,000 anaphylactic episodes due to food allergy each year, resulting in about 150–200 deaths [7,8].

There are four types of hypersensitivity reactions based on the type of mechanism that produces the immune response. Type I hypersensitivity reactions result from IgE-mediated immune response causing mast cell activation. Most food allergy, allergic rhinitis, wheal and flare, asthma, and systemic anaphylaxis falls under type I category. Type II and III hypersensitivity are IgG-mediated while the immune reactants in type IV are T_H1 or T_H2 cells. Type I hypersensitivity is triggered by antigen cross-linking of preexisting IgE antibodies associated with the Fc receptors on mast cells. The activated mast cells undergo degranulation and release inflammatory molecules. The primary inflammatory mediators stored in the granules include histamine, serotonin, chemotactic peptides (ECF-A, NCF-A), and proteases, while the secondary mediators are synthesized *de novo* and comprise of arachidonic acid metabolites (prostaglandins and leukotrienes) and proteins (cytokines and enzymes). The immunologic reactions in food hypersensitivity/allergy may be due to specific immunoglobulin IgE, or other Ig classes, or by T cells. Cell-mediated hypersensitivity is predominant in infants and children, and responsible for several gastrointestinal disorders including food protein-induced protocolitis, enterocolitis, enteropathy syndrome, and celiac disease [9]. IgE-mediated food allergy occurs when the immune system produces specific immunoglobulin (IgE) antibodies against the allergen.

The first encounter of the food-containing allergen begins the sensitization process by inducing the production of IgE antibodies. Subsequent exposure of the allergen triggers the IgE cross-linking on mast cells to release histamines and other biological mediators. Though the molecular mechanism involved in the release of these mediators after allergen-IgE binding has occurred is reasonably well understood, the very reason why the allergen binds the IgE remains unanswered. The IgE antibody may recognize a linear epitope (continuous linear sequence of amino acids) or conformational epitope (discontinuous sequence of amino acids produced by three-dimensional structural motifs). The amount of allergen required for sensitization and subsequent allergic reaction is very low and differs among allergic individuals. Based on the sensitization by the allergen, there are two classes of food allergy. The class 1 food allergy is characterized by sensitization occurring in the gastrointestinal tract when allergens stable toward gastric digestion reach intestinal mucosa relatively intact where they are absorbed in intestinal mucosa for sensitization [10]. This form of allergy occurs predominantly in children. On the other hand, class 2 food allergy results from cross-reactivity with inhalant allergens and mainly occurs in adults.

Food allergens are typically proteins and may occur in trace amounts or may constitute a major portion of total food protein. Many food allergens appear to be heat-stable proteins with monomer molecular mass 10–70 kDa [11]. A number of the major food allergens are multimeric. Homology observed among allergens of same class/type is not uncommon. However, such homology is not sufficient to distinguish allergenic proteins from nonallergenic ones *a priori*. Among the allergens of plant origins, seed storage proteins (legumins, vicilins, albumins, conglutins, glycinins, and β-conglycinins, etc.), pathogenesis-related proteins (β-1,3-glucanases, chitinases, thaumatin-like proteins, Bet v 1 homologous proteins, and lipid transfer proteins), protease and α-amylase inhibitors, peroxidases, profilins, proteases, and lectins have been reported to induce IgE-mediated allergic reactions [12–15]. The molecular structure of these proteins plays an important role in the allergic response. IgE-binding epitopes may be present on the surface of the native protein or buried within the three-dimensional structure to be released or exposed during protein digestion by digestive tract proteases *in vivo*. Although stable epitopes may remain unaffected, susceptible epitopes may be altered or destroyed and/or new epitopes may be formed as a consequence of food processing [16].

Theoretically, any food protein is a potential allergen. Food allergens have been found in a variety of food products ranging from commonly consumed foods like milk and egg to exotic foods such as lychee and mango [17–19]. The most common food allergens are "The Big Eight" foods that include milk, egg, peanut, tree nuts, wheat, soy, fish, and shellfish [20]. Milk and egg allergy is common in children and is outgrown by many. The mechanism involved in developing tolerance to cow's milk allergy and egg allergy is not known. However, decreased level of food-specific IgE has been observed in patients and may partly explain development of clinical tolerance by some patients [21,22]. The major allergens in eggs are ovomucoid (Gal d 1) and ovalbumin (Gal d 2), while the minor allergens include ovotransferrin (Gal d 3) and lysozyme (Gal d 4) [23]. The most commonly reported fish allergies are to salmon and tuna, while reaction to shellfish typically involves shrimp and crab [24]. The major allergens in fish and shellfish are parvalbumin and tropomyosin, respectively [25–27]. Among legumes, peanut and soybean proteins have been extensively investigated and many allergens characterized [28]. Soybean allergens are derived from cupin family (β-conglycinin, glycinin, and Gly m Bd 28k), 2S albumin, profilin (Gly m 3), Bet v 1 homologue (Gly m 4), cysteine protease C1 (Gly m Bd 30k), and trypsin inhibitor. Since there are homologous proteins in peanut and soy, it is not uncommon to have coallergy (cross-reactivity) among them or other legumes [29]. Similarly cross-reactivity exists among the closely related species of tree nuts [30,31]. Allergy to wheat has gained recognition in recent years due to occupational respiratory allergy (baker's asthma) to inhaled wheat flour and wheat dependent exercise-induced anaphylaxis [32–34]. The water- and salt-soluble wheat proteins, albumins, and globulins are the common cause for baker's asthma while the insoluble gliadins and glutenin are often responsible for IgE-mediated allergy to ingested wheat [35,36].

4.1.3 Prevalence of Tree Nut Allergy

Globally, food allergy incidences are not subject to mandatory reporting requirements and therefore accurate statistics on food allergy is difficult to obtain. Attempts have begun in recent years to estimate prevalence of food allergy. For instance, according to some estimates, prevalence of food allergy in the United States is about 6% in children (≤5 years) and 3.7% in adults [2,37–39]. More than 75% of total food allergies in young children are attributed to milk (41%), egg (21%), and peanuts (13%), whereas shellfish (54%), peanuts (16%), and tree nuts (13%) account for about 85% of the total food allergies in adults [38]. Wheat and soy allergies are not well documented although soybean allergies have been recently reviewed [40] and attempts are being made to determine the number and type of allergens present [41,42]. An international survey based on self-reported questionnaire suggests that the prevalence of food allergy/intolerance among adults varies significantly among different countries with estimates ranging from 4% to 20% [43]. Reported allergies toward a particular food also vary significantly depending on the food, study design, and the population sampled [44]. Since most surveys are based on questionnaires, accurate estimates are often difficult to obtain as symptoms of food allergy and (nonimmunological) food intolerance may be similar, resulting in overestimates of true allergy incidence. A lack of accurate and reliable diagnostic methods to determine food allergies and the inability of such tests to distinguish food allergy from food intolerance may further complicate such estimates. Inadequate consumer awareness and education may exacerbate the difficulties in obtaining accurate and reliable data on food allergies. A study by Armstrong and Rylance [45] involving 96 acute allergic children aged 18 months to 15 years and referred from general practitioners or accident and emergency department doctors concluded that skin prick testing and radioallergosorbent test for IgE measurement are inadequate tests for nut allergy and are not comparable to oral challenge. Allergy to any particular food depends on many factors such as geographical location, diet, and the individual's current immune status and immunological history.

Majority of tree nut allergies are associated with one or more of nine widely consumed tree nuts: almond, Brazil nut, cashew nut, hazelnut, macadamia, pecan, pine nut, pistachio, and walnut. Almond, pecan, and walnut together account for about 64% of total tree nut consumption in the United States [46,47]. Cashew, pine nut, Brazil nut (together 25%), pistachio (8%), and hazelnut (3%) follow. The most commonly consumed tree nuts in the European countries include hazelnut, almond, and walnut [48]. Though not as frequent as milk or egg allergy, tree nut allergies are more typically associated with severe or fatal reactions and, therefore, are of serious concern. For example, in a study by Bock et al. [8] peanuts (63%) and tree nuts (31%) were responsible for more than 90% of the 32 reported fatalities during 1994–1999. The authors found that fatal anaphylactic reactions occurred in both sexes equally, and most victims were adolescents or young adults with known food allergy. Most severe reactions occur due to accidental exposure outside the home [49]. A telephone interview of 129 subjects who had allergic reactions to peanut, tree nuts, or both in restaurant and other food establishments showed that 67% of reactions were caused by peanut and 24% by tree nuts. The establishments most commonly associated with food allergy incidents include restaurants and establishments serving or catering Asian foods, ice cream shops, and bakeries. Desserts are the common meal course associated with 43% of total reactions in the study [50].

Overall, tree nut allergy accounts for about 0.7% in the United States, where adults are more sensitive (0.5%) than young children (0.2%) [38,51]. It was thought that tree nuts tend to induce "persistent sensitivity" in the vast majority of patients, but recent reports suggest that about 10% of young patients may outgrow tree nut allergies [3,7,52]. Sicherer et al. [53] conducted a random digit dial telephone survey with standardized questionnaire involving 4374 households, representing 12,032 individuals to determine the prevalence of peanut and tree nut allergy in the United States. Peanut or tree nut allergy was reported in 151 households (164 individuals) with a corrected prevalence rate of 1.1%. Specific allergy to peanut and tree nuts was found in 0.6% and 0.5% of the total participants, respectively. Also, the incidence of tree nut allergy was high in adults (0.7%) than children below 18 years of age (0.2%). Of 118 patients who were able to provide detailed information, 58 patients were allergic to peanut followed by walnut (24), cashew (8), Brazil nut (8), almond (7), pecan (7), hazelnut (3), macadamia (2), and unspecified mixed nuts (6). Allergy to more than one tree nut was observed in five adults, while four adults reported both peanut and tree nut allergy.

Allergy to multiple tree nuts is not uncommon, which may be attributed to cross-reactivity of similar proteins as discussed further below. A more recent survey by Sicherer et al. [37] was conducted to estimate the prevalence of peanut and tree nut allergy in the United States by self-reported random digit dial telephone and compared with the earlier survey carried out in 1999. The survey included 4855 households representing 13,493 individuals. Though the total allergic cases attributed to any nut including peanut (1.04%) were not significantly different from earlier report (1.1%), the incidence doubled from 0.6% to 1.2% in children below 18 years of age. Also, there was no significant change in the rate of tree nut allergy in children and peanut or tree nut allergy in adults, but allergy to peanut among children increased from 0.4% to 0.8%. Additional food allergies were observed in 50% of individuals with peanut allergy, tree nut allergy, or both. Thirty-two respondents (0.3%) had allergy to both peanut and tree nut as compared to four individuals in previous report. Information about allergy to specific tree nut was available from 82 individuals of total 89 tree nut allergic patients. Forty-nine individuals indicated allergy to one tree nut, 15 to two nuts, and 28 to three or more different tree nuts. Among the tree nuts, walnut allergy was most frequent cause (51 individuals), followed by cashew (36), almond (32), pecan (29), Brazil nut (25), hazelnut (24), macadamia (21), pistachio (18), and pine nut (16). The noteworthy point is the increase in incidence of allergy to infrequently consumed nuts. The severity of the tree nut allergies can be surmised from the fact that self-injectable epinephrine was prescribed for 46% of the children and 23% of the adults evaluated for allergy by the physician.

With the aid of Food Allergy and Anaphylaxis Network, a voluntary peanut and tree nut allergy registry was established for better understanding of the prevalence and clinical features associated with them. Out of a total of 5149 registrants, allergy to peanut was reported by 3842 (68%), tree nut by 464 (9%), and to both by 1203 (23%) [49]. Most registrants in the study were children with an average age of 8.5 years, who also suffered from atopic disorders like atopic dermatitis (50%), asthma (46%), and allergic rhinitis (27%). The initial reaction to tree nuts occur at a median age of 36 months, with ingestion being the most common mode of exposure (88%) followed by skin contact (9%) and airborne exposure (3%). More than half of the tree nut allergic patients (54%) listed a single tree nut as a cause of reaction. Walnut was responsible for 34% of reactions followed by cashew (20%), almond (15%), pecan (9%), and pistachio (7%). Other tree nuts (hazelnut, Brazil nut, pine nut, macadamia, and hickory) accounted for less than 5% each.

A similar rise in allergic cases has been observed in other countries. A study involving 62 nut-allergic patients from allergy clinic at Addenbrooke's Hospital (in the United Kingdom) had peanut as most common cause of allergy (47 patients) [54]. Among the tree nuts, Brazil nut was responsible for allergy in 18 patients, followed by almond (14), hazelnut (13), walnut (8), and cashew (3). Thirty-seven patients were allergic only to a single specific tree nut or peanut, while 25 had allergy to multiple nut seeds. Reports from Isle of Wight have shown 1.2% of children were sensitized to peanut or tree nuts by age of 4 years [55]. Out of 1218 children reviewed at age 4 years, six had allergic reactions to peanut, seven had positive skin prick test or detectable IgE to peanut without clinical symptoms, one had allergy to hazelnut, and one to cashew. A questionnaire survey of 33,110 Frenchmen reported the rate of allergies to peanut and tree nuts to be 1% and 3%, respectively [56]. As is apparent from the foregoing discussion, there is a global and critical need to accurately document incidence and statistics for food allergies.

4.2 TREE NUT ALLERGENS

4.2.1 Tree Nut Allergen Identification

Most food allergies may be clinically identified through double-blind placebo-controlled crossover food challenge studies. However, due to inherent risks involved in such procedures, food allergies are often identified, at least initially, using a variety of tests either alone or in combination. For example, skin prick tests alone or in combination with patient IgE reactivity assays may be used for such purpose. Regardless of the method(s) used, the protein(s) responsible for human allergies are difficult to identify. Consequently, only a few proteins have been demonstrated to be true food allergens. Tree nut proteins that have been identified to be human allergens are summarized in Table 4.1.

4.2.2 Epitope Identification and Characterization

The IgE-binding epitopes typically involve 8–10 amino acids, which may be continuous (linear) or discontinuous (conformational), that react with patient IgE. Epitope identification and characterization is one of the most critical steps in efforts that seek to reduce protein allergenicity. Once the epitope is identified, it can be modified such that IgE is unable to recognize the targeted epitope. Such modifications may include denaturation of conformational epitopes, mutations, epitope hydrolysis, or a combination thereof. The solid-phase peptide synthesis (SPPS) technology is one of the most common techniques used for identification of linear epitopes. Characterization of conformational epitopes, as compared to defining the linear epitopes, is more challenging. Many immune-reactive linear epitopes of major allergens in peanut have been characterized in the last 10 years [57–59].

TABLE 4.1
Allergens Identified in Widely Consumed Tree Nuts

Source	Allergen	Protein Class/Family	kDa	Ref.
Almond (*Prunus dulcis*)	Amandin	14S legumin	63.02	[91,92]
	2S albumin	2S albumin	12	[131]
	Conglutin γ	7S vicilin	45	[131]
	Pru du 4	Profilin	14.06	[118]
	Pru du 5	60s ribosomal protein	10	[132][a]
Brazil nut (*Bertholetia excelssa*)	Ber e 1	2S albumin	12.2	[133]
	Ber e 2	11S legumin	52.3	[134][a]
Cashewnut (*Anacardium occidentale*)	Ana o 1	7S vicilin	50	[60]
	Ana o 2	13S legumin	53	[61]
	Ana o 3	2S albumin	12.6	[62]
Chestnut (*Castanea sativa*)	Cas a 1	Pathogenesis-related	22	[135][a]
	Cas a 5	Class I chitinase	32	[136]
	Cas a 8	Lipid transfer protein	9.7	[137]
Hazelnut (*Corylus avellana*)	Cor a 1	Bet v 1 homologue	17–18	[138]
	Cor a 2	Profilin	14	[139][a]
	Cor a 8	Lipid transfer protein	9	[140]
	Cor a 9	11S legumin	40	[108]
	Cor a 11	7S vicilin	48	[141]
Pecan (*Carya illinoinensis*)	Car i 1	2S albumin	10.73	[142][a]
Pistachio (*Pistacia vera*)	Pis v 1	2S albumin	7	[b]
	Pis v 2	11S legumin	32	[b]
	Pis v 3	7S vicilin	55	[148]
Walnut (*Juglans regia*)	Jug r 1	2S albumin	15–16	[143]
	Jug r 2	7S vicilin	47	[144]
	Jug r 3	Lipid transfer protein	9	[145]
	Jug r 4	11S legumin	58.1	[146]

[a] Direct submission to National Center for Biotechnology Information.

[b] Extracted from International Union of Immunological Societies, Allergen Nomenclature Subcommittee, www.allergen.org.

In order to develop effective immunotherapy candidates, clinically relevant details of epitope–IgE interactions are needed.

Epitope identification in tree nuts has begun in recent years with identification of linear epitope in cashew vicilin allergen, Ana o 1 [60]. Sixty-six overlapping peptides, each constituting of 15 amino acids and offset by eight amino acids, were synthesized using solid phase peptide synthesis (SPPS) technique. The immunoblotting of these peptides with three pools of cashew allergic patient sera revealed 11 peptides to be reactive to the IgE. Three immunodominant linear epitopes (shown in Table 4.2) were identified by all three of the tested pools of human sera. A comparison of linear epitopes of Ana o 1 and Ara h 1, the peanut homologue, shows positional overlap at four epitope regions, but overall there was no significant conservation found between their epitopes [57,60]. Similar method for cashew legumin allergen, Ana o 2, revealed 22 IgE-binding epitopes distributed throughout the length of protein [61]. The acidic chain region of Ana o 2 was found to be more

TABLE 4.2
Immunodominant Epitopes of Seed Storage Proteins Identified in Nut Seeds

Protein Type/ Function	Source	Allergen	Protein Accession No.	Immunodominant IgE Binding Epitopes	Amino Acid Residues[a]	Ref.
2S albumin	Brazil nut	Ber e 1	BAA96554.1	EGLRMMMRMMQQKEMQPRG	63–100	[66]
				EQMRRMMRLAENIPSRCNL[b]		
				QMQRQQMLSHCRMY	11–24	
	Cashew	Ana o 3	AAL91665.1	SGREQSCQRQFE	33–44	[62]
				KQEVQRGGRYNQ	57–68	
				SLRECCQELQEV	72–83	
				QEQIKGEEVREL	102–113	
	Peanut	Ara h 2	AAN77576.1	DRRCQSQLER	27–36	[58]
				YERDPYSPSQ	57–66	
				SQDPYSPSPY	65–74	
	Walnut	Jug r 1	AAB41308.1	QGLRGEEMEEMV	33–44	[63]
7S vicilin	Cashew	Ana o 1	AAM73730.2	AIMGPPTKFSFSLFL	1–15	[60]
				KECEKYYKEKKGRER	57–71	
				EEFFFQGPEWRKEKE	521–535	
	Peanut	Ara h 1	AAB00861.1	AKSSPYQKKT	25–34	[57]
				LEYDPRLVYD	65–74	
				GERTRGRQPG	89–98	
				RRYTARLKEG	498–507	
11S legumin	Cashew	Ana o 2	AAN76862.1	SRQEWQQQDECQIDR	15–29	[61]
				YQAPQQGRQQGQSGR	105–119	
				VFQQQQQHQSRGRNL	185–199	
				KVKDDELRVIRPSRS	233–247	
				VIRPSRSQSERGSES	241–255	
				EESEDEKRRWGQRDN	257–271	
				FQISREDARKIKFNN	425–439	
	Peanut	Ara h 3	ABI17154.1	IETWNPNNQEFECAG	33–47	[59]
				GNIFSGFTPEFLEQA	240–254	
				VTVRGGLRILSPDRK	279–293	
				DEDEYEYDEEDRG	303–317	

[a] Amino acid residues of epitopes are based on the published references and may be different from the accession no. provided here.

[b] Structural epitope.

reactive (15 epitopes) compared to the basic subunit. Moreover, six out of seven immunodominant epitopes (Table 4.2) are present in the acidic chain, with the exception of FQISREDARKIKFNN in the basic chain. Only two immunodominant epitopes show significant positional overlap with peanut homologous allergen, Ara h 3, epitope-bearing region, but interestingly, nine of the 11 G2 glycinin epitopes show significant overlap with Ana o 2 epitopes. Robotham et al. [62] recently cloned a 12.6 kDa 2S albumin allergen in cashew, designated as Ana o 3. The authors identified four strongly reactive epitopes, one of which (KQEVQRGGRYNQ) was bound to the IgE of all four pools of allergic patients sera used in the study. Sequence alignment of Ana o 3 with 2S albumins from walnut, sesame, and yellow mustard seed indicated one immunodominant epitope of Ana o 3 had significant sequential and positional overlap with walnut major epitope of Jug r 1, and positional overlap with yellow mustard Sin a 1 epitope, while the other three immunodominant epitopes exhibited considerable positional overlap with an extended segment of sesame seed 2S albumin protein [63–65].

The linear epitope mapping and mutational analysis of immunodominant epitope of walnut 2S albumin Jug r 1 was done by Robotham et al. [63]. Out of the 25 overlapping peptides (each containing 13 amino acids) used for immunoblotting with four pools of 20 walnut allergic patient sera, three neighboring peptides bearing a common sequence of GLRGEEM were found to be reactive. Based on these peptides, an IgE-reactive peptide (QGLRGEEMEEMV) was synthesized and mutational analysis was performed by substituting each amino acid with alanine. The core amino acid sequence RGEE and a glutamic acid residue at position 113 of Jug r 1 were found to be critical for IgE binding. These critical amino acids also share the epitope in Ana o 3, showing 100% similarity and 80% identity. A structural and linear epitope has been shown to be important for IgE binding to the major allergen Ber e 1 in Brazil nut [66]. Recent efforts in understanding conformational epitopes and role of protein homology in IgE binding among different food proteins suggest that cross-reactivity may be observed in botanically distant plants [67].

4.3 EFFECTS OF FOOD PROCESSING ON TREE NUT ALLERGENS

A recent study by Varshney et al. [68] on mountain cedar allergen, Jun a 1, illustrates the importance and complexity of investigating effects of processing on conformational and linear epitopes. The authors reported that upon heating (75°C, 1 h), exposure to 6 M guanidine hydrochloride or reductive alkylation denatured the conformational epitope, resulting in statistically significant reduction (31%–36% reduction) in patient serum IgE binding (inhibition assay format). Interestingly, the same treatments increased the IgE binding (enzyme-linked immunosorbent assay [ELISA]) to the linear epitopes. The authors attributed the increased binding to the linear epitopes to increased epitope exposure upon protein denaturation. For the affected patient, the combined effect of IgE reactivity is relevant. Similarly, Kleber et al. [69] noted a remarkable increase in antigenicity (ELISA assays) when milk whey β-globulin was heated to 90°C, which could be attributed to improved exposure of buried epitopes. However, when heated to higher temperatures (>90°C and up to 150°C), antigenicity decreased by as much as 90% due to protein aggregation. Particle size measurement in conjunction with ELISA was used to support this observation. These two examples indicate that effects of processing on allergenicity should be evaluated carefully as foods are consumed in both raw and processed forms.

Most foods are processed prior to consumption to improve sensory qualities and shelf life. Processing also ensures the safety of food against microbes and toxins. Processing may be done at home, food service facility, or in an industry. The main difference in these three settings is usually the scale and degree of sophistication of the processing machinery handling the food. Food processing could be broadly classified into two categories: thermal (drying, roasting, frying, blanching, infrared heating, microwave heating, and other heat treatments) and nonthermal (fermentation, germination, irradiation, and filtration). These processing methods can cause enzymatic as well as nonenzymatic biochemical changes in the processed foods, resulting in the altered availability of some nutrients and structural changes in others, especially proteins that are sensitive to denaturation. The modified structure of proteins can result in an increased or decreased level of IgE binding by formation of new allergen epitopes or inactivation of the existing allergen epitopes, respectively. Generalizations about the relationship of the allergen with the processing method are difficult to establish. Very few studies have been done to evaluate the effect of processing techniques on allergen modification.

The effect of food processing on allergenicity depends on various factors such as type and intensity of processing, role of food matrix, protein type and stability, and epitopes identified by an individual's IgE. Since thermal processing is the most common method of food preservation, more studies have been conducted on their role in allergenicity [16,70,71]. Also, they have been found to be more effective than nonthermal processing in reducing the allergenicity in plant foods such as celery [72]. Conformational epitopes, which are more sensitive to protein denaturation and/or aggregation than linear epitopes, may be the first to get altered during processing. This could be one of the reasons for loss of IgE binding directed against conformational epitopes [73,74]. Sometimes the processing conditions needed to eliminate/reduce allergenicity would be so harsh as to adversely affect the sensory qualities of the food [75]. There have been many such cases reported where a patient allergic to raw food can consume the processed food without any reactions due to inactivation or destruction of the allergen [76–78]. On the other hand, patients who can consume raw food but are allergic to processed foods are not uncommon [79,80]. This could be attributed to the formation of neoallergens where epitopes hidden in the native structure of protein get exposed to react with IgE upon processing. Initially, it was believed that food allergens may exhibit higher gastric stability than nonallergens [10], but a recent study on protein digestibility in simulated gastric and intestinal fluid did not support this hypothesis and thus cannot be used to distinguish between food allergens and nonallergens [81]. It should be noted that in these studies, high enzyme to protein ratio (13:1, w/w) was used. The complexity involved in the relationship between processing and allergenicity can be further elucidated from the fact that some allergens remain unaltered and maintain their antigenicity even after subjecting to processing treatment [82,83].

Effects of thermal processing on biophysical changes and IgE binding of peanut major allergens have been investigated [84]. Most allergens in peanut are thermostable and resistant to *in vitro* digestion [28,81]. Mondoulet et al. [85] observed a twofold decrease in allergenicity of boiled peanut, which can be attributed to the loss of some low-molecular-weight soluble proteins in the cooking water since these proteins were also recognized by the IgE of peanut allergic patients. Similar reduction in allergenicity was observed by Beyer et al. [86] when peanut was subjected to frying or boiling. The IgE binding capacity of major peanut allergens has been shown to increase by thermal processes such as roasting [85–88] and curing at high temperature [87]. Though the heat treatment causes an irreversible change in secondary structure and decreased solubility of Ara h 1, its allergenicity remains unaffected [89]. Moreover, roasted mature peanut exhibits a higher IgE binding compared to roasted immature peanut [87]. Interestingly, peroxidase treatment in the presence of hydrogen peroxide showed a significant decrease in the levels of Ara h 1 and Ara h 2 in roasted peanut, but had no effect on raw peanut [90]. The peroxidase induced cross-linking of Ara h 1 and Ara h 2 from roasted peanut may be a helpful tool in reducing the allergenicity of roasted peanut.

The majority of almond protein is composed of 14S legumin-like protein called amandin, which is one of the major allergen identified by almond allergic patients [91,92]. The stability of amandin to blanching, roasting, and autoclaving has been confirmed by Western blots using almond-allergic human sera [92]. Similarly, Venkatachalam et al. [93] reported thermal stability of protein extracts prepared from almonds subjected to different thermal processes including roasting, blanching, autoclaving, and microwave heating when probed with anti-almond rabbit polyclonal antibodies and human IgE in Western blots and ELISA. A significant decrease in antigenicity as measured by inhibition and sandwich ELISA using polyclonal antibodies raised against whole almond extract was observed only after prolonged roasting (160°C for 20 and 30 min) and microwave heating (3 min). These extreme conditions, however, are not suitable for processing as they would adversely affect product sensory qualities, resulting in unacceptable products. Even almonds treated with γ-irradiation alone or in combination with thermal treatments was found to have antigenically stable proteins [83]. The antigenic proteins in cashew and walnut were also stable toward γ-irradiation. The major proteins in cashew and walnut have been studied for *in vitro* digestibility. Sze-Tao and Sathe [94] investigated the *in vitro* digestibility of walnut glutelin fraction using chymotrypsin, trypsin, and pepsin and found that the proteins were readily hydrolyzed within a few minutes. Similarly, cashew globulin was easily hydrolyzed by pepsin and the polypeptide composition of hydrolyzed globulin remained unaffected by heat denaturation treatment [95]. The major allergen in Brazil nut (Ber e 1), a 2S albumin, has been shown to have poor digestibility when assessed *in vitro* by proteolysis with pepsin, and is thermally stable up to 110°C at neutral pH [96,97]. Venkatachalam et al. [98] assessed the antigenicity of pecan against thermal processes and *in vitro* digestion. Pecan antigens were more sensitive to moist heat than dry heat processing treatments and stable toward digestion. Neoallergens that are capable of causing fatal anaphylactic reactions have been shown to occur in pecan during thermal processing and storage [77,99]. More recently, decreased allergenicity in chestnuts that were boiled or that were exposed to simulated gastric and intestinal fluid has been reported [100]. The effect of processing on hazelnut allergenicity has been investigated more thoroughly as compared to other tree nuts. Wigotzki et al. [101] examined the effect of heat and storage on allergenicity of four different varieties of hazelnut by immunoblotting and enzyme-allergosorbent test inhibition experiments. Microwave treatment (630W/10min) and storage (19 weeks at room temperature) of hazelnut did not influence its allergenicity, but the IgE-binding activity of the main hazelnut allergens was destroyed after 15 min heat treatment at 170°C. Hazelnut allergens are more labile than peanut allergens when subjected to gastric digestion for 2h followed by 45 min treatment under duodenal conditions [102]. A more recent study examined the effect of roasting hazelnut (140°C/40 min) by double-blind placebo-controlled food challenges in 17 birch pollen-allergic patients with confirmed food allergy to raw hazelnuts [103]. In contrast to the effect on peanut allergenicity, the roasting of hazelnut reduced the allergenicity (only 5 of the 17 patients were positive to roasted hazelnut).

4.4 CROSS-REACTIVITY

The sensitivity of some patients to multiple tree nuts has been known for some time [4,28,104,105]. At times, individuals who have not been exposed to a certain food or a food ingredient exhibit allergic sensitivity upon their very first exposure. Occasionally, sensitive individual sensitized to a food or food component may react to an unrelated non-food item. Often such sensitivity is described as cross-reactivity. Cross-reactivity has been reported between tree nuts [15,30,106,147], seeds and tree nuts [107], tree nuts and other legumes [116], peanut and tree nuts [28,67,108–111], fruits and pollen [112–114], cereals and tree nuts [115], tree nut and pollen [116,118]. Additional articles providing overview of the topic have recently appeared [28,104,119,120]. Goetz et al. [121] noted the strongest cross-reactivity among tree nuts was between related tree nuts: walnut and pecan of the *Juglandaceae* family, and cashew and pistachio of the *Anacardiaceae* family. On the other hand, it is important to note that, with the exception of peanut cross-reactivity with pistachio, peanut antigens did not exhibit significant serological cross-reactivity with the tree nuts tested in the study. One of the recent reported examples of cross-reactivity is the case report involving citrus seed and peanut allergy [122]. A 26-year-old male experienced an episode of anaphylaxis upon showering and washing with soap containing lemon. The individual was known to be allergic to citrus (lemon, orange, and mandarin) seeds, but not to citrus fruit juices. Skin prick tests revealed the individual to be sensitive to peanut and tree nuts (hazelnut, Brazil nut, macadamia, pine nut, and almond), experiencing laryngeal edema, generalized urticaria, and asthma symptoms within minutes of exposure. Additionally, the authors reported that peanut proteins exhibited cross-reactivity with three out of five IgE-binding orange seed proteins (22, 42, and 48 kDa). Based on molecular mass, orange seed 22 kDa IgE-binding protein was suggested to be citrin. Citrin appears to have amino acid homology with Ara h 3 and Ara h 4 (49%) and with Ara h 1, all of which are known to be major allergens in peanuts. Ara h 3 and Ara h 4 belong to glycinin (legumin family) while Ara h 1 is a vicilin protein.

A recent European study on apple allergens, Mal d 1, 2, 3, and 4, in relation to cross-reactivity with birch pollen Bet v 1 is particularly interesting as the study focused on parameters that may help explain cross-sensitization [123]. The study investigated allergy sufferers from the Netherlands, Italy, Spain, Austria, and the United Kingdom. The main findings of the study were as follows: The analysis indicated (a) apple allergy in the Netherlands, Austria, and Italy was mild and related to birch pollinosis and sensitization to Bet v 1 and its apple homologue Mal d 1; (b) in Spain, apple allergy was severe and related to peach allergy and sensitization to nonspecific lipid transfer protein (nsLTP) Mal d 3. Apple consumption in the Netherlands and Austria is not known to be less than in Spain. Therefore, the study did raise an important question. Why was Mal d 3 sensitization not seen in Netherlands or Austria, while apple allergies were observed to be severe in Spain? The investigators suggested that such differential sensitivity may be due to early and frequent exposure to peach fruit, mugwort pollen and plane tree pollens, and the lack of birch pollen in Spain. Allergy to birch pollen Bet v, 1 is quite common in Europe. Using a T-cell activation assay, Jahn-Schmid et al. [124] have shown that Bet v 1 contains a linear epitope (amino acid sequence 142–156, TLLRAVESYLLAHSD) that exhibits 60%–80% linear peptide sequence similarity and 57%–79% overall protein sequence similarity with diverse food allergens such as apple (Mal d 1), cherry (Pru av 1), hazelnut (Cor a 1), celery (Api g 1), carrot (Dau c 1), and soybean (Gly m 4). Such streches of conserved amino acid sequence in cross-reactive allergens may explain cross-reactivity in some but not in other instances. Some of the possible causes of observed cross-reactivity among various food allergens and between food and nonfood sources include:

1. Conserved epitopes (linear, conformational, or both),
2. Sensitization to epitopes from a single allergen upon initial exposure, followed by subsequent exposure to other foods containing at least one cross-reactive epitope,
3. Genetic predisposition,
4. Significant individual differences with respect to threshold level of exposure to elicit allergic response. This concept has often been described as "no observable effect level (NOEL)",
5. Individual patient characteristics (age, sex, immunological maturity and status, medical condition, fitness, asthma, and other diseases),
6. Time (chronology, duration, or both) and level of allergen exposure,
7. Environmental factors (location, presence or absence of certain antigens in the locality, climatic conditions, and others),
8. Exposure to raw versus processed food,
9. Transmission of body fluids from a sensitized person to a person free of allergies (e.g., mother to infant, nonallergic patient receiving blood transfusion from allergy-suffering donor),
10. Formation of allergens during food processing (e.g., neoallergen generation) either due to protein modification, protein interaction with nonprotein components (e.g., sugars or lipids), or both,
11. Other unexplained mechanisms.

Attempts to understand cross-reactivity and their clinical implications have been reviewed [125]. Of particular interest is the compilation of relative risks of cross-reactivity based on relatedness of foods and the list of special risk factors summarized by Sicherer [125]. The author correctly pointed out that the risk of cross-reactivity was quite variable within as well as between different foods. The high risk (92%) of reaction by melon-allergenic patients to other fruits such as banana and avocado is of particular note in this regard. Stadler and Stadler [126] analyzed the then available databases for allergen sequences (total of 779 sequences) and found various sequence motifs were a better predictor (95.5% precision), compared to the Food and Agricultural Organization (FAO)–World Health Organization (WHO) decision tree (35.5% precision), of allergen potential of a particular food protein. A recent paper has, however, concluded that amino acid sequence homology is a poor predictor of cross-reactivity for profilins [127]. Whether sequence homology is equally unpredictive for other classes of proteins is not fully understood. The inherent difficulty in predicting cross-reactivity stems from several considerations including individual sensitivity, epitope conservation and variability in epitope recognition by IgE from different patients. The fact that some individuals exhibit clinical manifestations and others do not, even though both groups may exhibit comparable mast cell triggering, may further exacerbate the situation. One of the major confounding factors in accurately determining cross-reactivity between different food groups, especially those of plant origin, is the cross-reactivity between pollens and fruits and vegetables. A recent review discusses importance of pollen cross-reactivity, especially Bet v 1, with apple (Mal d 1), cherry (Pru a 1), and pear (Pyr c 1) [106]. The review points out that cross-reactivity between Bet v 1 and food allergens are independently mediated by T cells and IgE. As new models are being developed based on bioinformatics and genomic approaches, improved understanding of cross-reactivity is anticipated [128–130].

4.5 CONCLUSION

Tree nut allergies are a serious health threat to sensitive individuals as anaphylactic shock may, in severe cases, be fatal. With increasing globalization of food production, processing, and supply, and the ensuing increased risk of inadvertent exposure to tree nuts, it is important to continue to investigate ways to improve fundamental understanding of tree nut allergies. In absence of proven medical treatments, avoidance of the offending agent remains the best choice for sensitive patients suffering from tree nut allergies. The mode of entry and fate of allergens in mucosal tissues, determination of mast cell triggering threshold levels, and further details on the nature of cross-reactivity between different allergens are some of the research areas deserving additional attention.

ACKNOWLEDGEMENT

Partial financial support from the Department of Nutrition, Food and Exercise Sciences and the USDA NRICGP (#9901530; #2004-35503-14117; #355503-14117) is gratefully acknowledged.

REFERENCES

1. Keet, C.A. and Wood, R.A., Food allergy and anaphylaxis, *Immunol. Allergy Clin. North Am.*, 27, 193–212, 2007.
2. Sicherer, S.H. and Sampson, H.A., Food allergy, *J. Allergy Clin. Immunol.*, 117 (Suppl. 2), 470S–475S, 2006.
3. Fleischer, D.M., Conover-Walker, M.K., Matsui, E.C., and Wood, R.A., The natural history of tree nut allergy, *J. Allergy Clin. Immunol.*, 116, 1087–1093, 2005.
4. Clark, A.T. and Ewan, P.W., The development and progression of allergy to multiple nuts at different ages, *Pediatr. Allergy Immunol.*, 16, 507–511, 2005.
5. FALCPA, Food and Drug Administration, Federal Register, 70, 76463, 2005.
6. European Commission (EC), (a) Commission Directive 2000/13/EC of the European Parliament and of the Council of March 20, 2000 on the approximation of the laws of the Member States relating to the labelling, presentation and advertising of foodstuffs, *Off. J. Eur. Commun. Legislation*, 109, P0029–P0042, 2000. (b) Commission Directive 2006/142/EC, Amending Annex IIIa of Directive 2000/13/EC, *Off. J. Eur. Commun. Legislation*, L368/110–L368/111, 2006.
7. Sampson, H.A., Anaphylaxis and emergency treatment, *Pediatrics*, 111, 1601–1608, 2003.

8. Bock, S.A., Munoz-Furlong, A., and Sampson, H.A., Fatalities due to anaphylactic reactions to foods, *J. Allergy Clin. Immunol.*, 107, 191–193, 2001.
9. Sampson, H.A., Food allergy, *J. Allergy Clin. Immunol.*, 111 (Suppl. 2), 540S–547S, 2003.
10. Astwood, J.D., Leach, J.N., and Fuchs, R.L., Stability of food allergens to digestion in vitro, *Nat. Biotechnol.*, 14, 1269–1273, 1996.
11. Sampson, H.A., Food allergy. Part 1: Immunopathogenesis and clinical disorders, *J. Allergy Clin. Immunol.*, 103, 717–728, 1999.
12. Breiteneder, H. and Ebner, C., Molecular and biochemical classification of plant-derived food allergens, *J. Allergy Clin. Immunol.*, 106, 27–36, 2000.
13. Asensio, T., Crespo, J.F., Sanchez-Monge, R., Lopez-Torrejon, G., Somoza, M.L., Rodriguez, J., and Salcedo, G., Novel plant pathogenesis-related protein family involved in food allergy, *J. Allergy Clin. Immunol.*, 114, 896–899, 2004.
14. Radauer, C. and Breiteneder, H., Evolutionary biology of plant food allergens, *J. Allergy Clin. Immunol.*, 120, 518–525, 2007.
15. Benito, C., Gonzalez-Mancebo, E., de Durana, M.D.A.D., Tolon, R.M., and Fernandez-Rivas, M., Identification of 7S globulin as a novel coconut allergen, *Ann. Allergy Asthma Immunol.*, 98, 580–584, 2007.
16. Sathe, S.K., Teuber, S.S., and Roux, K.H., Effects of food processing on the stability of food allergens, *Biotechnol. Adv.*, 23, 423–429, 2005.
17. Raap, U., Schaefer, T., Kapp, A., and Wedi, B., Exotic food allergy: Anaphylactic reaction to lychee, *J. Investig. Allergol. Clin. Immunol.*, 17, 199–201, 2007.
18. Eriksson, N.E., Werner, S., Foucard, T., Möller, C., Berg, T., Kiviloog, J., Norrlind, K., Söderberg, M., and Wihl, J., Self-reported hypersensitivity to exotic fruit in birch pollen-allergic patients, *Allergol. Int.*, 52, 199–206, 2003.
19. Paschke, A., Kinder, H., Zunker, K., Wigotzki, M., Webbecher, R., Vieluf, S., and Steinhart, H., Characterization of allergens in mango fruit and ripening dependence of the allergenic potency, *Food Agric. Immunol.*, 13, 51–61, 2001.
20. Hefle, S.L., Hidden food allergens, *Curr. Opin. Allergy Clin. Immunol.*, 1, 269–271, 2001.
21. Vila, L., Beyer, K., Jarvinen, K.M., Chatchatee, P., Bardina, L., and Sampson, H.A., Role of conformational and linear epitopes in the achievement of tolerance in cow's milk allergy, *Clin. Exp. Allergy*, 31, 1599–1606, 2001.
22. Shek, L.P.C., Soderstrom, L., Ahlstedt, S., Beyer, K., and Sampson, H.A., Determination of food specific IgE levels over time can predict the development of tolerance in cow's milk and hen's egg allergy, *J. Allergy Clin. Immunol.*, 114, 387–391, 2004.
23. Jarvinen, K.M., Beyer, K., Vila, L., Bardina, L., Mishoe, M., and Sampson, H.A., Specificity of IgE antibodies to sequential epitopes of hen's egg ovomucoid as a marker for persistence of egg allergy, *Allergy*, 62, 758–765, 2007.
24. Sicherer, S.H., Munoz-Furlong, A., and Sampson, H.A., Prevalence of seafood allergy in the United States determined by a random telephone survey, *J. Allergy Clin. Immunol.*, 114, 159–165, 2004.
25. Lehrer, S.B., Ayuso, R., and Reese, G., Seafood allergy and allergens: A review, *Mar. Biotechnol.*, 5, 339–348, 2003.
26. Ishikawa, M., Ishida, M., Shimakura, K., Nagashima, Y., and Shiomi, K., Tropomyosin, the major oyster *Crassostrea gigas* allergen and its IgE binding epitopes, *J. Food Sci.*, 63, 44–47, 1998.
27. Swoboda, I., Bugajska-Schretter, A., Verdina, P., Keller, W., Sperr, W.R., Valent, P., Valenta, R., and Spitzauer, S., Recombinant carp parvalbumin, the major cross-reactive fish allergen: A tool for diagnosis and therapy of fish allergy, *J. Immunol.*, 168, 4576–4584, 2002.
28. Sathe, S.K., Kshirsagar, H.H., and Roux, K.H., Advances in seed protein research: A perspective on seed allergens, *J. Food Sci.*, 70, R93–R120, 2005.
29. Sicherer, S.H., Sampson, H.A., and Burks, A.W., Peanut and soy allergy: A clinical and therapeutic dilemma, *Allergy*, 55, 515–521, 2000.
30. Teuber, S.S., Sathe, S.K., Roux, K.H., and Peterson, W.R., Cross-reactivity of walnut and other tree nuts by IgE immunoblot inhibition, *J. Allergy Clin. Immunol.*, 105, 140S, 2000.
31. Fernandez, C., Fiandor, A., Martinez-Garate, A., and Quesada, J.M., Allergy to pistachio: Crossreactivity between pistachio nut and other *Anacardiaceae*, *Clin. Exp. Allergy*, 25, 1254–1259, 1995.
32. Brisman, J., Baker's asthma, *Occup. Environ. Med.*, 59, 498–502, 2002.
33. Romano, A., Di Fonso, M., Giuffreda, F., Papa, G., Artesani, M.C., Viola, M., Venuti, A., Palmieri, V., and Zeppilli, P., Food-dependent exercise-induced anaphylaxis: Clinical and laboratory findings in 54 subjects, *Int. Arch. Allergy Immunol.*, 125, 264–272, 2001.

34. Lehto, M., Palosuo, K., Varjonen, E., Majuri, M.L., Andersson, U., Reunala, T., and Alenius, H., Humoral and cellular responses to gliadin in wheat-dependent, exercise-induced anaphylaxis, *Clin. Exp. Allergy*, 33, 90–95, 2003.
35. Palosuo, K., Varjonen, E., Kekki, O.M., Klemola, T., Kalkkinen, N., Alenius, H., and Reunala, T., Wheat ω-5 gliadin is a major allergen in children with immediate allergy to ingested wheat, *J. Allergy Clin. Immunol.*, 108, 634–638, 2001.
36. Mittag, D., Niggemann, B., Sander, I., Resse, I., Fiedler, E.M., Worm, M., Vieths, S., and Reese, G., Immunoglobulin E-reactivity of wheat-allergic subjects (baker's asthma, food allergy, wheat-dependent, exercise-induced anaphylaxis) to wheat protein fractions with different solubility and digestibility, *Mol. Nutr. Food Res.*, 48, 380–389, 2004.
37. Sicherer, S.H., Muñoz-Furlong, A., and Sampson, H.A., Prevalence of peanut and tree nut allergy in the United States determined by means of a random digit dial telephone survey: A 5-year follow-up study, *J. Allergy Clin. Immunol.*, 112, 1203–1207, 2003.
38. Sampson, H.A., Update on food allergy, *J. Allergy Clin. Immunol.*, 113, 805–816, 2004.
39. Sampson, H.A., Food allergy—accurately identifying clinical reactivity, *Allergy*, 60, 19–24, 2005.
40. Wilson, S., Blaschek, K., and de Mejia, E.G., Allergenic proteins in soybeans: Processing and reduction of P34 allergenicity, *Nutr. Rev.*, 63, 47–58, 2005.
41. Fu, C.J., de Leon, M.P., Rolland, J.M., and O'Hehir, R.E., Identification, characterization, epitope mapping, and three-dimensional modeling of the α-subunit of β-conglycinin of soybean, apotential allergen for young pigs, *J. Agric. Food Chem.*, 55, 4014–4020, 2007.
42. Xu, C., Caperna, T.J., Garrett, W.M., Cregan, P., Bae, H., Luthria, D.L., and Natarajan, S., Proteomic analysis of the distribution of the major seed allergens in wild, landrace, ancestral, and modern soybean genotypes, *J. Sci. Food Agric.*, 87, 2511–2518, 2007.
43. Woods, R.K., Abramson, M., Bailey, M., and Walters, E.H., International prevalences of reported food allergies and intolerances. Comparisons arising from the European Community Respiratory Health Survey (ECRHS) 1991–1994, *Eur. J. Clin. Nutr.*, 55, 298–304, 2001.
44. Rona, R.J., Keil, T., Summers, C., Gislason, D., Zuidmeer, L., Sodergren, E., Sigurdardottir, S.T., Lindner, T., Goldhahn, K., Dahlstrom, J., McBride, D., and Madsen, C., The prevalence of food allergy: A meta-analysis, *J. Allergy Clin. Immunol.*, 120, 638–646, 2007.
45. Armstrong, D. and Rylance, G., Definitive diagnosis of nut allergy, *Arch. Dis. Child.*, 80, 175–177, 1999.
46. U.S. Department of Agriculture (USDA), *Tree Nut Supply Bountiful, Economic Research Service, January 2000*, Published online at: http://www.ers.usda.gov/publications/agoutlook/jan2000/ao268c.pdf (accessed December 19, 2007).
47. Pollack, S. and Perez A., *Fruit and Tree Nuts Situation and Outlook Yearbook, Economic Research Service, 2007*, Published online at: http://www.ers.usda.gov/publications/FTS/2007/Yearbook/FTS2007.pdf (accessed December 19, 2007).
48. Jenab, M., Sabaté, J., Slimani, N., Ferrari, P., Mazuir, M., Casagrande, C., Deharveng, G., Tjønneland, A., Olsen, A., Overvad, K., Boutron-Ruault, M., Clavel-Chapelon, F., Boeing, H., Weikert, C., Linseisen, J., Rohrmann, S., Trichopoulou, A., Naska, A., Palli, D., Sacerdote, C., Tumino, R., Mattiello, A., Pala, V., Bueno-de-Mesquita, H., Ocke, M.C., Peeters, P.H., Engeset, D., Skeie, G., Jakszyn, P., Ardanaz, E., Quiros, J.R., Chirlaque, M.D., Martinez, C., Amiano, P., Berglund, G., Palmqvist, R.,Guelpen, B., Bingham, S., Key, T., and Riboli, E., Consumption and portion sizes of tree nuts, peanuts and seeds in the European Prospective Investigation into Cancer and Nutrition (EPIC) cohorts from 10 European countries, *Br. J. Nutr.*, 96 (Suppl. 2), 12S–23S, 2006.
49. Sicherer, S.H., Furlong, T.J., Muñoz-Furlong, A., Burks, A.W., and Sampson, H.A., A voluntary registry for peanut and tree nut allergy: Characteristics of the first 5149 registrants, *J. Allergy Clin. Immunol.*, 108, 128–132, 2001.
50. Furlong, T.J., DeSimone, J., Sicherer, S.H., Peanut and tree nut allergic reactions in restaurants and other food establishments, *J. Allergy Clin. Immunol.*, 108, 867–870, 2001.
51. Food and Drug Administration (FDA), Approaches to Establish Thresholds for Major Food Allergens and for Gluten in Food, Threshold Working Group, Food and Drug Administration, Washington, DC, 2006.
52. Fleischer, D.M., The natural history of peanut and tree nut allergy, *Curr. Allergy Asthma Rep.*, 7, 175–181, 2007.
53. Sicherer, S.H., Muñoz-Furlong, A., Burks, A.W., and Sampson, H.A., Prevalence of peanut and tree nut allergy in the US determined by a random digit dial telephone survey, *J. Allergy Clin. Immunol.*, 103, 559–562, 1999.

54. Ewan, P.W., Clinical study of peanut and nut allergy in 62 consecutive patients: New features and associations, *Br. Med. J.*, 312, 1074–1076, 1996.
55. Tariq, S.M., Stevens, M., Matthews, S., Ridout, S., Twiselton, R., and Hide, D.W., Cohort study of peanut and tree nut sensitization by age of 4 years, *Br. Med. J.*, 313, 514–517, 1996.
56. Kanny, G., Moneret-Vautrin, D., Flabbee, J., Beaudouin, E., Morisset, M., Thevenin, F., Population study of food allergy in France, *J. Allergy Clin. Immunol.*, 108, 133–140, 2001.
57. Burks, A.W., Shin, D., Cockrell, G., Stanley, J.S., Helm, R.M., and Bannon, G.A., Mapping and mutational analysis of the IgE-binding epitopes on Ara h 1, a legume vicilin protein and a major allergen in peanut hypersensitivity, *Eur. J. Biochem.*, 245, 334–339, 1997.
58. Stanley, J.S., King, N., Burks, A.W., Huang, S.K., Sampson, H., Cockrell, G., Helm, R.M., West, C.M., and Bannon, G.A., Identification and mutational analysis of the immunodominant IgE binding epitopes of the major peanut allergen Ara h 2, *Arch. Biochem. Biophys.*, 342, 244–253, 1997.
59. Rabjohn, P., Helm, E.M., Stanley, J.S., West, C.M., Sampson, H.A., Burks, A.W., and Bannon, G.A., Molecular cloning and epitope analysis of the peanut allergen Ara h 3, *J. Clin. Invest.*, 103, 535–542, 1999.
60. Wang, F., Robotham, J.M., Teuber, S.S., Tawde, P., Sathe, S.K., and Roux, K.H., Ana o 1, a cashew (*Anacardium occidentale* L.) allergen of the vicilin seed storage protein family, *J. Allergy Clin. Immunol.*, 110, 160–166, 2002.
61. Wang, F., Robotham, J.M., Teuber, S.S., Sathe, S.K., and Roux, K.H., Ana o 2, a major cashew (*Anacardium occidentale* L.) nut allergen of the legumin family, *Int. Arch. Allergy Immunol.*, 132, 27–39, 2003.
62. Robotham, J.M., Wang, F., Seamon, V., Teuber, S.S., Sathe, S.K., Sampson, H.A., Beyer, K., Seavy, M., and Roux, K.H., Ana o 3, an important cashew nut (*Anacardium occidentale* L.) allergen of the 2S albumin family, *J. Allergy Clin. Immunol.*, 115, 1284–1290, 2005.
63. Robotham, J.M., Teuber, S.S., Sathe, S.K., and Roux, K.H., Linear IgE epitope mapping of the English walnut (*Juglans regia*) major food allergen, Jug r 1, *J. Allergy Clin. Immunol.*, 109, 143–149, 2002.
64. Wolff, N., Cogan, U., Admon, A., Dalal, I., Katz, Y., Hodos, N., Karin, N., and Yannai, S., Allergy to sesame in humans is associated primarily with IgE antibody to a 14kDa 2S albumin precursor, *Food Chem. Toxicol.*, 41, 1165–1174, 2003.
65. Menendez-Arias, L., Moneo, I., Dominguez, J., and Rodriguez, R., Epitope mapping of the major allergen from yellow mustard seeds, Sin a I, *Mol. Immunol.*, 27, 143–150, 1990.
66. Alcocer, M.J.C., Murtagh, G.J., Wilson, P.B., Progias, P., Lin, J., and Archer, D.B., The major human structural IgE epitope of the Brazil nut allergen Ber e 1: A chimaeric and protein microarray approach, *J. Mol. Biol.*, 343, 759–769, 2004.
67. Barre, A., Jacquet, G., Sordet, C., Culerrier, R., and Rougé, P., Homology modeling and conformational analysis of IgE binding epitopes of Ara h 3 and other legumin allergens with a cupin fold from tree nuts, *Mol. Immunol.*, 44, 3243–3255, 2007.
68. Varshney, S., Goldblum, R.M., Kearney, C., Watanabe, M., and Midoro-Horiuti, T., Major mountain cedar allergen, Jun a 1, contains conformational as well as linear IgE epitopes, *Mol. Immunol.*, 44, 2781–2785, 2007.
69. Kleber, N., Krause, I., Illgner, S., and Hinrichs, J., The antigenic response of β-lactoglobulin is modulated by thermally induced aggregation, *Eur. Food Res. Technol.*, 219, 105–110, 2004.
70. Maleki, S.J., Food processing: Effects on allergenicity, *Curr. Opin. Allergy Clin. Immunol.*, 4, 241–245, 2004.
71. Wal, J.M., Thermal processing and allergenicity of foods, *Allergy*, 58, 727–729, 2003.
72. Jankiewicz, A., Baltes, W., Bogl, K.W., Dehne, L.I., Jamin, A., Hoffmann, A., Haustein, D., and Vieths, S., Influence of food processing on the immunochemical stability of celery allergens, *J. Sci. Food. Agric.*, 75, 359–370, 1997.
73. Cooke, S.K. and Sampson, H.A., Allergenic properties of ovomucoid in man, *J. Immunol.*, 159, 2026–2032, 1997.
74. Hajos, G., Polgar, M., and Farkas, J., High-pressure effects on IgE immunoreactivity of proteins in a sausage batter, *Innovative Food Sci. Emerging Technol.*, 5, 443–449, 2004.
75. Sancho, A.I., Rigby, N.M., Zuidmeer, L., Asero, R., Mistrello, G., Amato, S., Gonzalez-Mancebo, E., Fernandez-Rivas, M., Ree, R., and Mills, E.N.C., The effect of thermal processing on the IgE reactivity of the non-specific lipid transfer protein from apple, Mal d 3, *Allergy*, 60, 1262–1268, 2005.
76. Urisu, A., Ando, H., Morita, Y., Wada, E., Yasaki, T., Yarnada, K., Komada, K., Torii, S., Goto, M., and Wakamatsu, T., Allergenic activity of heated and ovomucoid depleted egg white, *J. Allergy Clin. Immunol.*, 100, 171–176, 1997.

77. Primavesi, L., Brenna, O.V., Pompei, C., Pravettoni, V., Farioli, L., and Pastorello, E.A., Influence of cultivar and processing on cherry (*Prunus avium*) allergenicity, *J. Agric. Food Chem.*, 54, 9930–9935, 2006.
78. Fiocchi, A., Restani, P., Bernardo, L., Martelli, A., Ballabio, C., D'Auria, E., and Riva, E., Tolerance of heat-treated kiwi by children with kiwifruit allergy, *Pediatr. Allergy Immunol.*, 15, 454–458, 2004.
79. Berrens, L., Neoallergens in heated pecan nut: Products of Maillard-type degradation? *Allergy*, 51, 277–278, 1996.
80. Ludec, V., Moneret-Vautrin, D.A., Guerin, L., Morisset, M., and Kanny, G., Anaphylaxis to wheat isolates: Immunochemical study of a case proved by means of double-blind, placebo-controlled food challenge, *J. Allergy Clin. Immunol.*, 111, 897–899, 2003.
81. Fu, T.J., Abbot, U.R., and Hatzos, C., Digestibility of food allergens and nonallergenic proteins in simulated gastric fluid and simulated intestinal fluids—a comparative study, *J. Agric. Food Chem.*, 50, 7154–7160, 2002.
82. Ballmer-Weber, B.K., Hoffmann, A., Wuthrich, B., Luttkopf, D., Pompei, C., Wangorsch, A., Kastner, M., and Vieths, S., Influence of food processing on the allergenicity of celery: DBPCFC with celery spice and cooked celery in patients with celery allergy, *Allergy*, 57, 228–235, 2002.
83. Su, M., Venkatachalam, M., Teuber, S.S., Roux, K.H., and Sathe, S.K., Impact of γ-irradiation and thermal processing on the antigenicity of almond, cashew nut and walnut proteins, *J. Sci. Food Agric.*, 84, 1119–1125, 2004.
84. Maleki, S.J. and Hurlburt, B.K., Structural and functional alterations in major peanut allergens caused by thermal processing, *J. AOAC Int.*, 87, 1475–1479, 2004.
85. Mondoulet, L., Paty, E., Drumare, M.F., Ah-leung, S., Scheinmann, P., Willemot, R.M., Wal, J.M., and Bernard, H., Influence of thermal processing on the allergenicity of peanut proteins, *J. Agric. Food Chem.*, 53, 4547–4553, 2005.
86. Beyer, K., Morrow, E., Li, X., Bardina, L., Bannon, G.A., Burks, A.W., and Sampson, H.A., Effects of cooking methods on peanut allergenicity, *J. Allergy Clin. Immunol.*, 107, 1077–1081, 2001.
87. Chung, S., Butts, C.L., Maleki, S.J., and Champagne, E.T., Linking peanut allergenicity to the processes of maturation, curing, and roasting, *J. Agric. Food Chem.*, 51, 4273–4277, 2003.
88. Pomes, A., Butts, C.L., and Chapman, M.D., Quantification of Ara h 1 in peanuts: Why roasting makes a difference, *Clin. Exp. Allergy*, 36, 824–830, 2006.
89. Koppelman, S.J., Bruijnzeel-Koomen, C., Hessing, M., and de Jongh, H., Heat-induced conformational changes of Ara h 1, a major peanut allergen, do not affect its allergenic properties, *J. Biol. Chem.*, 274, 4770–4777, 1999.
90. Chung, S., Maleki, S.J., and Champagne, E.T., Allergenic properties of roasted peanut allergens may be reduced by peroxidase, *J. Agric. Food Chem.*, 52, 4541–4545, 2004.
91. Sathe, S.K., Wolf, W.J., Roux, K.H., Teuber, S.S., Venkatachalam, M., and Sze-Tao, K.W.C., Biochemical characterization of amandin, the major storage protein in almond (*Prunus dulcis* L.), *J. Agric. Food Chem.*, 50, 4333–4341, 2002.
92. Roux, K.H., Teuber, S.S., Robotham, J.M., and Sathe, S.K., Detection and stability of the major almond allergen in foods, *J. Agric. Food Chem.*, 49, 2131–2136, 2001.
93. Venkatachalam, M., Teuber, S.S., Roux, K.H., and Sathe, S.K., Effects of roasting, blanching, autoclaving, and microwave heating on antigenicity of almond (*Prunus dulcis* L.) proteins, *J. Agric. Food Chem.*, 50, 3544–3548, 2002.
94. Sze-Tao, K.W. and Sathe, S.K., Walnuts (*Juglans regia* L): Proximate composition, protein solubility, protein amino acid composition and protein in vitro digestibility, *J. Sci. Food Agric.*, 80, 1393–1401, 2000.
95. Sathe, S.K., Sze-Tao, K.W.C., Wolf, W.J., and Hamaker, B.R., Biochemical characterization and in vitro digestibility of the major globulin in cashew nut (*Anacardium occidentale*), *J. Agric. Food Chem.*, 45, 2854–2860, 1997.
96. Koppelman, S.J., Nieuwenhuizen, W.F., Gaspari, M., Knippels, L.M., Penninks, A.H., Knol, E.F., Hefle, S.L., and de Jongh, H.H., Reversible denaturation of Brazil nut 2S albumin (Ber el) and implication of structural destabilization on digestion by pepsin, *J. Agric. Food Chem.*, 53, 123–131, 2005.
97. Murtagh, G.J., Archer, D.B., Dumoulin, M., Ridout, S., Matthews, S., Arshad, S.H., and Alcocer, M.J., In vitro stability and immunoreactivity of the native and recombinant plant food 2S albumins Ber e 1 and SFA-8, *Clin. Exp. Allergy*, 33, 1147–1152, 2003.
98. Venkatachalam, M., Teuber, S.S., Peterson, W.R., Roux, K.H., and Sathe, S.K., Antigenic stability of pecan [*Carya illinoinensis* (Wangenh.) K. Koch] proteins: Effects of thermal treatments and in vitro digestion, *J. Agric. Food Chem.*, 54, 1449–1458, 2006.
99. Malanin, K., Lundberg, M., and Johansson, S.G.O., Anaphylactic reaction caused by neoallergens in heated pecan nut, *Allergy*, 50, 988–999, 1995.

100. Lee, S., Yoon, S., Kim, S., Choi, J., and Park, H., Chestnut as a food allergen: Identification of major allergens, *J. Korean Med. Sci.*, 20, 573–578, 2005.
101. Wigotzki, M., Steinhart, H., and Paschke, A., Influence of varieties, storage and heat treatment on IgE-binding proteins in hazelnuts (*Corylus avellana*), *Food Agric. Immunol.*, 12, 217–229, 2000.
102. Vieths, S., Reindl, J., Müller, U., Hoffmann, A., and Haustein, D., Digestibility of peanut and hazelnut allergens investigated by a simple in vitro procedure, *Eur. Food Res. Technol.*, 209, 379–388, 1999.
103. Hansen, K.S., Ballmer-Weber, B.K., Luttkopf, D., Skov, P.S., Wuthrich, B., Bindslev-Jensen, C., Vieths, S., and Poulsen, L.K., Roasted hazelnuts—allergenic activity evaluated by double-blind, placebo-controlled food challenge, *Allergy*, 58, 132–138, 2003.
104. Roux, K.H., Teuber, S.S., and Sathe, S.K., Tree nut allergens, *Int. Arch. Allergy Immunol.*, 131, 234–244, 2003.
105. Sutherland, M., O'Hehir, R., Czarny, D., and Suphioglu, C., Macadamia nut anaphylaxis: Demonstration of specific IgE reactivity and partial cross reactivity with hazelnut, *J. Allergy Clin. Immunol.*, 104, 889–890, 1999.
106. Bohle, B., The impact of pollen-related food allergens on pollen allergy, *Allergy*, 62, 3–10, 2007.
107. Crespo, J.F., James, J.M., Fernandez-Rodriguez, C., and Rodriguez, J., Food allergy: Nuts and tree nuts, *Br. J. Nutr.*, 96 (Suppl. 2), 95S–102S, 2006.
108. Beyer, K., Grishina, G., Bardina, L., Grishin, A., and Sampson, H.A., Identification of an 11S globulin as a major hazelnut food allergen in hazelnut-induced systemic reactions, *J. Allergy Clin. Immunol.*, 110, 517–523, 2002.
109. de Leon, M.P., Drew, A.C., Glaspole, I.N., Suphioglu, C., O'Hehir, R.E., and Rolland, J.M., Functional analysis of cross-reactive immunoglobulin E antibodies: Peanut specific immunoglobulin E sensitizes basophils to tree nut allergens, *Clin. Exp. Allergy*, 35, 1056–1064, 2005.
110. de Leon, M.P., Drew, A.C., Glaspole, I.N., Suphioglu, C., O'Hehir, R.E., and Rolland, J.M., IgE cross reactivity between the major peanut allergen Ara h 2 and tree nut allergens, *Mol. Immunol.*, 44, 463–471, 2007.
111. de Leon, M.P., Glaspole, I.N., Drew, A.C., Rolland, J.M., O'Hehir, R.E., and Suphioglu, C., Immunological analysis of allergenic cross-reactivity between peanut and tree nuts, *Clin. Exp. Allergy*, 33, 1273–1280, 2003.
112. Florido-Lopez, J.F., Quiralto Enriquez, J., Arias de Saavedra Alia, B., Saencz de San Pedro, B., and Martin Casañez, E., An allergen from *Olea europaea* pollen (Ole e 7) is associated with plant-derived food anaphylaxis, *Allergy*, 57, 53–59, 2002.
113. López-Torrejón, G., Ibáñez, M.D., Ahrazem, O., Sánchez-Monge, R., Sastre, J., Lombardero, M., Barber, D., and Salcedo, G., Isolation, cloning and allergenic reactivity of natural profilin Cit s 2, a major orange allergen, *Allergy*, 60, 1424–1429, 2005.
114. Marzban, G., Mansfeld, A., Herndl, A., Jäger, S., Stoyanova, M.F., Hemmer, W., Katinger, H., and Laimer, M., Direct evidence for the presence of allergens in *Rosaceae* fruit tree pollen, *Aerobiologia*, 22, 237–245, 2006.
115. Lee, S.H., Benmoussa, M., Sathe, S.K., Roux, K.H., Teuber, S.S., and Hamaker, B.R., A 50 kDa maize gamma-zein has marked cross-reactivity with the almond major protein, *J. Agric. Food Chem.*, 53, 7965–7970, 2005.
116. Müller, U., Lüttkopf, D., Hoffmann, A., Petersen, A., Becker, W.M., Schocker, F., Niggemann, B., Altmann, F., Kolarich, D., Haustein, D., and Vieths, S., Allergens in raw and roasted hazelnuts (*Corylus avellana*) and their cross-reactivity to pollen, *Eur. Food Res. Technol.*, 212, 2–12, 2000.
117. Hirschwehr, R., Valenta, R., Ebner, C., Ferreira, F., Sperr, W.R., Valent, P., Rohac, M., Rumpold, H., Scheiner, O., and Kraft, D., Identification of common allergenic structures in hazel pollen and hazelnuts—a possible explanation for sensitivity to hazelnuts in patients allergic to tree pollen, *J. Allergy Clin. Immunol.*, 90, 927–936, 1992.
118. Tawde, P., Venkatesh, Y., Wang, F., Teuber, S.S., Sathe, S.K., and Roux, K.H., Cloning and characterization of profilin (Pru du 4), a cross-reactive almond (*Prunus dulcis*) allergen, *J. Allergy Clin. Immunol.*, 118, 915–922, 2006.
119. Sicherer, S.H. and Leung, D.Y.M., Advances in allergic skin disease, anaphylaxis, and hypersensitivity reactions to foods, drugs, and insects, *J. Allergy Clin. Immunol.*, 119, 1462–1469, 2007.
120. Breiteneder, H. and Radaur, C., A classification of plant food allergens, *J. Allergy Clin. Immunol.*, 113, 821–830, 2004.
121. Goetz, D.W., Whisman, B.A., and Goetz, A.D., Cross-reactivity among edible nuts: Double immunodiffusion, crossed immunoelectrophoresis, and human specific IgE serologic surveys, *Ann. Allergy Asthma Immunol.*, 95, 45–52, 2005.

122. Glaspole, I.N., de Leon, M.P., Rolland, J.M., and O'Hehir, R.E., Anaphylaxis to lemon soap: Citrus seed and peanut allergen cross-reactivity, *Ann. Allergy Asthma Immunol.*, 98, 286–289, 2007.
123. Fernández-Rivas, M., Bolhaar, S., González-Mancebo, E., Asero, R., van Leeuwen, A., Bohle, B., Ma, Y., Ebner, C., Rigby, N., Sancho, A.I., Miles, S., Zuidmeer, L., Knulst, A., Breiteneder, H., Mills, C., Hoffmann-Sommergruber, K., and van Ree, R., Apple allergy across Europe: How allergen sensitization profiles determine the clinical expression of allergies to plant foods, *J. Allergy Clin. Immunol.*, 118, 481–488, 2006.
124. Jahn-Schmid, B., Radkovis, A., Lüttkopf, D., Scheurer, S., Vieths, S., Ebner, C., and Bohle, B., Bet v 1142–156 is the dominant T-cell epitope of the major birch pollen allergen and important for cross reactivity with Bet v 1-related food allergens, *J. Allergy Clin. Immunol.*, 116, 213–219, 2005.
125. Sicherer, S.H., Clinical implications of cross-reactive food allergens, *J. Allergy Clin. Immunol.*, 108, 881–890, 2001.
126. Stadler, M.B. and Stadler, B.M., Allergenicity prediction by protein sequence, *FASEB J.*, Express article 10.1096.fj.021052fje, pp. 17, 2003.
127. Sankian, M., Varasteh, A., Pazouki, N., and Mahmoudi, M., Sequence homology: A poor predictive value for profilins cross reactivity, *Clin. Mol. Allergy*, 3, 13, doi:10.1186/1476-7961-3-13, pp. 9, 2005.
128. Bowyer, P., Fraczek, M., and Denning, D.W., Comparative genomics of fungal allergens and epitopes shows widespread distribution of closely related allergen and epitope orthologues, *BMC Genomics*, 7, 1–14, 2006.
129. Schein, C.H., Ivanciuc, O., and Braun, W., Bioinformatics approaches to classifying allergens and predicting cross-reactivity, *Immunol. Allergy Clin. North Am.*, 27, 1–27, 2007.
130. Marti, P., Truffer, R., Stadler, M.B., Keller-Gautschi, E., Crameri, R., Mari, A., Schmid Grendelmeier, P., Miescher, S.M., Stadler, B.M., and Vogel, M., Allergen motifs and the prediction of allergenicity, *Immunol. Lett.*, 109, 47–55, 2007.
131. Poltronieri, P., Cappello, M.S., Dohmae, N., Conti, A., Fortunato, D., Pastorello, E.A., Ortolani, C., and Zacheo, G., Identification and characterization of the IgE-binding proteins 2S albumin and conglutin γ in almond (*Prunus dulcis*) seeds, *Int. Arch. Allergy Immunol.*, 128, 97–104, 2002.
132. Abolhassani, M. and Roux, K.H., Cloning and expression of an allergenic 60S acidic ribosomal protein of almond (*Prunus dulcis*) (unpublished), Direct Submission to NCBI, Accession no. ABH03379.1, 2006.
133. Pastorello, E.A., Farioli, L., Pravettoni, V., Ispano, M., Conti, A., Ansaloni, R., Rotondo, F., Incorvaia, C., Bengtsson, A., Rivolta, F., Trambaioli, C., Previdi, M., and Ortolani, C., Sensitization to the major allergen of Brazil nut is correlated with the clinical expression of allergy, *J. Allergy Clin. Immunol.*, 102, 1021–1027, 1998.
134. Beyer, K., Bardina, L., Grishina, G., Ashraf, A., Teuber, S.S., Niggemann, B., and Sampson, H.A., Identification of an 11S globulin as a Brazil nut food allergen (unpublished), Direct Submission to NCBI, Accession no. AAO38859.1, 2003.
135. Hofbauer, B., Radauer, C., Jaeger, S., and Hoffmann-Sommergruber, K., Molecular characterization of Cas s 1, the major allergen from European chestnut *Castanea sativa* (unpublished), Direct Submission to NCBI, Accession no. CAD10374.1, 2001.
136. Diaz-Perales, A., Collada, C., Blanco, C., Sánchez-Monge, R., Carrillo, T., Aragoncillo, C., and Salcedo, G., Class I chitinases with hevein-like domain, but not class II enzymes, are relevant chestnut and avocado allergens, *J. Allergy Clin. Immunol.*, 102, 127–133, 1998.
137. Diaz-Perales, A., Lombardero, M., Sanchez-Monge, R., Garcia-Selles, F.J., Pernas, M., Fernandez-Rivas, M., Barber, D., and Salcedo, G., Lipid-transfer proteins as potential plant panallergens: Cross-reactivity among proteins of Artemisia pollen, Castanea nut and Rosaceae fruits, with different IgE-binding capacities, *Clin. Exp. Allergy*, 30, 1403–1410, 2000.
138. Lüttkopf, D., Mlüler, U., Skov, P.S., Ballmer-Weber, B.K., Wuthrich, B., Hansen, K.S., Poulsen, L.K., Kastner, M., Haustein, D., and Vieths, S., Comparison of four variants of a major allergen in hazelnut (*Corylus avellana*) Cor a 1.04 with the major hazel pollen allergen Cor a 1.01, *Mol. Immunol.*, 38, 515–525, 2001.
139. Lüttkopf, D., Schocker, F., Scheurer, S., Haustein, D., and Vieths, S., Characterization of minor allergen hazelnut profilin (unpublished), Direct Submission to NCBI, Accession no. AAK01236, 2000.
140. Schocker, F., Lüttkopf, D., Scheurer, S., Peterson, A., Cistero-Bahima, A., Enrique, E., Miguel-Monicin, M.S., Akkerdaas, J., Ree, R., Vieths, S., and Becker, W.M., Recombinant lipid transfer protein Cor a 8 from hazelnut: A new tool for in vitro diagnosis of potentially severe hazelnut allergy, *J. Allergy Clin. Immunol.*, 113, 141–147, 2004.

141. Lauer, I., Foetisch, K., Kolarich, D., Ballmer-Weber, B.K., Conti, A., Altmann, F., Vieths, S., and Scheurer, S., Hazelnut (*Corylus avellana*) vicilin Cor a 11: Molecular characterization of a glycoprotein and its allergenic activity, *Biochem. J.*, 383, 327–334, 2004.
142. Hu, Y., Chen, Z., Tang, M., Ni, T., Wang, J., and Lin, Z., Molecular cloning of a seed storage protein cDNA (unpublished), Direct submission to NCBI, Accession no. AAO32314.1, 2002.
143. Teuber, S.S., Dandekar, A.M., Peterson, W.R., and Sellers, C.L., Cloning and sequencing of a gene encoding a 2S albumin seed storage protein precursor from English walnut (*Juglans regia*), a major food allergen, *J. Allergy Clin. Immunol.*, 101, 807–814, 1998.
144. Teuber, S.S., Jarvis, K.C., Dandekar, A.M., Peterson, W.R., and Ansari, A.A., Identification and cloning of a complementary DNA encoding a vicilin-like proprotein, Jug r 2, from English walnut kernel (*Juglans regia*), a major food allergen, *J. Allergy Clin. Immunol.*, 104, 1311–1320, 1999.
145. Pastorello, E.A., Farioli, L., Pravettoni, V., Robino, A.M., Scibilia, J., Fortunato, D., Conti, A., Borgonovo, L., Bengtsson, A., and Ortolani, C., Lipid transfer protein and vicilin are important walnut allergen in patients not allergic to pollen, *J. Allergy Clin. Immunol.*, 114, 908–914, 2004.
146. Wallowitz, M., Peterson, W.R., Uratsu, S., Comstock, S.S., Dandekar, A.M., and Teuber, S.S., Jug r 4, a legumin group food allergen from walnut (*Juglans regia* Cv. Chandler), *J. Agric. Food Chem.*, 54, 8369–8375, 2006.
147. Mittag, D., Akkerdaas, J., Ballmer-Weber, B.K., Vogel, L., Wensing, M., Becker, W., Koppelman, S.J., Knulst, A.C., Helbling, A., Hefle, S.L., van Ree, R., and Vieths, S., Ara h 8, a Bet v 1–homologous allergen from peanut, is a major allergen in patients with combined birch pollen and peanut allergy, *J. Allergy Clin. Immunol.*, 114, 1410–1417, 2004.
148. Willison, L.N., Tawde, P., Robotham, J.M., Penney IV, R.M., Teuber, S.S., Sathe, S.K., and Roux, K.H., Pistachio vicilin, Pis v 3, is immunoglobulin E-reactive and cross-reacts with the homologous cashew allergen, Ana o 1, *Clin. Exp. Allergy*, 38, 1229–1238, 2008.

5 Sphingolipids in Tree Nuts

Yu Wang, Di Tan, and Chi-Tang Ho

CONTENTS

5.1 INTRODUCTION

Sphingolipids are found in all eucaryotic cells, but are especially abundant in the plasma membrane and related cell membranes, such as endoplasmic reticulum, golgi membranes, and lysosomes. They are critical for the maintenance of membrane structure and play an important role in intracellular signaling pathways [1]. As receptors and ligands, they are involved in interactions between cells, and cells and matrix; they also serve as a binding site for toxins of bacterial and nonbacterial origin, hormones, and virus, among others [2,3].

Sphingolipids are a major topic of current research for several reasons. Firstly, sphingolipids are mediators of the signaling pathway of growth factors (e.g., platelet-derived growth factor), cytokines (e.g., tumor necrosis factor), and chemotherapeutics and play an important role in the regulation of cell growth, differentiation, and death. The hydrolysis products of sphingolipids, ceramides, sphingosine, and sphingosine-1-phosphate are highly bioactive compounds that can, as a potent mitogen (sphingosine-1-phosphate and sphinganine-1-phosphate), act as lipid "second messengers" in the signal transduction pathways that either induce apoptosis (sphingosine, sphinganine, and ceramides) or inhibit apoptosis [2,4]. Secondly, disruption of sphingolipids metabolism is implicated in several animal diseases and possibly human cancer. Finally, although relatively little is known about the mechanism of sphingolipids as dietary components, it is reasonable to believe that they contribute to disease prevention.

Studies about dietary sphingolipids found that they can suppress colon carcinogenesis. Milk sphingolipids were fed to female CF1 mice, which were previously administered 1,2-dimethylhydrazine. It was found that sphingolipids reduced the number of aberrant colonic crypt foci and aberrant crypts per focus, both of which are early indicators of colon carcinogenesis, by 70% and 30%, respectively. A longer term study found that sphingolipids had no effect on colon tumor incidence, but up to 31% of the tumors of mice fed with sphingolipids were adenomas, while all of the tumors of mice fed without sphingolipids were adenocarcinomas [5–7]. Different classes of sphingolipids, containing different head groups (sphingomyelin, glycosphingolipids, and ganglioside), showed similar effects [7]. Symolon et al. [8] also showed that dietary soy sphingolipids suppressed tumorigenesis and gene expression in 1,2-dimethylhydrazine-treated CF1 mice and *Apc*$^{Min/+}$ mice. In their study, the number of aberrant colonic crypt foci could be reduced by 38% and 52% at 0.025% and 0.1% of soy sphingolipids in the diet (w/w), respectively, and cell proliferation in the upper half of the crypts could be reduced by 50% and 56% at the same concentrations of soy sphingolipids in the diets. Adenomas in the *APC*$^{min/+}$ mice were decreased by 22% and 37% when the mice were fed with 0.025% and 0.1% (w/w) soy sphingolipids diets. For gene expression confirmation, they found soy sphingolipids decreased two transcription factors (hypoxia-induced factor α and transcription factor) of mRNA expression, which are associated with cancer.

A model of tumor suppression (Figure 5.1) by dietary sphingolipids was suggested [4]. The digestion, uptake, and subsequent metabolism of these compounds were studied. A significant part of orally administered sphingolipids was found in the small intestine and colon of mice [9]. The enzymes (sphingomyelinase, glucosylceramidase, and ceramidase) that are normally found in the small intestine digest these compounds. The sphingolipids that are not digested and adsorbed into the blood stream through the small intestine are transported to the colon, where they are hydrolyzed, mainly by colonic microflora. The released bioactive molecules, ceramides, and the sphingoid bases have biological functions involved in regulating cell growth, induction of cell differentiation, and apoptosis. More studies are underway to determine if sphingolipids can be used as chemopreventive agents.

Ceramides, products of hydrolysis of sphingolipids, induce apoptosis by activating extracellular signal regulated kinase (ERK) pathway, stress-activated protein kinase/c-Jun N-terminal kinase (SAPK/JNK) pathway, and p38 a mitogen-activated protein kinase (MAPK) pathway [10]. Ceramide-activated protein kinase (CAPK) stimulated by ceramide could phosphorylate and activate Raf1, which is associated with ERK1 and ERK2. ERK1 and ERK2 are stimulated by TNF-α and IL-1β. Both SPARK/JNK and p38 pathways are stress response signaling cascades and closely linked. Rac1 (a G protein) stimulated by ceramide activates both pathways. Ahn and Schroeder [11] illustrated early activation of JNK and p38 MAPK in HT-29 human colon cancer cells by sphingolipids. When HT-29 cells were incubated with sphinganine, the active phosphorylated forms of JNK2/JNK1 and p38 MAPK were clearly increased after 15, 30, and 60 min of treatment. However, sphinganine had minimal effects on activation of ERK1/ERK2 and little or no effect on the protein expression level of any of the kinases.

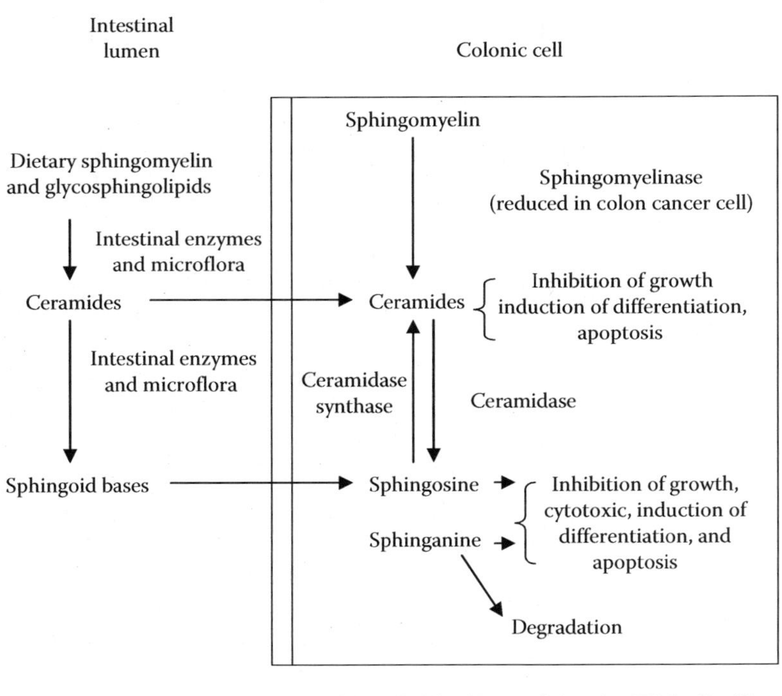

FIGURE 5.1 Tumor suppression by dietary sphingolipids. (From Schmelz, E.M., *Br. Nutr. Found. Nutr. Bull.*, 25, 135, 2000. With permission.)

In addition to tumor suppression activity, sphingolipids were found to have other beneficial effects. In short-term [12] and long-term [13] animal studies (feeding experiments with rats), sphingolipids were found to reduce plasma cholesterol, a risk factor for atherosclerosis. Also, the sphingolipids in foods may protect humans against bacterial toxins and viruses [14]. Many microorganisms, microbial toxins, and viruses bind to cell membranes through sphingolipids, therefore, sphingolipids in food can compete for cellular binding sites and facilitate the elimination of pathogenic microorganisms or toxins through the intestines.

5.2 STRUCTURE OF SPINGOLIPIDS

Sphingolipids were first discovered by J.L.W. Thudichum in 1884 while studying the chemical constituents of the brain. The first classes of sphingolipids, such as sphingomyelin and cerebrosides, were named by the tissue from which they were isolated. Today, there are over 300 known sphingolipids with considerable structural variation, but they all have in common a sphingoid base backbone, an amide-linked nonpolar aliphatic tail, and a polar head group. There are over 60 different sphingoid base backbones [15] that vary in alkyl chain lengths (from 14 to 22 carbon atoms), degree of saturation and position of double bonds, presence of a hydroxyl group at position 4, and branching of the alkyl chain. The amino group of the sphingoid base is often substituted by a long-chain fatty acid to produce ceramides. The fatty acids vary in chain length (from 14 to 30 carbon atoms), degree of saturation (but are normally saturated), and presence or absence of a hydroxyl group on the α- (or the ω-, in the case of ceramides of skin) carbon atom. More complex sphingolipids are formed when a polar head group is added at position 1 of sphingoid bases [14]. Figure 5.2 shows the general structure of sphingolipids.

As mentioned earlier, the structure of sphingolipids is of a complex nature; there is considerable variation among different organisms with respect to the type of sphingoid backbone, the polar group, and fatty acids. The sphingoid backbones of most mammalian sphingolipids consist mainly of sphingosine (*trans*-4-sphingenine, d18:1Δ^4), and a lesser amount of sphinganine (d18:0) and 4-hydroxysphinganine (t18:0), whereas plants contain sphinganine, 4-hydroxysphinganine, and *cis* and *trans* isomers of 8-sphingenine (d18:1Δ^8), 4,8-sphingadienine (d18:2$\Delta^{4,}\Delta^8$), and 4-hydroxy-8-sphingenine (t18:1Δ^8). The core structure of the sphingoid base is 2-amino-1,3-dihydroxyoctadecane, named sphinganine or d18:0, where d denotes a dihydroxy base. It can be substituted by an additional hydroxyl group at position 4, named 4-hydroxysphinganine or t18:0, where t denotes a trihydroxy base. If it has double bonds at position 4, 8, or 4 and 8, it is called sphingosine, d18:1Δ^4; 8-sphingenine, d18:1Δ^8; or 4,8-sphingadienine, d18:2$\Delta^{4,}\Delta^8$, respectively. Unlike mammalian sphingolipids, which consist of many different polar head groups (phosphocholine, glucose, galactose, *N*-acetylneuraminic acid, fructose, and other carbohydrates), plant sphingolipids have mainly glucose (to a lesser extent oligosaccharides containing glucose and mannose, and inositol) as their head groups. Depending on the head groups, sphingolipids are divided into two major classes—phosphosphingolipids, with a phosphoric acid linked to the position 1 of a ceramide through an ester bond (sphingomyelin with phosphocholine as the head group is the major component) and glycosphingolipids, with a glycosidic bond to a sugar moiety. The latter are further divided into neutral and acid glycosphingolipids. The neutral sphingolipids are cerebrosides with glucose, galactose, lactose, or oligosaccharides as the head group. The acid sphingolipids include gangliosides and sulfides. Gangliosides have oligoglycosidic head groups containing one or more sialic acid group (*N*-acyl, especially acetyl derivatives of neuraminic acid). Sulfides have sulfate ester bound with the sugar moiety. In the case of fatty acids, it appears that plants typically contain mostly 2-hydroxy fatty acids, saturated or monoenoic, ranging from C14 to C26, while sphingolipids from animals have fatty acids both with and without a 2-hydroxy group.

5.3 SPHINGOLIPIDS IN NUTS

Sphingolipids are components of a variety of foods. The amounts vary considerably. There is no evidence to indicate that sphingolipids are required for growth or survival. Generally, foods of

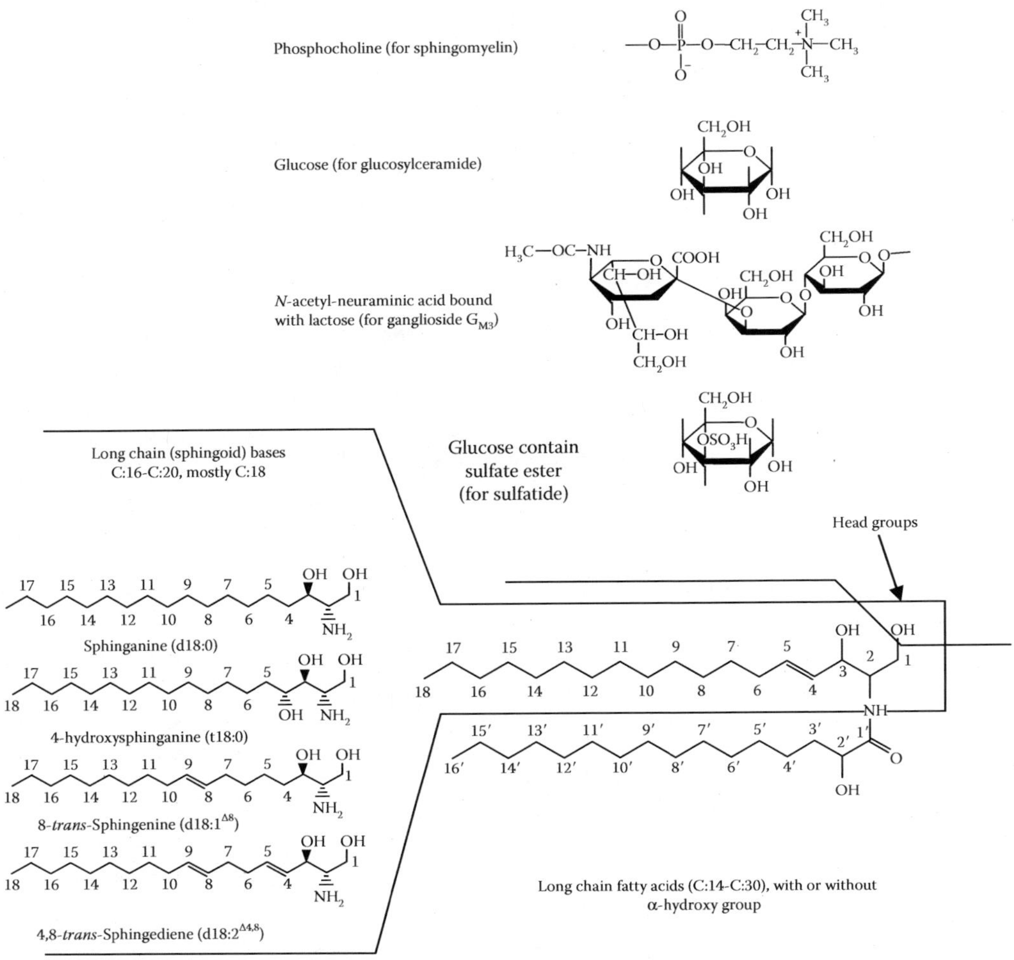

FIGURE 5.2 General structure of sphingolipids. (From Merrill, A.H. Jr., Schmelz, E.M., Dillehay, D.L., Spiegel, S., Shayman, J.A., and Schroeder, J.J., *Toxicol. Appl. Pharmacol.*, 142, 208, 1997. With permission.)

mammalian origin (dairy products and eggs) are rich in sphingolipids. On the other hand, a study found high amounts of cerebrosides in soybean [16]. Considerable amounts of sphingolipids were also found in cereals, fruits, and vegetables [17–20]. Recently, Sang et al. [21] reported the constituents of almond nuts, and one of the cerebrosides was found to be its major component for the first time. Actually, studies about sphingolipids in nuts are limited. However, nuts could be produced and consumed in a large amount every year in the United States. Nuts are one of the most popular snacks and ingredients in processed foods, especially in bakery and confectionery products.

Structural analysis of sphingolipids in nuts was carried out on a limited number of species. Cerebrosides and ceramides have been isolated from almond, cashew, hazelnut, peanut, and walnut (Table 5.1) [21,22]. In almond, Sang et al. [21] firstly reported on 1-*O*-α-D-glucopyranosyl-(2*S*,3*R*,4*E*,8*Z*)-2-[(2*R*)-2-hydroxyhexadecanoylamino]-4,8-octadecadiene-1,3-diol, a sphingolipid. They showed the presence of two aliphatic long chains. One was 4-*trans*-8-*cis*-sphingenine backbone. The other one was 2-hydroxypalmitic acid. The head group glucose was connected with C-1 at backbone. Furthermore, t18:1Δ^4 and d18:2Δ^4,Δ^8 were found as sphingoid backbones of ceramides, which were associated with C16:0h, C22:0-C25:0, and C22:0h-C25:0h fatty acids. Only glucose cerebrosides were found in almond, with d18: 2Δ^4,Δ^8 and t18: 1Δ^4 as the major backbones, and

TABLE 5.1
Identified Ceramides and Cerebrosides in Nuts

Sphingolipid	Almond	Cashew	Hazelnut	Peanut	Walnut
Ceramides [M+1]+					
T18:0-C16:0 (556)					*
T18:0-C22:0 (640)					*
T18:0-C23:0 (654)					*
T18:0-C24:0 (668)					*
T18:1-C22:0 (638)	*		*	*	*
T18:1-C23:0 (652)	*		*		*
T18:1-C24:0 (666)	*		*	*	*
T18:1-C25:0 (680)	*		*		*
T18:1-C26:0 (694)			*		
D18:2-C16:0h (552)	*	*	*	*	
T18:1-C22:0h (654)	*		*		*
T18:1-C23:0h (668)	*		*		*
T18:1-C24:0h (682)	*		*		*
T18:1-C25:0h (696)	*		*		*
T18:1-C26:0h (710)			*		
Cerebrosides [M+1]+					
D18:2-C16:0-Glu (698)	*	*	*	*	*
D18:2-C18:0-Glu (726)		*			
D18:2-C16:0h-Glu (714)	*	*	*	*	*
D18:2-C18:0h-Glu (742)		*			
D18:2-C20:0h-Glu (770)		*			
D18:2-C22:0h-Glu (798)	*	*	*	*	*
D18:2-C24:0h-Glu (826)	*	*	*	*	*
T18:1-C22:0h-Glu (816)	*	*	*	*	*
T18:1-C23:0h-Glu (830)	*		*		*
T18:1-C24:0h-Glu (844)	*	*	*	*	*
T18:1-C25:0h-Glu (858)	*				*
T18:1-C26:0h-Glu (872)			*		

Source: From Fang, F., Ho, C.-T., Sang, S., and Rosen, R.T., *J. Food Lipids*, 12, 327, 2005. With permission.

Note: * represents the presence of the concerned sphingolipid in the respective nuts.

palmitic acid (C16:0) and 2-hydroxypalmitic acid (C16:0h) as the main fatty acids. The remaining fatty acids of C22:0h-C25:0h were connected with 4-hydroxy-8-sphingenine (t18:1) backbone [22]. Peanut and cashew also contained mainly 4-hydroxysphingadienine (d18:2) and 4-hydroxysphingenine (t18:1) as the predominant backbones. In cashew, only C16:0h fatty acid was in the ceramide, while in the peanut, except for α-hydroxypalmitic acid (C16:0h), behenic acid (C22:0), and lignoceric acid (C24:0) also took an important part. For cerebrosides, both peanut and cashew contained C16:0, C16:0h, C22:0h, and C24:0h fatty acids, but in cashew C18:0, C18:0h, and C20:0h were also present. Similar to the nuts mentioned above, 4-hydroxysphingadienine (d18:2) and 4-hydroxysphingenine (t18:1) were still the major sphingoid backbones with C22:0-C25:0 or C22:0h-C25:0h fatty acids in hazelnut. However, ceramides and cerebrosides consisted of t18:1-C26:0h and t-18:1-C26:0-Glu, respectively, in hazelnut. The difference of walnut was t18:0 as one of backbones and the amide-linked fatty acids were also similar to other nuts [22]. From the above nut samples mentioned, it is found that the cerebroside consisting of 4,8-sphingadienine, α-hydroxypalmitic acid, and glucose as head group is the major cerebroside in almond, cashew, hazelnut, peanut, and walnut.

The structures of the sphingolipids in other foods vary considerably. The sphingolipids of mammalian tissues, lipoproteins, and milk typically contain ceramides, sphingomyelins, cerebrosides, and gangliosides; plants, fungi, and yeast mainly have cerebrosides and phosphoinositides [14]. For example, sphingomyelin is the major mammalian sphingolipid, and it is rarely present in plants or microorganisms. Milk contains many kinds of sphingolipids, including sphingomyelin, glucosylceramide, lactosylceramide, and gangliosides.

In spite of the biological activities of these compounds, studies about the sphingolipid content in nuts are quite sparse. Fang et al. [22] summarized the content of sphingolipids in some nuts (almond, cashew, hazelnut, peanut, and walnut) by using liquid chromatography/mass spectrometry (LC/MS) and found that the concentration cerebroside (d18:2-C16:0h-Glu) ranged from 0.021 to 0.068 mg/g, being lowest in hazelnut and highest in almond. As far as known, this is the only collection of data on the sphingolipid content in nuts. For other foods, milk, egg, and soybeans have the highest sphingolipid content, followed by meat (chicken, beef, and pork) and cereal (wheat). Fruits and vegetables have relatively low sphingolipid content.

Recent research stresses the importance of sphingolipids in biological systems and their preventive effect on colon carcinogenesis. Due to the enormous structural diversity described above, it is important to establish structure–function relationships. There are no studies about the biological effect of nuts sphingolipids, but studies with human adenocarcinoma cell line (HT29 cells) found that the toxicities of soy and wheat ceramides were comparable to brain ceramide [23]. Studies are underway using nuts sphingolipids as chemopreventive material. Knowledge about the structures and concentrations of sphingolipids in nuts and their metabolic pathways in human body are very important. Also, little is known about variations of sphingolipid amounts over season and during processing. Easy and sensitive methods are needed to determine sphingolipids in fresh and processed nuts, both qualitatively and quantitatively.

5.4 ANALYSIS OF SPHINGOLIPIDS

Sphingolipids pose a big challenge to analytical chemists for the following reasons. Firstly, it is difficult to quantitatively isolate them in completely pure form, since the content in biological materials is very low. Secondly, it is difficult to separate and identify the molecular species because of the structural variations mentioned in the above section. Thirdly, the lack of a chromophore makes it impossible to use ultraviolet visible (UV) detection. Different techniques were used to analyze this class of compounds, including thin layer chromatography (TLC) [24], gas chromatography/mass spectrometry (GC/MS) [19,25], and high-performance liquid chromatography (HPLC) [26–28] after derivatization. However, most of these methods are very cumbersome and time-consuming, and characterization of different species had often been carried out by analysis of different lipid residues after hydrolysis.

One of the most powerful techniques used in lipid analysis today is HPLC coupled with mass spectrometry (HPLC/MS). Several mass spectrometric ionization techniques, such as fast atom bombardment (FAB) [23], electrospray ionization (ESI) [29,30], ionspray ionization (ISI) [31], and atmospheric pressure chemical ionization (APCI) [22,30,32] have been used. By using HPLC/MS, one can get information on the molecular structure of the intact lipids, which helps differentiate molecular species within different lipid classes. By using tandem mass spectrometry (MS/MS), identification of molecular species of different sphingolipids can be achieved in an easier and more sensitive way. There are many other advantages of using MS, such as small sample size, minimal sample preparation, and lack of need for derivatization, speeds, and sensitivity. In the literature, sphingolipids of both animal and plant origin were analyzed by MS.

Ceramide profile of a bovine brain extract and a lipid extract of cultured T-cells were analyzed by ESI/MS and ESI/MS/MS [29]. The sample, either directly or after clean up, was infused into the MS. Collision-induced fragmentation (CIF) results in characteristic product ions, *m/z* at 264 and 282 for sphingosine and *m/z* at 266 and 284 for sphinganine, regardless of the length of the fatty acid chain. By using precursor ion scan analysis, sphingosine- and sphinganine-based ceramide species were detected. The change of ceramide levels in complex biological mixtures was measured quantitatively by comparison with mass intensity of an internal standard. Ceramides were also analyzed by using APCI/MS [32]. Ceramide species from the cells were separated by reverse phase HPLC and detected by APCI/MS. Selected ion monitoring (SIM) was used to detect sphingosine-based ceramides by monitoring the common fragment ion *m/z* at 264 at high-cone voltage. Quantification was carried out by comparing with known amounts of authentic samples. Bovine milk is a good source of sphingolipids, which include glucosyl ceramides, lactosyl ceramides, and sphingomyelins. Molecular species of these sphingolipids were analyzed by LC/MS [33], a method based on normal phase HPLC online with discharge assisted thermospray (plasmaspray) mass spectrometry. Through MS/MS using collision-induced dissociation (CID), specific long chain base and fatty acid compositions of the ceramide units can be revealed. In a paper a year later, the same authors, Karlsson et al. [30] discussed the analysis of a molecular species of sphingomyelin from bovine milk. Both ESI/MS and APCI/MS were used for structural determination. The sphingomyelin fraction was separated by normal phase HPLC. Firstly, using ESI, protonated molecules were detected; secondly, using APCI, fragmentation was achieved in the ion source. With the ceramide ions as precursors, ions representing both the long chain bases and fatty acids were identified via collision induced decomposition using APCI/MS/MS.

The determination of plant origin sphingolipids by MS has also been reported. The molecular species of the major sphingolipid, glucosylceramides, of soybean and wheat were analyzed by low- and high-resolution MS/MS using positive ion FAB [23]. The glucosylceramides were purified from soybean and wheat, and directly introduced into MS using a FAB probe. By analyzing the fragmentation pattern, different glucosylceramides were identified, but the principle used in the analysis was quite complicated. Fang et al. [22] showed that a single quadrupole LC/APCI/MS method using in-source CID was developed to separate and identify plant cerebrosides and free ceramides in most nuts and seeds at the same run for the first time; the major sphingolipids were quantified by external standard calibration with good recovery. The Zorbax NH_2 column was used with gradient separation, which was from 100% A (acetonitrile with 10 mM ammonium acetate buffer) to 34% A/64% B (methanol with 10 mM ammonium acetate buffer) in 20 min, and from 34% A/64 to 100% B in another 10 min. Full scan APCI/MS was used to show peaks of the protonated molecule and fragment ions at different sampling cone potentials. At low potential (15 V), the protonated molecule and an ion corresponding to the loss of water were obtained. For cerebrosides, the mass spectra also consisted of small peaks, which represent ceramide ions after loss of glucose. Fragment ions of long chain bases and fatty acids could be illustrated after potential increased to 45 V. For the backbones, ions at *m/z* 298, 280, and 262 demonstrated 4,8-sphingadienine (d18:2) with loss of one and two molecules of water, and *m/z* 316, 298, 280, and 262 represented 4-hydroxy-8-sphingenine (t18:1) by losing one, two, or three molecules of water. In spectra of d18:2, *m/z* 262 was the most significant peak at 45 V, while in that of t18:1, ion *m/z* 262 is not the biggest due to lack of double bond at position 4. Ion fragments of fatty acids were also found for the nonhydroxyl group, *m/z* 340, 354, and 368, representing behenic acid (C22:0), hexacosanoic acid (C23:0), and lignoceric acid (C24:0), respectively, while in the hydroxyl group, C16:0h, C22:0h, and C24:0h shown by *m/z* 272, 356, and 384, respectively, were noted.

5.5 CONCLUSION

Studies with experimental animals have shown that dietary sphingolipids significantly reduce the early stages of colon carcinogenesis and tumor formation, but knowledge about the amount and structure of sphingolipids in foods, especially foods of plant origin, is sparse. Recently, a developed LC/MS method with APCI positive ion mode has been used to separate, identify, and quantify ceramide and cerebroside species simultaneously in nut samples including almond, cashew, hazelnut, peanut, and walnut. The cerebroside consisting of 4,8-sphingadienine, α-hydroxypalmitic acid, and glucose as the head group was found to be the most abundant cerebroside in nuts.

REFERENCES

1. Hakomori, S.I., Bifunctional role of glycosphingolipids. Modulators for the transmembrane signaling and mediators for cellular interactions, *J. Biol. Chem.*, 265, 18713–18716, 1990.
2. Merrill, A.H. Jr., Schmelz, E.M., Dillehay, D.L., Spiegel, S., Shayman, J.A., and Schroeder, J.J., Sphingolipids—The enigmatic lipid class: Biochemistry, physiology, and pathophysiology, *Toxicol. Appl. Pharmacol.*, 142, 208–225, 1997.
3. Riboni, L., Viani, P., Bassi, R., Prinetti, A., and Tettamanti, G., The role of sphingolipids in the process of signal transduction, *Prog. Lipid Res.*, 36, 153–195, 1997.
4. Schmelz, E.M., Dietary sphingomyelin and other sphingolipids in health and disease, *Br. Nutr. Found. Nutr. Bull.*, 25, 135–139, 2000.
5. Dillehay, D.L., Webb, S.J., Schmelze, E.M., and Merrill, A.H. Jr., Dietary sphingomyelin inhibits 1,2-dimethylhydrazine-induced colon cancer in CF1 mice, *J. Nutr.*, 124, 615–620, 1994.
6. Schmelz, E.M., Dillehay, D.L., Webb, S.K., Reiter, A., Adams, J., and Merrill, A.H. Jr., Sphingomyelin consumption suppresses aberrant colonic crypt foci and increases the proportion of adenomas versus adenocarcinomas in CF1 mice treated with 1,2-dimethylhydrazine: Implications for dietary sphingolipids and colon carcinogenesis, *Cancer Res.*, 56, 4936–4941, 1996.
7. Schmelz, E.M., Sullards, M.C., Dillehay, D.L., and Merrill, A.H. Jr., Colonic cell proliferation and aberrant crypt foci formation are inhibited by dairy glycosphingolipids in 1,2-dimethylhydrazine treated CF1 mice, *J. Nutr.*, 130, 522–527, 2000.
8. Symolon, H., Schmelz, E.M., Dillehay, D.L., and Merrill, A.H. Jr., Dietary soy sphingolipids suppress tumorigenesis and gene expression in 1,2-dimethylhydrazine-treated CF1 mice and ApcMin/+ mice, *J. Nutr.*, 134, 1157–1161, 2004.
9. Schmelz, E.M., Crall, K.J., Larocque, R., Dillehay, D.L., and Merril, A.H. Jr., Uptake and metabolism of sphingolipids in isolated intestinal loops of mice, *J. Nutr.*, 124, 702–712, 1994.
10. Liu, G., Kleine, L., and Hebert, R.L., Advances in the signal transduction of ceramide and related sphingolipids, *Crit. Rev. Clin. Lab. Sci.*, 36, 511–573, 1999.
11. Ahn, E.H. and Schroeder, J.J., Sphinganine causes early activation of JNK and p38 and inhibition of AKT activation in HT-29 human colon cancer cells, *Anticancer Res.*, 26, 121–127, 2006.
12. Imaizumi, K., Tominaga, A., Sato, M., and Sugano, M., Effects of dietary sphingolipids on levels of serum and liver lipids in rats, *Nutr. Res.*, 12, 543–548, 1992.
13. Kobayashi, T., Schimizugawa, T., Osakabe, T., Watanabe, S., and Okuyama, H., A long-term feeding of sphingolipids affected the levels of plasma cholesterol and hepatic triacylglycerol but not tissue phospholipids and sphingolipids, *Nutr. Res.*, 17, 111–114, 1997.
14. Vesper, H., Schmelz, E.M., Nilolova-Karakashian, M.N., Dillehay, D.L., Lynch, D.V., and Merrill, A.H. Jr., Sphingolipids in food and the emerging importance of sphingolipids to nutrition, *J. Nutr.*, 129, 1239–1250, 1999.
15. Karlsson, K.A., On the chemistry and occurrence of sphingolipid long-chain bases, *Chem. Phys. Lipids*, 5, 6–43, 1970.
16. Ohnishi, M. and Fujino, Y., Sphingolipids in immature and mature soybeans, *Lipids*, 17, 803–810, 1982.
17. Laine, R.A. and Renkonen, O., Ceramide di- and trihexosides of wheat flour, *Biochemistry*, 13, 2837–2843, 1974.
18. Ohnishi, M., Ito, S., and Fujino, Y., Characterization of sphingolipids in sphinach leaves, *Biochim. Biophys. Acta*, 752, 416–422, 1983.
19. Whitaker, B.D., Cerebrosides in mature-green and red-ripe bell pepper and tomato fruits, *Phytochemistry*, 42, 627–632, 1996.
20. Zeisel, S.H., Choline, in *Modern Nutrition in Health and Disease*, Vol. 1, 8th ed., Shils, M.E., Olson, J.A., and Shike, M. (Eds.), Williams & Wilkins, Baltimore, MD, 1994, pp. 451–465.

21. Sang, S., Kikuzaki, H., Lapsley, K., Rosen, R.T., Nakatani, N., and Ho, C.-T., Sphingolipid and other constituents from almond nuts (*Prunus amygdalus* Batsch), *J. Agric. Food Chem.*, 50, 4709–4712, 2002.
22. Fang, F., Ho, C.-T., Sang, S., and Rosen, R.T., Determination of sphingolipids in nuts and seeds by a single quadrupole LC/MS method, *J. Food Lipids*, 12, 327–343, 2005.
23. Sullards, M.C., Lynch, D.V., Merrill, A.H. Jr., and Adams, J., Structure determination of soybean and wheat glucosylceramide by tandem mass spectrometry, *J. Mass Spect.*, 35, 347–353, 2000.
24. Motta, S., Monti, M., Sesana, S., Caputo, R., Carelli, S., and Ghidoni, R., Ceramide composition of the psoriatic scale, *Biochim. Biophys. Acta*, 1182, 147–151, 1993.
25. Cahoon, E.B. and Lynch, D.V., Analysis of glucocerebrosides of rye (*Secale cereale* L. cv Puma) leaf and plasma membrane, *Plant Physiol.*, 95, 58–68, 1991.
26. Iwamori, M., Costello, C., and Moser, H.W., Analysis and quantitation of free ceramide containing nonhydroxy and 2-hydroxy fatty acids, and phytosphingosine by high-performance liquid chromatography, *J. Lipid Res.*, 20, 86–96, 1979.
27. Previati, M., Bertolaso, L., Tramarin, M., Bertagnolo, V., and Capitan, S., Low nanogram range quantitation of diglycerides and ceramide by high performance liquid chromatography, *Anal. Biochem.*, 233, 108–114, 1996.
28. Snada, S., Uchida, Y., Anraku, Y., Izawa, A., Iwamori, M., and Nagai, Y., Analysis of ceramide and monohexaosyl glycolipid derivatives by high-performance liquid chromatography and its application to the determination of the molecular species in tissues, *J. Chromatogr.*, 400, 223–231, 1987.
29. Gu, M., Kerwin, J.L., Watts, J.D., and Aebersold, R., Ceramide profiling of complex lipid mixtures by electrospray ionization mass spectrometry, *Anal. Biochem.*, 244, 347–356, 1997.
30. Karlsson, A.Å., Michélsen, P., and Odham, G., Molecular species of sphingomyelin: Determination by high-performance liquid chromatography/mass spectrometry with electrospray and high-performance liquid chromatography/tandem mass spectrometry with atmospheric pressure chemical ionization, *J. Mass Spect.*, 33, 1192–1198, 1998.
31. Mano, N., Oda, Y., Yamada, K., Asakawa, N., and Katayama, K., Simultaneous quantitative determination method for sphingolipid metabolites by liquid chromatography/ionspray ionization tandem mass spectrometry, *Anal. Biochem.*, 244, 291–300, 1997.
32. Couch, L.H., Churchwell, M.I., Doerge, D.R., Tolleson, W.H., and Howard, P.C., Identification of ceramides in human cells using liquid chromatography with detection by atmospheric pressure chemical ionization-mass spectrometry, *Rapid Commun. Mass Spect.*, 11, 504–512, 1997.
33. Karlsson, A.Å., Arnoldsson, K.C., Westerdahl, G., and Odham, G., Common molecular species of glucosyl ceramides, lactosyl ceramides and sphingomyelins in bovine milk determined by high-performance liquid chromatography-mass spectrometry, *Milchwissenschaft*, 52, 554–559, 1997.

6 Health Aspects and Antiaflatoxigenic Activity of Phytochemicals in Tree Nuts

Russell J. Molyneux, Noreen Mahoney, Jong H. Kim, and Bruce C. Campbell

CONTENTS

6.1 INTRODUCTION

Tree nuts, while not generally regarded as a staple food, are particularly appreciated by consumers for their flavor and convenience. Snack foods in the United States such as chips, pretzels, and popcorn are often highly processed, but in many other countries, nuts and seeds play a more prevalent role and these undergo little or no postharvest treatment, although occasionally they may be smoked or flavored with spices. Even when incorporated into baked goods or savory dishes, or utilized as nut butters or marzipan, they are subject to minimal processing, consisting primarily of chopping, grinding, or blanching. Because of such factors, most consumers regard nuts as "natural" foods and have little concern for adverse or deleterious effects other than occasional allergenicity in some individuals. Development of off-flavors from rancidity and infection with spoilage microorganisms (generally *Aspergillus niger*, *Penicillium* spp., and *Rhizopus* spp.) are so obvious that such nuts are usually discarded and not eaten.

Nuts contain high levels of protein, fiber, and dietary fats, which in association with their pleasant flavor and convenience, has led to the recommendation that they should be an essential part of a healthy diet. This was recently endorsed by allowance of a qualified health claim for a relationship between the consumption of nuts and reduced risk of coronary heart disease (CHD) by the Food and Drug Administration (FDA) [1]. Most tree nuts have low levels of saturated fats but high levels of desirable unsaturated fats. Monounsaturated fatty acids (MUFA) such as oleic acid (18:1ω9) occur in almonds and hazelnuts and polyunsaturated fatty acids (PUFA) such as linoleic acid (18:2ω6) and α-linolenic acid (18:3ω3) predominate in walnuts [2]. A consistent decrease in serum cholesterol levels and reduced risk of CHD in humans was established by meta-analysis of five controlled diet clinical intervention trials with walnuts [3] and analogous effects have been found with other nuts [4].

Although they are not a primary nutrient source, and may sometimes be perceived almost as a "condiment" crop, tree nuts are an extremely valuable agricultural commodity in national and international trade. In California, where almost all almonds, pistachios, and walnuts in the United States are produced, they had an aggregate value of $3.42 billion in 2005, with 40%–60%, valued at $2 billion, being exported [5,6]. In spite of this, there is one serious constraint on the marketing of tree nuts, namely the potential presence of aflatoxins, which are highly regulated because of food safety concerns. The European Community (EC) in particular applies an extremely low tolerance level of 2 ng/g for aflatoxin B_1 and 4 ng/g total aflatoxins [7]; in contrast, the FDA has a domestic maximum guidance level for tree nuts intended for human consumption of 20 ng/g (e.g., 20 ppb) [8]. In 2005, 94% of the rapid alerts or notifications from the EC Rapid Alert System for Food and Feed (RASFF) for mycotoxins in tree nuts were for aflatoxins, with 28 for almonds and 13 for pistachios from the United States [9]. While this is a small number relative to the total of 827 alerts for all mycotoxins in tree nuts, it represents considerable economic loss to California producers and exporters, due to lost revenue, return, or reprocessing of the shipment and increased sampling of subsequent imports. Additional costs are incurred for preshipment quality control and from the rigorous sampling protocol mandated by the EC [7].

6.2 AFLATOXINS IN TREE NUTS

Aflatoxins are metabolites produced by many strains of the fungi *Aspergillus flavus* and *A. parasiticus*, which commonly infect major agricultural crops such as corn, peanuts, cotton, and tree nuts. Contamination of human foods and animal feeds by these compounds is of great concern because they are classified as carcinogens, particularly in humans infected with hepatitis [10,11]. There are also episodes of acute toxicity, the most recent being caused by contaminated maize in eastern Kenya in 2004 with 317 diagnosed cases and 125 deaths; analysis of maize samples revealed aflatoxin B_1 concentrations as high as 4400 ppb [12].

Structurally, aflatoxins are polyketide derivatives and can be classified into B and G groups, which share a common difurochromanone core but with furanone and δ-lactone moieties appended, respectively (Figure 6.1). As a general rule, *A. flavus* produces the B group aflatoxins whereas *A. parasiticus* produces both B and G groups, although not all strains of either species are capable of producing aflatoxins and these are regarded as atoxigenic. The most common metabolites are aflatoxins B_1 and G_1, accompanied to a lesser extent by aflatoxins B_2 and G_2, which are 8,9-dihydro derivatives of B_1 and G_1, respectively. Aflatoxins B_1 and G_1 are of greatest concern because the presence of the furan double bond permits biotransformation into the 8,9-epoxide through *in vivo*

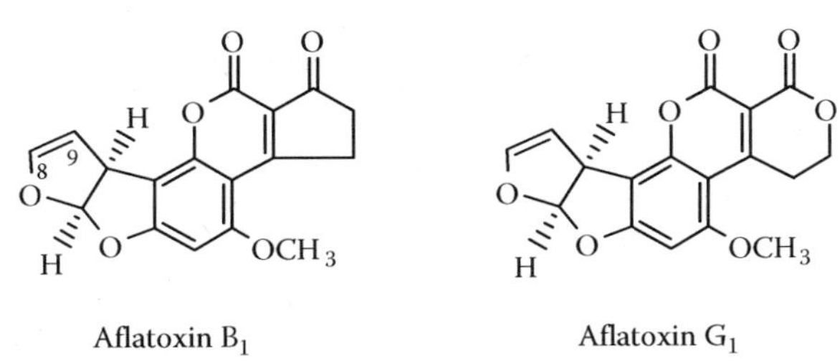

FIGURE 6.1 Structures of aflatoxin B_1 and aflatoxin G_1.

hepatic oxidation by cytochrome P450 [13]. Aflatoxin B_1-*exo*-8,9-epoxide has been shown to be the proximate carcinogen, intercalating into double-stranded DNA [14].

Since tree nuts are subjected to minimal processing, there are few postharvest opportunities to reduce aflatoxin levels and these are primarily influenced by nut species or cultivar and their proper handling during harvesting, drying, and packing. Because of the perception of tree nuts as a "natural" food, treatment with fungicides to reduce fungal infection or processing to destroy aflatoxins is unlikely to be acceptable to consumers or regulators. Such methods would involve the use of toxic antifungal compounds and could change the organoleptic characteristics of the product. Furthermore, the additional expense would add to the cost of an already relatively expensive product, leading to reduced consumption. There is an association of aflatoxins in tree nuts with damage by insect pests such as codling moth and navel orange worm, permitting ingress of fungal spores, and sorting to remove obviously damaged nuts can reduce the overall aflatoxin load. One method of preventing such damage would be to breed new cultivars with increased physical barriers to microorganisms, including the husk or hull, the shell, and the pellicle or seed coat (papery tissue surrounding the kernel). However, this could result in undesirable characteristics such as difficulty in cracking or shattering and even then the toxins can often be found in nuts which show no evidence that the integrity of the shell has been breached. It appears that the fungus may enter through the relatively less impenetrable suture or stem end, or be present in the flowers at the time of fertilization and thus incorporated into the fruit as it matures.

An alternative general strategy to limit aflatoxin formation is therefore to investigate natural factors within each nut species that might confer resistance to *Aspergillus* colonization and growth and/or aflatoxin biosynthesis. Such factors could then be incorporated and enhanced during breeding for new cultivars. The toughness of the shell and its lignin-derived composition suggests that it primarily presents a physical barrier, but the protective abilities of the husk, pellicle, and possibly the kernel itself are more likely to be due to the presence of natural chemical constituents. A bioactivity-directed strategy of determining the most aflatoxin-resistant nut species and varieties and focusing upon the natural product composition of these to isolate and identify specific resistance factors has proved fruitful. Obviously, reduction of aflatoxin biosynthesis is the most desirable outcome because merely eliminating aflatoxigenic fungi may result in their replacement by other toxic or spoilage microorganisms. Circumstantial evidence that aflatoxin biosynthesis can be influenced lies in the fact that mutant strains of *Aspergillus* exist that either do not produce the mycotoxins or their precursors, or that possess a disrupted biosynthetic pathway, suggesting that aflatoxigenesis is not absolutely essential to the fungus. For example, one mutant of *A. parasiticus* produces only the bright orange-colored norsolorinic acid, the initial product resulting from polyketide cyclization. Analogous mutants have been discovered for each intermediate product of aflatoxin biosynthesis [15].

6.3 AFLATOXIN RESISTANCE PHYTOCHEMICALS IN TREE NUTS

Phytochemical metabolites with antifungal activity fall into two distinct classes, phytoalexins and phytoanticipins [16]. Whereas phytoalexins are biosynthesized in significant amounts only in response to fungal attack, phytoanticipins are constituents always present in the plant, either in their active form or as precursors (e.g., glycosides) from which they can be generated. Phytoalexins are usually produced and located only in close proximity to the point of fungal attack and in variable amounts. This latent process means that even if the *Aspergillus* infection is ultimately overcome, there may have been sufficient time for significant amounts of aflatoxin to have been produced. In contrast, phytoanticipins are a characteristic of any particular plant species or variety, and therefore capable of genetic control for optimal effect. Identification of specific anticipins in tree nuts that confer resistance to aflatoxigenesis should be possible by first determining the relative inhibitory activity of different species and cultivars. Individual tissues of the most active with respect to aflatoxin suppression can then be examined for those having the highest potency, and such materials extracted and analyzed for content and concentrations of bioactive compounds. Since such constituents should be highest in the most active tissues, isolation in sufficient amounts for biological testing should thus be facilitated. The general approach to identifying antiaflatoxigenic compounds in tree nuts is therefore to determine the most resistant nut species, then the varieties within that species, and finally the location of the factors within the individual nuts.

6.3.1 Varietal Resistance of Tree Nut Species

Both anecdotal evidence, and the RASFF notification system for aflatoxins in products entering the EC [9], suggests that there is a species-related propensity for tree nuts to become contaminated. Thus, in 2005, almonds were subject to the greatest number of notifications, followed by pistachios, whereas there were none for walnuts. Furthermore, from experience, walnut producers and exporters are far more concerned with spoilage microorganisms, having little concern for aflatoxin contamination. *In vitro* experiments with selected cultivars and breeding lines of all three species confirmed this assumption. Comparison of the ability of *A. flavus* to produce aflatoxins when grown on agar medium containing 5% by weight of ground kernels (including seed coat) from 23 varieties of almonds (*Prunus dulcis*), 26 varieties of English walnuts (*Juglans regia*), single varieties each of pistachio (*Pistacia vera,* 'Kerman'), and black walnut (*Juglans hindsii*, 'Rawlins') showed that the English walnut was considerably more active in suppressing aflatoxigenesis [17]. Specifically, aflatoxin levels for almond ranged from 34 to 179 µg/plate (average 91 µg/plate) versus 0 to 28 µg/plate (average 4.2 µg/plate) for English walnuts, after 7 days of incubation at 30°C. The pistachio and black walnut kernels were intermediate in activity with values of 40 and 44 µg/plate, respectively. Most significantly, as shown in Figure 6.2, the 'Tulare' walnut allowed no aflatoxin formation, indicating that this variety would be the most rewarding to investigate, since it would most probably have the highest level of antiaflatoxigenic compounds.

6.3.2 Antiaflatoxigenic Constituents in Specific Tissues

The potent antiaflatoxigenic activity of 'Tulare' walnuts established that the bioactive constituents were located either throughout the edible portion of the nut, or alternatively specifically within the

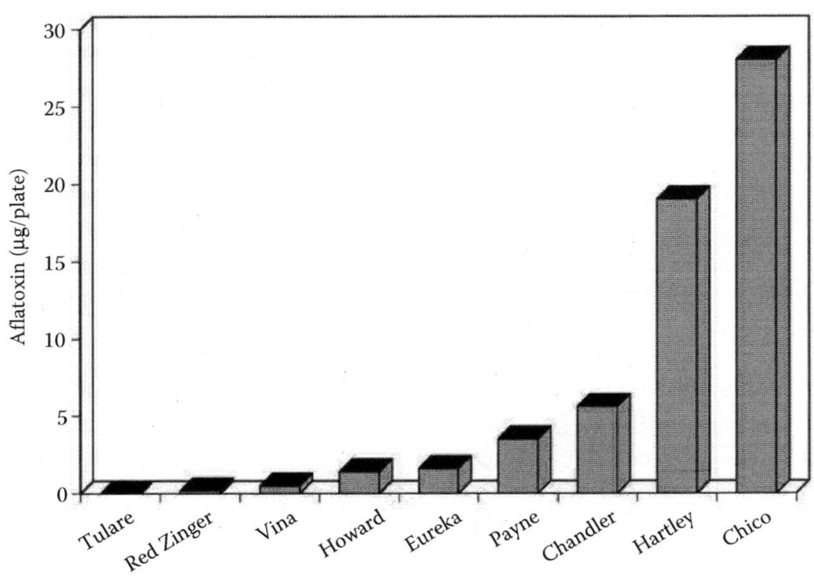

FIGURE 6.2 Differential production of aflatoxin B_1 by selected walnut varieties on agar media containing 5% by weight of ground kernels.

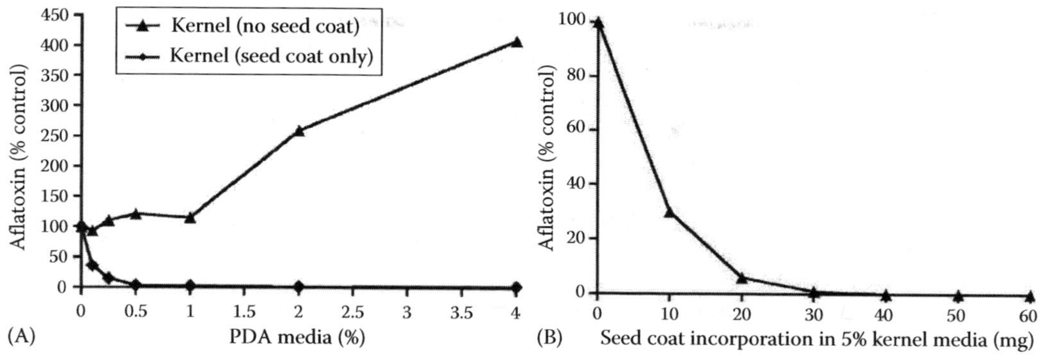

FIGURE 6.3 (A) Aflatoxin B_1 production on potato dextrose agar (PDA) media incorporating various amounts of kernel only or seed coat only of 'Tulare' walnut. (B) Decrease in aflatoxin B_1 production by incorporation of increasing amounts of seed coat material into agar media containing 5% by weight of ground kernels.

kernel itself (endosperm) or the pellicle (seed coat), the latter being the thin tissue surrounding the kernel. Samples of each tissue from 'Tulare' nuts were therefore obtained by peeling the pellicle from the whole kernel and aflatoxin production was measured *in vitro* with endosperm or pellicle incorporated into potato dextrose agar (PDA) medium as previously done for the complete kernel. As shown in Figure 6.3, there was no inhibition of aflatoxin by the endosperm alone, and at incorporation levels greater than 1% in PDA media, the aflatoxin production rapidly increased to approximately 4-fold at 4% incorporation. This effect is probably caused by an increase in levels of nutrients such as sugars and fatty acids. In contrast, incorporation of pellicle alone is completely different, with aflatoxin suppressed to 3% of control at an incorporation level of 1%. At levels as low as 0.1%, aflatoxin production was reduced to about one-third that of control [17].

These results established that antiaflatoxigenic constituents were located entirely in the pellicle and this was confirmed by adding back pellicle material to 5% endosperm media in agar. Under these conditions, the ability of kernel to enhance aflatoxin production was negated. Aflatoxin production was inversely proportional to the amount of added pellicle, with an incorporation level approximating the weight proportion of seed coat to endosperm in whole kernels reducing aflatoxin to 0.8% of control (Figure 6.3) [17].

6.4 ISOLATION AND IDENTIFICATION OF WALNUT ANTIAFLATOXIGENIC CONSTITUENTS

The identification of 'Tulare' walnut pellicle as the source of antiaflatoxigenic activity greatly simplified procedures for isolation of the bioactive constituents since it made it unnecessary to extract the whole kernel, which would have introduced large quantities of extraneous material. Attention was therefore focused on the constituents of pellicle alone.

6.4.1 Bioassay-Directed Fractionation

The extraction and fractionation of 'Tulare' pellicle adopted conventional bioassay-directed procedures commonly used for isolation of natural products. Sequential Soxhlet extraction of finely ground pellicle with solvents of increasing polarity gave extracted material, after evaporation of the solvent, which was suitable for testing *in vitro* in a similar way to the unextracted tissues. The results were unusual in that acetone, methanol, and water extracts all exhibited activity, as did the extracted pellicle material; only the hexane extract lacked activity [18]. This suggested that hexane-extractable material such as lipids did not have any effect on aflatoxin formation but that compounds extractable by more polar solvents consisted of a complex of constituents with variable solubilities. Analysis of extracts by gas chromatography/mass spectrometry (GC/MS) showed only trace amounts of methyl gallate in the water extract, indicating that a series of polar, relatively high molecular weight substances were the active compounds.

These results were considered in relation to literature on composition of walnut kernels and it was concluded that they were consistent with the established presence of hydrolysable tannins as a major component of the pellicle. In the 1950s, these compounds were extracted from pellicle of an unidentified walnut variety obtained from a commercial processor, and one of these was purified from the acetone-soluble fraction and named juglanin [19,20]. The latter was shown to be isomeric with, but not identical to, corilagin, a constituent of the seed pods of the legume *Caesalpinia coriaria* [21]. Corilagin possesses one of the simplest representative structures of a hydrolysable tannin and it is assumed that juglanin must have a different configuration with regard to substitution pattern or stereochemistry because of the differences in melting point and optical rotation. Hydrolysable tannins in walnuts consist of a glucose core, esterified with hexahydroxydiphenic acid and gallic acids, and as 1-*O*-galloyl-3,6-(*R*)-hexahydroxydiphenoyl-β-D-glucopyranose, corilagin contains only one of each moiety. A series of 17 more complex tannins have recently been isolated from walnut pellicle of the Chandler variety [22,23].

This information, in concert with the fact that no aflatoxin-suppressing activity is extractable with hexane, but with the demonstrated occurrence of activity in all other solvent extracts and even in exhaustively extracted residue, is consistent with activity being associated with the hydrolysable tannin content or composition of the pellicle.

6.4.2 Analysis of Hydrolysable Tannin Content in Walnuts

The extraordinary structural complexity and diversity of hydrolysable tannins, especially in walnuts [22,23], would appear to make their correlation with antiaflatoxigenic activity exceptionally difficult. However, in most cases, acid hydrolysis of almost all of the tannins gives only three products, namely glucose, gallic acid, and hexahydroxydiphenic acid, although the latter cannot be isolated as such because it spontaneously lactonizes to ellagic acid. This is illustrated in Figure 6.4 for strictinin, one of

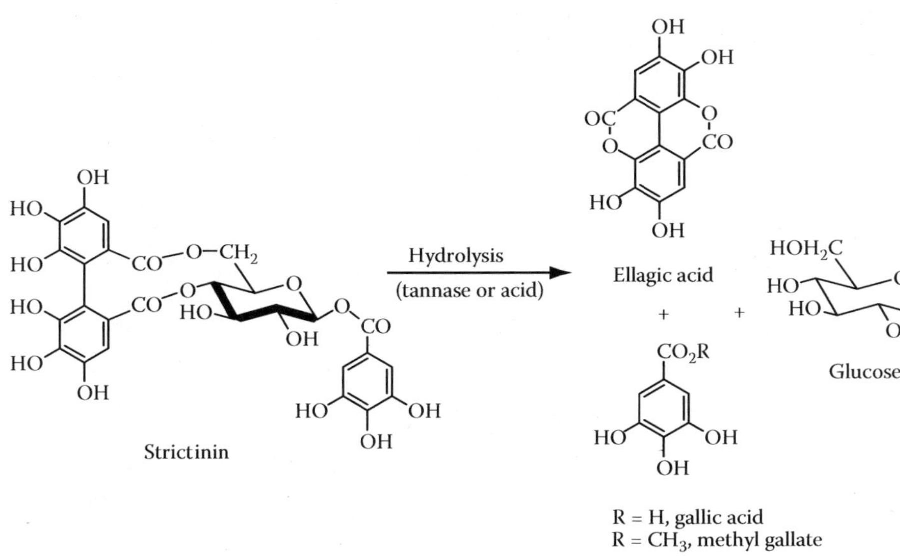

FIGURE 6.4 Structure of strictinin, a simple hydrolysable tannin present in walnuts and its hydrolysis products, ellagic acid, gallic acid, and glucose.

the major simple tannins in walnuts [22]. Interestingly, *A. flavus* is known to possess an extracellular tannase that is capable of hydrolyzing the tannins in an identical manner [24]. Furthermore, the presence of this tannase in any particular fungal strain can be demonstrated by observing a zone of clearing when the particular fungus is grown on media consisting of commercial tannic acid in agar [25].

Since gallic and ellagic acids are the ultimate products of the action of the fungus on hydrolysable tannins, these compounds maybe used as surrogates for hydrolysable tannin content. This analysis can be easily achieved by hydrolysis of plant tissue samples with methanolic hydrochloric acid to give methyl gallate and ellagic acid (Figure 6.4). Both of these products can be measured in a single run by high-performance liquid chromatography (HPLC) on a reversed phase C_{18} column with simultaneous monitoring at 252 and 280 nm, using a diode array detector (DAD), at concentrations as low as 0.1 µg/mL [26].

6.4.3 Gallic and Ellagic Acid Content in Tree Nuts

6.4.3.1 Variation of Gallic and Ellagic Acids with Maturity

The rapidity and convenience of the HPLC procedure has permitted analysis of a large number of samples, so that changes in gallic and ellagic acid levels with maturity could be compared for walnuts throughout the growing season. 'Tulare' was chosen as the aflatoxin-resistant cultivar and "Chico" as a more susceptible variety. During the 2002 growing season, samples were taken monthly, but on the basis of these results, this was changed to biweekly and collections started earlier during 2003 [18]. As shown in Figure 6.5, even before the kernel became firm ("jelly" stage), high levels of gallic acid were present, increasing by 3-fold over the first month of development, followed by a rapid decline and then reaching a relatively constant level throughout the rest of growth. Ellagic acid showed a contrasting pattern, increasing fairly steadily throughout kernel development. This is quite consistent with hydrolysable tannin biosynthesis, in which pentagalloyl glucose, the initially product, loses a proportion of gallic acid moities either via hydrolysis or by dimerization to form hexahydroxydiphenic esters; as a consequence, the ellagic acid content, derived from the latter, would be expected to increase. Further examination of the results showed that the gallic acid content of 'Tulare' was consistently higher than that of 'Chico,' indicating a higher overall level of hydrolysable tannins in the pellicle of the former. Ellagic acid levels were also higher overall in 'Tulare' relative to 'Chico' but the effect was less remarkable.

The ability of hydrolysable tannins to inhibit aflatoxin production, therefore, correlates well with the relative amounts of gallic acid contained within the structures of all of the individual tannins, considered as a whole, but the ellagic acid content appears to have little influence. Since the proportion of ellagic acid to gallic acid generally increases as biosynthesis of the tannins proceeds. It proceeds, it would be expected that antiaflatoxigenic activity should decline with maturity of the nut, but this may be offset by the fact that kernel is much less vulnerable than in early growth due to the physical barrier to infection provided by hardening of the shell. However, maintenance of gallic acid content above a specific, not yet determined, level should enable inhibitory activity to be preserved.

6.4.3.2 Variation of Gallic and Ellagic Acids with Cultivar

In addition to analyzing gallic and ellagic acid content throughout the growing season, the HPLC method can be used to compare levels of these compounds between cultivars and with other nut species. As shown in Table 6.1 for tree nuts harvested in the 2003 season, English walnut varieties have gallic acid contents ranging from 1.4% to 3.4% of the dry weight (dw) of the pellicle. These values generally correlate with the ability of the varieties to suppress aflatoxin production *in vitro* (Figure 6.2), although it should be recognized that such activity is not solely related to gallic acid but must also depend to some extent on ellagic acid content and on the activity of the intact hydrolysable tannins prior to tannase hydrolysis. Comparison of the ellagic acid/gallic acid ratio for the walnuts

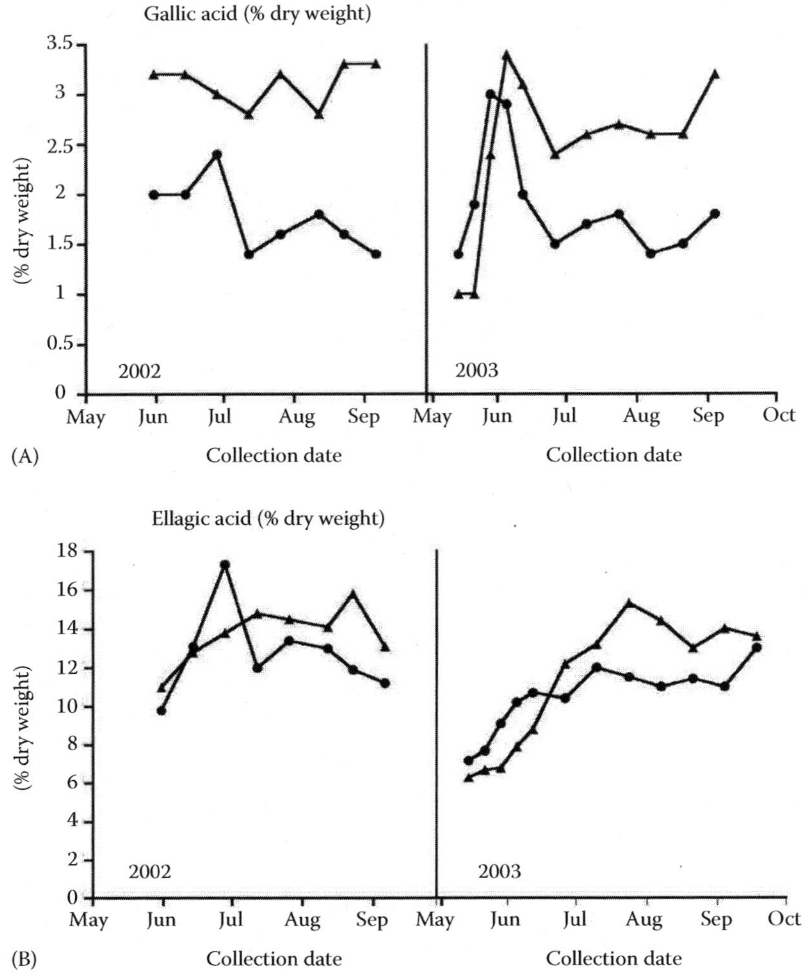

FIGURE 6.5 Variation in gallic acid and ellagic acid content in kernels of walnut varieties 'Tulare' (▲) and 'Chico' (●) during the growing seasons 2002 and 2003.

showed that there was a general trend to lower values for those varieties most resistant to aflatoxigenesis, with 'Tulare' having a ratio of 4.4 and 'Chico' a ratio of 6.1.

Black walnut cultivars 'Rawlins' and 'Thomas' had considerably lower gallic acid levels of ~1% dw, quite close to that of the pistachio cultivar 'Kerman' (0.5% dw), and *in vitro* aflatoxin production for the walnut 'Thomas' and pistachio 'Kerman' were also similar at 45 and 40 μg/plate, respectively [17]. The hydrolysable tannin in pistachios has a core of quinic acid, rather than glucose (unpublished results), and its stereochemistry is such that dimerization of gallic acid moieties to hexahydroxydiphenic esters, and consequent formation of ellagic acid, cannot occur. Any aflatoxin inhibitory activity is, therefore, dependent on either the tannin itself or gallic acid.

In contrast to walnuts and pistachios, almond pellicle contains no hydrolysable tannin and therefore no gallic acid was detectable in the cultivars 'Nonpareil' and 'Mission.' This is consistent with their propensity to accumulate aflatoxins [9] and their ability to support aflatoxin biosynthesis. However, there are considerable differences in aflatoxin production *in vitro* between cultivars and these must be due to compounds other than hydrolysable tannins. Almond pellicle has been shown to contain other phenolic constituents, primarily phenolic acids and flavonoids, which may suppress aflatoxin production although less effectively than the tannins [27–29].

TABLE 6.1
Gallic Acid and Ellagic Acid Content of Seed Coat at Maturity from Selected Nut Species Harvested in California in 2003

Species/Cultivar	Gallic Acid (% dw)	Ellagic Acid (% dw)	Ellagic/Gallic Ratio
Walnut, English (Juglans regia)			
Red Zinger	3.4	15.9	4.7
Tulare	3.2	14.0	4.4
Tehama	2.6	11.0	4.2
Hartley	2.2	13.3	6.0
Payne	2.0	12.3	6.2
Serr	2.0	11.8	5.9
Chico	1.8	11.0	6.1
Chandler	1.4	10.0	7.1
Walnut, Black			
Juglans nigra 'Thomas'	1.1	2.6	2.4
Juglans hindsii 'Rawlins'	1.0	3.1	3.1
Pistachio (Pistacia vera)			
Kerman	0.5	nd	—
Almond (Prunus dulcis)			
Nonpareil	<0.1	nd	—
Mission	<0.1	nd	—

Note: nd, not detected.

6.4.3.3 Structure–Activity Relationships

Although overall gallic acid content of the pellicle has been shown to be a primary indicator of the ability of any particular cultivar to resist aflatoxin formation, incorporation of this finding into breeding programs will depend on a much more complete analysis of structural relationships to the hydrolysable tannins from which it derives. Furthermore, a fundamental understanding of the mechanism by which this and other compounds suppress aflatoxin biosynthesis should open new directions to control of the contamination problem. Gallic acid in hydrolysable tannins is bound primarily as gallate esters of the carbohydrate core or as depsides of the underlying gallate or hexahydroxydiphenate moieties. Hydrolysis of such ester linkages is not likely to proceed at the same rate and may also be dependent on the substitution position on the carbohydrate and degree of steric hindrance. In addition, the fungal tannase consists of several isozymes with individual esterase and depsidase activities [30], while some strains of *Aspergillus* may not even possess such enzymes; in the latter situations, aflatoxin suppression will be dependent on the activity of the tannins themselves. *In vitro* bioassays therefore cannot be regarded as a comprehensive model for the situation *in vivo*, although they are convenient as a first approximation for evaluating and comparing antiaflatoxigenic potential. *Aspergillus* growing on the kernel of the nut will be directly exposed to the hydrolysable tannin in the seed coat and gallic acid will be generated where metabolic activity, including tannase production and aflatoxin biosynthesis, is the greatest. As a consequence, the effective gallic acid concentration at this point source may be much higher than in the *in vitro* bioassays where it is distributed throughout the media. It is, therefore, necessary to evaluate different structural types of hydrolysable tannins and other types of pellicle constituents, such as those in almonds, in order to develop a theoretical basis for understanding aflatoxin biosynthesis and control.

6.5 MECHANISM OF ANTIAFLATOXIGENIC ACTIVITY

The discovery of hydrolysable tannins and their derivatives as natural antiaflatoxigenic constituents is significant with regard to practical control of aflatoxin contamination in tree nut crops. Enhancement of levels of these compounds in new cultivars should enable growers to provide a product to wholesalers and exporters that, together with postharvest screening of nuts likely to be infected, will generally pass regulatory inspections. However, the information and compounds can also be used as a scientific tool to investigate the fundamental question of the role of aflatoxins in the life cycle of the fungus itself. Although many natural products are known to be fungicidal, the ability to affect metabolite biosynthesis without inhibiting fungal growth is unusual and raises the question as to how essential aflatoxins are to survival of the fungus. Initial observations suggest that it is not merely the penultimate stages of aflatoxin formation that are disrupted but rather that the whole biosynthetic pathway is affected, including the genes controlling fatty acid or polyketide synthases involved in the earliest stages of aflatoxin biosynthesis. If this is so, a basic physiological and biological response in the fungus must be involved.

6.5.1 Effect of Phenolic Antioxidants on Fungal Oxidative Stress Response

It has been suggested that aflatoxin production in *A. parasiticus* is a response to oxidative stress [31]. This implies that the aflatoxins themselves serve as antioxidants. However, it is noteworthy that although many of their precursors are phenolic compounds and therefore structurally feasible antioxidants, aflatoxins are not phenolic; it may be that the total biosynthetic network needs to be evaluated from this perspective. Defense responses [32] and environmental conditions such as drought [33] generate reactive oxygen species (ROS) and hydrogen peroxide (H_2O_2) in the plant that result in oxidative stress induced in the fungus, which may be alleviated by induction of aflatoxin biosynthesis. The presence of hydrolysable tannins and hydrolytic products such as gallic acid, which are potent antioxidants and free radical scavengers, could relieve the fungus of the costs associated with such biosynthetic necessity. This suggests that any phytochemical antioxidant, especially plant phenolics, could to some extent be used to reduce or eliminate aflatoxigenesis.

Comparative tests of phenolic compounds known to occur in walnuts, pistachios, and almonds have shown that these reduce aflatoxin levels *in vitro* (unpublished results). The most effective were pentagalloyl glucose, representative of a walnut-hydrolysable tannin, and 3,4-digalloylquinic acid, representative of a pistachio tannin, together with caffeic acid, all of which reduced aflatoxin production by >98% relative to the control. Gallic acid caused an aflatoxin reduction of 84% and ellagic acid was the least effective at 60%; phenolic acids and catechin, typical almond constituents, fell between these two values. The fact that all of these compounds show a significant degree of antiaflatoxigenic activity is further circumstantial evidence that aflatoxin biosynthesis is somehow involved in fungal oxidative stress response, but studies at the genetic level are necessary to establish the precise nature of this effect.

6.5.2 Functional Genomics of Fungal Oxidative Stress Response

The phenolic compounds shown to reduce aflatoxin production in response to oxidative stress have potential as tools to investigate functional elucidation of the genes involved but this is limited by the absence of a practical gene transformation system. In the absence of a direct approach with *A. flavus*, use can be made of the yeast *Saccharomyces cerevisiae*, for which numerous stress response pathways have been characterized [34]. Selected *S. cerevisiae* strains, including mutants with single gene deletions, have been used as a model to evaluate phenotypic response to oxidative stress induced by H_2O_2 [35].

With the exception of undiluted (~10^6) cells, growth of wild-type *S. cerevisiae* cells was completely inhibited for 6- to 10-fold serial dilutions on exposure to 3.3 mM H_2O_2, whereas under

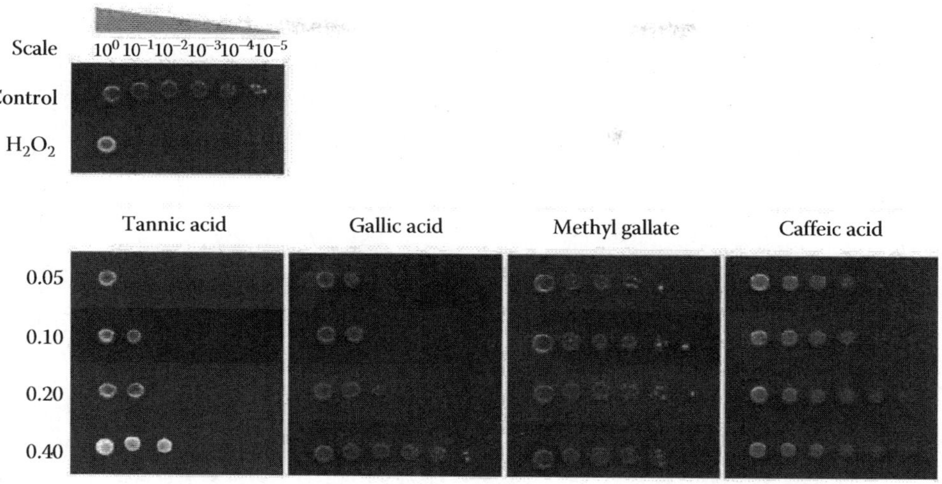

FIGURE 6.6 Effect of tannic acid, gallic acid, methyl gallate, and caffeic acid on growth recovery of serial dilutions of the yeast *S. cerevisiae* subjected to oxidative stress with H_2O_2.

control conditions in the absence of H_2O_2, all colonies from serially diluted cells were visible (Figure 6.6). When grown only in the presence of tannic acid, a hydrolysable gallotannin readily obtainable from commercial sources, or on gallic acid, at 0.4% (w/v), there was no inhibition of growth, demonstrating that these compounds alone are not toxic to the yeast. However, as shown in Figure 6.6, growth inhibition produced in the presence of H_2O_2 was overcome by culturing in the presence of the antioxidants tannic acid, gallic acid, methyl gallate, and caffeic acid in a concentration-dependant manner at 0.05%, 0.1%, 0.2%, and 0.4% incorporation, respectively. It is noteworthy that caffeic acid was much more effective than gallic acid, consistent with their differences in antiaflatoxigenic activity.

This approach was extended to the identification of the function of specific *S. cerevisiae* genes in antioxidative stress response using 22 deletion mutants. These encompassed strains defective in antioxidative stress response, gene regulation, DNA damage control, and signal transduction. When exposed to 2.5 mM H_2O_2 in the presence and absence of 0.4% antioxidant, the mutant *yap1Δ*, defective in a transcription factor for gene regulation of the antioxidative stress response, showed little appreciable recovery on treatment with the antioxidants, indicating that it is extremely sensitive to oxidative stress, while *rad54Δ*, deficient in DNA-dependent ATPase, showed partial recovery. In contrast, the strains *sho1Δ* and *cta1Δ*, deficient in transmembrane osmosensor signal transduction and catalase-dependent antioxidant stress response, respectively, exhibited complete recovery [35].

Using the expressed sequence tag (EST) database for *A. flavus* [36], 43 orthologs of *S. cerevisiae* genes involved in gene regulation, signal transduction, and antioxidation have been identified and the effect of oxidative stress on aflatoxin biosynthesis has been investigated in more detail. Functional complementation of the mitochondrial superoxide dismutase gene, *sodA*, an antioxidation stress gene from *A. flavus*, in a *sod2Δ* yeast mutant lacking the ortholog, demonstrated the utility of this approach [35]. The combination of knockout mutants and functional complementation analysis should thus enable the relationship between oxidative stress and aflatoxin biosynthesis to be elucidated. The recent availability of *A. flavus* genomic microarrays has enabled differential expression microarray analysis to show that treatment of *A. flavus* with antioxidants affects genes far upstream from the aflatoxin biosynthetic gene cluster (unpublished results). This approach will undoubtedly provide extremely detailed knowledge of the antioxidative stress response/aflatoxin biosynthesis relationship in the future.

6.6 CONCLUSION

It has been established that phenolic antioxidants and especially hydrolysable tannins, naturally present in tree nuts, can play an important role in preventing the formation of aflatoxins. Application of this knowledge should not only reduce exposure of consumers to these mycotoxins but also increase the marketability of tree nuts, especially through exports. A distinct advantage of enhancement of these natural products, as opposed to other approaches to aflatoxin elimination, is that there are no major contraindications to their presence in the product. In fact, there is abundant evidence that natural antioxidants in foods have significant health benefits. Not only can they reduce oxidative deterioration but hydrolysable tannins have also been reported to have antiviral, bacteriocidal, anthelmintic, and antihepatoxic properties [37], while phenolic antioxidants in food may limit diseases associated with aging, such as CHD, neurological degeneration, and various types of cancer [38].

REFERENCES

1. Food and Drug Administration (FDA), Qualified Health Claims: Letter of Enforcement Discretion—Nuts and Coronary Heart Disease, Docket No. 02P-0505, Food and Drug Administration, Washington, DC, 2003.
2. U.S. Department of Agriculture (USDA), *National Nutrition Database for Standard Reference, Release 19*, National Technical Information Service, USDA, Springfield, VA, 2006.
3. Feldman, E.B., The scientific evidence for a beneficial health relationship between walnuts and coronary heart disease, *J. Nutr.*, 132, 1062S–1101S, 2002.
4. Kris-Etherton, P.M., Zhao, G., Binkoski, A.E., Coval, S.M., and Etherton, T.D., The effects of nuts on coronary heart disease risk, *Nutr. Rev.*, 59, 103–111, 2001.
5. USDA-National Agricultural Statistics Service, *Noncitrus Fruits and Nut*, Published online at: http://usda.mannlib.cornell.edu/usda/nass/NoncFruiNu//2000s/2006/NoncFruiNu-07-06-2006_final.pdf (accessed July 25, 2007).
6. USDA-Foreign Agricultural Service, *Horticultural and Tropical Crops Division*, 2006. Published online at: http://www.fas.usda.gov/htp/horticulture/nuts.html (accessed July 25, 2007).
7. Commission of the European Community Commission Directive 98/53/EC of 16 July 1998 laying down the sampling methods and the methods of analysis for the official control of the levels of certain contaminants in foodstuffs, *Off. J. Eur. Commun. Legislation*, L201, 93–101, 1998.
8. Food and Drug Administration (FDA), Compliance Policy Guides Manual, Sec.555.400, 268; Sec. 570.500, 299, 1996.
9. European Commission (EC), The Rapid Alert System for Food and Feed (RASFF) Annual Report 2005, Published online at: http://ec.europa.eu/food/food/rapidalert/report2005_en.pdf (accessed July 25, 2007).
10. Henry, S.H., Bosch, F.X., Troxell, T.C., and Bolger, P.M., Reducing liver cancer: Global control of aflatoxin, *Science*, 286, 2453–2454, 1999.
11. Wang, J.-S. and Tang, L., Epidemiology of aflatoxin exposure and human liver cancer, *J. Toxicol. Toxin Rev.*, 23, 249–271, 2004.
12. Lewis, L., Onsongo, M., Njapau, H., Schurz-Rogers, H., Luber, G., Kieszak, S., Nyamongo, J., Backer, L., Dahiye, A.M., Misore, A., DeCock, K., Rubin, C., and the Kenya Aflatoxicosis Investigation Group, Aflatoxin contamination of commercial maize products during an outbreak of acute aflatoxicosis in eastern and central Kenya, *Environ. Health Persp.*, 113, 1763–1767, 2005.
13. Eaton, D.L. and Groopman, J.D., *The Toxicology of Aflatoxins. Human Health, Veterinary and Agricultural Significance*, Academic Press, San Diego, CA, 1994.
14. Gopalakrishnan, S., Harris, T.M., and Stone, M.P., Intercalation of aflatoxin B_1 in two oligodeoxynucleotide adducts: Comparative ^{1}H NMR analysis of d(ATC^{AFB}GAT).d(ATCGAT) and d(AT^{AFB}GCAT)$_2$, *Biochemistry*, 29, 10438–10448, 1990.
15. Minto, R.E. and Townsend, C.A., Enzymology and molecular biology of aflatoxin biosynthesis, *Chem. Rev.*, 97, 2537–2555, 1997.
16. Mansfield, J.W., Antimicrobial compounds and resistance: The role of phytoalexins and phytoanticipins, in *Mechanisms of Resistance to Plant Diseases*, Slusarenko, A.J., Fraser, R.S.S., and Van Loon, L.C., Eds., Kluwer Academic Publishers, Dordrecht, The Netherlands, 2000, pp. 325–370.

17. Mahoney, N., Molyneux, R.J., McKenna, J., Leslie, C.A., and McGranahan, G., Resistance of 'Tulare' walnut (*Juglans regia* cv.Tulare) to aflatoxigenesis, *J. Food Sci.*, 68, 619–622, 2003.
18. Mahoney, N. and Molyneux, R.J., Phytochemical inhibition of aflatoxigenicity in *Aspergillus flavus* by constituents of walnut (*Juglans regia*), *J. Agric. Food Chem.*, 52, 1882–1889, 2004.
19. Jurd, L., Plant polyphenols. I. The polyphenolic constituents of the pellicle of the walnut (*Juglans regia*), *J. Am. Chem. Soc.*, 78, 3445–3448, 1956.
20. Jurd, L., Plant polyphenols. III. The isolation of a new ellagitannin from the pellicle of the walnut, *J. Am. Chem. Soc.*, 80, 2249–2252, 1958.
21. Schmidt, O.Th., Schmidt, D.M., and Herok, J., Natural tannins. XIX. Constitution and configuration of corilagin, *Justus Liebigs Ann. Chem.*, 587, 67–74, 1954.
22. Fukuda, T., Ito, H., and Yoshida, T., Antioxidative polyphenols from walnuts (*Juglans regia* L.), *Phytochemistry*, 63, 795–801, 2003.
23. Ito, H., Okuda, T., Fukuda, T., Hatano, T., and Yoshida, T., Two novel dicarboxylic acid derivatives and a new dimeric hydrolyzable tannin from walnuts, *J. Agric. Food Chem.*, 55, 672–679, 2007.
24. Yamada, H., Adachi, O., Watanabe, M., and Sato, N., Studies on fungal tannase. Part 1. Formation, purification and catalytic properties of tannase of *Aspergillus flavus*, *Agric. Biol. Chem.*, 32, 1070–1078, 1968.
25. Bradoo, S., Gupta, R., and Saxena, R.K., Screening of extracellular tannase-producing fungi: Development of a rapid and simple plate assay, *J. Gen. Appl. Microbiol.*, 42, 325–329, 1996.
26. Lei, Z., Jervis, J., and Helm, R.F., Use of methanolysis for the determination of total ellagic and gallic acid contents of wood and food products, *J. Agric. Food Chem.*, 49, 1165–1168, 2001.
27. Frison, S. and Sporns, P., Variation in the flavonol glycoside composition of almond seedcoats as determined by MALDI–TOF mass spectrometry, *J. Agric. Food Chem.*, 50, 6818–6822, 2002.
28. Sang, S., Lapsley, K., Jeong, W.S., Lachance, P.A., Ho, C.T., and Rosen, R.T., Antioxidative phenolic compounds isolated from almond skins (*Prunus amygdalus* Batsch), *J. Agric. Food Chem.*, 50, 2459–2463, 2002.
29. Milbury, P.E., Chen, C.-Y., Dolnikowski, G.G., and Blumberg, J.B., Determination of flavonoids and phenolics and their distribution in almonds, *J. Agric. Food Chem.*, 54, 5027–5033, 2006.
30. Haslam, E. and Stangroom, J.E., The esterase and depsidase activities of tannase, *Biochem. J.*, 99, 28–31, 1966.
31. Jayashree, T. and Subramanyam, C., Oxidative stress as a prerequisite for aflatoxin production by *Aspergillus parasiticus*, *Free Radic. Biol. Med.*, 29, 981–985, 2000.
32. Levine, A., Tenhaken, R., Dixon, R., and Lamb, C., H_2O_2 from the oxidative burst orchestrates the plant hypersensitive disease resistance response, *Cell*, 79, 583–593, 1994.
33. Wang, W., Vinocur, B., and Altman, A., Plant responses to drought, salinity and extreme temperatures: Towards genetic engineering for stress tolerance, *Planta*, 218, 1–14, 2003.
34. Toone, W.M. and Jones, N., Stress-activated signalling pathways in yeast, *Genes Cells*, 3, 485–498, 1998.
35. Kim, J.H., Campbell, B.C., Yu, J., Mahoney, N., Chan, K.L., Molyneux, R.J., Bhatnagar, D., and Cleveland, T.E., Examination of fungal stress response genes using *Saccharomyces cervisiae* as a model system: Targeting genes affecting aflatoxin biosynthesis by *Aspergillus flavus* Link, *Appl. Microbiol. Biotechnol.*, 67, 807–815, 2005.
36. Yu, J., Whitelaw, C.A., Nierman, W.C., Bhatnagar, D., and Cleveland, T.E., *Aspergillus flavus* expressed sequence tags for identification of genes with putative roles in aflatoxin contamination of crops, *FEMS Microbiol. Lett.*, 237, 333–340, 2004.
37. Haslam, E., Natural polyphenols (vegetable tannins) as drugs: Possible modes of action, *J. Nat. Prod.*, 59, 205–215, 1996.
38. Young, I.S. and Woodside, J.V., Antioxidants in health and disease, *J. Clin. Pathol.*, 54, 176–186, 2001.

7 Flavor and Volatile Compounds in Tree Nuts

Keith R. Cadwallader and Sirima Puangpraphant

CONTENTS

7.1 INTRODUCTION

Tree nuts are appreciated worldwide for their desirable flavor attributes and are used extensively in confectionary, bakery, culinary, and other food product applications. Consumption of tree nuts has grown in recent years due to reports on the health benefits of a diet rich in nuts and nut oils [1,2]. The world's most popular tree nut is the almond followed by the walnut and the hazelnut (filbert). Other economically important tree nuts include pecan, Brazil nut, cashew, chestnut, pistachio, macadamia, and pine nut. Despite their economic importance and growing popularity, little information is available regarding the characteristic flavor and aroma constituents of most tree nuts. In fact, there has been little addition to the published literature on the subject since Maga's 1991 overview on the volatiles in nuts [3]. Hazelnut is an exception and extensive studies have been conducted to identify the volatile components of both the raw and roasted forms of this nut. This chapter presents an overview of the literature on the flavor and volatile compounds in tree nuts and discusses their possible origins.

7.2 FLAVOR AND VOLATILE COMPONENTS OF TREE NUTS

7.2.1 ALMOND

As mentioned above, almond (*Prunus dulcis*, synonyms *P. amygdalus*, *Amygdalus communis* L., and *A. dulcis* Mill.) is, from a commercial standpoint, the most important tree nut. Almond is commonly consumed in either its natural (raw) or roasted form. A multitude of other products are also produced from almond, including almond butter, almond milk, etc. Very few studies have been published on the volatile components of roasted almond [4,5] and to our knowledge, no reports have been published on the volatiles of raw almond. Takei et al. [4] isolated volatiles from roasted almond by acetone extraction–vacuum carbon dioxide distillation and conducted gas chromatography–mass spectrometry (GC-MS) analysis of the basic fraction. Takei and Yamanishi [5] applied this same procedure for the GC-MS analysis of basic, carbonyl, and noncarbonyl fractions. The combined results of these studies are shown in Table 7.1. In roasted almond, the

TABLE 7.1
Flavor and Volatile Compounds in Roasted Almond

Hydrocarbons	*Furans/Furanones*	*Pyrazines/Pyridines/Amines*
n-Hexane[c]	Furfurylmethylether[b]	Methylpyrazine[a,b]
3-Methylcyclopentane[c]	2-Furfural[b]	2,5- and 2,6-Dimethylpyrazine[a,b]
n-Heptane[c]	5-Methylfurfural[b]	2-Ethyl-6-methylpyrazine[a,b]
Toluene[c]	2-Acetylfuran[b]	Trimethylpyrazine[a,b]
Alcohols	2-Methyltetrahydrofuran-3-one[b]	2,5- and 2,6-Dimethyl-3-ethylpyrazine[a,b]
n-Butanol[b]	Furanylacetate[b]	2,5- and 2,6-Diethyl-3-methylpyrazine[a,b]
Phenol[c]	Methylfuroate[a,b]	2,5-Dimethyl-3-vinylpyrazine[a,b]
n-Hexanol[c]	Ethylfuroate[b]	Acetylpyrazine[b]
Carbonyls	Furfuryl alcohol[b]	5-Methyl-6,7-dihydro-5*H*-cyclopentapyrazine[a,b]
4-Methyl-3-pentanone[b]	5-Methylfurfuryl alcohol[b]	6,7-Dihydro-5*H*-cyclopentapyrazine[a,b]
Diacetone alcohol[b]	5-Hydroxymethylfurfural[b]	2 (or 3), 5-Dimethyl-6,7-dihydro-5*H*-cyclopentapyrazine[a,b]
Decanal[b]	Cyclotene[c]	5-Methyl-2-acetylpyrazine[a,b]
Benzaldehyde[b]	2,5-Dimethyl-4-hydroxy-3(2*H*)furanone[c]	2-Methyl-6,7-dihydro-5*H*-cyclopentapyrazine[b]
Decadienal (isomer)[b]	β-Angelicalactone[c]	6-Ethyl-2-acetylpyrazine[b]
(*E*,*E*)-2,4-Decadienal[b]	2-Hydroxy-2-methoxy-5-methyl-3(2*H*) furanone[c]	2,3-Dimethyl-6,7-dihydro-5*H*-cyclopentapyrazine[a,b]
α-Ionone[b]	*Pyrroles*	2-Acetyl-5-methoxypyrazine[b]
β-Ionone[b]	1-Methyl-2-acetylpyrrole[b]	2-Acetyl-6-allylpyrazine[b]
o-Hydroxyacetophenone[b]	1-Methyl-2-formylpyrrole[b]	2,5-Diacetylpyrazine[b]
2-Phenyl-2-butenal[b]	1-Acetyl-2-methylpyrrole[b]	2-(2′-Furyl) pyrazine[a,b]
Methylallylketone[c]	1-Ethyl-2-formylpyrrole[b]	3-Methyl-2-(2′-furyl)pyrazine[a,b]
p-Hydroxybenzaldehyde[c]	1-Acetyl-2-methylpyrrole[b]	2-Acetyl-3-hydroxy-5-methyl-6,7-dihydro-5H-cyclopentapyrazine[c]
	1-Furfurylpyrrole[b]	Pyrrylthiophenylketone[c]
	2-Acetylpyrrole[b]	5-Methylquinoxaline[a,b]
	2-Formylpyrrole[b]	2,4-Dimethoxy-3-hydroxypyridine[b]
	5-Methyl-2-formylpyrrole[b]	2-Methoxy-3-hydroxy-4-formylpyridine[b]
	5-Vinyl-2-formylpyrrole[a,b]	*N*-Acetylpropylamine[c]
	1-Furfuryl-2-formylpyrrole[b]	
	2-Carboxyethoxypyrrole[c]	

Source: From Takei, Y., Shimada, K., Watanabe, S., and Yamanishi, T., *Agric. Biol. Chem.*, 38, 645, 1974; Takei, Y. and Yamanishi, T., *Agric. Biol. Chem.*, 38, 2329, 1974.

[a] Acetone/carbon dioxide distillation extracts [4].

[b] Acetone/carbon dioxide distillation extracts [5].

[c] Methanol extracts [5].

TABLE 7.2
Flavor and Volatile Compounds in Almond Oil

Hexanal[b]	Methylbenzene[a]	*n*-Octane[a]
Benzaldehyde[a,b]	Ethylbenzene[a]	*n*-Tridecane[a]
Nonanal[b]	Propylbenzene[a]	*n*-Tetradecane[a]
Decadienal (isomer)[b]	1,2-Dimethylbenzene[a]	*n*-Hexadecane[a]
(*E*,*E*)-2,4-decadienal[b]	1,3-Dimethylbenzene[a]	1-Methylindene[a]
2 or 3-Methylcyclopent-2-en-1-one[a]	1,4-Dimethylbenzene[a]	Cyclopentadienes[a]
5-Ethyldihydrofuran-3-one[a]	Benzyl alcohol[a]	
5-Propyldihydrofuran-3-one[a]	Methylphenol[a]	

Source: From Pićurić-Jovanović, K. and Milovanović, M., *J. Am. Oil Chem. Soc.*, 70, 1101, 1993; Caja, M.M., Ruis del Castillo, M.L., Martínez Alvarez, R., Herraiz, M., and Blanch, G.P., *Eur. Food Res. Technol.*, 211, 45, 2000.

[a] From Pićurić-Jovanović, K. and Milovanović, M., Analysis of volatile compounds in almond and plum kernel oils, *J. Am. Oil Chem. Soc.*, 70, 1101–1104, 1993.

[b] Caja, M.M., Ruis del Castillo, M.L., Martínez Alvarez, R., Herraiz, M., and Blanch, G.P., Analysis of volatile compounds in edible oils using simultaneous distillation-solvent extraction and direct coupling of liquid chromatography with gas chromatography, *Eur. Food Res. Technol.*, 211, 45–51, 2000.

majority of the volatiles are generated by the Maillard reaction during the roasting process. In particular, 4-hydroxy-2,5-dimethyl-3(2*H*)-furanone was considered to contribute a strong and sweet aroma to roasted almond [5]. Pyrazines would also be expected to contribute nutty notes to roasted almond. In addition to Maillard reaction compounds, lipid-derived volatiles also were identified in roasted almond (Table 7.1). Lipid-derived compounds such as aldehydes also have been reported in almond oil [6,7] (Table 7.2) and are also likely to be found in significant quantities in raw almond, since, like most tree nuts, almonds contain a high level of polyunsaturated lipid that is prone to oxidation [8–10].

7.2.2 Walnut

Walnut (*Juglans* sp.) cultivars of commercial importance are of the Persian walnut (*Juglans regia* L.), a species which is favored because of its large kernel and thin shell. Walnut is appreciated as a snack and is used extensively in baking and confectionary products. Walnut contains a high amount of lipid (~62%–70%) [11–13] and not surprisingly, the majority of the volatile compounds identified in walnut is derived via breakdown of unsaturated fatty acids (Table 7.3). The volatile composition of walnut was first examined by Clark and Nursten [14] (Table 7.3). They identified 29 compounds, including eight carbonyls, four alcohols, and two terpenes. Among 44 volatiles later identified in walnut by the same authors [15], hexanal, pentanal, 2-methyl-2-pentenal, and 2,3-pentanedione were suggested as the most important aroma contributors (Table 7.3). These same four compounds were among 118 volatiles detected by dynamic headspace analysis (DHA)-GC-MS in walnuts from three geographical locations (Table 7.3) [16]. Considerable differences in volatile profiles were observed among the walnut samples studied; however, all contained an abundance of volatiles originating via oxidative decomposition of linoleic acid [16]. Likewise, the volatile components of walnut oil also is primarily products of lipid oxidation (Table 7.4) [7,17,18]. Moderate levels of lipid-derived compounds maybe important for the generation of typical walnut aroma; however, excessive oxidation may have a negative impact on flavor [13,17,19]. To date, there are no detailed reports on the characteristic aroma components of walnut.

TABLE 7.3
Flavor and Volatile Compounds in Natural Walnut

Hydrocarbons	*Alcohols*	*Aldehydes*	*Ketones*
n-Pentane[b]	Ethanol[a]	2-Methylpropanal[c]	Acetone[a,b]
n-Heptane[b]	*n*-Propanol[a,b,c]	Butanal[c]	2-Butanone[c]
n-Octane[b]	2-Propanol[b]	2- and 3-Methylbutanal[c]	3-Buten-2-one[c]
n-Nonane[a,b]	*n*-Butanol[b,c]	Pentanal[b,c]	2,3-Butanedione[b]
n-Decane[a,b]	*n*-Pentanol[a,b,c]	Hexanal[a,b,c]	2-Pentanone[c]
n-Undecane[a,b]	2-Pentanol[c]	2-Methyl-2-butenal[c]	4-Methyl-2-pentanone[c]
n-Dodecane[a,b]	(*Z*)-2-Penten-1-ol[c]	(*E*)-2-Pentenal[c]	1-Penten-3-one[c]
n-Tridecane[a]	2-Methyl-1 (and 2)-propanol[c]	2-Methyl-2-pentenal[b,c]	2,3-Pentanedione[b]
n-Tetradecane[a]	2-Methyl-1-propanol[c]	(*E*)-2-Hexenal[c]	5-Methylhexan-2-one[b]
Terpenes	1-Penten-3-ol[c]	Heptanal[c]	2-Heptanone[a,b,c]
Limonene[a,b,c]	2-Methyl-2-butanol[a]	(*E*)-2-Heptenal[b,c]	Cyclopentanone[c]
α-Pinene[a,b,c]	2- and 3-Methyl-1-butanol[b,c]	Octanal[c]	2-Octanone[b,c]
β-Pinene[c]	3-Methyl-1-pentanol[c]	(*E*)-2-Octenal[a,b,c]	1-Octen-3-one[c]
Sabinene[c]	*n*-Hexanol[a,b]	Nonanal[c]	6-Methyl-5-hepten-2-one[c]
3-Carene[b]	1-Octen-3-ol[c]	5-Ethyl-1-formylcyclopentene[c]	3-Octen-2-one[c]
Aromatic hydrocarbons	*n*-Heptanol[c]	(*E,E*)-2,4-Heptadienal[c]	(*E,E*)-3,5-Octadien-2-one[c]
Benzene[b,c]	*n*-Octanol[c]	(*E*)-2-Nonenal[c]	*Miscellaneous*
Toluene[b,c]		Decanal[b,c]	Ethyl acetate[a,c]
Ethylbenzene[a,c]		Benzaldehyde[a,b,c]	1-Methylpropyl acetate[c]
1,2-Dimethylbenzene[c]		2,4-Decadienal (isomers)[a]	Chloroform[a,b,c]
1,3-Dimethylbenzene[b,c]		(*E,E*)-2,4-Decadienal[b]	Dichlorobenzene[b]
1,4-Dimethylbenzene[a,b,c]		*Furans/Lactones*	Pentyl acetate[b]
Styrene[b,c]		2-Ethylfuran[b,c]	Pentyloxirane[c]
Isopropylbenzene[c]		2-Butylfuran[b,c]	2-Methylpyrrole[c]
Propylbenzene[c]		2-Pentylfuran[b,c]	
1-Ethyl-2(3, and 4)-ethylbenzene[c]		Tetrahydrofuran[c]	
[1 (and 2)-Methylpropyl]benzene[c]		γ-Butyrolactone[a]	
1,3,5-Trimethylbenzene[a]		γ-Caprolactone[c]	
1,2,4-Trimethylbenzene[a,b,c]		4,5-Dimethyldihydro-2(3*H*)-furanone[c]	
1,2,3-Trimethylbenzene[a,b]		5-Ethenyldihydro-5-methyl-2(3*H*)-furanone[c]	
o-, *m*- and/or *p*-Cymene[c]			
1,3-Diethylbenzene[c]			
α-Methylpropylbenzenes (3 isomers)[c]			
1,4-Diethylbenzene[c]			
Butylbenzene[c]			
α-Dimethylethylbenzenes (4 isomers)[c]			
Indane[c]			
1,2,3,4-Tetramethylbenzene[c]			
1,2,3,5-Tetramethylbenzene[c]			
Naphthalene[a,c]			
1-Methylnaphthalene[a]			
2-Methylnaphthalene[a]			

Source: From Clark, R.G. and Nursten, H.E., *J. Sci. Food Agric.*, 27, 902, 1976; Clark, R.G. and Nursten, H.E., *J. Sci. Food Agric.*, 28, 69, 1977; Elmore, J.S., Nisyrios, I., and Mottram, D.S., *Flavour Fragr. J.*, 20, 501, 2005.

[a] Clark, R.G. and Nursten, H.E., Volatile flavour components of walnuts (*Juglans regia* L.), *J. Sci. Food Agric.*, 27, 902–908, 1976.

[b] Clark, R.G. and Nursten, H.E., The sensory analysis and identification of volatiles from walnut (*Juglans regia* L.) headspace, *J. Sci. Food Agric.*, 28, 69–77, 1977,

[c] Elmore, J.S., Nisyrios, I., and Mottram, D.S., Analysis of the headspace aroma compounds of walnuts (*Juglans regia* L.), *Flavour Fragr. J.*, 20, 501–506, 2005.

TABLE 7.4
Flavor and Volatile Compounds in Walnut Oil

n-Pentane[c]	*n*-Octanol[c]	Hexanal[a,b,c]	Octanal[c]
n-Octane[c]	1-Acetylcyclohexene[a]	2-Hexanal[c]	1-Octen-3-ol[b]
n-Nonane[c]	Furancarboxaldehyde[a]	2-Hexanone[c]	Nonanal[a,b,c]
Ethanol[c]	Furfuryl alcohol[a]	Heptanal[c]	2-Nonenal[c]
Cyclobutanol[c]	2-Pentylfuran[c]	(*E*)-2-Heptenal[b]	Decanal[c]
n-Pentanol[c]	2-Octylfuran[c]	2,4-Heptadienal (isomers)[a,c]	2-Decenal[c]
n-Hexanol[c]	Pentanal[c]	(*E,Z*)-2,4-Heptadienal[b]	2,4-Decadienal (isomer)[a,c]
n-Heptanol[c]	(*E*)-2-Pentenal[b]	(*E,E*)-2,4-Heptadienal[b]	(*E,E*)-2,4-Decadienal[a,b]

Source: From Caja, M.M., Ruis del Castillo, M.L., Martínez Alvarez, R., Herraiz, M., and Blanch, G.P., *Eur. Food Res. Technol.*, 211, 45, 2000; Crowe, T.D. and White, P.J., *J. Am. Oil Chem. Soc.*, 80, 569, 2003; Torres, M.M., Martínez, M.L., and Maestri, D.M., *J. Am. Oil Chem. Soc.*, 82, 105, 2005.

[a] Caja, M.M., Ruis del Castillo, M.L., Martínez Alvarez, R., Herraiz, M., and Blanch, G.P., Analysis of volatile compounds in edible oils using simultaneous distillation-solvent extraction and direct coupling of liquid chromatography with gas chromatography, *Eur. Food Res. Technol.*, 211, 45–51, 2000.

[b] Crowe, T.D. and White, P.J., Oxidation, flavor, and texture of walnuts reduced in fat content by supercritical carbon dioxide, *J. Am. Oil Chem. Soc.*, 80, 569–574, 2003.

[c] Torres, M.M., Martínez, M.L., and Maestri, D.M., A multivariate study of the relationship between fatty acids and volatile flavor components in olive and walnut oils, *J. Am. Oil Chem. Soc.*, 82, 105–110, 2005.

7.2.3 Hazelnut

Among the tree nuts, hazelnut (*Corylus avellana* L.) has received the most attention in regards to the study of its characteristic flavor and aroma. Hazelnut can be consumed raw, but the roasted counterpart is generally the preferred form. A well-known characteristic odorant of roasted hazelnut is the compound filbertone [5-methyl-(*E*)-2-hepten-4-one], which has a characteristic "roasted hazelnut-like" note and is formed during the roasting process [20,21]. This potent odorant, with a low odor detection threshold of only 5 ng/kg (ppb) in oil, was first identified by Emberger [22]. Güntert et al. [23] later demonstrated that the (*E*)-S form had the lowest threshold and possessed the most roasted hazelnut-like odor quality among the four possible isomers. In addition to filbertone, both raw and roasted hazelnuts contain other important aroma components (Table 7.5) [24–29]. The volatile compounds in hazelnut oil has also been reported (Table 7.6) [7,30]. Alasalvar et al. [28] compared the flavor of natural (raw) and roasted hazelnuts by sensory analysis and DHA-GC-MS. A greater number of volatile compounds were found in the roasted nut (71) than in the raw nut (39), and both types of hazelnuts contained appreciable amounts of lipid-derived aldehydes, ketones, and alcohols. Thermally derived compounds, such as Strecker aldehydes and pyrazines, were found in much higher abundance in the roasted nut. In particular, filbertone was found at 10-fold greater abundance in the roasted hazelnut compared to raw hazelnut. This is in agreement with the results of Pfnuer et al. [20] who reported that heating of hazelnuts was necessary to generate sufficient quantities of filbertone.

7.2.4 Pecan

The pecan tree (*Carya illinoinensis*) is native to North America and has been commercially exploited for production of pecan nuts for nearly a century, mainly in the southeastern United States. Pecan can be consumed either raw or roasted and is used extensively in confectionary, bakery, culinary, and other food product applications. Like most other tree nuts, pecan contains high amounts of lipid (55%–75%) [31–33], thus the majority of volatile compounds identified in this nut are derived via breakdown of unsaturated fatty acids. Only a few studies have been published on the volatile constituents of pecan

TABLE 7.5
Flavor and Volatile Compounds in Roasted Hazelnut

Hydrocarbons
n-Hexane[a]
n-Heptane[a]
n-Octane[a]
n-Nonane[a]
n-Decane[a,b]
n-Undecane[a]
n-Dodecane[a]
n-Tridecane[a]

Terpenes
α-Pinene[a,e]
β-Pinene[a]
Sabinene[a]
Myrcene[a]
β-Phellandrene[a]
α-Terpinene[a]
p-Cymene[a]
4-Terpineol[a]
3-Carene[e]
Limonene[a]
β-Caryophyllene[e]

Aromatic hydrocarbons
Benzene[a,b]
Toluene[a,b,e,f]
o-Xylene[a,e,f]
m-Xylene[a,c,f]
p-Xylene[a,c,e,f]
Styrene[a]
1,2,3-Trimethylbenzene[e,f]
1,2,4-Trimethylbenzene[a,c,e,f]
Ethylbenzene[a]
Butylbenzene[a]
Propylbenzene[a]
2,3-Dihydroindene[a]
Indene[e]
Naphthalene[a,d]
1-Methylnaphthalene[a]
2-Methylnaphthalene[a]
Ethylnaphthalene[a]
1,6,7-Trimethylnaphthalene[e]
Biphenyl[a]
Methylethylbenzene[a]

Alcohols
Methanol[b]
Ethanol[b]
n-Pentanol[d,e,f]
2-Pentanol[e,f]
1-Penten-3-ol[f]
3-Penten-2-ol[d]
4-Methylpentan-2-ol[d]
n-Hexanol[d,e,f]
n-Heptanol[d,e,f]
3-Methyl-1-butanol[e,f]
n-Octanol[e]
1-Octen-3-ol[a,e]
Benzyl alcohol[a]
Phenethyl alcohol[a]

Phenols
2-Ethylphenol[a]
m-Cresol[a]
p-Cresol[a]
o-Methoxyphenol[a]
Butylated hydroxytoluene[c]

Esters/Acids
Methyl formate[b]
Ethyl formate[a]
Methyl acetate[b]
Ethyl acetate[a,b]
Butyl acetate[a]
Isopropyl acetate[d]
Allyl crotonate[c]
2-Methylbutyl 2-methylbutyrate[a]
Ethyl heptanoate[a]
Ethyl decanoate[a]
Methyl undecanoate[a]
Benzyl acetate[a]
1-Hydroxy-2-propanone acetate[a]
Acetic acid[a,c,e]
3-Butenoic acid[e]
Pentanoic acid[a,d]
Hexanoic acid[c,d]
Heptanoic acid[d]
Octanoic acid[d]
Nonanoic acid[d]

Aldehydes
Acetaldehyde[a,b]
Propanal[b]
Pentanal[a,d]
Hexanal[a,b,d,e,f]
Heptanal[a,b,d,e,f]
Octanal[a,c,d]
Nonanal[a,d,e,f]
Decanal[a,d]
Methylpropanal[a,b,e,f]
2-Methylbutanal[a,b,e,f]
3-Methylbutanal[a,b,e,f]
2-Methyl-2-butenal[a,e]
2-Hexanal[e,f]
2-Heptenal[a,d]
2-Octenal[a,d]
2-Nonenal[a,d]
2-Decenal[a,d]
2-Undecenal[a,d]
Benzaldehyde[a,c,d,e]
Phenylacetaldehyde[a,c]
4-Methylbenzaldehyde[a,e]
(*E*,*E*)-2,4-Hexadienal[e,f]
2,4-Decadienal (isomer)[a]
(*E*,*E*)-2,4-Decadienal[a]
2-Phenyl-2-butenal[a]
4-Methyl-2-phenyl-2-pentenal[a]
5-Methyl-2-phenyl-2-hexenal[a]

Ketones
Acetone[a,b]
2-Butanone[a]
2-Pentanone[a,e,f]
2-Hexanone[a]
2-Heptanone[a,d,e,f]
2-Octanone[a,d]
2-Nonanone[a,d,f]
2-Decanone[a,d]
2-Undecanone[a,d]
3-Methyl-2-pentanone[a,e,f]
5-Methyl-2-hexanone[a]
3-Penten-2-one[a,c,d,e,f]
5-Methyl-3-hexen-2-one[a]
5-Methyl-5-hexen-2-one[e]
5-Methyl-2-heptanone[e,f]
6-Methyl-5-hepten-2-one[e,f]
5-Methy-(*E*)-2-hepten-4-one[e,f]
2-Hepten-4-one[a]
3-Hepten-2-one[a]
3-Octen-2-one[a,d]
1-Octen-3-one[e,f]
3-Hexanone[a]
2,3-Butanedione[a,b]
2,3-Pentanedione[a,e,f]
2,5-Hexanedione[a]
6-Methyl-3,5-heptadien-2-one[a]
p-Mentha-6,8-dien-2-one[a]
2-Cyclopentenone[a]
2-Cyclohexenone[a]
3-Methyl-2-cyclohexen-1-one[a,e]
3,5,5-Trimethyl-2-cyclohexen-1-one[a]
2-Hydroxy-3-methyl-2-cyclopentanone[a]
1-Cyclopentylethanone[e,f]
Acetophenone[a]
1-Phenyl-1-propanone[a]
1-Phenyl-2-propanone[a]

Lactones
γ-Butyrolactone[a,c]
γ-Crotonolactone[a]
γ-Valerolactone[a]
γ-Hexalactone[a]
γ-Heptalactone[a]
γ-Octalactone[a,d]
γ-Nonalactone[a,d]
δ-Verolactone[a]
δ-Octalactone[a]

Furans/Furanones/Pyranones
Furan[a]
2-Methylfuran[a]
2-Ethylfuran[a]
2-Butylfuran[a]
2-Pentylfuran[a,c,e,f]
2-Hexylfuran[a]
2-Heptylfuran[a]
2-Octylfuran[a]
Dimethylfuran[a,e,f]
2-Ethyl-5-methylfuran[e,f]
2,3,5-Trimethylfuran[e,f]
2-Vinylfuran[a]
Phenylfuran[a]
2-Furfural[a,c]
5-Methylfurfural[a,c]
2-Methyltetrahydrofuran-3-one[a,c]
2-Acetylfuran[a,c]
5-Methyl-2-acetylfuran[a]

(*continued*)

TABLE 7.5 (continued)
Flavor and Volatile Compounds in Roasted Hazelnut

Fufural acetate[a]
Methyl furoate[a]
Fufuryl alcohol[a,c]
2-Furancarboxaldehyde[e]
3-Furanmethanol[e]
3-Hydroxy-2-methyl-4-pyrone[c]
2,5-Dimethyl-4-hydroxy-3(2H)-furanone[c]

Pyrazines
Pyrazine[a,c]
Methylpyrazine[a,e]
Ethylpyrazine[c,e]
Isopropylpyrazine[a]
Propylpyrazine[a]
2,5-Dimethylpyrazine[a,c,e]
2,6-Dimethylpyrazine[a,e]
2,3-Dimethylpyrazine[a,c]
2-Ethyl-6-methylpyrazine[a,e,f]
2-Ethyl-5-methylpyrazine[a,c,e]
2-Ethyl-3-methylpyrazine[a,e]
2,6-Diethylpyrazine[a]
2,5-Diethylpyrazine[a]
2,3-Diethylpyrazine[a]
2-Methyl-5-pentylpyrazine[a]
Trimethylpyrazine[a]
2-Ethyl-3,6-dimethylpyrazine[a,c,e]
2-Ethyl-3,5-dimethylpyrazine[a,c,e]
2,3-Diethyl-5-methyl-pyrazine[a]
2,5-Diethyl-3-methyl-pyrazine[a,c]
2,6-Diethyl-3-methyl-pyrazine[a,c]
Dimethyl isobutylpyrazine[a]
Triethylpyrazine[a,e]
Tetramethylpyrazine[a]
Diethyldimethylpyrazine[a]
Vinylpyrazine[a]
2-Methyl-5-acetylpyrazine[a]
Ethylacetylpyrazine[a]
6,7-Dihydro-*5H*-cyclopenta-pyrazine[a]
2-Methyl-6,7-dihydro-*5H*-cyclo-penta-pyrazine[a]
5-Methyl-6,7-dihydro-*5H*-cyclo-pentapyrazine[a]
2-Ethyl-6,7-dihydro-*5H*-cyclo-pentapyrazine[a]
2,5-Dimethyl-6,7-dihydro-*5H*-cyclopentapyrazine[a]
3,5-Dimethyl-6,7-dihydro-*5H*-cyclopentapyrazine[a]
2,3-Dimethyl-6,7-dihydro-*5H*-cyclopentapyrazine[a]
Quinoxaline[a]
5,6,7,8-Tetrahydroquinoxaline[a]
2-Methyl-5,6,7,8-tetrahydro-quinoxaline[a]
(2′-Furyl)pyrazine[a]

Pyrroles
Pyrrole[e]
2-Pentylpyrrole[a]
2-Isobutylpyrrole[a]
1-Methylpyrrole[a,e]
1-Acetylpyrrole[a]
2-Acetylpyrrole[a,c]
2-Propionylpyrrole[a]
1-Methylpyrrole-2-carboxaldehyde[a]
5-Methylpyrrole-2-carbox-aldehyde[a]
Pyrrole-2-carboxaldehyde[a,c]
1-Furfurylpyrrole[a,c]
1-Methyl-2-pyrrolidinone[a]
Indole[a]
N-Methy-2-pyrrole aldehyde[c]
2,5-Dihydropyrrole[e]

Pyridines
2-Pentylpyridine[a]
2-Acetylpyridine[a]
3-Acetylpyridine[a]
Methyl nicotinate[a]

Sulfur compounds
Methanethiol[a]
Dimethyl sulfide[b]
Dimethyl disulfide[a]
Diethyl disulfide[a]
Dimethyl trisulfide[a]
Methional[a]
Dihydro-1*H*-thiophen-3-one[a]
Thiophene-2-carbox-aldehyde[a]
4-Methyl-4-vinylthiazole[a]
Benzothiazole[a]
3,5-Dimethyl-1,2,4-trithiolan[a]
Dimethyltrithiolan (isomer)[a]

Miscellaneous
Chloroform[a]
Dichlorobenzene[a]
Trichloroethylene[f]
Benzonitrile[a]
Ionole[a]
Diethyl phthalene[a]
Pentyloxirane[a]
2-Methylpyrrole[a]
Thiazole[c]

Source: From Kinlin, T.E., Maralidhara, R., Pittet, A.O., Sanderson, A., and Walradt, J.R., *J. Agric. Food Chem.*, 20, 1021, 1972; Sheldon, R.M., Lindsay, R.C., and Libbey, L.M., *J. Food Sci.*, 37, 313, 1972; Kinderlerer, J.L. and Johnson, S., *J. Sci. Food Agric.*, 58, 89, 1992; Alasalvar, C., Shahidi, F., and Cadwallader, K.R., *J. Agric. Food Chem.*, 51, 5067, 2003; Alasalvar, C., Odabasi, A.Z., Demir, N., Balaban, M.Ö., Shahidi, F., and Cadwallader, K.R., *J. Food Sci.*, 69, SNQ99, 2004.

[a] Volatile isolated by four different techniques [24].
[b] Headspace analysis [25].
[c] Volatiles isolated by vacuum distillation [25].
[d] Volatiles isolated by Likens and Nickerson distillation [26].
[e] Volatiles isolated by dynamic headspace analysis [28].
[f] Volatiles isolated by dynamic headspace analysis [29].

TABLE 7.6
Flavor and Volatile Compounds in Hazelnut Oil

Limonene[b]	Decanal[a]	(*E*)-5-Methyl-2-hepten-4-one (filbertone)[a,b]
Sabinene[a]	(*Z*)-2-Octenal[b]	Fufural[a]
Terpineol[a]	(*E*)-2-Nonenal[b]	Furfuryl alcohol[a]
n-Octanol[a]	(Z)-2-Decenal[a]	Pentylfuran[a]
Ethyl 2-methylbutanoate[b]	(*E*)-2-Decenal[b]	Pyrazine[a]
Ethyl isobutyrate[b]	(*E*)-2-Undecenal[a,b]	2-Acetylpyrrole[a]
Hexanoic acid[b]	(*E,Z*)-2,4-Nonadienal[b]	2-Ethyl-3-methylpyrazine[b]
3-Methylbutanoic acid[b]	(*E,E*)-2,4-Nonadienal[b]	2-ethyl-3,5-Dimethylpyrazine[b]
(*E*)-β-Damascenone[b]	2,4-Decadienal (isomer)[a]	3-Ethyl-2,5-dimethylpyrazine[b]
4-Vinylguaiacol[b]	(*E,Z*)-2,4-Decadienal[b]	2,3-Diethyl-5-methylpyrazine[b]
Vanillin[b]	(*E,E*)-2,4-Decadienal[a,b]	2,5-(or 2,6)-Diethylpyrazine[b]
Haxanal[a,b]	Phenylacetaldehyde[a,b]	
Nonanal[a]	*trans*-4,5-Epoxy-(*E*)-2-decenal[b]	

Source: From Caja, M.M., Ruis del Castillo, M.L., Martínez Alvarez, R., Herraiz, M., and Blanch, G.P., *Eur. Food Res. Technol.*, 211, 45, 2000; Matsui, T., Guth, H., and Grosch, W., *Lipid*, 100, 51, 1998.

[a] Caja, M.M., Ruis del Castillo, M.L., Martínez Alvarez, R., Herraiz, M., and Blanch, G.P., Analysis of volatile compounds in edible oils using simultaneous distillation-solvent extraction and direct coupling of liquid chromatography with gas chromatography, *Eur. Food Res. Technol.*, 211, 45–51, 2000.

[b] Matsui, T., Guth, H., and Grosch, W., A comparative study of potent odorants in peanut, hazelnut, and pumpkin seed oils on the basis of aroma extract dilution analysis (AEDA) and gas chromatography–olfactometry of headspace samples (GCOH), *Lipid*, 100, 51–56, 1998.

[34–37] and only the Stuart variety has been studied. Wang and Odell [34], in roasted pecan, identified 19 carbonyl compounds including eight pyrazines, seven acids, five alcohols, one pyridine, and one lactone (Table 7.7). They concluded that most of these components were formed by the roasting process. In a different study, the decline in sensory quality of natural pecan was closely correlated with increases in the contents of hexanal, dodecane, tridecane, and tetradecane [36]. Horvat and Senter [37] characterized the volatiles associated with mildly oxidized pecan oil and identified 20 typical lipid degradation compounds (Table 7.7). The same laboratory examined the relationship of the phenolic acid content with the storage stability of natural pecan [38]. In a study that did not specifically analyze mature edible pecan, Mody et al. [35] identified the volatile compounds in pecan leaf and immature pecan kernel. They reported seven monoterpene hydrocarbons, four oxygenated mono-terpenoids, seven sesquiterpene hydrocarbons, and three sesquiterpene alcohols. The oils associated with the leaf and immature nut contained approximately 78% terpenoids.

7.2.5 Brazil Nut

Brazil nut (*Bertholletia excelsa* L.) is consumed either raw or roasted as a snack or used in a variety of confectionary products. Brazil nut contains a high amount of fat (65%–70%), which is rich in unsaturated fatty acids [39,40]. The only report on the volatile composition of Brazil nut was published by Clark and Nursten [41]. They compared both high vacuum distillation and Likens and Nickerson steam distillation for the extraction of the volatile compounds, and found the latter technique was more susceptible to artifact formation as evidenced by the presence of the lipid oxidation products 2,4-nonadienal and 2,4-decadienal in that extract. Both extracts were analyzed by gas chromatography–olfactometry (GCO) on both polar (Carbowax 20M) and nonpolar

TABLE 7.7
Flavor and Volatile Compounds in Pecan and Pecan Oil

Immature Pecan[a]	**Pecan Oil**[c]	**Roasted Pecan**[d]	
p-Cymene	*n*-Pentane	*Alcohols*	2-Ethyl-5-methylpyrazine
α-Phellandrene	*n*-Hexane	Ethanol	2,3,5-Trimethylpyrazine
α-Pinene	*n*-Heptane	*n*-Pentanol	2,5-Dimethyl-3-ethylpyrazine
Camphene	*n*-Octane	*n*-Hexanol	*Aldehydes/Ketones*
Sabinene	1-Hexene	*n*-Heptanol	Acetaldehyde
α-Terpinene	1-Dodecene	*n*-Octanol	Propanal
Limonene	1-Tetradecene	*Acids*	Butanal
Borneol	Toluene	Acetic acid	Pentanal
α-Terpineol	*n*-Propylbenzene	Propanoic acid	Hexanal
1,8-Cineole	*n*-Pentanol	Pentanoic acid	Heptanal
4-Terpineol	Acetaldehyde	4-Methylpentanoic acid	Octanal
α-Santalene	Butanal	Hexanoic acid	2-Hexenal
β-Caryophyllene	Pentanal	Heptanoic acid	2-Heptenal 2-decenal
Humulene	Hexanal	Octanoic acid	2-Undecenal
β_2-Bisabolene	Heptanal	*Basic compounds*	Acrolein
α-Ferulene	Octanal	Pyridine	2,4-Heptadienal
Cubenol	Decanal	Methylpyrazine	2,4-Decadienal
Natural Pecan[b]	Benzaldehyde	2,5-Dimethylpyrazine	Furfural
Hexanal	2-Butenal	2,6-Dimethylpyrazine	Glyoxal
Dodecane	Acetone	2,3-Dimethylpyrazine	Pyruvaldehyde
Tridecane	2-Heptanone	2-Ethyl-6-methylpyrazine	2,3-Butanedione
Tetradecane	Ethylphenylketone		2,3-Pentanedione

Source: From Wang, P.-S. and Odell, G.V., *J. Agric. Food Chem.*, 20, 206, 1972; Mody, N.V., Hedin, P.A., and Neel, W.W., *J. Agric. Food Chem.*, 24, 175, 1976; Forbus Jr., W.R., Senter, S.D., Lyon, B.G., and Dupuy, H.P., *J. Food Sci.*, 45, 1376, 1980; Horvat, R.J. and Senter, S.D., *J. Am. Oil Chem. Soc.*, 57, 111, 1980.

[a] Mody, N.V., Hedin, P.A., and Neel, W.W., Volatile components of pecan leaves and nuts, *Carya illinoensis* Koch, *J. Agric. Food Chem.*, 24, 175–177, 1976.

[b] Forbus Jr., W.R., Senter, S.D., Lyon, B.G., and Dupuy, H.P., Correlation of objective and subjective measurements of pecan kernel quality, *J. Food Sci.*, 45, 1376–1379, 1980.

[c] Horvat, R.J. and Senter, S.D., Identification of some volatile products from mildly oxidized pecan oil, *J. Am. Oil Chem. Soc.*, 57, 111, 1980.

[d] Wang, P.-S. and Odell, G.V., Characterization some volatile constituents of roasted pecans, *J. Agric. Food Chem.*, 20, 206–210, 1972.

(SE30) columns. A summary of the compounds identified is provided in Table 7.8. The compounds that were specifically linked to compounds detected by GCO included hexanal (*strong green* note), heptanal (*green* note), 2,4-nonadienal (*fatty/lard* note), and 2,4-decadienal (*fatty* note). The predominance of these volatiles can be explained by the oxidation of the abundant polyunsaturated fatty acids (PUFA) in Brazil nut [40].

7.2.6 Cashew

Cashew, the seed of the cashew tree (*Anacardium occidentale* L), is consumed as a snack and used as a substitute for peanut and almond in some confectionary applications. Cashew contains a high amount of fat (~43%–50%), which is composed mainly of oleic acid (~57%–65%) and appreciable levels of linoleic acid (~15%–19%) [42]. Information on the volatile components of cashews is very

TABLE 7.8
Flavor and Volatile Compounds in Brazil Nut

Alcohols	*Carbonyls*	*Hydrocarbons*	*Terpenes*
n-Ethanol (VD)[a]	2-Heptanone (VD)	*n*-Nonane (VD)	Limonene (LN, VD)
n-Butanol (LN, VD)	2-Nonanone (LN, VD)	*n*-Decane (LN, VD)	*Miscellaneous*
n-Pentanol (LN, VD)	2-Decanone (VD)	*n*-Undecane (LN, VD)	Ethyl acetate (VD)
2-Methyl-2-butanol (VD)	2-Undecanone (VD)	*n*-Dodecane (LN, VD)	Benzofuran (LN)
n-Hexanol (LN, VD)	2-Dodecanone (VD)	*n*-Tridecane (LN, VD)	Methylbenzofuran (LN, VD)
n-Octanol (LN)	Hexanal (LN, VD)	*n*-Tetradecane (LN, VD)	Cyanobenzene (VD)
2-Octanol (VD)	Heptanal (LN, VD)	Toluene (LN)	Chloroform (VD)
2-Nonanol (VD)	2,4-Nonadienal (LN)	Ethylbenzene (LN)	
Phenol (VD)	2,4-Decadienal (isomers) (LN)	Styrene (LN)	
Cresol (LN)	Benzaldehyde (LN, VD)	*n*-Butylbenzene (VD)	
		1,2,3-Trimethylbenzene (VD)	
		1,2,4-Trimethylbenzene (LN, VD)	
		1,3,5-Trimethylbenzene (VD)	
		Naphthalene (LN, VD)	
		2-Methylnaphthalene (LN)	

Source: From Clark, R.G. and Nursten, H.E., *J. Sci. Food Agric.*, 27, 713, 1976.
[a] Compound identified in Likens and Nickerson (LN) or vacuum distillation (VD) extracts.

limited. Coleman et al. [43] reported 18 headspace volatiles in heat-treated aqueous extracts of cashew (Table 7.9). Because of their low odor detection thresholds, the compounds methylpropanal, 2-methylbutanal, 3-methylbutanal, hexanal, and phenylacetaldehyde may contribute to the aroma of cashew.

7.2.7 Chestnut

Chestnut (*Castanea* sp.) is consumed in a number of preparations including roasted, boiled, and candied. The sensory quality and volatile composition of roasted chestnut has been studied [44–47]. In contrast to the other tree nuts discussed in this chapter, chestnut is comparatively low in fat and protein and high in carbohydrate [44]. This, at first, might seem to have important ramifica-

TABLE 7.9
Flavor and Volatile Compounds in Heated Aqueous Extract of Cashew

Propanal (7.59)[a]	3-Methylbutanal (11.49)	Furfural (7.44)
Acetone (1.69)	2-Methylbutanal (8.08)	Dimethylpyrazines (0.44)
Methylpropanal (30.48)	1-Hydroxy-2-propanone (1.9)	Acetylfuran (0.49)
2-Methylfuran (13.95)	Hexanal (0.35)	Benzaldehyde (0.24)
Butanone (1.62)	Methylpyrazine (0.68)	5-Methylfurfural (0.42)
2-Pentanone (7.62)	Dihydro-2-methyl-3(2H)-furanone (0.45)	Phenylacetaldehyde (0.46)

Source: From Coleman III, W.M., White Jr., J.L., and Perfetti, T.A., *J. Agric. Food Chem.*, 42, 190, 1994.
[a] Percent peak area.

tions in regards to volatile composition, but despite the relatively low fat content, chestnut still has appreciable levels of linoleic and linolenic acids [45,46] that can undergo oxidative and thermal degradation to form many of the same volatiles found in other tree nuts. The published literature on the volatile composition of roasted chestnut is limited [45,47] and no study has gone into enough depth to provide much insight into the important aroma constituents of this nut. The volatile compounds identified to date in roasted chestnuts are listed in Table 7.10. It is difficult to speculate as to which compounds contribute to roasted chestnut aroma since no quantitative data are available. However, certain compounds with burnt sugar and caramel-like

TABLE 7.10
Flavor and Volatile Compounds in Roasted Chestnut

Roasted and Boiled Chinese Chestnuts[a]		Italian Chestnut[b,c]	
Roasted chestnut (only)	*Common to both types*	α-Pinene[c]	Methyl 2-butenoic acid[b]
3-Hydroxy-2-butanone	Cyclohexene	3-Carene[c]	Acetic acid[b]
2,4-Hexadien-1-ol	2-Ethenyl-2-butenal	α-Thujene[b]	Hexanoic acid[c]
4-Hydroxy-2-methyl-1,3-cyclopentane-dione	Cyclohexanol	Sabinene[b]	Furfural[b,c]
3-Penten-3-one	3-Cyclohexen-1-ol	α-Phellandrene[b]	2-Pentylfuran[b,c]
2,5-Dimethyl-4-hydroxy-3(2*H*)-furanone	2-Cyclohexen-1-one	α-Terpinene[b]	Hexanal[b,c]
3-Hydroxy-2-methyl-4-pyrone (maltol)	2-(2-Methoxyethoxy)ethanol	*p*-Cymene[b]	Heptanal[b,c]
1-(3-Methoxyphenyl)ethanone	2,2′-Oxybis(ethanol)	γ-Terpinene[b]	Octanal[c]
Dodecanoic acid	5-Ethyl-2(5H)-furanone	α-Terpinolene[b]	Benzaldehyde[b,c]
2,4-Dimethylpyrrole	2,3-Dihydro-3,5-dihydroxy-6-methyl-4H-pyran-4-one	Carvone[b]	2-Heptenal[b]
7-Hyroxy-6-methoxy-1-benzopyran-2(2*H*)-one	5-(Hydroxymethyl)furfural	Ethylbenzene[b]	Phenylacetaldehyde[b]
Methyl 9,12-octadecadienoate	2,3-Dihydrobenzofuran	Styrene[b]	3-Penten-2-one[b]
9-Octadecanoic acid	4-Hydroxy-4-(3oxo-1-butenyl)-3,5,6-trimethyl-2-cyclohexen-1-one	2,3-Butanediol[c]	3-Hydroxy-2-butanone[c]
Docosane	Hexadecanoic acid	*n*-Hexanol[b,c]	4-Methyl-2-pentanone[c]
Dioctyl hexanedioate	9,12-Octadecadienoic acid	*n*-Heptanol[c]	2-Heptanone[b,c]
Boiled chestnut (only)	Octadecanoic acid	1,6-Heptadien-4-ol[b]	3-Hepten-2-one[b,c]
n-Heptane	Eicosanoic acid	1-Octen-3-ol[b]	γ-Butyrolactone[b,c]
Hexanal	Docosanoic acid	2-Phenylethanol[b]	2,3-Dimethylpyrazine[b]
2-Decenal	Heptacosane	Ethyl acetate[b]	2-Acetylthiazoline[b]
2-Undecenal	Cyclotetracosane	Hexyl acetate[b]	
n-Tridecane	1-Docosanol		
Eicosene			
2,5-Dihydro-3-methylfurandione			
2-Methoxy-6-methylpyrazine			
6,7-Dimethoxy-2H-1-benzopyran-2-one			
Mono(2-ethylhexyl) hexanedioate			

Source: From Morini, G. and Maga, J.A., *Lebensm. Wiss. u.-Technol.*, 28, 638, 1995; Krist, S., Unterweger, H., Bandion, F., and Buchbauer, G., *Eur. Food Res. Technol.*, 219, 470, 2004.

[a] Volatiles extracted by CH_2Cl_2 and analyzed by GC-MS [45].

[b] Headspace volatiles analyzed by solid-phase microextraction (SPME)-GC-MS [47].

[c] Volatiles extracted by CH_2Cl_2 and the resulting oily extract analyzed by SPME-GC-MS [47].

notes [e.g., 2,5-dimethyl-4-hydroxy-3(*2H*)-furanone and maltol] may have some impact on the characteristic aroma of roasted chestnut.

7.2.8 Pistachio

Pistachio (*Pistacia vera* L.) is consumed either raw or roasted. It is also used as ingredients in desserts, such as baklava and ice cream, and for the production of natural flavors and flavored oils. Like many tree nuts, pistachio is high in fat and a rich source of linoleic and linolenic acids. It is, therefore, prone to lipid oxidation and development of rancid flavors. There are several reports on the chemical composition of pistachio [48–51]. The effects of processing and storage on the composition, fatty acids, and some quality attributes of pistachio also have been reported [52–55]. However, the published literature on the volatile components of pistachio is scarce and only one such study has been published [56]. In this study, the volatiles were extracted from roasted pistachio using the procedure of Takei et al. [4], which was previously described in the earlier section devoted to almond (Section 7.2.1). The resulting aroma extract was separated into neutral/acidic and basic factions and analyzed by GC-MS. The neutral/acidic fraction was reported to have a sweet, roast aroma, while the basic fraction had a strong, nutty note. A total of 18 volatile compounds were identified (Table 7.11). Among the 12 neutral/acid compounds, furfural was in highest abundance followed by 2-hexenal, ethanol, and 2-methyl-2-hexenal. Some minor components such as 2,4-decadienal and 2-nonenal might be important in pistachio aroma because these compounds have low odor detection thresholds. The basic fraction contained mainly 2,5-dimethylpyrazine, and low amounts of some other pyrazines (2-methylpyrazine, 2-ethyl-5-acetylpyrazine, and 2,6-dimethylpyrazine) and pyrroles (pyrrole and 2-pentylpyrrole). All of the reported basic components have relatively high odor thresholds and might make only minor contributions to the overall aroma of roasted pistachio.

7.2.9 Macadamia

Generally consumed roasted, macadamia (*Macadamia integrifolia*) was once considered a gourmet item due to limited availability. Macadamia is now readily available as a snack and is used extensively in baking and confectionary applications. To our knowledge, only one study has been published on the volatile component of macadamia. Crain and Tang [57] employed two techniques for the isolation of the volatile compounds from dry roasted (177°C, 20min) macadamia; (1) direct solvent extraction with CH_2Cl_2 followed compound class fractionation into neutral and basic components and (2) headspace analysis. Forty volatiles were identified by GC-MS (Table 7.12). These consisted of 11 highly volatile headspace, 22 neutral, and seven basic compounds. Methyl sulfide, methylpropanal, methyl disulfide, and 2- and 3-methylbutanal are noteworthy among the headspace volatiles since these compounds have

TABLE 7.11
Flavor and Volatile Compounds in Roasted Pistachio

Neutral/Acidic Extract		Basic Extract
2-Fufural (53.30)[a]	(*E,E*)-2,4-Decadienal (1.06)	2,5-Dimethylpyrazine (92.91)
2-Hexenal (14.20)	2-Nonenal (0.70)	2-Ethyl-5-acetylpyrazine (2.50)
Ethanol (13.53)	2-Acetylfuran (0.14)	2-Methylpyrazine (2.30)
2-Methyl-2-hexenal (8.85)	5-Methylfurfural (0.14)	2,6-Dimethylpyrazine (0.80)
Furfuryl alcohol (2.24)	Nonenal *isomer* (trace)	Pyrrole (0.56)
Acetic acid (1.13)		2-Pentylpyrrole (0.07)

Source: From Mervat, M., Soliman, A., Osman, F., and El-Sawy, A., *Agric. Biol. Chem.*, 45, 2123, 1981.
[a] Percent peak area.

TABLE 7.12
Flavor and Volatile Compounds in Roasted Macadamia

Headspace	Neutral Extract	Basic Extract
n-Hexane (S)[a]	*p*-Xylene (S)	2-Methylpyrazine (S)
n-Heptane (M)	*p*-Cymene (S)	2,5-Dimethylpyrazine (L)
n-Octane (M)	2-Pentylfuran (S)	2,3-Dimethylpyrazine (S)
Benzene (S)	2-Methyltetrahydrofuran-3-one (S)	2-Ethyl-5-methylpyrazine (S)
Toluene (S)	2-Fufural (S)	2,3,5-Trimethylpyrazine (M)
Methylfuran (S)	*n*-Pentanol (S)	2-Ethyl-3,5-dimethyl-pyrazine (L)
Methyl sufide (L)	*n*-Hexanol (M)	2,5-Diethyl-3-methyl-pyrazine (S)
Methyl disulfide (S)	*n*-Heptanol (M)	
Methylpropanal (L)	*n*-Octanol (M)	
2-Methylbutanal (L)	Phenol (M)	
3-Methylbutanal (L)	2-Phenylethanol (S)	
	2,3-Pentanedione (S)	
	2-Heptanone (S)	
	3,5,5-Trimethyl-2-cyclo-hexen-1-one (S)	
	1-Phenyl-2-propa-none (M)	
	n-Hexanal (M)	
	n-Heptanal (M)	
	n-Octanal (M)	
	Methional (S)	
	Benzaldehyde (S)	
	Phenylacetaldehyde (L)	
	2-Phenyl-2-butanal (M)	

Source: From Crain Jr., W.O. and Tang, C.S., *J. Food Sci.*, 40, 207, 1975.

[a] Letters in parentheses represent the relative size of the GC-MS peak (S = small; M = medium; L = large).

been implicated as important aroma compounds in various kinds of heated foods. Methyl sulfide, in particular, was found at a high level in the headspace and could be important in macadamia aroma. The neutral fraction contained several aldehydes and ketones that might, depending upon its concentrations, contribute to the overall aroma of roasted macadamia. Among the seven pyrazines identified in the basic fraction, 2-ethyl-3,6-dimethylpyrazine and 2,5-diethyl-3-methylpyrazine are reported to have low odor detection thresholds [58] and are, therefore, probably important in the aroma of roasted macadamia.

7.2.10 Pine Nut

Some species of pine (*Pinus* sp.) are capable of producing seeds or nuts that are large enough for practical food use. Pine nut can be eaten as snacks or used as an essential ingredient of many foods, including Italian pesto and some Middle Eastern dishes such as kibbeh. Pine nut is high in PUFA [59–61] and is thus prone to oxidative rancidity. Pinolenic acid, a positional isomer of γ-linolenic acid, is found exclusively in pine nut and at appreciable levels in Siberian pine (*Pinus sibirica*). The oxidation of this unsaturated fatty acid might give rise to some unique volatile compounds in pine nut. Maga [3], in his review of the volatile compounds in nuts, cited one study that reported that pyrazines increased from 15 μg/g in raw nut to 29 μg/g after roasting [62]. There appear to be no other reports on the volatile components of pine nut.

7.3 FLAVOR AND VOLATILE FORMATION IN TREE NUTS

7.3.1 Natural Nuts

Soon after harvest, tree nuts, because of their high concentrations of unsaturated fatty acids, may undergo development of oxidative rancidity. This leads to the formation of undesirable rancid flavors and a decline in both unsaturated fatty acids (e.g., oleic, linoleic, and linolenic acids) and natural antioxidants (e.g., tocopherols) [63]. The postharvest stability and sensory quality of tree nuts are influenced by several factors such as chemical composition (e.g., fatty acid composition and presence of antioxidants such as tocopherols), moisture content, oxygen concentration, and temperature, among others. There is some evidence that lipid oxidation is at least in part due to the action of oxidative enzymes, such as lipoxygenases. This is supported by the fact that mild to moderate heat treatment of some nuts, such as pecan, retards the development of rancid flavors during storage [64,65]. Nevertheless, mild oxidation is probably necessary for the development of the characteristic volatile flavor components of natural tree nuts [63]. In addition to the lipid oxidation volatiles, some other compounds, such as terpenes, lactones, and short-chain volatile acids, may impact the aroma profiles of some types of natural tree nuts.

7.3.2 Roasted Nuts

The roasting of tree nuts catalyzes the formation of numerous volatile compounds by three main pathways: (1) Maillard reaction (including Strecker degradation) between amino acids and sugars; (2) thermal breakdown or carmelization of sugars; and (3) lipid degradation, including both thermal (pyrolysis) and oxidative reactions [43]. These reactions work in concert to produce the characteristic volatile compounds that comprise the aroma profiles of the various types of roasted tree nuts. Many factors influence the thermal generation of volatiles in tree nuts. The chemical composition of the nut is very important. This includes types and levels of aroma precursors (sugars, amino acids, fatty acids, etc.), presence of prooxidants and antioxidants, moisture content, and pH. In addition, the conditions of roasting (time and temperature) are also important.

The Maillard reaction between reducing sugars and amino acids under specific conditions (pH, water activity, and temperature) is primarily responsible for the production of heterocyclic volatile compounds such as pyrazines, pyridines, pyrroles, furans, and the Strecker aldehydes. Maillard reactions produce many potent aroma compounds identified in some roasted tree nuts, including 3-methylbutanal, 2,3-butanedione, methional, phenylacetaldehyde, 2-ethyl-3,5-dimethylpyrazine, and 2,5-dimethyl-4-hydroxy-3(2*H*)-furanone, among others.

In addition to Maillard reactions, caramelization reactions, involving the dehydration and decomposition of sugars, can form a variety of volatile compounds found in roasted tree nuts such as heterocyclic oxygen-containing furan derivatives and maltol.

Lipid oxidation is a major pathway for the formation of volatile compounds in roasted tree nuts. After the enzymes in the nuts (e.g., lipoxygenase) are inactivated by the high temperatures used for roasting, autoxidation becomes the principal source of lipid breakdown [66]. Lipid degradation reactions are not necessarily deleterious to flavor [67], and some aldehydes and ketones (e.g., *n*-aldehydes and 2-alkanones) produced during roasting may impart desirable sweet, fruity, and pungent aroma notes to roasted tree nuts.

7.4 CONCLUSION

This review demonstrates that over the past four decades, considerable effort has gone into the determination of the volatile compounds in natural and roasted tree nuts. It is also evident that lipid oxidation plays a predominant role in the generation of the characteristic volatile components of all types of natural tree nuts. Unfortunately, there is insufficient information available to clearly define the key aroma components of most of the natural tree nuts reviewed in this chapter. In general, roasted tree nuts have received more attention than their natural counterparts. In the case of roasted nuts, the volatile profiles are highly complex and are composed of compounds arising not only from lipid oxidation, but also from Maillard reaction, Strecker degradation, and caramelization of sugars. With the exception of hazelnut, there is not enough information available to accurately define the characteristic aroma components of roasted tree nuts. Additional studies are needed in order to fully characterize the aroma components of both natural and roasted tree nuts.

REFERENCES

1. Dreher, M.L., Maher, C.V., and Kearney, P., The traditional and emerging role of nuts in healthful diets, *Nutr. Rev.*, 54, 241–245, 1996.
2. Mukuddem-Peterson, J., Oosthuizen, W., and Jerling, J.C., A systematic review of the effects of nuts on blood lipid profiles in humans, *J. Nutr.*, 135, 2082–2089, 2005.
3. Maga, J.A., Nuts, in *Volatile Compounds in Foods and Beverages*, Maarse, H., Ed., Marcel Dekker Inc., New York, 1991, pp. 671–688.
4. Takei, Y., Shimada, K., Watanabe, S., and Yamanishi, T., Volatile components of roasted almonds: Basic fraction, *Agric. Biol. Chem.*, 38, 645–648, 1974.
5. Takei, Y. and Yamanishi, T., Flavor components of roasted almond, *Agric. Biol. Chem.*, 38, 2329–2336, 1974.
6. Pićurić-Jovanović, K. and Milovanović, M., Analysis of volatile compounds in almond and plum kernel oils, *J. Am. Oil Chem. Soc.*, 70, 1101–1104, 1993.
7. Caja, M.M., Ruis del Castillo, M.L., Martínez Alvarez, R., Herraiz, M., and Blanch, G.P., Analysis of volatile compounds in edible oils using simultaneous distillation-solvent extraction and direct coupling of liquid chromatography with gas chromatography, *Eur. Food Res. Technol.*, 211, 45–51, 2000.
8. García-López, C., Grané-Turuel, N., Berenguer-Navarro, V., García-García, J.E., and Martín-Carratalá, M.L., Major fatty acid composition of 19 almond cultivars of different origins. A chemometric approach, *J. Agric. Food Chem.*, 44, 1751–1755, 1996.
9. Severini, C., Gomes, T., De Pilli, T., Romani, S., and Massini, R., Autoxidation of packed almonds as affected by Maillard reaction volatile compounds derived from roasting, *J. Agric. Food Chem.*, 48, 4635–4640, 2000.
10. Karantzis, I., Nanos, G.D., and Stavroulakis, G.G., Effect of harvest time and storage conditions on almond kernel oil and sugar composition, *J. Sci. Food Agric.*, 83, 354–359, 2003..
11. Sze-Tao, K.W.C. and Sathe, S.K., Wanuts (*Jugans regia* L.): Proximate composition, protein solubility, protein amino acids composition and protein in vitro digestibility, *J. Sci. Food Agric.*, 80, 1393–1401, 2000.
12. Savage, G.P., Chemical composition of walnuts (*Juglans regia* L.) grown in New Zealand, *Plant Foods Human Nutr.*, 56, 75–82, 2001.
13. Crowe, T.D., Crowe, T.W., Johnson, L.A., and White, P.J., Impact of extraction method on yield of lipid oxidation products from oxidized and unoxidized walnuts, *J. Am. Oil Chem. Soc.*, 79, 453–456, 2002.
14. Clark, R.G. and Nursten, H.E., Volatile flavour components of walnuts (*Juglans regia* L.), *J. Sci. Food Agric.*, 27, 902–908, 1976.
15. Clark, R.G. and Nursten, H.E., The sensory analysis and identification of volatiles from walnut (*Juglans regia* L.) headspace, *J. Sci. Food Agric.*, 28, 69–77, 1977.
16. Elmore, J.S., Nisyrios, I., and Mottram, D.S., Analysis of the headspace aroma compounds of walnuts (*Juglans regia* L.), *Flavour Fragr. J.*, 20, 501–506, 2005.
17. Crowe, T.D. and White, P.J., Oxidation, flavor, and texture of walnuts reduced in fat content by supercritical carbon dioxide, *J. Am. Oil Chem. Soc.*, 80, 569–574, 2003.
18. Torres, M.M., Martínez, M.L., and Maestri, D.M., A multivariate study of the relationship between fatty acids and volatile flavor components in olive and walnut oils, *J. Am. Oil Chem. Soc.*, 82, 105–110, 2005.
19. Al-Bachir, M., Effect of gamma irradiation on fungal load, chemical and sensory characteristics of walnuts (*Juglans regia* L.), *J. Stored Prod. Res.*, 40, 355–362, 2004.
20. Pfnuer, P., Matsui, T., Grosch, W., Guth, H., Hofmann, T., and Schieberle, P., Development of a stable isotope dilution assay for the quantification of 5-methyl-(*E*)-2-hepten-4-one: Application to hazelnut oils and hazelnuts, *J. Agric. Food Chem.*, 47, 2044–2047, 1999.

21. Langourieux, S., Perren, R., and Escher, F., Infuence of processing parameters on the aroma of dry-roasted hazelnuts, in *Frontiers of Flavour Science,* Schieberle, P. and Engel, K.H., Eds., Deutsche Forschungsanstalt für Lebensmittelchemie, Garching, Germany, 2000, pp. 527–581.
22. Emberger, R., The role of nature-identical aroma compounds in food flavorings today (in German), in *Proceedings of the 43rd Meeting of the Forschungskreis der Ernährungsindustrie,* Bremen, Germany, 1985, pp. 41–60.
23. Güntert, M., Emberger, R., Hopp, R., Köpsel, M., Silberzahn, W., and Werkhoff, P., Chirospecific analysis in flavor and essential oil chemistry. Part A. Filbertone—the character impact odorant of hazelnuts, *Z. Lebesm. Unters. Forsch.*, 192, 108–110, 1991.
24. Kinlin, T.E., Maralidhara, R., Pittet, A.O., Sanderson, A., and Walradt, J.R., Volatile components of roasted filbers, *J. Agric. Food Chem.*, 20, 1021–1028, 1972.
25. Sheldon, R.M., Lindsay, R.C., and Libbey, L.M., Identification of volatile flavor compounds from roasted filberts, *J. Food Sci.*, 37, 313–316, 1972.
26. Kinderlerer, J.L. and Johnson, S., Rancidity in hazelnuts due to volatile aliphatic aldehydes, *J. Sci. Food Agric.*, 58, 89–93, 1992.
27. Wickland, S.E., Johnston, J.J., and Stone, M.B., Evaluation of roasted and natural hazelnut volatiles by purge-and-trap/gas chromatography/mass spectrometry, Presented at the Institute of Food Technologists Annual Meeting and Food Expo, New Orleans, LA, June 23–27, 2001, Abstract 88B-17.
28. Alasalvar, C., Shahidi, F., and Cadwallader, K.R., Comparison of natural and roasted Turkish Tombul hazelnut (*Corylus avellana* L.) volatiles and flavor by DHA/GC/MS and descriptive sensory analysis, *J. Agric. Food Chem.*, 51, 5067–5072, 2003.
29. Alasalvar, C., Odabasi, A.Z., Demir, N., Balaban, M.Ö., Shahidi, F., and Cadwallader, K.R., Volatiles and flavor of five Turkish hazelnut varieties as evaluated by descriptive sensory analysis, electronic nose, and dynamic headspace analysis/gas chromatography–mass spectrometry, *J. Food Sci.*, 69, SNQ99–SNQ106, 2004.
30. Matsui, T., Guth, H., and Grosch, W., A comparative study of potent odorants in peanut, hazelnut, and pumpkin seed oils on the basis of aroma extract dilution analysis (AEDA) and gas chromatography–olfactometry of headspace samples (GCOH), *Lipid*, 100, 51–56, 1998.
31. Wakeling, L.T., Mason, R.L., D'Arcy, B.R., and Caffin, N.A., Composition of pecan cultivars Wichita and Western Schley [*Carya illinoinensis* (Wangenh.) K. Koch] grown in Australia, *J. Agric. Food Chem.*, 49, 1277–1281, 2001.
32. Singanusong, R., Mason, R.L., and D'Arcy, B.R., Compositional changes of Australia-grown Western Schley pecans [*Carya illinoinensis* (Wangenh.) K. Koch] during maturation, *J. Agric. Food Chem.*, 51, 406–412, 2003.
33. Venkatachalam, M. and Sathe, S.K., Chemical composition of selected edible nut seeds, *J. Agric. Food Chem.*, 54, 4705–4714, 2006.
34. Wang, P.-S. and Odell, G.V., Characterization some volatile constituents of roasted pecans, *J. Agric. Food Chem.*, 20, 206–210, 1972.
35. Mody, N.V., Hedin, P.A., and Neel, W.W., Volatile components of pecan leaves and nuts, *Carya illinoensis* Koch, *J. Agric. Food Chem.*, 24, 175–177, 1976.
36. Forbus Jr., W.R., Senter, S.D., Lyon, B.G., and Dupuy, H.P., Correlation of objective and subjective measurements of pecan kernel quality, *J. Food Sci.*, 45, 1376–1379, 1980.
37. Horvat, R.J. and Senter, S.D., Identification of some volatile products from mildly oxidized pecan oil, *J. Am. Oil Chem. Soc.*, 57, 111, 1980.
38. Senter, S.D., Horvat, R.J., and Forbus Jr, W.R., Relation between phenolic acid content and stability of pecans in accelerated storage, *J. Food Sci.*, 45, 1380–1382, 1392, 1980.
39. Woodroof, J.G., *Tree Nuts: Production, Processing, Products,* 2nd ed., AVI Publishing Company, Inc., Westport, CT, 1979.
40. Assunção, F.P., Bentes, M.H.S., and Serruya, H., A comparison of the stability of oils from Brazil nut, Para rubber and passion fruit seeds (*Bertholletia excelsa, Hevea brasiliensis, Passiflora edulis*), *J. Am. Oil Chem. Soc.*, 61, 1031–1036, 1984.
41. Clark, R.G. and Nursten, H.E., Volatile flavour components of Brazil nuts *Bertholletia exelsa* (Humb. and Bonpl.), *J. Sci. Food Agric.*, 27, 713–720, 1976.
42. Gallina Toschi, T., Caboni, M.F., Penazzi, G., Lercker, G., and Capella, P., A study of cashew nut oil composition, *J. Am. Oil Chem. Soc.*, 70, 1017–1020, 1993.
43. Coleman III, W.M., White Jr., J.L., and Perfetti, T.A., Characteristics of heat-treated aqueous extracts of peanuts and cashews, *J. Agric. Food Chem.*, 42, 190–194, 1994.

44. Senter, S.D., Payne, J.A., Miller, G., and Anagnostakis, S.L., Comparison of total lipids, fatty acids, sugars and nonvolatile organoleptic acids in nuts from four *Castanea* species, *J. Sci. Food Agric.*, 65, 223–227, 1994.
45. Morini, G. and Maga, J.A., Volatile compounds in roasted and boiled Chinese chestnuts (*Castanea molissima*), *Lebensm. Wiss. u.-Technol.*, 28, 638–640, 1995.
46. Künsch, U., Schärer, H., Patrian, B., Höhn, E., Conedera, M., Sassella, A., Jermini, M., and Jelmini, G., Effects of roasting on chemical composition and quality of different chestnut (*Casanea sativa* Mill) varieties, *J. Sci. Food Agric.*, 81, 1106–1112, 2001.
47. Krist, S., Unterweger, H., Bandion, F., and Buchbauer, G., Volatile compound analysis of SPME headspace and extract samples from roasted Italian chestnuts (*Castanea sativa* Mill.) using GC-MS, *Eur. Food Res. Technol.*, 219, 470–473, 2004.
48. Kamangar, T. and Farsam, H., Composition of pistachio kernels of various Iranian origins, *J. Food Sci.*, 42, 1135–1136, 1977.
49. Shokraii, E.H., Chemical composition of the pistachio nuts (*Pistacia vera* L.) of Kerman, Iran, *J. Food Sci.*, 42, 244–245, 1977.
50. Luh, B.S., Wong, W.S., and El-Shimi, N.E., Effects of processing on some chemical constituents of pistachio nuts, *J. Food Qual.*, 5, 33–41, 1981.
51. Garcia, J.M., Agar, I.T., and Steif, J., Fat content and fatty acid composition in individual seeds of pistachio varieties grown in Turkey, *Gartenbauwissenschaft*, 57, 130–133, 1992.
52. Kadar, A.A., Heintz, C.M., Labavitch, J.M., and Rae, H.L., Studies related to the description and evaluation of pistachio nut quality, *J. Am. Soc. Hort. Sci.*, 107, 812–816, 1982.
53. Kashani, G.G. and Valadon, L.R.G., Effect of salting and roasting on the lipids of Iranian pistachio kernels, *J. Food Technol.*, 18, 461–467, 1983.
54. Kashani, G.G. and Valadon, L.R.G., Effect of salting and roasting on the lipids carbohydrates and proteins of Iranian pistachio kernels, *J. Food Technol.*, 19, 247–253, 1984.
55. Maskan, M. and Karataş, Ş., Fatty acid oxidation of pistachio nuts stored under various atmospheric conditions and different temperatures, *J. Sci. Food Agric.*, 77, 334–340, 1998.
56. Mervat, M., Soliman, A., Osman, F., and El-Sawy, A., Volatile components of roasted pistachio-nut, *Agric. Biol. Chem.*, 45, 2123–2125, 1981.
57. Crain Jr., W.O. and Tang, C.S., A research note: Volatile components of roasted macadamia nuts, *J. Food Sci.*, 40, 207–208, 1975.
58. Guadagni, D.G., Butter, R.G., and Turnbaugh, J.G., Odour thresholds and similarity ratings of some potato chip components, *J. Sci. Food Agric.*, 23, 1435–1444, 1972.
59. Kim, M., Rhee, S.H., and Cheigh, H.S., Lipid contents and fatty acid composition of three lipid classes in Korean pine nut, *J. Kor. Soc. Food Nutr.*, 13, 406–412, 1984.
60. Yoon, T.-H., Im, K.-J., Koh, E.T., and Ju, J.-S., Fatty acid composition of *Pinus koraiensis* seed, *Nutr. Res.*, 9, 357–361, 1989.
61. Sagrero-Nieves, L., Fatty acid composition of Mexican pine nut (*Pinus cembroides*) oil from three seed coat phenotypes, *J. Sci. Food Agric.*, 59, 413–414, 1992.
62. Carballido, A., Villanua, L., Valdehita, M.T., and Robisco, M.A., Quantitative determination of pyrazines in foods, *Anal. Bromatol.*, 26, 391–419, 1974 (cited by Maga, J.A., Nuts, in *Volatile Compounds in Foods and Beverages*, Maarse, H., Ed., Marcel Dekker Inc., New York, 1991, pp. 671–688).
63. Rudolph, C.J., Odell, G.V., Hinrichs, H.A., Thomson, H.J., and Kays, S.J., Chemical changes in pecan oils during oxidation, *J. Food Qual.*, 15, 279–293, 1992.
64. McGlamery, J.B. and Hood, M.P., Effect of two heat treatments on rancidity development in unshelled pecans, *Food Res.*, 16, 80–84, 1951.
65. Forbus Jr., W.R. and Senter, S.D., Conditioning pecans with steam to improve shelling efficiency and storage stability, *J. Food Sci.*, 41, 794–798, 1976.
66. St. Angelo, A.J., Kuck, J.C., and Ory, R.L., Role of lipoxygenase and lipid oxidation in quality of oilseeds, *J. Agric. Food Chem.*, 27, 229–234, 1979.
67. Ho, C-T. and Chen, Q., Lipids in food flavors: An overview, in *Lipids in Food Flavors*, Ho, C.T. and Hartman, T.G. Eds., ACS Symposium Series No. 558, American Chemical Society, Washington, DC, 1994, pp. 2–14.

8 Almond and Almond Products: Nutraceutical Components and Health Effects

Fereidoon Shahidi, Ying Zhong, Subhashinee S.K. Wijeratne, and Chi-Tang Ho

CONTENTS

8.1 INTRODUCTION

Almond, scientifically known as *Prunus dulcis* (Mill.) D.A. Webb (synonyms: *Prunus amygdalus* or *Amygdalus communis*) belongs to the family Rosaceae and is related to stone fruits such as peaches, plums, and cherries. It is the number one tree nut produced on a global basis. According to the Food and Agriculture Organization (FAO), the world production of almond is 1,725,638 MT [1], and the top five almond producers in 2004 were the United States at 761,286 MT (45% of the world's production), followed by Spain, Syria, Italy, and Iran. Almond seeds, whole nuts or with skin peeled, are consumed as snacks or used as ingredients for processed foods such as various bakeries, confectioneries, and chocolates, whereas the inedible counterparts, including hull, shell, and skin are discarded or used as fuel material or livestock feed [2].

Almond is highly nutritious, rich in monounsaturated fatty acids (MUFA), and packed with a wide variety of vitamins and minerals (see Chapter 2). The U.S. Food and Drug Administration (FDA) has defined almond as an excellent source of vitamin E and manganese [3]. Energy provided by almond is basically from its fat content, which is mainly composed of MUFA, particularly oleic acid (18:1ω9). Almond is also a good source of protein, which is high in arginine and possesses good digestibility [4]. In addition to its nutritional value, almond also contains a number of phytochemicals that are responsible for a variety of bioactivities, among which antioxidant activity has been frequently studied. Almond is one of the most important natural sources of antioxidants that are known to possess health-promoting properties. Phenolic compounds, including phenolic acids and flavonoids, are believed to play an important role in antioxidant effectiveness of almond. Almond and its by-products such as hull, shell, and skin have been reported to have powerful free radical scavenging capacities [5]. Almond skin, while representing only approximately 4% of the total weight, contains 70%–100% of total phenolics present in the nut [6]. Almond hulls, which are produced as a by-product in large quantities, can also serve as a potential source of antioxidants. Other phytochemicals, such as terpenoids, sterols, and sphingolipids as well as their related functional properties have also been reported.

Not only regarded as a good source of nutrients, almond has recently been associated with a number of health benefits. These include cholesterol-lowering effect, anticancer activity, prevention of cardiovascular disease (CVD), and other chronic diseases. More details on health benefits of almond will be given later in this chapter.

8.2 PHENOLICS IN ALMOND

Phenolics are commonly present in both edible and inedible parts of plants. They act as antioxidants in food systems in order to minimize rancidity and protect cells from oxidative stress in the body. With the growing interest in replacing synthetic antioxidants with natural alternatives, many plant materials including tree nuts have been explored for their phenolic contents and antioxidant efficacies. A variety of phenolic compounds such as phenolic acids, flavonoids, and other polyphenols have been isolated from almond and identified (Table 8.1).

8.2.1 Solvent Extraction of Almond Phenolics

An effective and reliable extraction method is important for studies of phenolics. Different solvent extractions may provide different types of compounds due to their variable chemical nature and sensitivity toward extraction or hydrolysis methods. For instance, phenolic compounds extracted from almond skins and hulls with diethyl ether [2], methanol [7], ethyl acetate, and *n*-butanol [8] may result in different yields and compositions of the extracts.

When 80% ethanol was used to extract phenolics from defatted whole seed, skins, and shells, the highest yield (w/w) was obtained from shells (41%), followed by seeds (19%) and skins (8%), whereas the highest quantity of total phenolics (as mg quercetin equivalents/g extract) was present in the skins (88 mg), followed by shells (71 mg) and seeds (8 mg) [5]. Generally, the hydrophobic phenolic contents were two to three times higher than those of their hydrophilic counterparts in all these extracts. However, when 80% acetone was used as the solvent, almond seed extracts had double the amount of phenolics compared to the 80% ethanol extracts [9], probably because acetone has a higher efficiency than alcohol in extracting high-molecular-weight phenolics such as tannins. Diethyl ether extracts of fresh almond hulls afforded a yield of 2% (w/w), whereas extraction with diethyl ether extractions followed by extraction with methanol contributed to a 45% (w/w) yield. Acids were added as extracting cosolvents in some modified cases to serve different purposes. Moure et al. [10] achieved high phenolic yields from almond shells (2 g gallic acid equivalents/100 g shells) by performing acid hydrolysis (2% sulfuric acid) prior to ethyl acetate extraction. An acidified methanolic extraction (HCl:H_2O:methanol, 1:19:80) of almond skin powder resulted in 8.2 μmol gallic acid equivalents/g skin [11]. Therefore, solvent systems could be manipulated to extract phenolics at variable efficiencies. Other factors may also influence the yield of phenolic compounds. Pinelo et al. [12] optimized extraction of phenolics by adjusting liquid–solid ratio, temperature, and contact time. Moreover, yield of phenolics can be enhanced by other means. For instance, gamma-irradiation has been reported to increase phenolic contents in various spices [13]. At doses of 12–12.7 kGy, γ-irradiation of almond skins increased the total content of phenolics by 20% in 40% ethanol extracts [14].

TABLE 8.1
Bioactive Phenolic Constituents of Almond

Almond Component	Major Solvent	Identified Compounds	References
Hull	Methanol	5-*O*-Caffeoylquinic acid (chlorogenic acid) 4-*O*-Caffeoylquinic acid (cryptochlorogenic acid) 3-*O*-Caffeoylquinic acid (neochlorogenic acid)	[18]
Seed coat	Methanol	Isorhamnetin rutinoside Isorhamnetin glucoside Kaempferol rutinoside Kaempferol glucoside	[6]
Skin	Ethyl acetate and *n*-butanol	3′-*O*-Methylquercetin 3-*O*-β -D-glucopyranoside 3′-*O*-Methylquercetin 3-*O*-β-D-galactopyranoside 3′-*O*-Methylquercetin 3-*O*-α-L-rhamnopyranosyl-(1→6)-β-D-glucopyranoside Kaempferol 3-*O*-α-L-rhamnopyranosyl-(1→6)-β-D-glucopyranoside Naringenin 7-*O*-β-D-glucopyranoside Catechins Protocatechuic acid Vanillic acid *p*-Hydroxybenzoic acid	[7]
Whole seed, skin, and shell	Ethanol	Caffeic acid, ferulic acid, *p*-coumaric acid, and sinapic acid	[15]
Whole seed, skin, and shell	Ethanol	Quercetin, isorhamnetin, quercitrin, kaempferol 3-*O*-rutinoside, isorhamnetin 3-*O*-glucoside, and morin	[30]
Hull	Diethyl ether	Betulinic acid, oleanolic acid, and ursolic acid	[2]
Seed	Acetone	Vanillic acid, caffeic acid, *p*-coumaric acid, ferulic acid, delphinidin, cyanidin, quercetin, kaempferol, isorhamnetin, procyanidin B2, and procyanidin B3	[8]

8.2.2 Phenolic Acids

Phenolic acids occur naturally in higher plants, usually as free acids, glycosides, esters, or in insoluble-bound form in a complex mixture of other phenolic compounds. The presence of phenolic acids in food products has been associated with astringency, discoloration, inhibition of enzyme activity, and antioxidant properties, among others [15]. A major portion of phenolic acids present in almond was found in the form of soluble esters. Wijeratne et al. [16] reported that the total amounts of identified free phenolic acids were 16.3, 14 μg/g and a trace amount in skin, shell, and whole seed extracts, respectively, whereas the amounts for total esterified phenolic acids were 279.6, 967.1, and 40.3 μg/g, respectively.

Substituted derivatives of hydroxybenzoic and hydroxycinnamic acids differing in hydroxylation and methoxylation patterns of their aromatic rings are the predominant phenolic acids present in many foods of plant origin [17]. In general, the phenolic acids tentatively identified in almond had the same profile, e.g., caffeic, *p*-coumaric, ferulic, sinapic, syringic, vanilic, gallic, *p*-hydroxybenzoic, and protocatechuic acids (Figures 8.1 and 8.2) [15], which are basically derivatives of either benzoic acid or cinnamic acid. Senter et al. [15] have demonstrated that protocatechuic acid is the predominant phenolic acid in nutmeat of American grown almond, followed by *p*-hydroxybenzoic

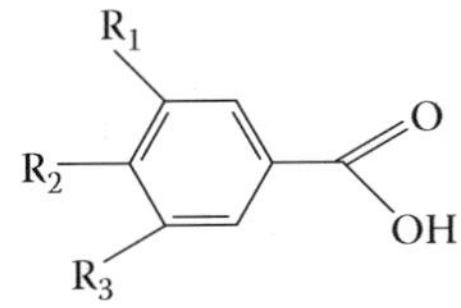

	R_1	R_2	R_3
p-Hydroxybenzoic acid	H	OH	H
Protocatechuic acid	OH	OH	H
Gallic acid	OH	OH	OH
Vanillic	OCH_3	OH	H
Syringic	OCH_3	OH	OCH_3

FIGURE 8.1 Chemical structures of hydroxybenzoic acid derivatives.

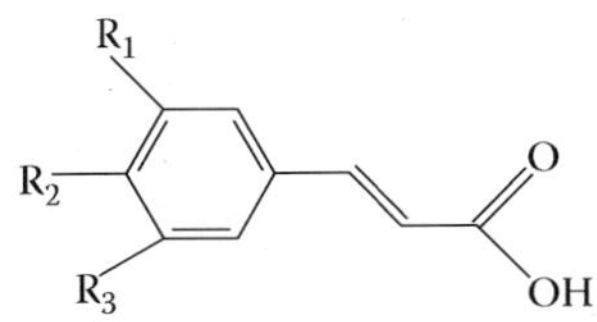

	R_1	R_2	R_3
p-Coumaric acid	H	OH	H
Caffeic acid	OH	OH	H
Ferulic acid	OCH_3	OH	H
Sinapic acid	OCH_3	OH	OCH_3

FIGURE 8.2 Chemical structures of hydroxycinnamic acid derivatives.

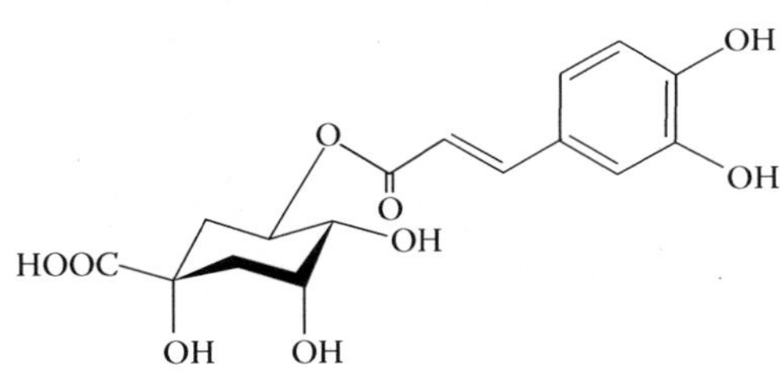

Chlorogenic acid

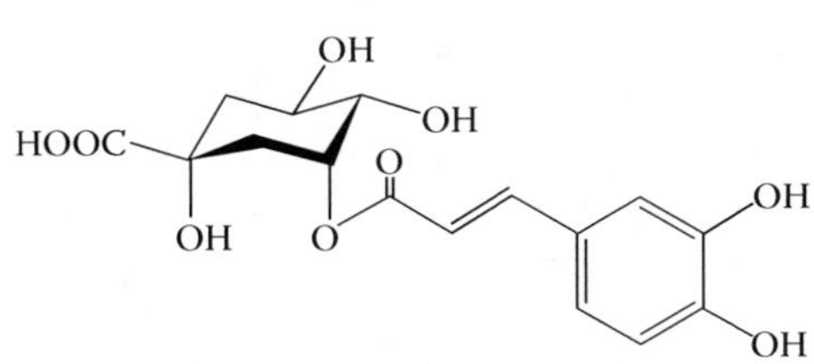

Neochlorogenic acid

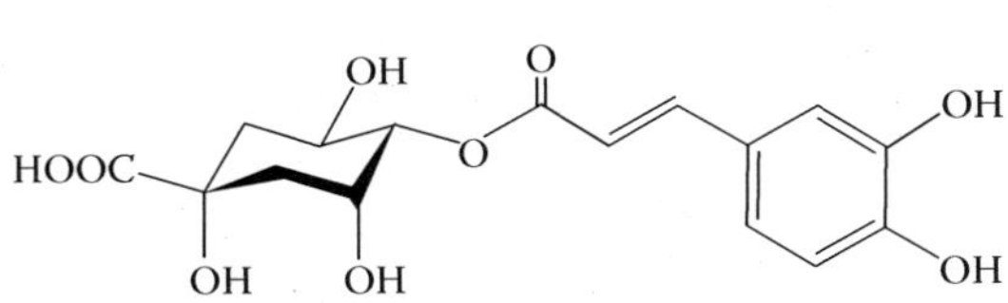

Cryptochlorogenic acid

FIGURE 8.3 Chemical structures of chlorogenic, cryptochlorogenic, and neochlorogenic acids.

acid and vanilic acid. Gallic and caffeic acids, which were present as major phenolic acids in most other tree nuts such as chestnut, pine nut, and walnut, however, were found only in trace amounts in almond in this study. Similar results were obtained for almond skins by Sang et al. [8] and Milbury et al. [6], who demonstrated that glycosides of protocatechuic, vanillic, and *p*-hydroxybenzoic acids dominated among all phenolics in almond skin. Ferulic, *p*-coumaric and caffeic acids were also identified in almond seed extract, but at very low concentrations [9]. However, these phenolic acids together with sinapic acids were isolated and quantified by Wijeratne et al. [16] as the major phenolic acids in almond whole seed, brown skin, and green shell cover extracts, possibly due to different extraction method employed. In almond hulls, the presence of a prenylated benzoic acid, 2-prenyl-4-*O*-β-D-glucopyranosyloxy-4-hydroxybenzoic acid, was reported [18]. More recently, improved technologies and methodologies for the analysis of phenolics have led to the identification of new phenolic acids. These include chlorogenic (5-*O*-caffeoylquinnic) acid and its isomers cryptochlorogenic (4-*O*-caffeoylquinnic), and neochlorogenic (3-*O*-caffeoylquinnic) acids (Figure 8.3) isolated from almond hulls [19].

Phenolic acids are strong antioxidants. For instance, protocatechuic acid, a simple phenolic acid present in almond, showed an antioxidant effect 10-fold higher than that of α-tocopherol [20]. The antioxidant activity of phenolic acids and their esters varies depending on their chemical structure, more specifically, the number of hydroxyl and methoxy groups in the molecule [21]. For example, caffeic acid with two hydroxyl groups was found to be more effective than *p*-coumaric acid with only one hydroxyl group in preventing oxidation of a stripped corn oil system [22]. Studies on *in vitro* human low-density lipoprotein (LDL) oxidation have revealed that the antioxidant activity of phenolic acids was improved as the number of hydroxyl and methoxy groups increased, and that the presence of *o*-dihydroxy groups in the phenolic ring, as in caffeic acid, enhanced the antioxidant activity [23]. Andreasen et al. [24] reported that human LDL cholesterol was protected against oxidation by phenolic acids in the order of caffeic acid > sinapic acid > ferulic acid > *p*-coumaric acid.

8.2.3 Flavonoids

Flavonoids, including flavanols, flavonols, flavononols, flavones, isoflavones, flavanones, and anthocyanidins, are another important group of phenolics in higher plants. They are abundant in plant sources and have been under intensive investigation owing to their health-promoting properties, partly attributed to their antioxidant activity. Flavonoid composition of plants generally depends upon genetic factors, such as variety, geographic origin, and cultivation methods, as well as environmental conditions, including exposure to fungi and bacteria, pest, weather, and UV light, and is also influenced by the ripeness status, processing, and storage of the plant-based materials [25–28]. The total amounts of flavonoids and their patterns vary in different parts of the plants. Studies on distribution of flavonoids in almond, for instance, revealed that most flavonoids were present exclusively in the skin, while nonflavonoids contributed to the majority of total phenolics in the kernel (Table 8.2). The flavonoids in almond skin layer act as phytoalexins protecting the seeds and nuts against bacterial, fungal, and other environmental stress factors [29].

The majority of flavonoids present in almond are in their glycosylated forms. The most common sugar moieties, which account for the hydrophilic property of glycosylated flavonoids, include glucose, rhamnose, rutinose, and galactose, among others. An unusual glycoside named amygdaloside (kauranoid diterpene glycoside) was also isolated from almond seeds [30]. Isorhamnetin, kaempferol, catechin, quercetin, and epicatechin (Figures 8.4 and 8.5) are the most prevalent flavonoid aglycones identified in almond [6,11]. The presence of morin [31], naringenin (Figure 8.6), and *O*-methylquercetin [8] has also been reported in almond phenolic extracts. Analysis of methanolic extracts of seed coats of 16 almond varieties using matrix-assisted laser desorption/ionization time-of-flight mass spectrometry (MALDI-TOF MS) showed the presence of four main flavonol glycosides, namely isorhamnetin rutinoside, isorhamnetin glucoside, kaempferol

TABLE 8.2
Distribution (%) of Identified Falvonoids in Skin and Kernel from Almond

Compound	Skin	Kernel
Catechin	35.7	8.8
Epicatechin	33.9	4.0
Quercetin-3-*O*-galactoside	41.4	nd
Naringenin-7-*O*-glucoside	16.1	10.5
Quercetin-3-*O*-rutinoside	43.7	nd
Quercetine-3-*O*-glucoside	24.5	nd
Dihydroxyhaempferol	50.8	nd
Kaempferol-3-*O*-galactoside	36.4	20.1
Isorhamnetin-3-*O*-galactoside	35.0	3.2
Kaempferol-3-*O*-glucoside	39.0	16.6
Kaempferol-3–*O*-rutinoside	40.4	7.0
Isorhamnetin-3-*O*-rutinoside (and glucoside)	28.5	2.9
Eriodictyol	48.7	nd
Quercetin	100	nd
Naringenin	34.1	28.1
Kaempferol	100	nd
Isorhamnetin	47.2	2.1

Source: From Siriwardhana, S.S.K.W. and Shahidi, F., *J. Am. Oil Chem. Soc.*, 79, 903, 2002. With permission.

Note: nd, not detected.

rutinoside, and kaempferol glucoside. In all almond varieties, isorhamnetin rutinoside was the most abundant flavonol glycoside [7,32]. In ethanolic extracts of almond seed, skin, and shell the main flavonoids detected were three flavonols (quercetin, isorhamnetin, and morin) and four flavonol glycosides (quercitrin, astragalin, kaempferol 3-*O*-rutinoside, and isorhamnetin 3-*O*-glucoside) [16,31]. Sang et al. [8] isolated naringenin glucoside, kaempferol rhamnoglucoside as well as galactoside, glucoside, and rhamnoglucoside of 3′-*O*-methylquercetin.

Similar to phenolic acids, the antioxidant activity of flavonoids depends on the number and location of the hydroxyl groups as well as the degree of structural conjugation [33]. For example, the antioxidant activity of flavonoids can be influenced by *o*-dihydroxylation in the B-ring, the

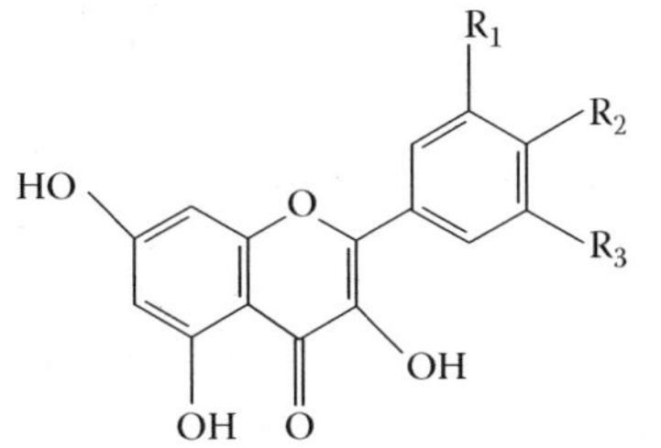

	R_1	R_2	R_3
Kaempferol	H	OH	H
Quercetin	OH	OH	H
Isorhamnetin	H	OH	OCH_3

FIGURE 8.4 Chemical structures of kaempferol, quercetin, and isorhamnetin.

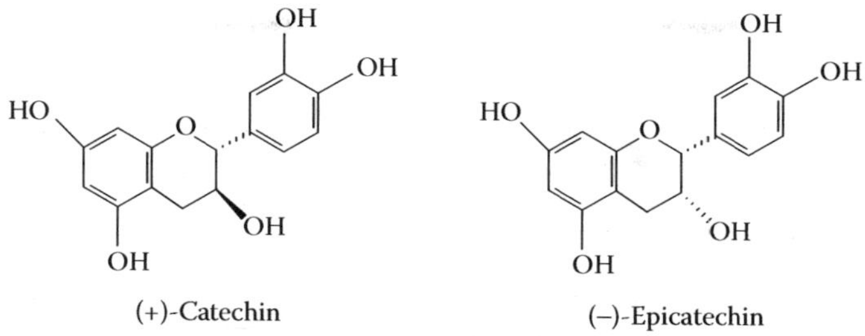

FIGURE 8.5 Chemical structures of catechin and epicatechin.

presence of a C2-3 double bond in conjunction with 4-oxo in the C-ring, and 3- and 5-hydroxyl groups and the 4-oxo function in the A- and C-rings [34–36]. Flavonoid glycosides exhibit lower radical scavenging activity than their corresponding aglycones [37] due to the loss of one or more hydroxyl groups. Isorhamnetin, in which a conjugated group is introduced in place of the dihydroxyl group in the B-ring, was less effective than quercetin in inhibiting copper-induced lipid peroxidation in human LDL cholesterol [38]. Quercetin is more powerful than morin in preventing oxidation of lipids, protein, and DNA [39], chelating iron [40], and quenching hydroxyl radical [41], possibly due to the more favorable *o*-diphenol structure of its B-ring.

8.2.4 Proanthocyanidins (Condensed Tannins)

Proanthocyanidins, also known as condensed tannins, are important flavonoid oligomers/polymers widely distributed in plants. They are structurally dimers, oligomers, or polymers of flavan-3-ols. Based on the structure of their monomers, e.g., the hydroxylation pattern of the A- and B-rings, proanthocyanidins can be classified into procyanidins, prodelphinidins, and propelargonidins (Figure 8.7) [17]. Procyanidins, of which (epi)catechins are the flavanol subunits, are the most commonly occurring proanthocyanidins from natural sources, although prodelphinidins and propelargonidins are also found [42,43]. The U.S. Department of Agriculture has a separate nutrient database of proanthocyanidins [44].

Information on proanthocyanidins in almond is limited. Two monomers, (+)-catechin and (−)-epicatechin, and 15 oligomeric procyanidins, (six dimers, seven trimers, and two tetramers) have been isolated from almond extract and identified [45]. Based on a hydrolysis method in the presence of phloroglucinol, the authors concluded that higher procyanidins present in the extract may have an average degree of polymerization of five. Brieskorn and Betz [46] identified procyanidins B1, B3, and B4 in almond phenolics. Presence of B-type procyanidins was confirmed by Lazarus et al. [47] and Amarowicz et al. [9]. Prodelphinidins and propelargonidins, which contain (epi)gallocatechin and (epi)afzelechin, respectively, as their monomers, are also present in almond. Gu et al. [48]

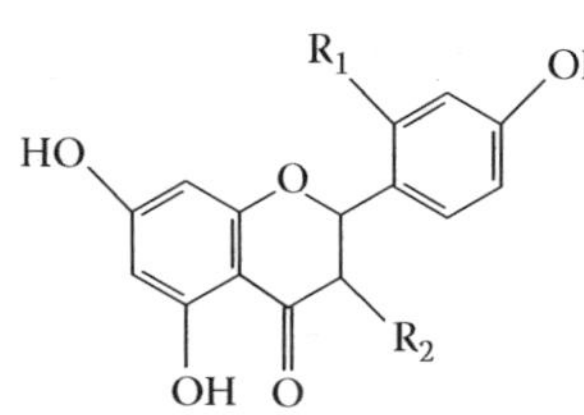

	R_1	R_2
Morin	OH	OH
Naringenin	H	H

FIGURE 8.6 Chemical structures of morin and naringenin.

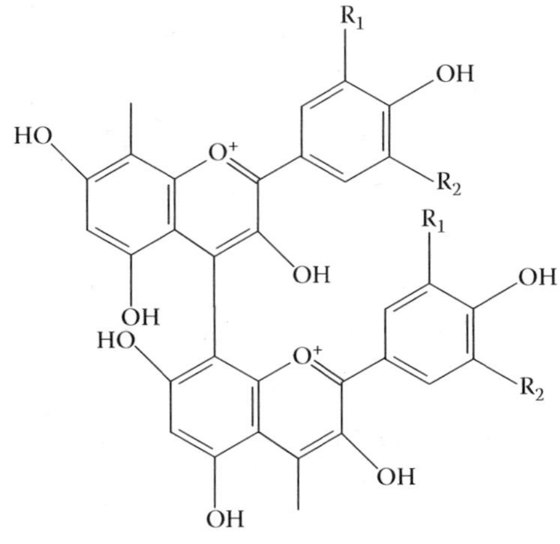

	R_1	R_2
Propelargonidin	H	H
Procyanidin	OH	H
Prodelphinidin	OH	OH

FIGURE 8.7 Chemical structures of propelargonidin, procyanidin, and prodelphinidin.

identified propelargonidins in almond. Amarowicz et al. [9,49] reported that the hydrolysis of extracts of almonds afforded delphinidin and cyanidin.

The tannin fraction of almond has the highest content of total phenolics and the highest antioxidant efficiency among all phenolics [9]. The total phenolics content of tannin fraction from almond extract was found to be 10 times higher than that of the low-molecular-weight phenolics [9]. The tannins possessed higher total antioxidant activity and scavenging capacity against 2,2-diphenyl-1-picrylhydrazyl (DPPH) radical as well as reducing power, and provided better protection to β-carotene–linoleate against oxidation than both crude extract and low-molecular-weight phenolics [9].

8.3 OTHER PHYTOCHEMICALS IN ALMOND

Almond contains ~50% lipid, and lipid-soluble substances are important components of its micronutrients and phytochemicals. These include tocopherols, terpenoids, and sphingolipids, which are considered functional components of foods. Tocopherols, including four homologues α-, β-, γ- and δ-tocopherol, are naturally occurring antioxidants widely distributed in plants as well as animal sources, and are one of the most extensively used antioxidants in food products. Distribution of tocopherols in plants varies with species. In tree nuts, for example, Brazil nut, cashew, pecan, and walnut have higher concentrations of γ- and δ-tocopherols than α-tocopherol, whereas opposite trend is true for hazelnut and almond [50]. α-Tocopherol is the most abundant homologue among all in almonds, while γ- and δ-tocopherols are present at similar levels, which are 10 times lower than α-tocopherol [51]. The concentrations of tocopherols in almond oil were reported by Kornsteiner et al. [50] to be 24.2 mg/100 g oil for α-tocopherol and 3.1 mg/100 g oil for β- and γ-tocopherols; δ-tocopherol was not detected.

Terpenoids such as some triterpenoids and phytosterols are also present in almond and contribute to the bioactivity and functional properties of almond. Triterpenoids are believed to possess anti-inflammatory, anti-human immunodeficiency virus (HIV), and anticancer activities [2]; phytosterols are able to reduce total cholesterol and LDL cholesterol concentrations [52,53]. Triterpenoids generally occur in the waxy coatings of leaves and on fruits such as apple and pear and serve as insect antifeedants and antimicrobial agents [54]. Almond hulls, containing ~1% triterpenoids, are a rich source of these phytonutrients [2]. Triterpenoids in almond hulls, mostly identified as their methyl esters, include betulinic, oleanolic, and ursolic acids [2,19], and their corresponding aldehydes as well as alphitolic, corosolic, and maslinic acids (Figure 8.8) [55]. The presence of amygdaloside, an unusual diterpene glycoside, has recently been reported [30]. In addition to triterpenoids, almond contains a variety of sterols, among which sitosterol, stigmasterol [19], 24-methylenecholesterol, fucosterol, campesterol, Δ^5-avenasterol [46,56], and daucosterol [57] have been identified. β-Sitosterol (Figure 8.9) is prevalent in almond skin, comprising 71% of the total sterols [46], while stigmasterol predominates in the hull [19].

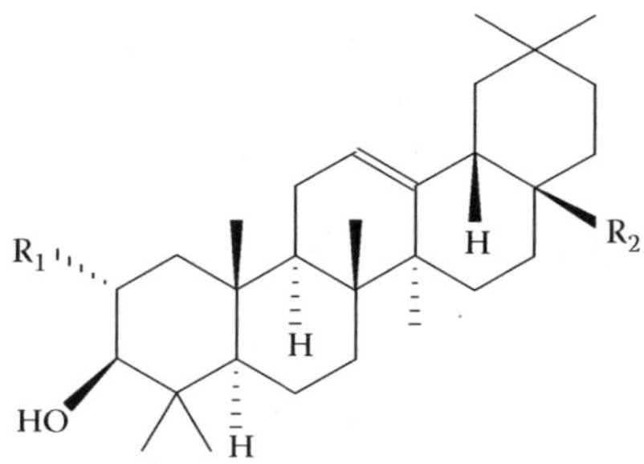

	R_1	R_2
Oleanolic acid	H	COOH
Maslinic acid	OH	COOH

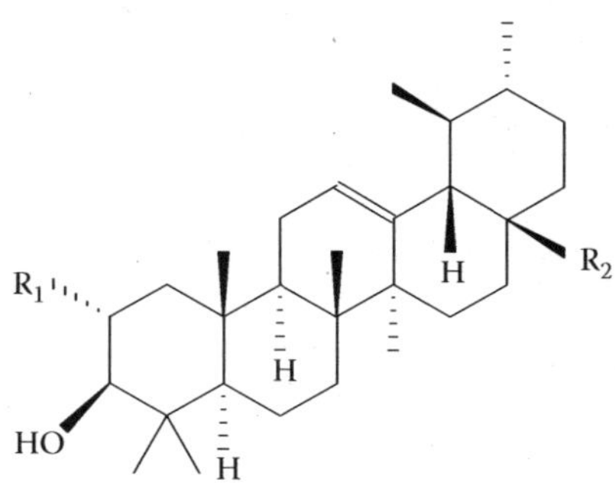

	R_1	R_2
Ursolic acid	H	COOH
Corosolic acid	OH	COOH

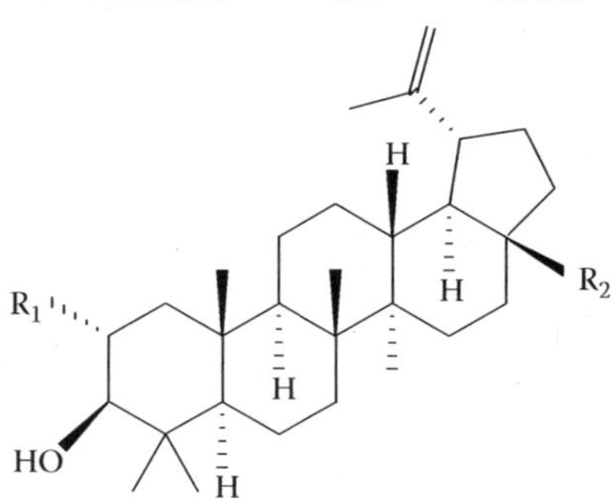

	R_1	R_2
Betulinic acid	H	COOH
Alphitolic acid	OH	COOH

FIGURE 8.8 Chemical structures of triterpenoids.

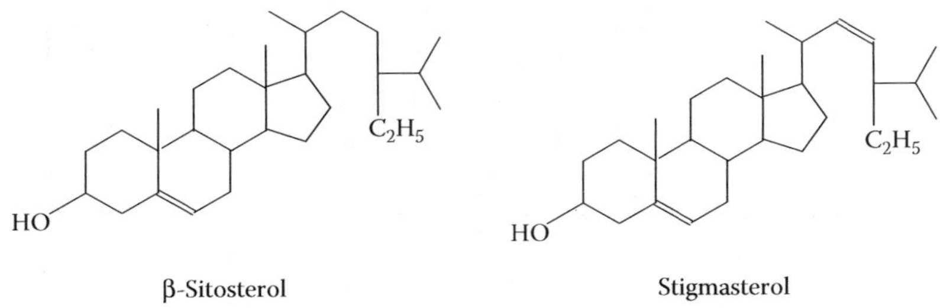

FIGURE 8.9 Chemical structures of β-sitosterol and stigmasterol.

Another lipid-soluble constituent in almond is sphingolipids, which are a group of compounds that have a sphingoid base backbone, an amide linked nonpolar aliphatic "tail" and a polar head-group. They are usually located in cellular membranes, lipoproteins, and other lipid-rich structures [57]. Sphingolipids of plant origin are mainly cerebrosides (mono- and oligohexosylceramides) with a sugar moiety such as glucose, galactose, mannose, and inositol [58]. Sphingolipids have been shown to reduce serum LDL cholesterol and evaluate high-density lipoprotein (HDL) concentrations [59,60]. Sphingomyelin and glycosphingolipids of milk were able to inhibit early stages of colon cancer in mice [61–63]. Almond consumption was found to provide protection from the risk of colon cancer in a rat model system [64], which was attributed to the presence of the sphingolipid 1-*O*-β-D-glucopyranosyl-(2*S*,3*R*,4*E*,8*Z*)-2-[(2*R*)-2-hydroxyhexadecanoylamino]-4,8-octadecadiene-1,3-diol (Figure 8.10) [57].

8.4 ANTIOXIDANT ACTIVITY OF ALMOND AND ITS BY-PRODUCTS

Fruits, herbs, vegetables, nuts, cereals, and other plant materials rich in phenolics are good sources of natural antioxidants. The antioxidant activity of many plants has been evaluated. Almond, one of the most popular tree nuts in the world, has been investigated for its antioxidant activity, and its antioxidant substances has been isolated and identified. Phenolic compounds play an important role in antioxidant effectiveness of almond, mostly due to their redox properties [8]. Phenolic compounds inhibit oxidation reaction mainly by scavenging free radicals, quenching reactive oxygen species (ROS), chelating metal ions, and decomposing peroxides [65]. Many assays have been employed to determine antioxidant efficacy of almond and its by-products. These include oxygen radical absorbance capacity (ORAC), DPPH radical scavenging activity, hydroxyl, and superoxide radicals scavenging capacity, reducing power, and metal chelating capability, among others.

Siriwardhana and Shahidi [5] reported that extracts of whole almond seed, brown skin, and shell rendered potent free radical scavenging capacities. Their studies showed a 100% scavenging

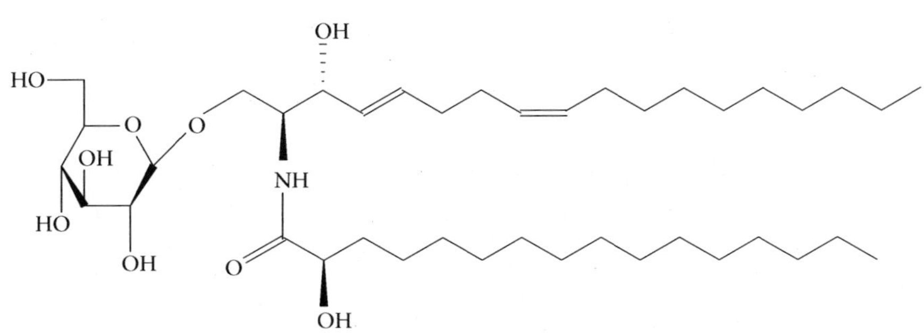

FIGURE 8.10 Chemical structure of sphingolipid identified in almond.

activity of the DPPH radical for skin and shell extracts at 100–200 ppm, whereas seed extracts scavenged 21%–73% of the DPPH radical. The scavenging activity of superoxide radical by these three almond extracts ranged from 76% to 99% with corresponding reduction of hydrogen peroxide concentration ranging from 59% to 91%. The hydroxyl radical-scavenging capacities ranged from 16% for seed to 100% for skin. The antiradical activity of almond extracts against free radicals examined followed the same order: skin > shell > seed. Sang et al. [8] isolated nine phenolic compounds from almond skin and evaluated its antioxidant activity by measuring their DPPH radical scavenging capacity. Their results showed that phenolics present were effective in scavenging DPPH radical and that almond skin can be a potential natural source of phenolic antioxidants. Another study on almond shell revealed that ethyl acetate extract of almond shell acid hydrolysates had a DPPH radical scavenging activity comparable to that of synthetic antioxidants such as butylated hydroxyanisole (BHA) and butylated hydroxytoluene (BHT) [10]. Amarowicz et al. [9] demonstrated antiradical property against DPPH radical and reducing power of almond seed extract and its fractions. In addition, metal ion chelation efficacy of different almond extracts has been reported [31].

Almond extracts have also been evaluated for their inhibitory effect on oxidation in lipid and/or nonlipid model systems. Almond whole seed, brown skin, and green shell extracts effectively inhibited lipid oxidation in a bulk stripped corn oil system as well as in a cooked comminuted pork model system, with shell extract exerting the highest antioxidant activity in both model systems [16]. The authors also demonstrated that these extracts were effective in retention of β-carotene in a β-carotene–linoleate model system, which is similar to those reported by Amarowicz et al. [9] that almond seed extract and its fractions exhibited antioxidant activity in a β-carotene–linoleate model system. According to Takeoka and Dao [19], almond hull extracts were able to inhibit oxidation of methyl linoleate better than α-tocopherol. The crude extracts from almond shells showed protection to labile lipid systems, such as fish oils and fish oil-in-water emulsion, from oxidation as efficiently as propyl gallate [10]. Almond skin extracts inhibited iron-catalyzed oxidation of soybean oil as measured by conjugated dienes and trienes formation and peroxide value [14]. The antioxidant activity of almond extracts in biological systems such as human LDL and DNA has been assessed. Flavonoids extracted from almond skin were found to possess antioxidant capacity and interact with vitamins E and C in a synergistic manner to protect LDL against oxidation both *in vitro* and *in vivo* in hamsters [11]. Wijeratne et al. [31] reported that almond shell, skin, and seed extracts inhibited copper-induced oxidation of human LDL cholesterol. The highest preventive effect was observed for skin extract. However, the shell cover extract arrested peroxyl radical–induced DNA scission more effectively than skin and seed extracts.

8.5 HEALTH BENEFITS

Almond consumption is known to afford health benefits, owing to the presence of various nutrients and phytochemicals including plant protein, unsaturated fatty acids, dietary fiber, tocopherols, phytosterols, and phenolic compounds, among others. Almond is a nutrient-dense food providing an excellent source of protein with good digestibility, MUFA, fiber with insoluble/soluble ratio at 4:1, and a variety of minerals and vitamins [3]. A daily supplement of almond can help build a healthful dietary pattern, which is favorable for chronic disease prevention. Jaceldo-Siegl et al. [66] demonstrated that long-term almond supplementation resulted in spontaneous nutrient modifications which closely matched the dietary recommendations for preventing CVD and other chronic diseases. Study by Jambazian et al. [67] revealed that incorporation of almond into diet simultaneously improved plasma α-tocopherol concentrations and reduced plasma lipid.

As a natural source of antioxidants, almond may be of great value in preventing the onset and/or occurrence of oxidative stress and its related diseases [68]. Almond appears to have beneficial effects on blood cholesterol level and lipoprotein profile in humans [69]. Diets containing almond meal or oil were shown to reduce plasma triacylglycerols (TAG) and total and LDL cholesterol concentrations, increase the resistance of LDL to oxidation, and increase the concentration of HDL cholesterol in humans [70–72]. Jenkins et al. [72] reported that incremental intake of 7 g/day of almond was able to reduce LDL cholesterol concentration by 1%. Almond consumption has been associated with decreased incidence of CVD [73–75]. Results from several epidemiological studies suggest that frequent almond consumption gives rise to a reduced risk of coronary heart disease (CHD) [74,76–78]. The cardioprotective effect of almond is possibly due to the less atherogenic plasma lipid profiles [69] as well as the decreased glycemic excursion [79]. In 2003, FDA promulgated a qualified health claim for tree nuts, including almond, that eating ~42.5 g daily as part of a diet low in saturated fat and cholesterol may reduce the risk of heart disease [80].

Anticancer property of almond and its extracts has been reported. *In vivo* studies have shown that whole almond and almond fractions reduce aberrant crypt foci (ACF) in a rat model of colon carcinogenesis [64]. Two phytochemicals in almond, quercetin and kaempferol, have been found to be strong suppressors of lung and prostate tumor cell growth [81]. Furthermore, triterpenoids present in almond, including betulinic, oleanolic, and ursolic acids, have been reported as antitumor promotion agents [82,83]. These compounds are also known to display other biological activities such as anti-HIV, anti-inflammatory, and *in vitro* antimalarial activities [83], as well as hepatoprotective and antihyperlipidemic properties [82]. While benefits of almond for human health appear promising, further research on bioavailability and bioaccessibility of almond phytochemicals as well as potential allergenic reaction is required for a better understanding of the role of almond in human health.

8.6 CONCLUSION

Almond is a nutrition-dense food providing a spectrum of macro- and micronutrients. Moreover, it is an excellent source of bioavailable phytochemicals that are believed to possess health promotion potentials. Among these phenolic compounds, including phenolic acids and flavonoids, which are the major antioxidant active substances in almond. Other components such as tocopherols and terpenoids also make contribution to the antioxidant activity of almond. The health benefits of almond have been explored. Almond appears to be effective in reducing the risk of heart disease and cancer prevention, and consumption of almond is recommended by FDA for better health conditions.

REFERENCES

1. Food and Agriculture Organization (FAO), FAOSTAT Statistics Database 2004, Published online at: http://faostat.fao.org/site/340/default.aspx (accessed July 20, 2007).
2. Takeoka, G., Dao, L., Teranishi, R., Wong, R., Flessa, S., Harden, L., and Edwards, R., Identification of three triterpenoids in almond hulls, *J. Agric. Food Chem.*, 48, 3437–3439, 2000.
3. Chen, C.Y., Lapsley, K., and Blumberg, J., A nutrition and health perspective on almonds, *J. Sci. Food Agric.*, 86, 2245–2250, 2006.
4. Ahrens, S., Venkatachalam, M., Mistry, A.M., Lapsley, K., and Sathe, S.K., Almond (*Prunus dulcis* L.) protein quality, *Plant Foods Hum. Nutr.*, 60, 123–128, 2005.
5. Siriwardhana, S.S.K.W. and Shahidi, F., Antiradical activity of extracts of almond and its by-products, *J. Am. Oil Chem. Soc.*, 79, 903–908, 2002.
6. Milbury, P.E., Chen, C.Y., Dolnikowski, G.G., and Blumberg, J.B., Determination of flavonoids and phenolics and their distribution in almonds, *J. Agric. Food Chem.*, 54, 5027–5033, 2006.
7. Frison-Norrie, S. and Sporns, P., Identification and quantification of flavonol glycosides in almond seedcoats using MALDI-TOF MS, *J. Agric. Food Chem.*, 50, 2782–2787, 2002.
8. Sang, S., Lapsley, K., Li, G., Jeong, W.-S., Lachance, P.A., Ho, C.-T., and Rosen, R.T., Antioxidative phenolic compounds isolated from almond skins (*Prunus amygdalus* Batsch), *J. Agric. Food Chem.*, 50, 2459–2463, 2002.
9. Amarowicz, R., Troszynska, A., and Shahidi, F., Antioxidant activity of almond seed extract and its fractions, *J. Food Lipids*, 12, 344–358, 2005.
10. Moure, A., Pazos, M., Medina, I., Dominguez, H., and Parajo, J.C., Antioxidant activity of extracts produced by solvent extraction of almond shells acid hydrolysates, *Food Chem.*, 101, 193–201, 2007.
11. Chen, C.Y., Milbury, P.E., Lapsley, K., and Blumberg, J.B., Flavonoids from almond skins are bioavailable and act synergistically with vitamins C and E to enhance hamster and human LDL resistance to oxidation, *J. Nutr.*, 135, 1366–1373, 2005.

12. Pinelo, M., Rubilar, M., Sineiro, J., and Nunez, M.J., Extraction of antioxidant phenolics from almond hulls (*Prunus amygdalus*) and pine sawdust (*Pinus pinaster*), *Food Chem.*, 85, 267–273, 2004.
13. Variyar, P., Bandyopadhyay, C., and Thomas, P., Effect of gamma-irradiation on the phenolic acids of some Indian spices, *Int. J. Food Sci. Technol.*, 33, 533–537, 1998.
14. Harrison, K. and Were, L.M., Effect of gamma irradiation on total phenolic content yield and antioxidant capacity of almond skin extracts, *Food Chem.*, 102, 932–937, 2007.
15. Senter, S.D., Horvat, R.J., and Forbus, W.R., Comparative GLC-MS analysis of phenolic acids of selected tree nuts, *J. Food Sci.*, 48, 798–799, 1983.
16. Wijeratne, S.S.K., Amarowicz, R., and Shahidi, F., Antioxidant activity of almonds and their by-products in food model systems, *J. Am. Oil Chem. Soc.*, 83, 223–230, 2006.
17. Shahidi, F. and Naczk, M., *Phenolics in Food and Nutraceuticals*, CRC Press, Boca Raton, FL, 2004.
18. Sang, S., Lapsley, K., Rosen, R.T., and Ho, C.-T., New prenylated benzoic acid and other constituents from almond hulls (*Prunus amygdalus* Batsch), *J. Agric. Food Chem.*, 50, 607–609, 2002.
19. Takeoka, G.R. and Dao, L.T., Antioxidant constituents of almond [*Prunus dulcis* (Mill.) D.A. Webb] hulls, *J. Agric. Food Chem.*, 51, 496–501, 2003.
20. Nakamura, Y., Torikai, K., and Ohigashi, H., A catechol antioxidant protocatechuic acid pententiates inflammatory leukocyte-derived oxidative stress in mouse skin via a tyrosinase bioactivation pathway, *Free Radical Biol. Med.*, 30, 967–978, 2001.
21. Dziedzic, S.Z. and Hudson, B.J.F., Polyhydroxy chalcones and flavonones as antioxidants for edible oils, *Food Chem.*, 12, 205–222, 1983.
22. Pratt, D.E. and Hudson, B.J.F., Natural antioxidant not exploited commercially, in *Food Antioxidants*, Hudson, B.J.F., Ed., Elsevier Applied Science, New York, 1990, pp. 171–192.
23. Natella, F., Nardini, M., Di Filici, M., and Caccini, C., Benzoic and cinnamic acid derivatives as antioxidants: Structure–activity relationship, *Biosci. Biotechnol. Biochem.*, 56, 324–325, 1999.
24. Andreasen, M.F., Landbo, A.-K., Christensen, L.P., Hansen, A., and Meyer, A.S., Antioxidant effects of phenolic rye (*Secale cereale* L.) extracts, monomeric hydroxycinnamates, and ferulic acid dehydrodimers on human low-density-lipoproteins, *J. Agric. Food Chem.*, 49, 4090–4096, 2001.
25. Herrman, K., Flavonols and flavones in food plants: A review, *J. Food Technol.*, 11, 433–448, 1976.
26. Mazza, G., Anthocyanins in grapes and grape products, *Crit. Rev. Food Sci. Nutr.*, 35, 341–371, 1995.
27. Parcerisa, J., Rafecas, M., Castellote, A.I., Codony, R., Farran, A., Garcia, J., Gonzalez, C., Lopez, A., Romera, A., and Boatella, J., Influence of variety and geographical origin on the lipid fraction of hazelnuts (*Corylus avellana* L.) from Spain: (III) Oil stability, tocopherol content and some mineral contents (Mn, Fe, Cu), *Food Chem.*, 53, 71–74, 1995.
28. Lavedrine, F., Ravel, A., Villet, A., Ducros, V., and Alary, J., Mineral composition of two walnut cultivars originating in France and California, *Food Chem.*, 68, 347–351, 2000.
29. Koes, R.E., Quattrocchio, F., and Mol, J.N.M., The flavonoid biosynthetic pathway in plants: Function and evolution, *BioEssays*, 16, 123–132, 1994.
30. Sang, S., Li, G., Tian, S., Lapsley, K., Stark, R.E., Pandey, R.K., Rosen, R.T., and Ho, C.-T., An unusual diterpene glycoside from the nuts of almond (*Prunus amygdalus Batsch*), *Tetrahedron Lett.*, 44, 1199–1202, 2003.
31. Wijeratne, S.S.K., Abou-Zaid, M.M., and Shahidi, F., Antioxidant polyphenols in almond and its coproducts, *J. Agric. Food Chem.*, 54, 312–318, 2006.
32. Frison, S. and Sporns, P., Variation in the flavonol glycoside composition of almond seedcoats as determined by MALDI-TOF mass spectrometry, *J. Agric. Food Chem.*, 50, 6818–6822, 2002.
33. Rice-Evans, C., Miller, N.J., and Paganga, G., Structure–antioxidant activity relationships of flavonoids and phenolic acids, *Free Radical Biol. Med.*, 20, 933–956, 1996.
34. Foti, M., Piattelli, M., Baratta, M.T., and Ruberto, G., Flavonoids, coumarins and cinnamic acids as antioxidants in a micellar system. Structure–activity relationship, *J. Agric. Food Chem.*, 44, 497–501, 1990.
35. Nieto, S., Garrido, A., Sanhueza, J., Loyola, L., Morales, G., Leighton, F., and Valenzuela, A., Flavonoids as stabilizers of fish oil: An alternative to synthetic antioxidants, *J. Am. Oil Chem. Soc.*, 70, 773–778, 1993.
36. Ho, C.-T., Antioxidant properties of plant flavonoids, in *Food Factors for Cancer Prevention*, Ohigashi, H., Osawa, T., Terao, J., Wanatabe, S., and Yoshikawa, T., Eds., Springer-Verlag, Tokyo, Japan, 1997, pp. 503–597.
37. Ioku, K., Tsushida, T., Takei, Y., Nakatani, N., and Terao, J., Antioxidative activity of quercetin and quercetin monoglucosides in solution and phospholipid bilayers, *Biochim. Biophys. Acta*, 1234, 99–104, 1995.
38. Yamaoto, N., Moon, J.H., Tsushida, T., Nagao, A., and Terao, J., Inhibitory effect of quercetin metabolites and their related derivatives on copper ion-induced lipid peroxidation in human low-density lipoprotein, *Arch. Biochem. Biophys.*, 372, 347–354, 1999.

39. Makris, D.R. and Rossiter, J.T., Comparison of quercetin and a non-orthohydroxy flavonoid as antioxidants by competing in vitro oxidation reactions, *J. Agric. Food Chem.*, 49, 3370–3377, 2001.
40. Brown, J.E., Khodr, H., Hider, R.C., and Rice-Evans, C.A., Structural dependence of flavonoid interactions with Cu^{2+} ions: Implications for their antioxidant properties, *Biochem. J.*, 330, 1173–1178, 1998.
41. Husain, S.R., Cillard, J., and Cillard, P., Hydroxyl radical scavenging activity of flavonoids, *Phytochemistry*, 26, 2489–2491, 1987.
42. Santos-Buelga, C., and Scalbert, A., Proanthocyanidins and tannin-like compounds—Nature, occurrence, dietary intake and effects on nutrition and health, *J. Sci. Food Agric.*, 80, 1094–1117, 2000.
43. Gu, L., Kelm, M.A., Hammerstone, J.F., Beecher, G., Holden, J., Haytowitz, D., Gebhardt, S., and Prior, R.L., Concentrations of proanthocyanidins in common foods and estimations of normal consumption, *J. Nutr.*, 134, 613–617, 2004.
44. U.S. Department of Agriculture (USDA), Database for the Proanthocyanidin Content of Selected Foods, August 2004, Published online at: http://www.nal.usda.gov/fnic/foodcomp (accessed September 28, 2006).
45. De Pascual-Teresa, S.Y., Gutiérrez-Fernández, J.C., Rivas-Gonzalo, C., Santos-Buelga, C., Characterization of monomeric and oligomeric flavan-3-ols from unripe almond fruits, *Phytochem. Anal.*, 9, 21–27, 1998.
46. Brieskon, C.H. and Betz, R., Die steroide der mandel-samenschale, *Fett. Wissenschaft. Technologie*, 89, 230–235, 1987.
47. Lazarus, S.A., Adamson, G.E., Hammerstone, J.F., and Schmitz, H.H., High performance liquid chromatography/mass spectrometry analysis of proanthocyanidins in food and beverages, *J. Agric. Food Chem.*, 47, 3693–3701, 1999.
48. Gu, L.W., Kelm, M.A., Hammerstone, J.F., Beecher, G., Holden, J., Haytowitz, D., and Prior, R.L., Screening of foods containing proanthocyanidins and their structural characterization using LC-MS/MS and thiolytic degradation, *J. Agric. Food Chem.*, 51, 7513–7521, 2003.
49. Amarowicz, R. and Pegg, R.B., Content of proanthocyanidins in selected plant extracts as determined via *n*-butanol/HCl hydrolysis and a colorimetric assay or by HPLC—A short report, *Polish J. Food Nutr. Sci.*, 15, 319–322, 2006.
50. Kornsteiner, M., Wagner, K.H., and Elmadfa, I., Tocopherols and total phenolics in 10 different nut types, *Food Chem.*, 98, 381–387, 2006.
51. Kodad, O., Socias-i-Company, R., Prats, M.S., and Lopez-Ortiz, M.C., Variability in tocopherol concentrations in almond oil and its use as a selection criterion in almond breeding, *J. Hortic. Sci. Biotechnol.*, 81, 501–507, 2006.
52. Nguyen, T.T., The cholesterol-lowering action of plant stanol esters, *J. Nutr.*, 129, 2109–2112, 1999.
53. Law, M., Plant sterol and stanol margarines and health, *Br. Med. J.*, 320, 861–864, 2000.
54. Harborne, J.B., *Phytochemical Methods: A Guide to Modern Techniques of Plant Analysis*, 3rd ed., Chapman & Hall, London, 1998.
55. Amico, V., Barresi, V., Condorelli, D., Spatafora, C., and Tringali, C., Antiproliferative terpenoids from almond hulls (*Prunus dulcis*): Identification and structure–activity relationships, *J. Agric. Food Chem.*, 54, 810–814, 2006.
56. Savage, G.P. and Dutta, P.C., Desmethylsterol composition of some common nuts, *Proc. Nutrition Soc. New Zealand*, 26, 87–91, 2001.
57. Sang, S., Kikuzaki, H., Lapsley, K., Rosen, R.T., Nakatani, N., and Ho, C.-T., Sphingolipid and other constituents from almond nuts (*Prunus anygdalus* Batsch), *J. Agric. Food Chem.*, 50, 4709–4712, 2002.
58. Vesper, H., Schmelz, E.M., Nilolova-Karakashian, M.N., Dillehay, D.L., Lynch, D.V., and Merrill, A.H.Jr, Sphingolipids in food and the emerging importance of sphingolipids to nutrition, *J. Nutr.*, 129, 1239–1250, 1999.
59. Imaizumi, K., Tominaga, A., Sato, M., and Sugano, M., Effects of dietary sphingolipids on levels of serum and liver lipids in rats, *Nutr. Res.*, 12, 543–548, 1992.
60. Kobayashi, T., Shimizugawa, T., Osakabe, T., Watanabe, S., and Okuyama, H., A long-term feeding of sphingolipids affected the levels of plasma cholesterol and hepatic triacylglycerol but not tissue phospholipids and sphingolipids, *Nutr. Res.*, 17, 111–114, 1997.
61. Dillehay, D.L., Webb, S.J., Schmelz, E.M., and Merrill, A.H. Jr., Dietary sphingomyelin inhibits 1,2-dimethylhydrazine-induced colon cancer in CF1 mice, *J. Nutr.*, 124, 615–620, 1994.
62. Schmelz, E.M., Dillehay, D.L., Webb, S.K., Reiter, A., Adams, J., and Merrill, A.H. Jr., Sphingomeylin consumption suppresses aberrant colonic crypt foci and increases the proportion of adenomas versus adenocarcinomas in CF1 mice treated with 1,2-dimethylhydrazine: Implications for dietary sphingolipids and colon carcinogenesis, *Cancer Res.*, 56, 4936–4941, 1996.

63. Schmelz, E.M., Sullards, M.C., Dillehay, D.L., and Merrill, A.H. Jr., Colonic cell proliferation and averrant crypt foci formation are inhibited by dairy glycosphngolipids in 1,2-dimethylhydrazine-treated CF1 mice, *J. Nutr.*, 130, 522–527, 2000.
64. Davis, P.A. and Iwahashi, C.K., Whole almonds and almond fractions reduce aberrant crypt foci in a rat model of colon carcinogenesis, *Cancer Lett.*, 165, 27–33, 2001.
65. Osawa, T., Novel natural antioxidants for utilization in food and biological system, in *Postharvest Biochemistry of Plant Food-Materials in the Tropics*, Uritani, I., Garcia, V.V., and Mendoza, E.M., Eds., Japan Scientific Societies Press, Tokyo, Japan, 1994, pp. 241–251.
66. Jaceldo-Siegl, K., Sabaté, J., Rajaram, S., and Fraser, G.E., Long-term almond supplementation without advice on food replacement induced favourable nutrient modifications to the habitual diets of free-living individuals, *Br. J. Nutr.*, 92, 533–540, 2004.
67. Jambazian, P.R., Haddad, E., Rajaram, S., Tanzman, J., and Sabaté, J., Almonds in the diet simultaneously improve plasma alpha-tocopherol concentrations and reduce plasma lipids, *J. Am. Diet. Assoc.*, 105, 449–454, 2005.
68. Willet, W.C., Diet and health-what should we eat? *Science*, 264, 532–537, 1994.
69. Spiller, G.A., Jenkins, D.J.A., Bosello, O., Gates, J.E., Cragen, L.N., and Bruce, B., Nuts and plasma lipids: An almond-based diet lowers LDL-C while preserving HDL-C, *J. Am. Coll. Nutr.*, 17, 285–290, 1998.
70. Spiller, G.A., Jenkins, D.J.A., and Cragen, L.N., Effect of a diet high in monounsaturated fat from almonds on plasma cholesterol and lipoproteins, *J. Am. Coll. Nutr.*, 1, 126–130, 1992.
71. Hyson, D.A., Schneeman, B.O., and Davis, P.A., Almonds and almond oil have similar effects on plasma lipids and LDL oxidation in healthy men and women, *J. Nutr.*, 132, 703–707, 2002.
72. Jenkins, D.J., Kendall, C.W., Marchie, A., Parker, T.L., Connelly, P.W., and Qian, W., Dose response of almonds on coronary heart disease risk factors: blood lipids, oxidized low-density-lipoproteins, lipoprotein(a), homocysteine, and pulmonary nitric oxide: A randomized, controlled, crossover trial, *Circulation*, 106, 1327–1332, 2002.
73. Fraser, G.E., Sabaté, J., Beeson, W.L., and Strahan, T.M., A possible protective effect of nut consumption on risk of coronary heart disease. The Adventist Health Study, *Arch. Intern. Med.*, 152, 1416–1424, 1992.
74. Hu, F.B., Stampfer, M.J., Manson, J.E., Rimm, E.B., Colditz, G.A., and Rosner, B.A., Frequent nut consumption and risk of coronary heart disease in women: Prospective cohort study, *Br. Med. J.*, 317, 1341–1345, 1998.
75. Albert, C.M., Gaziano, J.M., Willett, W.C., and Manson, J.E., Nut consumption and decreased risk of sudden cardiac death in the physician's health study, *Arch. Intern. Med.*, 162, 1382–1387, 2002.
76. Kendall, C.W.C., Jenkins, D.J.A., Marchie, A., Parker, T., and Connelly, P., Dose response of almonds on coronary heart disease risk factors-blood lipids, oxidized LDL, Lp(a), homocysteine and pulmonary nitric oxide: A randomized controlled cross-over trial, *Circulation*, 106, 1327–1332, 2002.
77. Kris-Etherton, P.M., Zhao, G., Binkoski, A.E., Coval, S.M., and Etheron, T.D., The effects of nuts on coronary heart disease risk, *Nutr. Rev.*, 59, 103–111, 2001.
78. Simopoulos, A.P., The Mediterranean diets: What is so special about the diet of Greece? The scientific evidence, *J. Nutr.*, 131, 3065S–3073S, 2001.
79. Jenkins, D.J., Kendall, C.W.C., Josse, A.R., Salvatore, S., Brighenti, F., Augustin, L.S.A., Ellis, P.R., Vidgen, E., and Venket-Rao, A., Almonds decrease postprandial glycemia, insulinemia, and oxidative damage in healthy individuals, *J. Nutr.*, 136, 2987–2992, 2006.
80. Food and Drug Administration (FDA), Qualified Health Claims: Letter of Enforcement Discretion—Nuts and Coronary Heart Disease, Docket No 02P-0505, Food and Drug Administration, Washington, DC, 2003.
81. Heasman, M. and Mellentin, J., Nuts about nutraceuticals, *World Ingredients*, 48, 50, 1998.
82. Liu, J., Pharmacology of oleanolic and ursolic acid, *J. Ethnopharm.*, 49, 57–68, 1995.
83. Cichewicz, R.H. and Kouzi, S.A., Chemistry, biological activity, and chemotherapeutic potential of betulinic acid for the prevention and treatment of cancer and HIV infection, *Med. Res. Rev.*, 24, 90–114, 2004.

9 Bioactives and Health Benefits of Brazil Nut

Fereidoon Shahidi and Zhuliang Tan

CONTENTS

9.1 INTRODUCTION

Brazil nut has been a popular snack food since the early 1600s. The world production of Brazil nut since mid-1960s is shown in Table 9.1. Brazil and Bolivia are the major producers of Brazil nut followed by Peru and Côte d'Ivoire [1]. Despite what is commonly believed, Brazil nut is not a nut; in botany, nuts are considered to be a specific kind of fruit. However, Brazil nut is the seed of a cannon ball-like, woody capsule fruit. The Brazil nut tree (*Bertholletia excelsa,* family Lecythidaceae) is native to Brazil, Guianas, Venezuela, eastern Colombia, eastern Peru, and eastern Bolivia and is one of the most important economic plants of the Amazon. Brazil nut tree produces fruit only in virgin forests (forests not previously disturbed by human activity), as forests that are not virgin usually lack orchid and orchid bees (*Euglossa* spp.), which are believed to be responsible for the pollination of the flowers and production of Brazil nut. Although avocado, Brazil nut, rubber, and vanilla all originate from tropical forest, only Brazil nut is widely traded and still harvested from the wild forest rather than from plantation [2]. Almost all Brazil nuts consumed around the world come from wild Brazil nut trees [3].

Bioactive plant food constituents (bioactives), including those from nuts in general and Brazil nut in particular, are defined as inherent nonnutritive constituents with generally anticipated health-promoting/beneficial effects when ingested [4,5]. Plant foods such as fruits, vegetables, grains, and nuts are common sources for plant bioactives. Thousands of plant bioactives have scientifically been identified [6]. The health benefits of some plant bioactives are due to their special functions when ingested, such as acting as antioxidants, inhibiting cholesterol absorption, blocking the activity of bacterial or viral toxins, decreasing platelet aggregation, or destroying harmful gastrointestinal bacteria, among others [7]. A wide range of chemical compounds with varying structures are classified as plant bioactives. These include phenolics, carotenoids, phytosterols, phytostanols,

TABLE 9.1
World Production Quantities of Brazil Nut (With Shell, 1000 T)

Year	Brazil	Bolivia	Peru	Côte d'Ivoire	Total
1965	40.8	6	1.7	3.0	51.5
1970	104	8.5	1.7	3.0	117.2
1975	51.7	11.8	1.4	3.0	67.9
1980	40.5	9.4	1.1	3.0	54
1985	45	12	1.4	3.0	61.4
1990	51	17	1.6	5.0	74.6
1995	40	15	1.6	5.2	61.8
2000	33.4	36	0.5	5.2	75.1
2001	28.5	38	0.5	5.2	72.2
2002	27.4	38	0.5	5.2	71.1
2003	28	38.5	0.5	5.2	72.2
2004	28.5	38.2	0.5	5.2	72.4
2005	30	na	0.5	5.2	35.7
2006	30	38.2	0.5	5.2	73.9

Source: From Food and Agricultural Organization (FAO), World Production Quantity (1000 tonnes) of Brazil Nut, Published online at: http://faostat.fao.org (accessed November 28, 2007).

Note: na, not available.

tocopherols, tocotrienols, and mono- and polyunsaturated fatty acids (MUFA and PUFA, respectively), among others.

The study of plant bioactives has been of considerable importance in recent years with the recognition that they are involved in many vital biological and metabolic processes in plants as well as in the human body when consumed as part of a normal human diet. This chapter deals mainly with the bioactives identified in Brazil nut. The potential health benefits of Brazil nut are also discussed. Several groups of plant bioactives have been identified in Brazil nut, including MUFA, PUFA, phenolics (mainly tocopherols), selenium, phytosterols, phytostanols, squalene, and other minor bioactives.

9.2 BIOACTIVES OF BRAZIL NUT

9.2.1 UNSATURATED FATTY ACIDS (MUFA AND PUFA)

As with most tree nuts, Brazil nut is rich in oil, which is clear yellowish, with a pleasant and sweet smell and taste [8]. In animal studies, Brazil nut oil has been shown to be similar to other vegetable oils and animal fats, both fresh and heat-treated, as measured by growth parameters and percentage digestibility in rats [9].

Among foods with favorable fatty acid profile, nuts have received particular attention because epidemiological studies have indicated that their increased intake is associated with protection from coronary heart disease (CHD) [10–13]. The total lipid content and fatty acid composition of Brazil nut oil from different sources are listed in Table 9.2. The total lipid content of Brazil nut ranges from 60.87% to 66.71% of the whole nut. Brazil nut is low in saturated fatty acids (SFA) and high in unsaturated fatty acids. There are great variations in fatty acid composition of Brazil nut from different sources. However, the available data all indicate that their PUFA and MUFA are mainly 18:2ω6 and 18:1ω9, respectively. The variation in the results from different research groups are nonetheless

TABLE 9.2
Fatty Acid Composition (%) of Oil Extracted from Brazil Nut

Fatty Acid	Ref. [14]	Ref. [36]	Ref. [37]	Ref. [71]	Ref. [96]
Total lipid (g/100 g nut)	68.7	60.8	66.43	66.71	—
6:0	—	—	0.00	0.14	—
11:0	—	—	0.00	0.11	—
14:0	—	0.06	0.08	0.05	—
16:0	14.1	13.50	15.09	15.11	12.0
16:1	0.32	0.33	0.38	0.29	—
17:0	tr	0.22	0.08	0.08	—
18:0	8.6	11.77	9.62	9.51	10.4
18:1ω9	29.0	29.09	40.22	28.75	41.2
18:2ω6	46.6	42.80	34.11	45.43	36.1
18:3ω3	tr	0.20	0.06	0.13	—
20:0	tr	0.54	0.27	0.25	—
20:1ω9	—	0.21	0.09	0.00	—
22:0	tr	0.12	0.00	0.06	—
22:1ω9	—	0.34	0.00	0.00	—
Total SFA	22.7	26.21	25.14	25.31	22.4
Total MUFA	29.32	29.97	40.69	29.04	41.2
Total PUFA	46.6	43.0	34.17	45.56	36.1

Note: tr, trace.

due to different factors that influence nut fatty acid composition, including changes in agronomic practices and differences in cultivar or species [14]. Although there are no direct investigations of the above factors for Brazil nut, similar research has been reported for other nuts, such as irrigation effects on hazelnuts that could retard oleic acid synthesis and reduce its proportion during all stages of ripening [14], application of nitrogen to pecan trees leading to decrease of 18:1ω9 and increase of 18:2ω6 in nut kernels [15], variations in fatty acid composition of pecan [16], pistachio [17], and macadamia [18] according to cultivar and year of production. Despite the existing differences in total MUFA and PUFA content among different research reports, the average total unsaturated fatty acids of Brazil nut oil is 74%–78% of total fatty acids from all reports. In 2003, Food and Drug Administration (FDA) promulgated a qualified health claim for tree nuts, that eating 1.5 ounces (~42.5 g) of nuts daily as part of a diet low in saturated fat and cholesterol may reduce the risk of heart disease [19]. Thus, Brazil nut, similar to other nuts, enjoys the same status for health benefits.

From a nutritional point of view, linoleic (LA, 18:2ω6), α-linolenic (ALA, 18:3ω3), and arachidonic (AA, 20:4ω6) acids are essential fatty acids [20]. LA can be elongated to AA while ALA can be elongated to eicosapentaenoic acid (EPA, 20:5ω3) and docosahexaenoic acid (DHA, 22:6ω3). AA and EPA play an important role in prostaglandin metabolism and may influence the thrombotic process [21]. DHA is vital for the proper development of infants' brain and retina [22]. Studies have also indicated that a high intake of ω6 fatty acids shifts the physiologic state to one that is prothrombotic and proaggregatory, characterized by an increase in blood viscosity, vasospasm, and vasoconstriction and decreases bleeding time. The ω3 fatty acids, however, have anti-inflammatory, antithrombotic, antiarrhythmic, hypolipidemic, and vasodilatory properties [20]. Therefore, Brazil nut may also be considered as a healthy lipid source based on its MUFA content, but not its ω3 and ω6 fatty acids.

The oxidative stability of tree nut oils was examined under Schaal oven conditions at 60°C for 12 days (equivalent to 1 year storage at room temperature) [23]. The stability of the oils, as measured by the headspace hexanal, showed that Brazil nut and pecan oils were most stable as compared to those of pine nut, walnut, and hazelnut. The relatively high resistance of Brazil nut and pecan and oils to autoxidation was explained to be due to their lower degree of unsaturation and high content of tocols and phospholipids.

9.2.2 Antioxidants

Antioxidants are compounds that may inhibit, retard, or delay the oxidation of other molecules by inhibiting the initiation or propagation of oxidizing chain reactions. According to the USDA [24] Code of Federal Regulations (21, CFR 170.3), antioxidants are defined as substances used to preserve food by retarding deterioration, rancidity, or discoloration due to oxidation. A more broader definition given by Halliwell [25] states any substance that, when present at low concentrations compared with those of an oxidizable substrate, significantly delays or prevents oxidation of that substrate. The term "oxidizable substrate" includes different molecules found both *in vitro* and *in vivo*. The major antioxidants identified in Brazil nut include phenolics (mainly tocopherols) and selenium.

9.2.2.1 Phenolics

Phenolics are ubiquitous secondary metabolites that are widely distributed in the plant kingdom. They play a very important role in plants and are one of the most important sources of plant bioactives in the human diet [26]. Phenolic compounds might participate in the prevention of various diseases associated with oxidative stress [27,28]. Tocopherols are the major phenolics identified in Brazil nut, thus more research is needed to determine other phenolics present in Brazil nut. This work is currently under investigation in our laboratories.

Phenolic compounds are associated with antioxidant activity and play an important role in preventing lipid peroxidation [29]. Table 9.3 summarizes the total phenolics content (TPC) of different nuts (expressed as mg of gallic acid equivalents (GAE)/100 g dry nut) investigated by different researchers [30,31]. The TPC of Brazil nut from two different sources were quite different, one being 112 mg GAE/100 g and the other being 310 mg GAE/100 g of nuts. This difference might be

TABLE 9.3
Total Phenolics Content (TPC) and Total Antioxidant Activity (TAA) of Different Nuts

	TPC (mg of GAE/100g)[a]		TAA (μmol of TE/g)[b]
Nut	**Ref. [30]**	**Ref. [31]**	**Ref. [31]**
Brazil nut	112	310	14.19
Cashew	137	274	19.97
Hazelnut	291	835	96.45
Macadamia	46	156	16.95
Pecan	1284	2016	179.4
Pistachio	867	1657	79.83
Walnut	1625	1556	135.41

[a] TPC, expressed as milligrams of gallic acid equivalents (GAE) per 100 gram of dry nut (mg of GAE/100 g).

[b] TAA, expressed as micromoles of Trolox equivalents (TE) per gram of dry nut (μmol of TE/g).

due to the variety or different contribution of skin of Brazil nut samples used. Brazil nut was found to have the lowest total antioxidant activity (TAA) among the listed nuts in Table 9.3, while its TPC was the second lowest in one study and the third lowest in another. TAA of Brazil nut did not correlate well with its TPC. Nuts are an important source of dietary lipids and have been suggested as a potential source of dietary antioxidants on the basis of epidemiological and cohort studies [12]. Both lipophilic and hydrophilic antioxidant capacities of different nuts have been determined using the oxygen radical absorbance capacity ($ORAC_{FL}$) assay with fluorescein as the fluorescent probe. The hydrophilic antioxidant capacity of Brazil nut (expressed as μmol of Trolox equivalents (TE)/g) accounted only for 60.7% of the TAA, while the ones from most other nuts (macadamia, pecan, almond, and hazelnut) ranged from 85% to 97.7% [31]. This finding was partially confirmed by the report of Miraliakbari and Shahidi [32]. Although no direct comparison could be made between these two studies due to different antioxidant extraction procedures and ORAC standards used, some striking similarities were observed; for example, lipidic extract of Brazil nut had a relatively high antioxidant activity [32]. Although TPC and TAA have been reported in both studies, no details were made available with respect to the identity of the phenolics contributing to the antioxidant activity of Brazil nut.

Tocopherols are a subgroup of phenolics with important antioxidant and nutritional properties [33]. The side chain of tocopherols is saturated [34]. Brazil nut is a good dietary source of α- and γ-tocopherols and can contribute to a balanced intake of vitamin E. As shown in Table 9.4, the content of tocopherols varies considerably as given in different relevant research reports. The γ-tocopherol content in oil extracted from Brazil nut ranged from 5.1 to 13.8 mg/100 g oil [30,35–38]. γ-Tocopherol was found to be prevalent in Brazil nut as given in several literature reports [30,36,38], while α-tocopherol seems to be predominant in another study [35]. The contents of both α-tocopherol and γ-tocopherol were moderate in the USDA report. δ-Tocopherol was present in trace amounts. γ-Tocopherol is a more effective antioxidant in oils than α- and β-tocopherols [39], but its vitamin E activity is lower than other tocopherols [40]. Since tocopherols have cardioprotective effects, due to their inhibition of low-density lipoprotein (LDL) oxidation, they are implicated to play a key role in the atherogenic process. Thus increased intake of tocopherols appears to be protective against the occurrence of these diseases [41]. Low quantities of tocopherols obtained from average nut consumption have been shown to have a beneficial effect on CHD [42]. However, Miller et al. [43] have shown that high dosage (>268 mg/day) of vitamin E supplementation may increase all-cause mortality.

9.2.2.2 Selenium

Brazil nut is one of the very few naturally occurring sources of exceptionally high level of selenium [44]. Chemically, selenium is a metalloid with properties of both metals and nonmetals. It has a

TABLE 9.4
Tocopherol Composition (mg/100 g) of Oil Extracted from Brazil Nut

Tocopherol	Ref. [30]	Ref. [35]	Ref. [36]	Ref. [37]	Ref. [38]
α-Tocopherol	1.0	11.0	8.29	5.73	1.28
γ-Tocopherol	13.2[a]	5.1	11.62	7.87	13.8
δ-Tocopherol	nd	2.6	—	0.77	1.76
Total	14.2	18.6	19.91	14.37	16.84

Note: nd, not detected.

[a] This value also include β-tocopherol.

similar structure and function to sulfur [45]. Selenium is an essential trace element in living organisms as it forms the active center of certain antioxidant enzymes that protect cells against adverse effects of free radicals produced during normal oxygen metabolism or other enzymes related to immune system and the thyroid gland [46]. However, it may render a toxic effect when consumed in large amounts. Selenium may exist in several chemical forms, both organic (such as selenites and selenates) and inorganic (such as selenoamino acids) in foods and nutritional supplements. Brazil nut has been reported in several studies as one of the richest sources of natural dietary selenium [47,48]. It contains approximately 2500 times as much selenium as any other nuts. The content of selenium in Brazil nut from different sources is assembled in Table 9.5 that shows that the content of selenium in Brazil nut is highly variable. Brazil nut with shells had a selenium content of 3500–4990 μg/100 g sample. The selenium content range for the shelled nut was 254–830 μg/100 g sample. The difference of selenium content between nuts with and without shells in studies carried out by Vonderheide et al. [49] was quite large; their research also analyzed the selenium content of the actual shells removed from the purchased Brazil nut samples with shells. Only 200 μg selenium/100 g was found in the shells alone. Large variations among selenium content of individual Brazil nuts have also been reported by other researchers. The average content (± standard deviation) of selenium in 72 Brazil nut samples was 1466 ± 3790 μg/100 g with a range of 200–25,300 μg/100 g [50]. In another study carried out by Chang et al. [51], 162 Brazil nuts from Acre to Rondonia region, on the upper Amazon in Brazil, where soil selenium levels were low, varied between 30 and 3170 μg selenium/100 g, while another 162 Brazil nuts from Manaus to Belem region of Brazil, lower reaches of the Amazon basin with high soil selenium, contained relatively higher content of selenium with a range of 125–51,200 μg/100 g. Several factors have been proposed to explain the variations in selenium content among individual Brazil nuts. These include the difference in soil and availability of selenium as determined by soil type, moisture content, and other factors [52], the efficiency of selenium taken up by the roots system from soil, or the efficiency of the vascular system at various locations in the branching system for accumulating selenium [50].

Medical surveys have shown that increased selenium intake decreases the risk of breast, colon, lung, and prostate cancer. In order to determine which species could be responsible for disease-preventive effects of selenium, several studies have characterized the selenium compounds in Brazil nut [46,49,53–56].

The selenium distribution in lipid, low-molecular-weight, and protein fractions was studied by Kannamkumarath et al. [53]. Results showed that the lipid fraction was devoid of selenium. The low-molecular-weight fractions, extracted with perchloric acid, were found to contain 3% of the total selenium in Brazil nut. Proteins were isolated by dissolving nut samples in sodium hydroxide, then precipitated with acetone and redissolved in phosphate buffer at pH 7.5 prior to selenium speciation analysis using size exclusion high-performance liquid chromatography (HPLC) with online inductively coupled plasma (ICP)-mass spectrometry (MS). Results obtained revealed that approximately 12% of total selenium was weakly protein bound, but the rest were firmly bound to proteins [53]. To release the firmly bound selenium compounds from proteins, enzymatic digestion with proteinase K [53] or protein hydrolysis with methanesulfonic acid [54] was carried out. The use of

TABLE 9.5
Selenium Content of (μg/100 g nut) Brazil Nut

Brazil Nut	Ref. [49]	Ref. [53]	Ref. [56]	Ref. [97]	Ref. [98]
Nut with shell	3500	3510	4990	—	3800
Nut without shell	800	830	510	254	—
Shell	200	—	—	—	—

ion-pair reversed-phase HPLC-ICP-MS for analysis of the enzymatic digests has shown that selenomethionine (SeMet) was the primary selenium species in Brazil nut (25% of total selenium). This conclusion was confirmed by two other research reports [54,55]. Better cleavage of SeMet (75% of total selenium) was observed using hydrolysis with methanesulfonic acid compared to the enzymatic hydrolysis. Results from these studies indicate that Brazil nuts could serve as a valuable dietary source of SeMet. A study by Chunhieng et al. [57] focused on the distribution of selenium in Brazil nut among the different protein fractions. Analysis by HPLC-MS showed that selenium was linked by a covalent bond to two amino acids to form SeMet and selenocysteine (SeCys). The SeMet represented less than 1% of the total amount of methionine.

Various sample preparation approaches, including microwave-assisted extraction and enzymatic digestion procedures, were examined to extract selenium from the defatted Brazil nut matrix among these approaches, enzymatic treatment with Proteinase K proved most effective [49]. SeMet was demonstrated to be the most abundant of these seleno-amino acids. In their study, another selenium species with *m/z* = 361 was also detected. By application of collision-induced dissociation (CID), there was evidence for this compound to be a dipeptide consisting of tyrosine and methionine with Se [49] as shown in Figure 9.1. It was proposed by the authors that this was a dipeptide with the structure: *p*-$HO(C_6H_4)CH_2CH(NH_2)CONHCH(COOH)CH_2CH_2SeCH_3$.

By combining liquid chromatography and ICP-MS and LC-electrospray ionization (ESI)-MS-MS, the major compounds determined were SeMet and SeCys [56]. This method was further extended to the analyses of *in vitro* gastrointestinal digests of the Brazil nut. CE-ICP-MS was also used, but did not unravel any other species [46].

Further studies should be pursued for identification of unknown selenium species in Brazil nuts. Although Brazil nut provides a good source of selenium for the human diet, it is not a commonly consumed foodstuff because the supply of Brazil nut is rather limited [56].

9.2.3 Phytosterols and Phytostanols

Phytosterols are nonnutrient bioactive substances and act as a structural component in the cell membranes, a role which in mammalian cells is played by cholesterol. The methyl or ethyl group at C-24 location makes them different from cholesterol [58]. They include plant sterols (unsaturated form) and plant stanols (saturated form). Both sterols and stanols are effective in lowering plasma total and LDL cholesterols and inhibit the absorption of cholesterol from the small intestine [59]. A wide spectrum of other biological activities in animals and humans has been reported, including anti-inflammatory [60], antibacterial [61], antioxidative [62], and anticancer activities [63]. β-Sitosterol, campesterol, stigmasterol, Δ^5-avenasterol, sitostanol, and campestanol are the most common representative members in this series. β-Sitosterol, campesterol, and stigmasterol are the major identified phytosterols in Brazil nut, with sitostanol, campestanol, and Δ^5-avenasterol present in trace amounts.

Table 9.6 presents the phytosterol and phytostanol composition of Brazil nut and Brazil nut oil reported by several research groups [36,38,64,65]. Phillips et al. [64] showed that β-sitosterol was

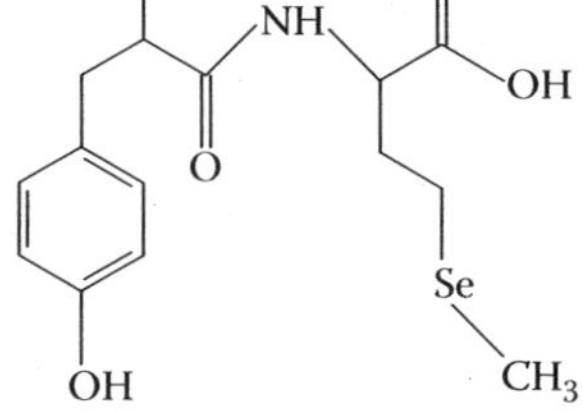

FIGURE 9.1 Proposed chemical structure of dipeptide from Brazil nut protein extraction.

TABLE 9.6
Phytosterol and Phytostanol Composition of Brazil Nut and Brazil Nut Oil

	Brazil Nut		Brazil Nut Oil (mg/100 g oil)		
Sterol and Stanol	Expressed as mg/100 g Nut [64]	Expressed as % Nut [65]	Extracted with H/I[a] [36]	Extracted with H[b] [38]	Extracted with C/M[c] [39]
β-Sitosterol	65.5	74.33	132.54	111	112
Campesterol	2.0	8.39	2.69	12	15
Stigmasterol	6.2	6.07	57.75	22	23
Δ^5-Avenasterol	13.6	—	—	10	11
Sitostanol	4.1	—	—	nd	nd
Campestanol	2.0	—	—	—	—
Other sterols	3.4	11.21	—	37	45
Total	95	100	192.98	192	206

Note: nd, not detected.
[a] H/I, hexane/isopropanol.
[b] H, hexane.
[c] C/M, chloroform/methanol.

the predominant phytosterol (95 mg/100 g) in Brazil nut, followed by Δ^5-avenasterol (13.6 mg/100 g), and stigmasterol (6.2 mg/100 g). This agrees, in part, with the result from Silva et al. [65], who reported that in the sterol fractions of Brazil nut, 74.33% were β-sitosterol, followed by campesterol (8.39%), stigmasterol (6.07%), and other minor sterols (11.21%). Compared with other nuts, Brazil nut is not a good source of dietary phytosterols, as it has the lowest total phytosterols content (95 mg/100 g nut) among nuts [64].

The total sterol content of Brazil nut oil was 192 mg/100 g, with β-sitosterol being the predominant one [38]. Comparison of two solvent extraction systems, namely hexane and chloroform/methanol, showed that the latter solvent render a 7% higher total sterols content in the extract than that extracted with hexane, but the difference was not significant at $P \leq 0.05$.

9.2.4 Squalene

Squalene is a naturally occurring triterpene and the precursor of steroids [66]. Oxidation (via squalene monooxygenase) of one of the terminal double bonds of squalene yields 2,3-squalene oxide, which undergoes enzyme-catalyzed cyclization to afford lanosterol, which is then elaborated into cholesterol and other steroids. Humans cannot live without squalene, because squalene is regarded as an essential building block for the production of hormones and other important substances in the human body. Squalene in olive oil may contribute to the low cholesterol levels of individuals consuming Mediterranean-style diets [67].

Brazil nut has been reported to have notably high levels of squalene (1377.8 μg/g) compared to other nuts, such as pine, cashew, pistachio, and pecan, which contain 39.5, 89.4, 91.4, and 151.7 μg squalene/g, respectively [36].

9.2.5 Other Minor Bioactives

Brazil nut contains small amounts of other minor bioactives and essential micronutrients such as essential minerals, phytic acid, dietary fiber, and thiamin. Brazil nut is an excellent source of magnesium, potassium, and calcium.

Phytic acid (known as inositol hexakisphosphate [IP_6] or phytate as a salt) is an important source of plant phosphorus. Due to its ability to complex with proteins and particularly with minerals, phytic acid has been regarded as an antinutrient [68]. However, it also serves as an important phytonutrient, providing antioxidant effect by binding of the excess free iron, thus preventing the formation of free radicals by Fenton reaction [69]. Vucenik and Shamsuddin [70] suggested that phytates may reduce the risk of colon cancer by reducing oxidative stress. Researchers now believe that phytic acid, found in the fiber of legumes, grains, and nuts, is the major factor responsible for preventing colon and other types of cancers. Brazil nut contains a moderate amount of phytate (1.9 mg/g) compared with other nuts [71]. The range of phytate content for other nuts such as macadamia, pecan, pistachio, walnut, and almond is 1.5 to 3.5 mg/g.

Dietary fiber is defined as the edible parts of plants or analogous carbohydrates that are resistant to digestion and absorption in the human small intestine with complete or partial fermentation in the large intestine. It includes nonstarch polysaccharides, oligosaccharides, lignin, and associated plant substances [72]. Dietary fiber has many health benefits, including improvement of gastrointestinal health, improvement in glucose tolerance and insulin response, reduction of hyperlipidemia, hypertension, and other CHD risk factors, reduction in the risk of developing some cancers, and increased satiety and hence some degree of weight management [73]. According to USDA [37], Brazil nut contains 7.5% total dietary fiber, which accounts for 61.12% of its total carbohydrates. Brazil nut is a food with medium fiber content [74].

The protein content of Brazil nut is 14%–16.5% [71,75]. Brazil nut kernel protein has been reported to be one of the richest sources of sulfur-containing amino acids [76]. Nevertheless, the cake, a by-product from Brazil nut oil extraction, in spite of its high protein quality, has so far only been used in animal feed formulation [75]. Detailed information about nutritional composition of Brazil nut is given in Chapter 2.

9.3 HEALTH BENEFITS OF BRAZIL NUT

Epidemiological evidences have demonstrated that frequent nut consumption has cardioprotective effect because they could lower total blood cholesterol, as well as LDL cholesterol and decrease the risk of CHD [10–13]. To the best of our knowledge, no clinical studies are yet available on cholesterol lowering effect of Brazil nut.

Since squalene is an intermediate in the endogenous synthesis of steroids, there are concerns that increased dietary intake of squalene may increase cholesterol synthesis and hence the risk for the development of atherosclerosis [77]. However, it has been reported that squalene significantly decreases the total cholesterol, LDL cholesterol, and triacylglycerol (TAG) levels in hypercholesterolemic patients [78]. It has also been suggested that squalene is a potential oxidation inhibitor since it can retard the degradation of unsaturated fatty acids and was found to limit the extent of polymerization of rapeseed oil [79]. It can protect cells against free radicals, strengthen the body's immune system, and decrease the risk of various types of cancer [80,81]. To the best of our knowledge, investigation on the bioavailability or biological activity of squalene from Brazil nut is lacking.

The health benefits of selenium have been intensively investigated, with few research directly related to Brazil nut. Sodium selenite, extracted Brazil nut meal, or dried mushroom powder were administered to weanling male rats, which were fed a selenium-deficient Torula yeast diet for 4–9 weeks followed by either continued depletion or repletion for 4–6 weeks [82]. The selenium in Brazil nut meal (*B. excelsa*) was as fully available as that of pure sodium selenite when judged by the ability of dietary selenium to restore plasma and liver glutathione peroxidase activities or selenium levels in rats. Ip and Lisk [83] investigated the bioactivity of selenium from Brazil nut for cancer prevention and selenoenzyme maintenance. Brazil nut was as effective as sodium selenite in preventing cancer when similar levels of Se were administered to rats. Mammary cancer protection by Brazil nut has been associated with increased selenium accumulation in the liver, kidney, mammary gland, and plasma [83]. The magnitude of tissue selenium accumulation was proportional to the amount of dietary Brazil nut. Supplementation with Brazil nut as the sole source of selenium to selenium-deficient rats produced an efficient gradient of two enzymes (selenoenzymes, glutathione peroxidase and type I 5′-deiodinase) restoration at 0.05–0.2 μ/g of dietary selenium [83].

The possible effectiveness of selenium in cancer therapy was considered as early as mid 1920s, with a paper published in 1915 on the therapeutic use of selenium in cancer [84]. Following the early findings, much research has been conducted on possible protective properties of selenium against several forms of cancer, including colon cancer [85,86], lung cancer [87], nonmelanoma skin cancer [88], breast cancer [89], and prostate cancer [90–92]. The preventive effect of selenium on the development of acquired immunodeficiency syndrome (AIDS) in human immunodeficiency virus (HIV)-positive individuals has also been reported [93].

There are other health benefits associated with Brazil nut and its related products. The indigenous tribes of Amazon rainforest brewed the husk of Brazil nut seed pods into tea to treat stomach aches. The Brazil nut tree bark is also brewed into tea to treat liver ailments [94]. In Venezuela, Brazil nut sometimes is used as insect repellent [94].

Brazil nut oil may be used as an ingredient in the manufacture of soap, shampoo, hair conditioner, skin moisturizer, and other cosmetic products. High levels of squalene maybe responsible for use of Brazil nut oil in these applications [94]. Squalene can quickly penetrate the skin, does not leave a greasy feeling on the skin, and blends well with other oils and vitamins [95].

9.4 CONCLUSION AND FUTURE PERSPECTIVES

In addition to being delicious, Brazil nut is a good source of bioactives, which are relevant to its many health beneficial attributes. These bioactive compounds include MUFA, PUFA, selenium, phytosterols, phytostanols, squalene, phenolics (mainly tocopherols), and other minor bioactive constituents.

More investigations are needed to explore the benefits of Brazil nut and its by-products for human health. Studies on the bioavailability and bioaccessibility of the known Brazil nut bioactive compounds as well as their potential allergenic reactions are required. Although very few publications have reported the TPC content of Brazil nut, more research work is necessary to determine the types of phenolics present in Brazil nut.

The chemical and functional characteristics of proteins in Brazil nut have not yet been fully explored and little information is available on the health effects of Brazil nut by-products. For example, the cake produced from Brazil nut oil extraction industries, which might be used as functional food ingredients, sources of nutraceutical extracts or dietary protein due to its high content of sulfur-containing amino acids, requires further research.

REFERENCES

1. Food and Agricultural Organization (FAO), World Production Quantity (1000 tonnes) of Brazil Nut, Published online at: http://faostat.fao.org (accessed November 28, 2007).
2. Silertown, J., Sustainability in a nut shell, *Trends Ecol. Evol.*, 19, 276–278, 2004.
3. Smith, N.J.H., Williams, J.T., Plucknett, D.L., and Talbot, J.P., *Tropical Forests and Their Crops*, Cornell University, Ithaca, NY, 1992.
4. Gry, J., Rosa, E., Rosner, H., Andrade, I., Gevers, E., Hallikainen, A., Hedley, C., Holm, S., Lambein, F., Laursen, P., Strigl, A., Sørensen, H., and Vidal-Valverde, C., *NETTOX List of Food Plants*, Danish Veterinary and Food Administration, Søborg, Denmark, 1998.
5. Gry, J., Black, L., Eriksen, F.D., Pilegaard, K., Plumb, J., Rhodes, M., Sheehan, D., Kiely, M., and Kroon, P., EuroFIR-BASIS—A combined composition and biological activity database for bioactive compounds in plant-based food, *Trends Food Sci. Technol.*, 18, 434–444, 2007.
6. Goldberg, G., *Plants: Diet and Health. Report of a British Nutrition Foundation Task Force*, Blackwell Publishing, Oxford, 2003.
7. Pennington, J.A.T., Food composition databases for bioactive food components, *J. Food Comp. Anal.*, 15, 419–434, 2002.

8. Shahidi, F. and Miraliakbari, H., Tree nut oils, in *Bailey's Industrial Oil and Fat Products, Edible Oil and Fat Products: Specialty Oils and Oil Products*, vol. 3, 6th ed., Shahidi F., Ed., Wiley Interscience, Hoboken, NJ, 2005, pp. 175–193.
9. Elias, L.G. and Bressani, R., The nutritive value of the Brazil nut oil, *J. Am. Oil Chem. Soc.*, 38, 450–452, 1961.
10. Hu, F.B., Stampfer, M.J., Manson, J.E., Rimm, E.B., Colditz, G.A., Rosner, B.A., Speizer, F.E., Hennekens, C.H., and Willett, W.C., Frequent nut consumption and risk of coronary heart disease in women: Prospective cohort study, *Br. Med. J.*, 317, 1341–1345, 1998.
11. Kris-Etherton, P.M., Yu-Poth, S., Sabaté, J., Ratcliffe, H.E., Zhao, G., and Etherton, T.D., Nuts and their bioactive constituents: Effects on serum lipids and other factors that affect disease risk, *Am. J. Clin. Nutr.*, 70, 504S–511S, 1999.
12. Kris-Etherton, P.M., Zhao, G., Binkoski, A.E., Coval, S.M., and Etherton, T.D., The effects of nuts on coronary heart disease risk, *Nutr. Rev.*, 59, 103–111, 2001.
13. Kelly, J.H. Jr. and Sabaté, J., Nuts and coronary heart disease: An epidemiological perspective, *Br. J. Nutr.*, 96 (Suppl. 2), 61S–67S, 2006.
14. Beuchat, L.R. and Worthington, R.E., Fatty acid composition of tree nut oils, *J. Food Technol.*, 13, 355–358, 1978.
15. Heaton, E.K., Marion, J.E., and Woodroof, J.G., Pecan oil is highly unsaturated, *Peanut J. Nut World*, 45, 36–38, 1966.
16. Heaton, E.K., Worthington, R.E., and Shewfelt, A.L., Pecan nut quality: Effect of time of harvest on the composition, sensory and quality characteristics, *J. Food Sci.*, 40, 1260–1263, 1975.
17. Shokraii, E.H., Chemical composition of the pistachio nuts (*Pistacia vera* L.) of Kerman, Iran, *J. Food Sci.*, 43, 244–245, 1977.
18. Saleeb, W.F., Yermanos, D.M., Huszar, C.K., Storey, W.B., and Labanauskas, C.K., The oil and protein in nuts of *Macadamia tetraphylla* L. Johnson, *Macadamia integrifolia* Maiden and Betche, and their F_1 hybrid, *J. Am. Soc. Hort. Sci.*, 98, 453–456, 1973.
19. Food and Drug Administration (FDA), Qualified Health Claims: Letter of Enforcement Discretion—Nuts and Coronary Heart Disease, Docket No 02P-0505, Food and Drug Administration, Washington, DC, 2003.
20. Simopoulos, A.P., Essential fatty acids in health and chronic disease, *Am. J. Clin. Nutr.*, 70, 560S–569S, 1999.
21. Hu, F.B., Stampfer, M.J., Manson, J.A.E., Rimm, E.B., Wolk, A., Colditz, G.A., Hennekens, C.H., and Willett, W.C., Dietary intake of α-linolenic acid and risk of fatal ischemia heart disease among women, *Am. J. Clin. Nutr.*, 69, 890–897, 1999.
22. Shahidi, F. and Finley, J.W., The chemistry, processing, and health benefits of highly unsaturated fatty acids: An overview, in *Omega-3 Fatty Acids: Chemistry, Nutrition, and Health Effects*, Shahidi, F. and Finley, J.W., Eds., ACS Symposium Series 788, Washington, DC, 2001, pp. 2–13.
23. Shahidi, F. and Miraliakbari, H., Oxidative stability of tree nut oils, submitted to the *J. Agric. Food Chem.*, 56, 4751–4759, 2008.
24. U.S. Department of Agriculture (USDA), Code of Federal Regulations, Title 21 Food and Drugs, PART 170—Food Additives, 2006, Published online at: http://www.cfsan.fda.gov/~lrd/FCF170.html (access January 2, 2008).
25. Halliwell, B., How to characterize a biological antioxidant, *Free Radical Res. Commun.*, 9, 1–32, 1990.
26. Ortuno, A., Garcia-Puig, D., Fuster, M.D., Perez, M.L., Sabater, F., Porras, I., Garcia-Lidon, A., and Del, R.J.A., Flavonone and flavone levels in different varieties of grapefruit and pumello, *J. Agric. Food Chem.*, 43, 1–5, 1995.
27. Temple, N.J., Antioxidants and disease: More questions than answers, *Nutr. Res.*, 20, 449–459, 2000.
28. Manach, C., Williamson, G., Morand, C., Scalbert, A., and Rémésy, C., Bioavailability and bioefficacy of polyphenols in humans. I. Review of 97 bioavailability studies, *Am. J. Clin. Nutr.*, 81, 230S–242S, 2005.
29. Shahidi, F. and Wanasundara, P.K.J.P.D., Phenolic antioxidants, *Crit. Rev. Food. Sci. Nutr.*, 32, 67–103, 1992.
30. Kornsteiner, M., Wagner, K.H., and Elmadfa, I., Tocopherols and total phenolics in 10 different nut types, *Food Chem.*, 98, 381–387, 2006.
31. Wu, X., Beecher, G.R., Holden, J.M., Haytowitz, D.B., Gebhardt, S.E., and Prior, R.L., Lipophilic and hydrophilic antioxidant capacities of common foods in the United States, *J. Agric. Food Chem.*, 52, 4026–4037, 2004.

32. Miraliakbari, H. and Shahidi, F., Antioxidant activity of minor components of tree nut oils, *Food Chem.*, 111, 421–427, 2008.
33. Brigelius-Flohe, R. and Traber, M.G., Vitamin E: Function and metabolism, *FASEB J*, 13, 1145–1155, 1999.
34. Shahidi, F., Antioxidants in food and food antioxidants, *Nahrung*, 44, 158–163, 2000.
35. Sheppard, A.J., Pennington, J.A.T., and Weihrauch, J.L., Analysis and distribution of vitamin E in vegetable oils and foods, in *Vitamin E in Health and Disease*, Packer, L. and Fuchs, J., Eds., Marcel Dekker, New York, 1993, pp. 9–31.
36. Ryan, E., Galvin, K., O'Connor, T.P., Maguire, A.R., and O'Brien, N.M., Fatty acid profile, tocopherol, squalene and phytosterol content of Brazil, pecan, pine, pistachio and cashew nuts, *Int. J. Food. Sci. Nutr.*, 57, 219–228, 2006.
37. U.S. Department of Agriculture (USDA), National Nutrition Database for Standard Reference, Release 20, National Technical Information Service, USDA, Springfield, VA, 2007.
38. Miraliakbari, H. and Shahidi, F., Lipid class composition, tocopherols and sterols of tree nut oils extracted with different solvents, *J. Food Lipids*, 15, 1–16, 2008.
39. Olcott, H.S. and Emerson, O.H., Antioxidants and the autoxidation of fats. IX. The antioxidant properties of the tocopherols, *J. Am. Oil Chem. Soc.*, 59, 1008–1009, 1937.
40. Burton, G.W. and Traber, M.G., Vitamin E: Antioxidant activity, biokinetics, and bioavailability, *Ann. Rev. Nutr.*, 10, 375–382, 1990.
41. Elmadfa, I. and Wagner, K.H., Non-nutritive bioactive food constituents of plants: Tocopherols (vitamin E), *Int. J. Vitam. Nutr. Res.*, 73, 89–94, 2003.
42. Bramley, P.M., Elmadfa, I., Kafatos, A., Kelly, F.J., Manios, Y., Roxboroug, H.E., Schun, W., Sheehy, P.J.A., and Wagner, K.H., Vitamin E, *J. Sci. Food Agric.*, 80, 913–938, 2000.
43. Miller, E.R. 3rd., Pastor-Barriuso, R., Dalal, D., Riemersma, R.A., Appel, L.J., and Gullar, E., Meta-analysis: High-dosage vitamin E supplementation may increase all-cause mortality, *Ann. Intern. Med.*, 142, 37–46, 2005.
44. Infante, H.G., Hearn, R., and Catterick, T., Current mass spectrometry strategies for selenium speciation in dietary sources of high-selenium, *Anal. Bioanal. Chem.*, 382, 957–967, 2005.
45. Reilly, C., Selenium: A new entrant into the functional food arena, *Trends Food Sci. Technol.*, 9, 114–118, 1998.
46. Kannamkumarath, S.S., Wrobel, K., and Wuilloud, R.G., Studying the distribution pattern of selenium in nut proteins with information obtained from SEC-UV-ICP-MS and CE-ICP-MS, *Talanta*, 66, 153–159, 2005.
47. Dumont, E., Vanhaecke, F., and Cornelis, R., Selenium speciation from food source to metabolites: A critical review, *Anal. Bioanal. Chem.*, 385, 1304–1323, 2006.
48. Reilly, C., *Selenium in Food and Health*, 2nd ed., Springer US, New York, 2006.
49. Vonderheide, A.P., Wrobel, K., Kannamkumarath, S.S., B'Hymer, C., Montes-Bayón, M., Ponce, D.E., León, C., and Caruso, J.A., Characterization of selenium species in Brazil nuts by HPLC-ICP-MS and ES-MS, *J. Agric. Food Chem.*, 50, 5722–5728, 2002.
50. Sector, C.L. and Lisk, D., Variation in the selenium content of individual Brazil nuts, *J. Food Safety*, 9, 279–281, 1989.
51. Chang, J.C., Gutenmann, W.H., Reid, C.M., and Lisk, D., Selenium content of Brazil nut from two geographic locations in Brazil, *Chemosphere*, 30, 801–802, 1995.
52. Reilly, C., Brazil nuts-a selenium supplement? *BNF Nutr. Bull.*, 24, 177–184, 1999.
53. Kannamkumarath, S.S., Wrobel, K., Wrobel, K., Vonderheide, A., and Caruso, J.A., HPLC-ICP-MS determination of selenium distribution and speciation in different types of nut, *Anal. Bioanal. Chem.*, 373, 454–460, 2002.
54. Wrobel, K., Kannamkumarath, S.S., Wrobel, K., and Caruso, J.A., Hydrolysis of proteins with methanesulfonic acid for improved HPLC-ICP-MS determination of seleno-methionine in yeast and nuts, *Anal. Bioanal. Chem.*, 375, 133–138, 2003.
55. Bodó, E.T., Stefánka, Z., Ipolyi, I., Sörös, C., Dernovics, M., and Fodor, P., Preparation, homogeneity and stability studies of a candidate LRM for Se speciation, *Anal. Bioanal. Chem.*, 377, 32–38, 2003.
56. Dumont, E., De Pauw, L., Vanhaecke, F., and Cornelis R., Speciation of Se in *Bertholletia excelsa* (Brazil nut): A hard nut to crack? *Food Chem.*, 95, 684–692, 2006.
57. Chunhieng, T., Petritis, K., Elfakir, C., Brochier, J., Goli, T., and Montet, D., Study of selenium distribution in the protein fractions of the Brazil nut, *Bertholletia excelsa*, *J. Agric. Food Chem.*, 52, 4318–4322, 2004.
58. Moreau, R.A., Whitaker, B.D., and Hicks, K.B., Phytosterols, phytostanols, and their conjugates in foods: Structural diversity, quantitative analysis, and health-promoting uses, *Prog. Lipid Res.*, 41, 457–500, 2002.

59. Piironen, V., Lindsay, D.G., Miettinen, T.A., Toivo, J., and Anna-Maija, L., Plant sterols: Biosynths, biological function and their importance to human nutrition, *J. Sci. Food Agric.*, 80, 939–966, 2000.
60. Bouic, P.J., Sterols and sterolins: New drugs for the immune system? *Drug Discov. Today*, 7, 775–778, 2002.
61. Zhao, C.C., Shao, J.H., Li, X., Xu, J., and Zhang, P., Antimicrobial constituents from fruits of *Ailanthus altissima* SWINGLE, *Arch. Pharmacol Res.*, 28, 1147–1151, 2005.
62. van-Rensburg, S.J., Daniels, W.M.U., van-Zyl, J.M., and Taljaard, J.J.F., A comparative study of the effects of cholesterol, beta-sitosterol, beta-sitosterol glucoside, dehydroepiandrosterone sulphate and melatonin on in vitro lipid peroxidation, *Metab. Brain Dis.*, 15, 257–265, 2000.
63. Awad, A.B., Downie, A., Fink, C.S., and Kim, U., Dietary phytosterols inhibits the growth and metastasis of MDA-MB-231 human breast cancer cells grown in SCID mice, *Anticancer Res.*, 20, 821–824, 2000.
64. Phillips, K.M., Ruggio, D.M., and Ashraf-Khorassani, M., Phytosterol composition of nuts and seeds commonly consumed in the United States, *J. Agric. Food Chem.*, 53, 9436–9445, 2005.
65. Silva, W.G.D., Cortesi, N., and Rovellini, N., The Brazilian nut (*Bertholletia excelsa* H.B.K.-Lecythidaceae group). II. Lipids: The chemical structure, *Rivista-Italiana-delle-Sostanze-Grasse*, 74, 311–314, 1997.
66. Bloch, K., Sterol structure and membrane function, *Curr. Top. Cell. Regul.*, 18, 289–299, 1981.
67. Smith, T.J., Squalene: Potential chemopreventive agent, Expert Opin. Investig. Drugs, 9, 1841–1848. 2000.
68. Urbano, G., Lopez-Jurado, M., Aranda, P., Vidal-Valverde, C., Tenorio, E., and Porres, J., The role of phytic acid in legumes: Antinutrient or beneficial function? *J. Physiol. Biochem.*, 56, 283–294, 2000.
69. Graf, E., Empson, K.L., and Eaton, J.W., Phytic acid: A natural antioxidant, *J. Biol. Chem.*, 262, 11647–11650, 1987.
70. Vucenik, I. and Shamsuddin, A.M., Cancer inhibition by inositol hexaphosphate (IP_6) and inositol: From laboratory to clinic, *J. Nutr.*, 133, 3778S–3784S, 2003.
71. Venkatachalam, M. and Sathe, S.K., Chemical composition of selected edible nut seeds, *J. Agric. Food Chem.*, 54, 4705–4714, 2006.
72. American Association of Cereal Chemists Report, The definition of dietary fiber, *Cereal Foods World*, 46, 112–126, 2001.
73. British Nutrition Foundation (BNF), Response to the FSA Regarding the Definition of Fiber-CODEX Committee on Nutrition and Foods for Special Dietary Uses, Oxford, 2003.
74. BC Cancer Agency, *Dietary Fiber Content of Common Foods*, Vancouver, BC, 2004.
75. Ramos, C.M.P. and Bora, P.S., Extraction and functional characteristics of Brazil nut (*Bertholletia excelsa* HBK) globulin, *Food Sci. Technol. Int.*, 9, 265–269, 2003.
76. Antunes, A.J. and Markakis, P., Protein supplementation of navy beans with Brazil nuts, *J. Agric. Food Chem.*, 25, 1096–1098, 1977.
77. Scolastici, C., Ong, T.P., and Moreno, F.S., Squalene does not exhibit a chemopreventive activity and increases plasma cholesterol in a Wistar rat hepatocarcinogenesis model, *Nutr. Cancer*, 50, 101–109, 2004.
78. Chan, P., Tomlinson, B., Lee, C.B., and Lee, Y.S., Effectiveness and safety of low dose pravastatin and squalene, alone and in combination, in elderly patients with hypercholesterolemia, *J. Clin. Pharmacol.*, 36, 422–427, 1996.
79. Malecka, M., The effect of squalene on the thermostability of rapeseed oil, *Nahrung*, 38, 135–140, 1994.
80. Newmark, H.L., Squalene, olive oil, and cancer risk: A review and hypothesis, *Cancer Epidemiol. Biomar. Prevent.*, 6, 1101–1103, 1997.
81. Rao, C.V., Newmark, H., and Reddy, B.S., Chemoprotective effect of squalene on colon cancer, *Carcinogenesis*, 19, 287–290, 1998.
82. Chansler, B.S.M.W., Mutanen, M.S.M., Morris, M.S.V.C., and Levander, Ph.D.O.A., Nutritional bioavailability to rats of selenium in Brazil nuts and mushrooms, *Nutr. Res.*, 6, 1419–1428, 1986.
83. Ip, C. and Lisk, D.J., Bioactivity of selenium from Brazil nut for cancer prevention and selenoenzyme maintenance, *Nutr. Cancer*, 21, 203–212, 1994.
84. Walker, C.H. and Klein, F., Selenium—its therapeutic value, especially in cancer, *American Med.* 628, 1915 as cited by Schrauzer, C.H., Trace elements in carcinogenesis, in *Advances in Nutritional Research*, vol. 2, Draper, H.H., Ed., Plenum, New York, 1979, pp. 219–244.
85. Findley, J.W., Davis, C.D., and Feng, Y., Selenium from high-selenium broccoli protects rats from colon cancer, *J. Nutr.*, 103, 2384–2389, 2000.
86. Findley, J.W. and Davis, C.D., Selenium (Se) from high-selenium broccoli is utilized differently than selenite, selenate and selenomethionine, but is more effective in inhibiting colon carcinogenesis, *Biofactors*, 14, 191–196, 2001.
87. Reid, M.E., Duffield-Lillico, A.J., Garland, L., Turnbull, B.W., Clark, L.C., and Marshall, J.R., Selenium supplementation and lung cancer incidence: An update of the nutritional prevention of cancer trial, *Cancer Epidemiol. Biomar. Prevent.*, 11, 1285–1291, 2002.

88. Duffield-Lillico, A.J., Slate, E.H., Reid, M.E., Turnbull, B.W., Wilkins, P.A., Combs, G.F. Jr., Park, H.K., Gross, E.G., Graham, G.F., Stratton, M.S., Marshall, J.R., and Clark, L.C., Selenium supplementation and the secondary prevention of non-melanoma skin cancer in a randomized trial, *J. Natl. Cancer Inst.*, 95, 1477–1481, 2003.
89. Medina, D., Thompson, H., Ganther, H., and Ip, C., Se-Methylselenocysteine; A new compound for the treatment of breast cancer, *Nutr. Cancer*, 40, 34–41, 2001.
90. Brooks, J.D., Metter, E.J., Chan, D.W., Sokoll, L.J., Landis, P., Nelson, W.G., Muller, D., Andres, R., and Carter, H.B., Plasma selenium level before diagnosis and the risk of prostate cancer development, *J. Urol.*, 166, 2034–2038, 2001.
91. Combs, G.F. Jr., Status of selenium in prostate cancer prevention, *Br. J. Cancer*, 91, 195–199, 2004.
92. Peters, U., Foster, C.B., Chatterjee, N., Schatzkin, A., Reding, D., Andriole, G.L., Crawford, E.D., Sturup, S., Chanock, S.J., and Hayes, R.B., Serum selenium and risk of prostate cancer—A nested case–control study, *Am. J. Clin. Nutr.*, 85, 209–217, 2007.
93. Rayman, M.P., The argument for increasing selenium intake, *Proc. Nutr. Soc.*, 61, 203–215, 2002.
94. BioTrade Facilitation Programme (BTFP), Market Brief in the European Union for Selected Natural Ingredients Derived from Native Species: Bertholletia excelsa, Compiled by ProFound for United Nations Conference on Trade and Development (UNCTAD), 2005, Published online at: http://r0.unctad.org/biotrade/docs/biotradebrief-bertholletiaexcelsa.pdf (accessed January 31, 2008).
95. Lu, H.T., Jiang, Y., and Chen, F., Preparative separation and purification of squalene from the microalga *Thraustochytrium* ATCC 26185 by high-speed counter-current chromatography, *J. Chromatogr. A.*, 994, 37–43, 2003.
96. Assuncão, F.P., Bentes, M.H.S., and Serruya, H., A comparison of the stability of oils from Brazil nut, para rubber and passion fruit seeds, *J. Am. Oil Chem. Soc.*, 61, 1031–1036, 1984.
97. Barclay, M.N., MacPherson, A., and Dixon, J., Selenium content of a range of UK foods, *J. Food Compos. Anal.*, 8, 307–318, 1995.
98. Manjusha, R., Dash, K., and Karunasagar, D., UV-photolysis assisted digestion of food samples for the determination of selenium by electrothermal atomic absorption spectrometry (ETAAS), *Food Chem.*, 105, 260–265, 2007.

10 Bioactive Compounds from Cashew Nut and Its Coproducts

Fereidoon Shahidi and Zhuliang Tan

CONTENTS

10.1 INTRODUCTION

Cashew (*Anacardium occidentale* L.) is a tropical evergreen plant known for its seeds that are consumed worldwide. It is one of the most economically important genera in the Anacardiaceae family and is presently cultivated in many regions of the world [1]. World production of cashew nut is about 2.8 million tons in 2005. Vietnam, Nigeria, and India are the major cashew nut producers, with 33.5%, 20.7% and 19.0%, respectively, of world production of cashew nut [2]. The cashew industry ranks third in the world production of edible nuts [3]. Cashew tree produces several products. Cashew nut is the seed of cashew. There is no doubt that the nut is the most important product of the cashew tree. It is consumed whole, roasted, shelled, and salted. Its fruit, also called as cashew apple, may be eaten raw, preserved as jam, made into a beverage, or fermented into a wine [4]. Cashew nut shell oil, also known as cashew nut shell liquid (CNSL), is extracted from the honey-combed shell of the cashew nut and has been used in several industrial applications.

For the last decades, there has been a surge of interest in searching for bioactive compounds originating from nuts. Bioactive compounds are beneficial components in functional/healthy foods that are responsible for their disease prevention properties. They may exert their effects by acting as antioxidants, inhibiting cholesterol absorption, blocking the activity of bacterial or viral toxins, decreasing platelet aggregation, or destroying harmful gastrointestinal bacteria. They include a wide range of chemical compounds with varying structures such as phenolics, carotenoids, phytosterols, phytostanols, tocopherols, and mono- and polyunsaturated fatty acids (MUFA and PUFA), among others. Nut consumption has been shown to lower total blood cholesterol, as well as low-density lipoprotein (LDL) cholesterol and has been associated with decreased risk of cardiovascular disease (CVD) [5]. Several groups of bioactive compounds have been identified in cashew nuts, including MUFA, PUFA, phenolics, phytosterols, phytostanols, tocopherols, phytates, and other minor components.

10.2 BIOACTIVE COMPOUNDS OF CASHEW NUT AND ITS COPRODUCTS

10.2.1 Unsaturated Fatty Acids (MUFA and PUFA)

Cashew nut contains a relatively high amount of fat. Its total lipids ranges from 40.4% to 46.5% on the whole seed. The fatty acid composition of cashew nut oil from different sources is shown in Table 10.1. It consists mainly of 18:1ω9 (59%–61%) and 18:2ω6 (16%–20%), with small amounts of 16:0 and 18:0, but with only trace amounts of 18:3ω3 [6–8]. The average content of total unsaturated fatty acids of cashew nut oil is around 78%–80% of total fatty acids. The content 18:1ω9 and 18:2ω6 for cashew nut are similar to that of canola oil [9]. Nutritionally, 18:2ω6 is considered as an essential fatty acid [75]. Moreover, 18:2ω6 can be elongated to 20:4ω6, which plays an important role in prostaglandin metabolism and may influence the thrombotic process [10]. In 2003, Food and Drug Administration (FDA) promulgated a qualified health claim for tree nuts, including cashew nut, that eating 1.5 ounces (~42.5 g) of nuts daily as part of a diet low in saturated fat and cholesterol may reduce the risk of heart disease [11].

10.2.2 Phenolic Compounds

Natural phenolic compounds are ubiquitous and widely distributed secondary metabolites in the plant kingdom possessing at least one aromatic ring with one or more hydroxyl groups attached to it.

TABLE 10.1
Fatty Acid Composition (%) of Oil Extracted from Cashew Nut

Fatty Acid	Ref.[a] [6]	Ref. [7]	Ref. [8]	Ref. [49]
Total lipid (g/100 g nut)	46.5	43.71	40.4	43.85
6:0	—	0.02	—	0.00
11:0	—	0.05	—	0.00
14:0	—	0.03	0.07	0.04
16:0	11.50	10.7	9.93	9.93
16:1	0.33	0.54	0.36	0.34
17:0	0.13	0.12	0.14	0.12
18:0	8.80	9.32	8.70	8.18
18:1ω9	61.44	61.14	57.24	59.67
18:2ω6	17.09	16.88	20.80	19.74
18:3ω3	0.20	0.32	0.23	0.16
20:0	0.51	0.63	0.97	0.67
20:1ω9	—	0.00	0.25	0.35
22:0	—	0.12	0.39	0.44
22:1ω9	—	0.00	0.28	0.00
24:0	—	0.13	—	0.26
Total SFA	20.94	21.12	20.20	19.74
Total MUFA	61.77	61.68	58.13	60.36
Total PUFA	17.29	17.20	21.03	19.90

[a] Values are average of fatty acid composition of eight cashew nut oil samples.

They play a very important role in plants and also in plant-derived foods. They are one of the most important sources of bioactive compounds in the human diet [12]. Accumulating clinical and epidemiological evidences have shown that phenolics might participate in prevention of various diseases associated with oxidative stress [13,14]. Phenolic compounds could serve as a major determinant of antioxidant potential of foods because of their ability to donate a hydrogen atom or an electron in order to form stable radical intermediates. The major phenolics found in cashew are anacardic acids, cardanols, cardols, tocopherols, and other minor phenolic constituents.

10.2.2.1 Total Phenolics Content and Total Antioxidant Activity

Phenolic compounds are associated with antioxidant activity and play an important role in preventing lipid peroxidation [15]. The total phenolics content (TPC) of different nuts (expressed as mg of gallic acid equivalents (GAE)/100 g dry nut) investigated by different researchers [16,17] is assembled in Table 10.2. The TPC of cashew nut from two different sources are quite different, one is 137 mg GAE/100 g and the other one is 274 mg GAE/100 g of nuts. The difference might be due to the variety or the skin of cashew nut sample. According to the study of Kamath and Rajini [18], the TPC of ethanolic extract of cashew nut skin is 243 mg GAE/100 g of cashew skin powder. It is indicated that the cashew nut skin/testa contains high amount of phenolics. Cashew nut has the second lowest TPC and total antioxidant activity (TAA) among the listed nuts in Table 10.2. TAA seems to correlate well with the TPC.

Significant antioxidant activity has also been observed in the hexane extracts of different cashew products (Table 10.3). The antioxidant activity was determined in the hypoxanthine/xanthine oxidase assay and expressed as inhibition of reactive oxygen species (ROS) attack on salicylic acid (%). CNSL has the highest antioxidant activity, followed by cashew fiber, cashew apple, raw cashew nut, and roasted cashew nut.

10.2.2.2 Anacardic Acids, Cardanols, Cardols, and 2-Methylcardols

Cashew nut and its coproducts are rich sources of long-chain alkyl substituted salicylic acid and resorcinol derivatives, namely anacardic acids, cardanols, cardols, and 2-methylcardols (Figure 10.1). The content of different alkyl substituted phenols, including anacardic acids, cardanols, and

TABLE 10.2
Total Phenolics Content and Total Antioxidant Activity of Different Nuts

	TPC (mg of GAE/100 g)[a]		TAA (μmol of TE/g)[b]
Nut	**Ref. [16]**	**Ref. [17]**	**Ref. [17]**
Cashew	137	274	19.97
Hazelnut	291	835	96.45
Macadamia	46	156	16.95
Pecan	1284	2016	179.4
Pistachio	867	1657	79.83
Walnut	1625	1556	135.41

[a] TPC, expressed as milligrams of gallic acid equivalents (GAE) per 100 g of dry nut (mg of GAE/100 g).

[b] TAA, expressed as micromoles of Trolox equivalents (TE) per gram of dry nut (μmol of TE/g).

TABLE 10.3
Antioxidant Activity of Hexane Extracts of Different Cashew Products as Determined in the Hypoxanthine/Xanthine Oxidase Assay

	Extract (10 mg/mL in Assay Buffer)				
				Cashew Nut	
	CNSL	Cashew Fiber	Cashew Apple	Raw	Roasted
Inhibition of ROS attack on salicylic acid (%)	100	94	53	41	37

Source: From Trevisan, M.T.S., Pfundstein, B., Haubner, R., Wurtele, G., Spiegelhalder, B., Bartsch, H., and Owen, R.W., *Food Chem. Toxicol.*, 44, 188, 2006. With permission.
Note: CNSL, cashew nut shell liquid; ROS, reactive oxygen species.

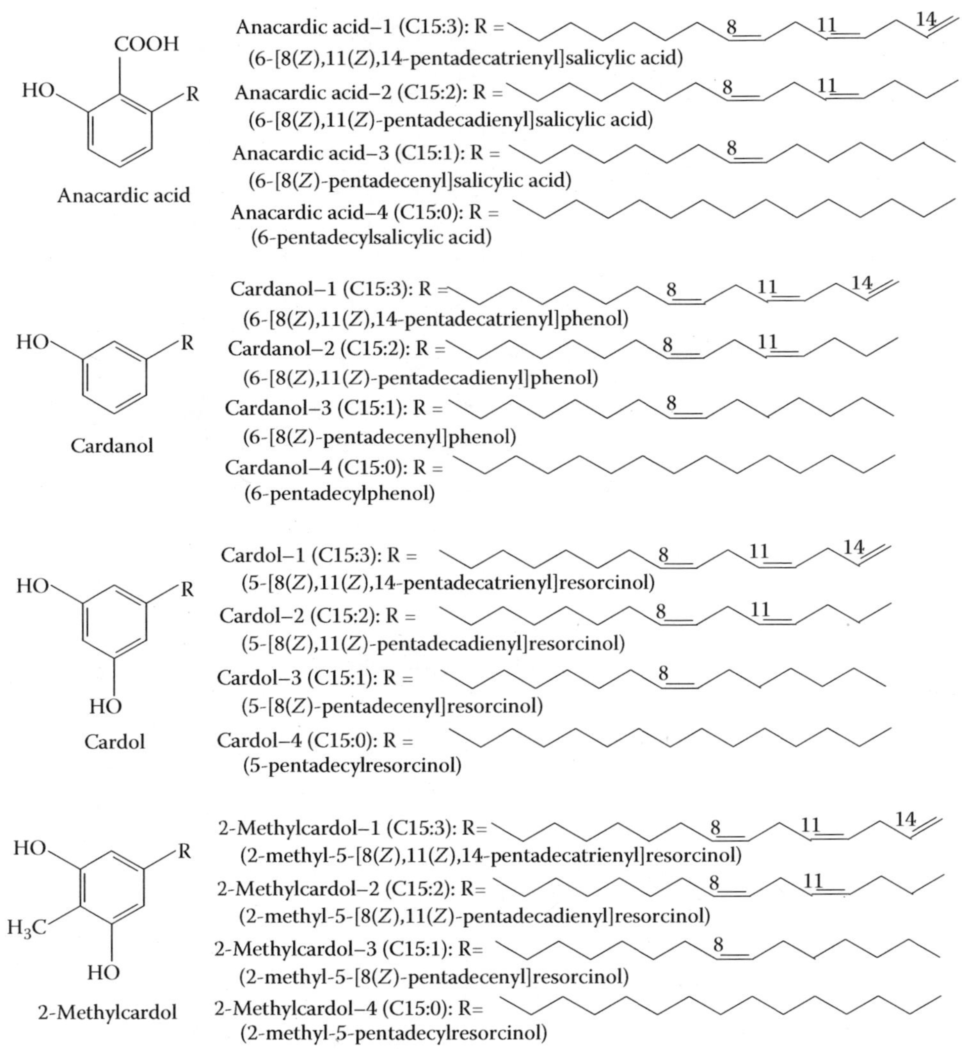

FIGURE 10.1 Chemical structures of anacardic acids, cardanols, cardols, and 2-methylcardols found in cashew nut and its coproducts.

TABLE 10.4
Content of Alkyl Phenols (Ancardic Acids, Cardanols, and Cardols) in Cashew Nut and Its Coproducts (mg/g)

Alkyl Phenols	Cashew Nut	CNSL	Cashew Apple
Anacardic acid-1	0.58	153.50	0.22
Anacardic acid-2	0.20	107.96	0.32
Anacardic acid-3	0.28	92.12	0.56
Cardanol-1	nd	97.16	nd
Cardanol-2	nd	63.57	nd
Cardanol-3	nd	55.31	nd
Cardol-1	0.22	98.18	nd
Cardol-2	0.07	46.65	nd
Total alkyl phenols	1.35	714.25	1.10

Source: From Trevisan, M.T.S., Pfundstein, B., Haubner, R., Wurtele, G., Spiegelhalder, B., Bartsch, H., and Owen, R.W., *Food Chem. Toxicol.*, 44, 188, 2006. With permission.
Note: CNSL, cashew nut shell liquid; nd, not detected.

cardols in cashew nut and its coproduct is shown in Table 10.4. Cashew apple contained only anacardic acids, while cashew nut contained both anacardic acids and cardols. CNSL contained an abundance of anacardic acids (50%), cardanols (30%), and cardols (20%). These results do not agree well with those of Shobha et al. [19], who reported that approximately 70% of CNSL were anacardic acids and 25% cardols.

Anacardic acids are 6-alkylsalicylic acids with different alkyl chain lengths and degrees of unsaturation. Cashew is the most important natural source of anacardic acids. Anacardic acids are found not only in cashew nut shell oil, but also in the nut and fruit juice made from cashew apple [20]. Anacardic acids are relatively nonpolar hydrophobic substances. Their structures are illustrated in Figure 10.1. They are insoluble in water, but well soluble in organic solvents such as ethanol or dimethyl sulfoxide (up to 10 mg/mL). The cardanols might be formed by thermal decarboxylation of the anacardic acids [19].

10.2.2.3 Tocopherols

Tocopherols are a class of plant phenolics often encountered in foods and possess important antioxidant and nutritional properties [21]. As shown in Table 10.5, γ-tocopherol is prevalent in cashew, while α- and δ-tocopherols are present in lower amounts. The γ-tocopherol content in oil extracted from cashew nut ranged from 5.10 to 6.18 mg/100 g oil [8,16]. The γ-tocopherol is a more potent antioxidant in oils than other tocopherols [22], but its poor vitamin E activity in biological systems has also been reported [23].

10.2.2.4 Other Minor Phenolic Constituents

The skin/testa of cashew nut is reddish-brown in color. It serves as a good source of catechol-type tannin [24]. (+)-Catechin and (−)-epicatechin are the main phenolic constituents of cashew kernel testa, which account for 6% and 7.5% of the dried weight of cashew kernel testa, respectively and they constitute approximately 40% of the total phenolics in cashew nuts. The polymeric proanthocyanidins account for 40% of total testa phenolics; other components of testa phenolics include

TABLE 10.5
Tocopherol Composition (mg/100 g) of Oil Extracted from Cashew Nut

Tocopherol	Ref. [6]	Ref. [8]	Ref. [16]	Ref. [a][49]
α-Tocopherol	0.51	3.60	nd	0.90
γ-Tocopherol	6.18	5.72	5.10	5.34[b]
δ-Tocopherol	0.41	nr	0.30	0.36

Note: nd, not detected; nr, not reported.
[a] Raw nut.
[b] This value also include β-tocopherol.

leucocyanidins and leucopelargonidins [25]. The presence of gallic, caffeic, and quinic acids apart from catechin and leucocyanidin was reported by Kantamoni [26]. Furthermore, nonylphenol has been identified in cashew apple headspace constituents and cinnamic acid in glycosidically bound components [27].

10.2.3 Phytosterols and Phytostanols

Phytosterols are triterpenes with a wide spectrum of biological activities in animal and humans such as anti-inflammatory [28], antibacterial [29], antioxidative [30], and anticancer [31] activities. They are bioactive nonnutrient substances and similar in structure to cholesterol, but include a methyl or ethyl group at C-24 [32]. They cover plant sterols (unsaturated form) and plant stanols (saturated form). Both plant sterols and plant stanols are effective in lowering plasma total and LDL cholesterols and in inhibiting the absorption of cholesterol from the small intestine [33]. The most common representative members in this series are sitosterol, campesterol, stigmasterol, Δ^5-avenasterol, sitostanol, and campestanol. The structures of identified phytosterols and phytostanols in cashew nut are shown in Figure 10.2. These mainly include sitosterol, campesterol, stigmasterol, sitostanol, campestanol, and Δ^5-avenasterol.

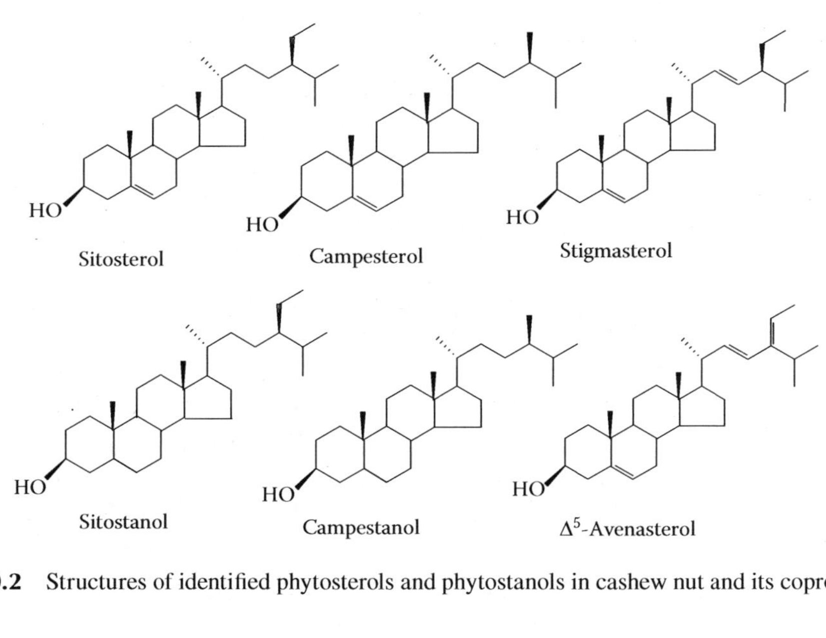

FIGURE 10.2 Structures of identified phytosterols and phytostanols in cashew nut and its coproducts.

TABLE 10.6
Phytosterol and Phytostanol Composition (mg/100 g) of Cashew Nut

	Ref. [76]		Ref. [49]
Sterol and Stanol	**Raw**	**Oil roasted**	**Dry roasted**
β-Sitosterol	112.6	119.3	nr
Campesterol	8.9	10.0	nr
Stigmasterol	<1.2	<1.7	nr
Δ^5-Avenasterol	13.7	13.6	nr
Sitostanol	<1.2	<1.7	nr
Campestanol	2.0	2.8	nr
Other sterols	13.3	12.6	nr
Total	150.0	158.0	158.0

Note: nr, not reported.

The phytosterol and phytostanol composition of cashew nut reported from different research groups [49,76] is listed in Table 10.6. β-Sitosterol is the predominant phytosterol (112.6–119.3 mg/100 g) in cashew nut, followed by campesterol (8.9–10 mg/100 g), and Δ^5-avenasterol (13.6–13.7 mg/100 g). Total content of phytosterols and phytostanols in cashew nut are around 150–158 mg/100 g. This agrees with the result from Toschi et al. [6], which reported that in the sterol fraction of cashew nut, 79.4% are β-sitosterol, then followed with Δ^5-avenasterol (9.4%), campesterol, 24-methylencholesterol (6.4%), and other minor sterols (4.7%).

10.2.4 Phytates

Phytic acid (known as inositol hexakisphosphate [IP_6] or phytate [as a salt]) is an important source of plant phosphorus. Its six reactive phosphate groups have strong chelating ability to complex with proteins and particularly with minerals, thus contributing to their antinutritional effects [34]. This is the main reason for mineral deficiency in young children in developing countries, particularly those with low intake of essential minerals and high intake of phytic acid [35]. Dietary phytates have attracted much interest in recent years due to their role in cancer prevention and their hypocholersterolemic effect [36]. Vucenik and Shamsuddin [37] have suggested that phytates may reduce the risk of colon cancer by reducing oxidative stress. Researchers now believe that phytic acid, found in the fiber of legumes and grains, is the major ingredient responsible for preventing colon and other types of cancer. They may be viewed as phytonutrients, providing antioxidant effect by binding the excess free iron thus preventing the formation of free radicals by Fenton reaction [38]. The content of IP_6 in cashew nut reported by Chen [39] and Venkatachalam and Sathe [7] is 4.99 and 2.9 mg/g, respectively. However, it is 12.29 mg/g from Harland et al. [40]. More research is necessary to investigate the role of phytic acid in cancer prevention and related biological activities.

10.2.5 Other Bioactive Micronutrients

Cashew fruit is also a rich source of vitamins, minerals, and other essential micronutrients. The cashew apple is an excellent source of ascorbic acid (averaging 269–287 mg/100 mL of juice), which is about four to seven times that of orange juice [41,42]. It is also a rich source of vitamin A precursors [43].

Vitamin K (phylloquinone) is an essential cofactor for the conversion of glutamic acid to gamma-carboxyglutamic acid residues in vitamin-K-dependent proteins, including hemostasis factors II, VII, IX, and X [44]. The high dietary intake of vitamin K and low risk of hip fracture is reported [45]. Cashew nut and pine nut are the two nuts reported to contain appreciable amounts of vitamin K.

Their contents are 34.8 and 53.9 μg/100 g nut, respectively [46], while most other commonly consumed nuts such as almond, Brazil nut, peanut, and walnut contained < 5 μg/100 g. However, the increased consumption of dietary vitamin K could reverse the anticoagulation effect of warfarin, an oral anticoagulant prescribed for primary and secondary prevention of thromboembolic disease, by acting through an alternate enzymatic pathway [47].

Folate is a generic term for a complex family of water-soluble B-group vitamins. It plays an important role in nucleotide synthesis, methylation, and gene expression, which makes it very necessary for the production and maintenance of new cells. This is especially important during periods of rapid cell division and growth such as infancy and pregnancy [48]. The content of folate in cashew nut is 25 μg/100 g nut [49]. The recommended dietary allowance (RDA) for folate equivalents for pregnant women is 600 μg, twice the normal RDA of 300 μg for unpregnant women [50]. Nutritional composition of cashew is given in detail in Chapter 2.

10.3 NUTRACEUTICAL AND PHARMACEUTICAL USAGES OF CASHEW NUT AND ITS COPRODUCTS

Cashew is a multipurpose tree. The leaves, stem bark, stem, and fruits of cashew have been used as traditional medicine for a long time in some countries [51]. Cashew tree leaves are used in South Cameroon by traditional practitioners as well as in other countries, as a folk remedy for diabetes mellitus [52,53]. The protective role of a cashew tree leaf extract against streptozotocin-induced diabetes in both mice and rats is reported [54,55]. The astringent cashew tree bark is rich in tannins and popularly used for healing wounds. It is also indicated for combating hypertension, in the treatment of gastric disturbances, and for anti-inflammatory and bactericidal treatment. An infusion of cashew stem bark presents analgesic and aphrodisiac properties and is also indicated against intestinal cramps, peptic ulcers, asthmas, and bronchitis [56,57]. Kubo et al. [58] reported that cashew fruit exhibited antibacterial activity against the Gram-negative bacterium *Helicobacter pylori*, which is now considered to cause acute gastritis and stomach ulcers. The same antibacterial compounds

TABLE 10.7
Worldwide Ethnomedical Usages of Cashew Products

Country	Medical Usages
Africa	For malaria
Brazil	For asthma, bronchitis, corns, cough, diabetes, dyspepsia, eczema, fever, genital disorders, impotence, intestinal colic, leishmaniasis, libido stimulation, muscular debility, pain, psoriasis, scrofula, syphilis, throat (sore), tonsillitis, ulcers (mouth), urinary disorders, urinary insufficiency, venereal disease, warts, wounds, and used as a gargle and mouthwash
Haiti	For cavities, diabetes, stomatitis, toothache, warts
Malaysia	For constipation, dermatosis, diarrhea, flu, nausea, thrush
Mexico	For diabetes, diarrhea, freckles, leprosy, skin, swelling, syphilis, ulcer, wart
Nigeria	For dysentery, diarrhea, piles, toothache, and pellagra
Panama	For asthma, colds, congestion, diabetes, diarrhea, hypertension, inflammation
Peru	For diarrhea, flu, infection, skin infections and used as an antiseptic and douche
Turkey	For diarrhea, fever, poisoning, warts
Venezuela	For dysentery, leprosy, sore throat and used as a gargle

Source: From Taylor, L., *The Healing Power of Rainforest Herbs*, Square One Publishers, Garden City Park, NY, 2005. With permission.

TABLE 10.8
Selected Biological Functions of Different Components Found in Cashew Nut and Its Coproducts

Alkyl Phenols	Molluscicidal Activity, Expressed as LD_{50} (ppm) [62]	Antitumor Activity Expressed as ED_{50} (μg/mL) [64]		Antimicrobial Activity MIC against Microorganisms (μg/mL) [60]	
		BT-20	HeLa	*Bacillus subtilis*	*Streptococcus mutans*
Anacardic acid-1	0.3	3.23	3.84	3.13	1.56
Anacardic acid-2	0.6	3.08	3.91	12.5	3.13
Anacardic acid-3	1	4.02	2.69	6.25	3.13
Anacardic acid-4	nr	7.42	4.94	12.5	3.13
Cardanol-1	80	>10	nt	50	1.56
Cardanol-2	80	>10	nt	>100	3.13
Cardanol-3	>100	>10	nt	>100	>100
Cardanol-4	nr	>10	nt	>100	>100
Cardol-1	15	2.08	2.56	1.56	0.78
Cardol-2	7	2.63	3.01	3.13	1.56
Cardol-3	7	1.72	2.74	6.25	1.56
Cardol-4	nr	6.25	4.02	>100	>100
2-Methylcardol-1	20	3.73	5.05	3.13	0.78
2-Methylcardol-2	15	2.64	7.38	6.25	1.56
2-Methylcardol-3	10	6.23	6.06	>100	>100
2-Methylcardol-3	nr	7.38	>10	>100	>100

Note: LD_{50}, the lethal doses for 50% mortality; ED_{50}, effective dose to produce 50% response; BT-20, BT-20 breast carcinoma cells; HeLa, Hela epithelioid cervix carcinoma cells; MIC, minimal inhibitory concentration; nr, not reported; nt, not tested.

have also been found to inhibit urease. The cashew fruit produces a resin with antiseptic, anthelmintic, and vesicant properties [59]. The worldwide ethnomedical usage of cashew products is shown in Table 10.7.

Various parts of the cashew plant are submitted to chemical and pharmacological screening. Diverse biological activities of anacardic acids, cardanols, cardols, and 2-methylcardols have been reported. These include antibacterial activity against Gram-positive bacteria [60,61], molluscididal activity [62], antitumor activities [63,64], weak antifungal activity against molds [60,65], tyrosinase [66], lipoxygenase [19], prostaglandin endoperoxidase synthase [67], and xanthine oxidase [68] inhibitory activities, and antioxidant activity [69,70]. The selected functions of different components isolated from cashew nut and its coproducts are summarized in Table 10.8.

Although CNSL does not have any nutraceutical and pharmaceutical usage, it is an important and versatile industrial raw material. It is used as a deterrent to pests of boat hulls, fishing nets, and stem borers of trees, or used to manufacture paints, synthetic rubber, varnishes, epoxy resins, oil soluble resins, surface active agents, automobile braking linings, wax compoundings, and foundry resins [71,72]. The cardanols, the main constituents of CNSL, are transformed for gasoline stabilization [73], and are also used for the synthesis of aryl glycolipids and their self-assembled nanostructures [74].

10.4 CONCLUSION

Cashew nut and its coproducts provide a complex food rich in macronutrients and micronutrients, as well as small quantities of various antioxidants and bioactive compounds that are relevant to many health beneficial attributes. These bioactive compounds include MUFA, PUFA, phenolics, phytosterols, phytostanols, tocopherols, and phytates. While benefits of cashew nut and its coproducts for human health appear promising, further research on bioavailability and bioaccessibility of cashew bioactive compounds as well as potential allergenic reaction is required for a better understanding of the role of cashew nut and its coproducts in human health.

REFERENCES

1. Mitchell, J.D. and Mori, S.A., *The Cashew and Its Relatives (Anacardium: Anavardiaceae)*, Memoirs of the New York Botanical Garden, New York, 1987.
2. FAOSTAT, World Production Quantity (1000 tonnes) of Cashew Nut in 2005, FAO Statistics Division, FAO, Rome, Italy, 2007.
3. Azam-Ali, S.H. and Judge, E.C., *Small-Scale Cashew Nut Processing*, FAO, Rome, Italy, 2004.
4. Mohanty, S., Ray, P., Swain, M.R., and Ray, R.C., Fermentation of cashew (*Anacardium occidentale* L.) "apple" into wine, *J. Food. Process. Pres.*, 30, 314–322, 2006.
5. John, H., Kelly, J.R., and Sabaté, J., Nuts and coronary heart disease: An epdidemiological perspective, *Br. J. Nutr.*, 96(Suppl. 2), 61S–67S, 2006.
6. Toschi, T.G., Caboni, M.F., Penazzi, G., Lercher, G., and Capella, P., A study on cashew nut oil composition, *J. Am. Oil Chem. Soc.*, 70, 1017–1020, 1993.
7. Venkatachalam, M. and Sathe, S.K., Chemical composition of selected edible nut seeds, *J. Agric. Food Chem.*, 54, 4705–4714, 2006.
8. Ryan, E., Galvin, K., O'Connor, T.P., Maguire, A.R., and O'Brien, N.M., Fatty acid profile, tocopherol, squalene and phytosterol content of Brazil, pecan, pine, pistachio and cashew nuts, *Int. J. Food. Sci. Nutr.*, 57, 219–228, 2006.
9. Kris-Etherton, P.M., Yu-Poth, S., Sabaté, J., Ratcliffe, H.E., Zhao, G., and Etherton, T.D., Nuts and their bioactive constituents: Effects on serum lipids and other factors that affect disease risk, *Am. J. Clin. Nutr.*, 70, 504S–511S, 1999.
10. Hu, F.B., Stampfer, M.J., Manson, J.A.E., Rimm, E.B., Wolk, A., Colditz, G.A., Hennekens, C.H., and Willett, W.C., Dietary intake of α-linolenic acid and risk of fatal ischemia heart disease among women, *Am. J. Clin. Nutr.*, 69, 890–897, 1999.
11. Food and Drug Administration (FDA), Qualified Health Claims: Letter of Enforcement Discretion—Nuts and Coronary Heart Disease, Docket No 02P-0505, FDA, Washington, DC, 2003.
12. Ortuno, A., Garcia-Puig, D., Fuster, M.D., Perez, M.L., Sabater, F., Porras, I., Garcia-Lidon, A., and Del, R.J.A., Flavonone and flavone levels in different varieties of grapefruit and pumello, *J. Agric. Food Chem.*, 43, 1–5, 1995.
13. Temple, N.J., Antioxidants and disease: More questions than answers, *Nutr. Res.*, 20, 449–459, 2000.
14. Manach, C., Williamson, G., Morand, C., Scalbert, A., and Rémésy, C., Bioavailability and bioefficacy of polyphenols in humans. I. Review of 97 bioavailability studies, *Am. J. Clin. Nutr.*, 81, 230S–242S, 2005.
15. Shahidi, F. and Wanasundara, P.K.J.P.D., Phenolic antioxidants, *Crit. Rev. Food. Sci. Nutr.*, 32, 67–103, 1992.
16. Kornsteiner, M., Wagner, K.H., and Elmadfa, I., Tocopherols and total phenolics in 10 different nut types, *Food Chem.*, 98, 381–387, 2006.
17. Wu, X., Beecher, G.R., Holden, J.M., Haytowitz, D.B., Gebhardt, S.E., and Prior, R.L., Lipophilic and hydrophilic antioxidant capacities of common foods in the United States, *J. Agric. Food Chem.*, 52, 4026–4037, 2004.
18. Kamath, V. and Rajini, P.S., The efficacy of cashew nut (*Anacardium occidentale* L.) skin extract as a free radical scavenger, *Food Chem.*, 103, 428–433, 2007.
19. Shobha, S.V., Ramadoss, C.S., and Ravindranath, B., Inhibition of soybean lipoxygenase-1 by anacardic acids, cardols, and vardanols, *J. Nat. Prod.*, 57, 1755–1757, 1994.
20. Toyomizu, M., Okamoto, K., Ishibashi, T., Chen, Z., and Nakatsu, T., Uncoupling effect of anacardic acids from cashew nut shell oil on oxidative phosphorylation of rat liver mitochondria, *Life Sci.*, 66, 229–234, 2000.

21. Brigelius-Flohe, R. and Traber, M.G., Vitamin E: Function and metabolism, *FASEB. J.*, 13, 1145–1155, 1999.
22. Olcott, H.S. and Emerson, O.H., Antioxidants and the autoxidation of fats. IX. The antioxidant properties of the tocopherols, *J. Am. Oil Chem. Soc.*, 59, 1008–1009, 1937.
23. Burton, G.W. and Traber, M.G., Vitamin E: Antioxidant activity, biokinetics, and bioavailability, *Ann. Rev. Nutr.*, 10, 375–382, 1990.
24. Pillai, M.K.S., Kedlaya, K.J., and Selvarangam, R., Cashew skin as a tanning material, *Leather Sci.*, 10, 317–318, 1963.
25. Mathew, A.G. and Parpia, H.A.B., Polyphenols of cashew kernel testa, *J. Food Sci.*, 35, 140–145, 1970.
26. Kantamoni, S., Isolation and identification of the constituents of cashew nut husk, *Leather Sci.*, 12, 396–398, 1965.
27. Bicalho, B., Pereira, A.S., Neto, F.R.A., Pinto, A.C., and Rezende, C.M., Application of high-temperature gas chromatography–mass spectrometry to the investigation of glycosidically bound components related to cashew apple (*Anacardium occidentale* L. Var. nanum) volatiles, *J. Agric. Food Chem.*, 48, 1167–1174, 2000.
28. Bouic, P.J., Sterols and sterolins: New drugs for the immune system? *Drug Discov. Today*, 7, 775–778, 2002.
29. Zhao, C.C., Shao, J.H., Li, X., Xu, J., and Zhang, P., Antimicrobial constituents from fruits of Ailanthus altissima SWINGLE, *Arch. Pharmacal Res.*, 28, 1147–1151, 2005.
30. van-Rensburg, S.J., Daniels, W.M.U., van-Zyl, J.M., and Taljaard, J.J.F., A comparative study of the effects of cholesterol, beta-sitosterol, beta-sitosterol glucoside, dehydroepiandrosterone sulphate and melatonin on *in vitro* lipid peroxidation, *Metab. Brain. Dis.*, 15, 257–265, 2000.
31. Awad, A.B., Downie, A., Fink, C.S., and Kim, U., Dietary phytosterols inhibits the growth and metastasis of MDA-MB-231 human breast cancer cells grown in SCID mice, *Anticancer Res.*, 20, 821–824, 2000.
32. Moreau, R.A., Whitaker, B.D., and Hicks, K.B., Phytosterols, phytostanols, and their conjugates in foods: Structural diversity, quantitative analysis, and health-promoting uses, *Prog. Lipid Res.*, 41, 457–500, 2002.
33. Piironen, V., Lindsay, D.G., Miettinen, T.A., Toivo, J., and Anna-Maija, L., Plant sterols: Biosynthesis, biological function and their importance to human nutrition, *J. Sci. Food Agric.*, 80, 939–966, 2000.
34. Urbano, G., Lopez-Jurado, M., Aranda, P., Vidal-Valverde, C., Tenorio, E., and Porres, J., The role of phytic acid in legumes: Antinutrient or beneficial function? *J. Physiol. Biochem.*, 56, 283–294, 2000.
35. Hurrell, R.F., Influence of vegetable protein sources on trace element and mineral bioavailability, *J. Nutr.*, 133, 2973S–2977S, 2003.
36. Shamsuddin, A.M., Inositol phosphates have novel anticancer function, *J. Nutr.*, 125 (Suppl. 3), 725S–732S, 1995.
37. Vucenik, I. and Shamsuddin, A.M., Cancer inhibition by inositol hexaphosphate [IP_6] and inositol: From laboratory to clinic, *J. Nutr.*, 133, 3778S–3784S, 2003.
38. Graf, E., Empson, K., and Eaton, J.W., Phytic acid: A natural antioxidant, *J. Biol. Chem.*, 262, 11647–11650, 1987.
39. Chen, Q., Determination of phytic acid and inositol pentakisphosphates in foods by high-performance ion chromatography, *J. Agric. Food Chem.*, 52, 4604–4613, 2004.
40. Harland, B.F., Smikle-Williams, S., and Oberleas, D., High performance liquid chromatography analysis of phytate [IP_6] in selected foods, *J. Food Comp. Anal.*, 17, 227–233, 2004.
41. Maciel, M.I., Hansen, J.T., Aldinger, S.B., and Labows, J.N., Flavor chemistry of cashew apple juice, *J. Agric. Food Chem.*, 34, 923–927, 1986.
42. Inyang, U.E. and Abah, U.J., Chemical composition and organoleptic evaluation of juice from steamed cashew apple blended with orange juice, *Plant Foods Hum. Nutr.*, 50, 295–300, 1997.
43. Cecchi, H.M. and Rodrignez-Amaya, D.B., Carotenoid composition and vitamin A value of fresh and pasteurized cashew apple juice, *J. Food Sci.*, 46, 147–149, 1981.
44. Furie, B., Bouchard, B.A., and Furie, B.C., Vitamin K dependent biosynthesis of gamma-carboxyglutamic acid, *Blood.*, 93, 1798–1808, 1999.
45. *Dietary Reference Intakes for Vitamin A, Vitamin K, Arsenic, Boron, Chromium, Copper, Iodine, Iron, Manganese, Molybdenum, Nickel, Silicon, Vanadium, and Zinc,* National Academy Press, Washington, DC, 2001.
46. Dismore, M., Haytowitz, D.B., Gebhardt, S., Peterson, J.W., and Booth, S., Vitamin K content of nuts and fruits in the U.S. diet, *J. Am. Diet. Assoc.*, 103, 1650–1652, 2003.
47. Booth, S.L., Charnley, J.M., Sadowski, J.A., Saltzman, E., Bovill, E.G., and Cushman, M., Dietary vitamin K_1 and stability of oral anticoagulation: Proposal of a diet with constant vitamin K_1 content, *Thromb. Haemost.*, 77, 504–509, 1997.

48. Stover, P.J., Physiology of folate and vitamin B_{12} in health and disease, *Nutr. Rev.*, 62(Suppl. June), 3S–12S. 2004.
49. U.S. Department of Agriculture (USDA), National Nutrition Database for Standard Reference, Release 20, National Technical Information Service, USDA, Springfield, VA, 2007.
50. Nordic Council of Ministers for Fishery, Agricultural, Forestry and Food, *Nordic Nutrition Recommendations: Integrating Nutrition and Physical Activity,* 4th ed., Nordic Council of Ministers, Copenhagen, Denmark, 2004.
51. Taylor, L., *The Healing Power of Rainforest Herbs*, Square One Publishers, Garden City Park, NY, 2005.
52. Gupta, M.P., Arias, T.D., Correa, M., and Lamba, S.S., Ethnopharmacognostic observations on Panamanian medicinal plants. Part I, *Q. J. Crude Drug Res.*, 17, 115–130, 1979.
53. Felcman, J. and Tristão-Bragança, M.L., Chromium in plants: Comparison between the concentration of chromium in Brazilian nonhypo and hypoglycemic plants, *Biol. Trace Elem. Res.*, 17, 11–16, 1988.
54. Swanston-Flatt, S.K., Day, C., Flatt, P.R., Gould, B.J., and Bailey, C.J., Glycaemic effects of traditional European plant treatments for diabetes. Studies in normal and streptozotocin diabetic mice, *Diabetes Res.*, 10, 69–73, 1989.
55. Kamtchouing, P., Sokeng, S.D., Moundipa, P.F., Watcho, P., Jatsa, H.B., and Lontsi, D., Protective role of *Anacardium occidentale* extract against streptozotocin-induced diabetes in rats, *J. Ethnopharmacol.*, 62, 95–99, 1998.
56. Mota, M.L.R., Thomas, G., and Barbosa Filho, J.M., Anti-inflammatory actions of tannins isolated from the bark of *Anacardium occidentale* L., *J. Ethnopharmacol.*, 13, 289–300, 1985.
57. Akinpelu, D.A., Antimicrobial activity of *Anacardium occidentale*, *Fitoterapia*, 72, 286–287, 2000.
58. Kubo, J., Lee, J.R., and Kubo, I., Anti-*Helicobacter pylori* agents from the cashew apple, *J. Agric. Food Chem.*, 47, 533–537, 1999.
59. Barcelos, G.R.M., Shimabukuro, F., Mori, M.P., Maciel, M.A.M., and Cólus, I.M.S., Evaluation of mutagenicity and antimutagenicity of cashew stem bark methanolic extract in vitro, *J. Ethnopharmacol.*, 114, 268–273, 2007.
60. Himejima, M. and Kubo, I., Antibacterial agents from the cashew *Anacardium occidentale* (Anacardiaceae) nut shell oil, *J. Agric. Food Chem.*, 39, 418–421, 1991.
61. Kubo, I., Muroi, H., Himejima, M., Yamagiwa, Y., Mera, H., and Tokushima, K., Structure-antibacterial activity relationships of anacardic acids, *J. Agric. Food Chem.*, 41, 1016–1019, 1993.
62. Kubo, I., Komatsu, S., and Ochi, M., Molluscicides from the cashew *Anacardium occidentale* and their large-scale isolation, *J. Agric. Food Chem.*, 34, 970–973, 1986.
63. Itokawa, H., Totsuka, N., Nakahara, K., Maezuru, M., Takeya, K., and Kondo, M., A quantitative structure-activity relationship for antitumor activity of long-chain phenols from *Ginko biloba* L., *Chem. Pharm. Bull.*, 37, 1619–1621, 1989.
64. Kubo, I., Ochi, M., Vieira, P.C., and Komatsu, S., Antitumor agents from the cashew *Anacardium occidentale* apple juice, *J. Agric. Food Chem.*, 41, 1012–1015, 1993.
65. Kubo, I., Nihei, K.I., and Tsujimoto, K., Antibacterial action of anacardic acids against methicillin resistant *Staphylococcus aureus* (MRSA), *J. Agric. Food Chem.*, 51, 7624–7628, 2003.
66. Kubo, I., Kinst-Hori, I., and Yokokawa, Y., Tyrosinase inhibitors from *Anacardium occidentale* fruits, *J. Nat. Prod.*, 57, 545–552, 1994.
67. Grazzini, R., Hesk, D., Heiminger, E., Hildenbrandt, G., Reddy, C.C., and Cox-Foster, D., Inhibition of lipoxygenase and prostaglandin endoperoxide synthase by anacardic acids, *Biochem. Biophys. Res. Commun.*, 176, 775–780, 1991.
68. Masuoka, N. and Kubo, I., Characterization of xanthine oxidase inhibition by anacardic acids, *Biochim. Biophys. Acta*, 1688, 245–249, 2004.
69. Kubo, I., Masuoka, N., Ha, T.J., and Tsujimoto, K., Antioxidant activity of anacardic acids, *Food Chem.*, 99, 555–562, 2006.
70. Trevisan, M.T.S., Pfundstein, B., Haubner, R., Wurtele, G., Spiegelhalder, B., Bartsch, H., and Owen, R.W., Characterization of alkyl phenols in cashew (*Anacardium occidentale*) products and assay of their antioxidant capacity, *Food Chem. Toxicol.*, 44, 188–197, 2006.
71. Rickson, F.R. and Rickson, M.M., The cashew nut, *Anacardium occidentale* (Anacardiaceae), and its perennial association with ants: Extrafloral nectary location and the potential for ant defense, *Am. J. Bot.*, 85, 835–849, 1998.
72. Patel, R.N., Bandyopadhyay, S., and Ganesh, A., Economic appraisal of supercritical fluid extraction of refined cashew nut shell liquid, *J. Chromatogr. A.*, 1124, 130–138, 2006.

73. Castro Dantas, T.N., Dantas, M.S.G., Dantas Neto, A.A., D'Ornellas, C.V., and Queiroz, L.R., Novel antioxidants from cashew nut shell liquid applied to gasoline stabilization, *Fuel*, 82, 1465–1469, 2003.
74. John, G., Masuda, M., Yase, K., and Shimizu, T., Nanotube formation from renewable resources via coiled nanofibers, *Adv. Mat.*, 13, 715–718, 2001.
75. Simopoulos, A.P., Essential fatty acids in health and chronic disease, *Am. J. Clin. Nutr.*, 70, 560S–569S, 1999.
76. Phillips, K.M., Ruggio, D.M., and Ashraf-Khorassani, M., Phytosterol composition of nuts and seeds commonly consumed in the United States, *J. Agric. Food Chem.*, 53, 9436–9445, 2005.

11 Chemical Composition and Health Aspects of Chestnut (*Castanea* spp.)

Figen Korel and Murat Ö. Balaban

CONTENTS

11.1 INTRODUCTION

According to the Food and Agricultural Organization [1] statistics, world's chestnut production was 1,164,959 MT in 2006 (Table 11.1). China was one of the largest producers, with 825,000 MT in 2006, followed by Korea (76,447 MT), Turkey (53,814 MT), Italy (52,000 MT), Bolivia (40,980 MT), Portugal (29,133 MT), and Japan (23,100 MT).

Chestnut has been part of the staple diet in many countries for centuries. It has a sweet and nutty flavor, and its texture is like a firm baked potato and not crunchy compared to other nuts. Chestnut, which is classified as a temperate nut, requires cold winters and warm summers. It has been grown on all continents in northern hemisphere and Australia [2,3]. In Asia, the Japanese chestnut (*Castanea crenata*) has been cultivated since the eleventh century and the Chinese chestnut (*Castanea mollissima*), possibly for 6000 years. In the Mediterranean region, chestnut has been cultivated for at least 3000 years [2]. In Europe, chestnut spreads from south to north and from Mediterranean to Sweden. It was brought to the Alps and Appennines, and new villages were built only where chestnut could grow and produce fruits [4].

TABLE 11.1
World Production of Chestnuts from 2000 to 2006 (in 10³ MT)

	Year													
Country	**2000**		**2001**		**2002**		**2003**		**2004**		**2005**		**2006**	
China	598	63[a]	599	63[a]	702	68[a]	797	71[a]	805	71[a]	825	71[a]	825	71[a]
South Korea	93	9.8	94	9.9	72	7.0	60	5.3	72	6.3	76	6.5	76	6.5
Italy	50	5.2	50	5.3	50	4.8	50	4.5	50	4.4	52	4.5	52	4.5
Turkey	50	5.2	47	5.0	47	4.6	48	4.3	49	4.3	50	4.3	54	4.6
Bolivia	34	3.6	35	3.7	35	3.4	35	3.1	39	3.4	41	3.5	41	3.5
Portugal	33	3.5	26	2.8	31	3.0	33	2.9	31	2.7	22	1.9	29	2.5
Japan	27	2.8	29	3.1	30	2.9	25	2.2	24	2.1	22	1.9	23	2.0
Russia	16	1.7	16	1.7	16	1.6	17	1.5	18	1.6	19	1.6	19	1.6
Greece	15	1.6	15	1.6	15	1.5	17	1.5	19	1.7	21	1.8	12	1.0
France	13	1.4	13	1.4	11	1.1	10	0.9	12	1.1	10	0.9	10	0.9
Spain	9	1.0	10	1.1	9	0.9	17	1.5	10	0.9	10	0.9	10	0.9
Total	953		945		1031		1122		1141		1162		1165	

Source: From Food and Agricultural Organization (FAO), Chestnut Production, Published online at: http://faostat.fao.org (accessed July 10, 2007). With permission.

[a] Percentages of total world production.

Chestnut contains 40–50 g/100 g carbohydrates and was consumed as a major source of complex carbohydrates until the introduction of potatoes [3]. It lost its importance in the diet of several communities since corn and potato plants proved to be more productive, and people started to consider chestnut as having less nutritional value. However, since the 1970s, the image of chestnut has changed from being viewed as the food for the poorest in society, and it became an ingredient of dishes and culinary items characterized by a high degree of sophistication due to its nutritional and dietary characteristics. Chestnut has a high content of polymeric carbohydrates, an acceptable content of lipids, and an adequate amount of minerals [5].

In this chapter, the taxonomy, origin, compositional and lipid characteristics, phenolic compounds, and allergens of chestnut as well as use of chestnut by-products are briefly discussed. The use of chestnut as a food or an ingredient in many recipes is mentioned. Health aspects of chestnut are particularly important due to its nutritional properties. These features as well as the consumption of chestnut are also summarized in this chapter.

11.2 TAXONOMY AND ORIGIN

Chestnut belongs to the *Fagaceae* family (with beeches and oaks) and the genus *Castanea*. The important nut-bearing species are *C. crenata*, *C. dentata*, *C. mollissima*, *C. sativa*, and the hybrids of *dentata* and *mollissima* [6].

C. crenata is a Japanese chestnut, cultivated for over 2000 years. Nuts of *C. crenata* are the largest in all the *Castanea* species (30 g), their quality is astringent, and nut flavor is considered inferior to other species. It is highly resistant to most known diseases, such as chestnut blight and gall wasp [2,7].

C. dentata is an American chestnut. Nuts of *C. dentata* are relatively small (3–12 g), but it is well known that it has the best flavor among the four known species. The Eastern half of the United States (from Maine to Georgia and as far west as Michigan and Louisiana) was once covered with chestnut trees, but the species is now nearly extinct due to the chestnut blight (*Cryphonectria parasitica*) [2,7]. Most of the current chestnut grown in the United States is usually of Chinese or Korean origin due to their resistance to blight and similarity to the native American chestnut.

C. mollissima is a Chinese chestnut. Nuts of *C. mollissima* are medium-large in size (10–30 g) with good quality. It is the most tolerant species to chestnut blight [2,7]. Hybrids of *dentata* and *mollisima* are supposedly blight-tolerant and have a quality similar to that of *C. dentata*.

C. sativa is a European chestnut. Nuts of *C. sativa* are medium-large (10–25 g) and their quality is variable from sweet to astringent depending on the cultivar. This species is susceptible to chestnut blight [7]. Chestnut grown in Turkey is of European origin (*C. sativa* Miller) and spreads from eastern Black Sea, Marmara, and Aegean to Antalya in the Mediterranean region [8].

11.3 USE OF CHESTNUT AS FOOD AND FOOD INGREDIENTS

Chestnut is consumed raw, or more commonly boiled or roasted to improve its flavor and digestibility. If the nuts are consumed immediately after harvesting, they may have an astringent taste; it is better to wait for a few weeks for the starch to slowly change to sugar [9,10]. Its distinctive flavor and texture give a unique taste to many dishes from appetizers to sweets. Chestnut is mainly used in candies, and in stuffing for meat and poultry. Many recipes are available, such as chestnut soup, cereals, ice cream, and stir fries as well as toasted, salted, buttered, candied, sautéed, stewed, and braised, among others [2,3,11]. Chestnut-chicken casserole, chestnut dressing, and Waldorf salad are other recipes [11]. In Italy, chestnut is ground to fine flour for confectionery. Before the introduction of maize to Europe, Italian polenta (like Italian grits) was made from chestnut flour [7]. The physical properties and sensory characteristics of a snack-like product produced from chestnut-rice flour blend are improved by the addition of chestnut flour due to its pleasant taste and texture. The density and color of the product also become more desirable [12].

Chestnut is dried, the pericarp and the endocarp are removed, and the nut is ground to obtain chestnut flour [13]. For an adequate conservation, the moisture content of chestnut should be around 50% [5]. Small nuts or nuts with double embryos are widely used for chestnut flour [13]. Chestnut flour is used as a confectionery paste, which is a basic ingredient for desserts [14]. Sacchetti et al. [12] used chestnut flour mixed with rice flour in the production of a snack-like product obtained through an extrusion-cooking process. A chestnut-rice flour blend with a 30% chestnut flour content extruded at 120°C had the best performance, resulting in a well-gelatinized and well-expanded product with good sensory attributes. Chestnut flour could be used as a functional ingredient in the formulation of snack-like products since it could improve the nutritional value of the extruded product. Sacchetti et al. [15] also stated that dried chestnut and chestnut flour could be easily preserved for several months because of their low water activity (a_w), but their fat content could limit their storage life because of oxidation processes.

11.4 COMPOSITIONAL AND LIPID CHARACTERISTICS

11.4.1 PROXIMATE COMPOSITION

Several researchers have examined the proximate composition of American, European, Chinese, and Japanese chestnuts [16–21]. The proximate composition of raw chestnuts of Chinese, European, and Japanese species is given in Table 11.2 [22]. According to this table, Japanese chestnut has the highest moisture content among the three sets of samples for different species. The protein content of Chinese, European, and Japanese chestnuts vary from 1.63 to 4.20 g/100 g of the edible portion. The fat content of the same chestnuts is in the range of 0.53–1.25 g/100 g of the edible portion. Chinese chestnut has the highest ash content, followed by Japanese and European chestnuts.

McCarthy and Meredith [23] analyzed the proximate composition of American and Chinese chestnuts grown in the United States and imported European chestnut. The moisture content of these chestnuts ranged from 43.70 to 54.88 g/100 g. The protein contents of American, Chinese, and European chestnuts were 4.83, 4.20, and 1.98 g/100 g, respectively. The fat content of European chestnut (1.63 g/100 g) was higher than those of American (1.32 g/100 g) and Chinese chestnuts (1.11 g/100 g). The crude fiber and ash contents of these nuts were between 1 and 2 g/100 g.

TABLE 11.2
Nutrient Contents of Raw Chestnuts in 100 g Edible Portion

Nutrient	Unit	Chinese Chestnut	European Chestnut	Japanese Chestnut
Proximate composition				
Water	g	43.95	52.00	61.41
Protein	g	4.20	1.63	2.25
Total lipid (fat)	g	1.11	1.25	0.53
Ash	g	1.67	0.96	0.91
Carbohydrates, by difference	g	49.07	44.17	34.91
Energy	kcal	224	196	154
Mineral				
Calcium	mg	18	19	31
Copper	mg	0.363	0.418	0.562
Iron	mg	1.41	0.94	1.45
Magnesium	mg	84	30	49
Manganese	mg	1.601	0.336	1.591
Phosphorus	mg	96	38	72
Potassium	mg	447	484	329
Sodium	mg	3	2	14
Zinc	mg	0.87	0.49	1.10
Vitamin				
Vitamin A (RAE)	μg	10	1	2
Vitamin C (total ascorbic acid)	mg	36.0	40.2	26.3
Thiamin (B_1)	mg	0.160	0.144	0.344
Riboflavin (B_2)	mg	0.180	0.016	0.163
Niacin (B_3)	mg	0.800	1.102	1.500
Pantothenic acid (B_5)	mg	0.555	0.476	0.206
Pyridoxine (B_6)	mg	0.410	0.352	0.283
Cobalamin (B_{12})	μg	0.00	0.00	0.00
Folate	μg	68	58	47
Amino acid				
Alanine	g	0.200	0.109	0.203
Arginine	g	0.430	0.116	0.148
Aspartic acid	g	0.852	0.281	0.474
Cystine	g	0.110	0.052	0.065
Glutamic acid	g	0.537	0.210	0.429
Glycine	g	0.184	0.084	0.114
Histidine	g	0.121	0.045	0.056
Isoleucine	g	0.157	0.064	0.111
Leucine	g	0.259	0.096	0.139
Lysine	g	0.228	0.096	0.147
Methionine	g	0.101	0.038	0.054
Phenylalanine	g	0.190	0.069	0.088
Proline	g	0.162	0.086	0.141
Serine	g	0.184	0.081	0.110
Threonine	g	0.167	0.058	0.090
Tryptophan	g	0.049	0.018	0.032
Tyrosine	g	0.125	0.045	0.064
Valine	g	0.220	0.091	0.134

TABLE 11.2 (continued)
Nutrient Contents of Raw Chestnuts in 100 g Edible Portion

Nutrient	Unit	Chinese Chestnut	European Chestnut	Japanese Chestnut
Fatty acid				
14:0	%	—	0.43	—
16:0	%	14.73	18.40	14.69
16:1 (undifferentiated)	%	0.78	1.04	0.82
18:0	%	1.07	1.04	1.02
18:1 (undifferentiated)	%	54.54	35.85	54.69
18:2 (undifferentiated)	%	25.17	38.19	25.10
18:3 (undifferentiated)	%	2.73	4.60	2.65
20:1	%	0.98	0.43	1.02
Total SFA	%	15.81	19.88	15.71
Total MUFA	%	56.29	37.33	56.53
Total PUFA	%	27.90	42.79	27.76

Source: From U.S. Department of Agriculture (USDA), Nutrient Data Lab, National Agricultural Library, Agricultural Research Service, 2006, Published online at: http://www.nal.usda.gov/fnic/foodcomp/cgi-bin/list_nut_edit.pl (accessed June 27, 2006).

Note: RAE, retinol activity equivalent.

Míguelez et al. [5] determined the nutrient composition of 15 common varieties of chestnuts from the region of Verín-Monterrei (Spain) and reported that the moisture content was in the range of 48.37 to 59.35 g/100 g. Protein content of these chestnut varieties was high, varying from 6.02 to 8.58 g/100 g of dry weight, and this was explained by the type of soil in which they were grown [5]. Chestnut grown in schistose soils showed a higher protein content than those of granite-based soils [24]. The mineral content of these varieties was low, ranging from 1.8 to 3 g/100 g [5]. Pereira-Lorenzo et al. [25] also determined the chemical composition of Spanish chestnut cultivars and reported that the average moisture, protein, fat, crude fiber, and ash contents were 54, 5.8, 3, 3, and 2.3 g/100 g, respectively.

Bellini et al. [26] investigated the compositional quality of Marrone del Mugello PGI (Protected Geographical Indication) chestnut of Mugello region in Italy. The moisture, carbohydrate, protein, fat, and insoluble fiber contents were 52.4, 41.3, 2.4, 2.1, and 1.4 g/100 g, respectively.

Korel and Balaban [27] determined the proximate composition of raw and candied chestnuts (European chestnut) and found that moisture content was in the range of 55.09 g/100 g for raw chestnut and 21.29–23.90 g/100 g for candied chestnut. The protein content was 2.33 for raw and 1.31–1.35 g/100 g for candied chestnuts. The lipid content for raw and candied chestnuts was 1.30 and 0.29–0.78 g/100 g of dry weight, respectively. The carbohydrate content was reported as 38.82 and 76.13 g/100 g for raw and candied chestnuts, respectively. The ash content was 2.46 g/100 g for raw and 0.68 g/100 g for candied chestnuts.

11.4.2 Minerals

The importance of the minerals and the consequences of their deficiency are well known [28]. The mineral contents of Chinese, European, and Japanese chestnuts are listed in Table 11.2 [22]. European chestnut has the highest potassium content compared to Chinese and Japanese chestnuts. Magnesium, phosphorus, and manganese contents are higher in Chinese chestnut than in European and Japanese chestnuts. On the other hand, Japanese chestnut contains the highest amounts of calcium, iron, sodium, zinc, and copper contents among these three varieties. The minimum and maximum percentages of recommended dietary allowances (RDA) or adequate intake (AI) of these minerals obtained by

TABLE 11.3
Mineral Content of Chestnut and Percentage of RDA or AI

Mineral	RDA or AI for Adult Males[a] (mg/day)	RDA or AI for Adult Females[a] (mg/day)	Mineral Content (mg/100 g) [b][22]	Percent of RDA or AI for Adult Males[c]	Percent of RDA or AI for Adult Females[c]
Calcium	1000*	1000*	18–31	0.77–1.32*	0.77–1.32*
Copper	0.9	0.9	0.363–0.562	16.67–26.67	16.67–26.67
Iron	8	18	0.94–1.45	5.00–7.75	2.22–3.44
Magnesium	400–420	310–320	30–84	3.04–8.93	3.98–11.52
Manganese	2.3*	1.8*	0.336–1.601	6.09–29.57*	7.78–37.78*
Phosphorus	700	700	38–96	2.31–5.83	2.31–5.83
Potassium	4700	4700	329–484	2.97–4.38	2.97–4.38
Sodium	1500	1500	2–14	0.06–0.40	0.06–0.40
Zinc	11	8	0.49–1.10	1.91–4.27	2.63–5.88

[a] Recommended dietary allowances (RDA) or adequate intake (AI) for adults (aged 19–50 years).
[b] Contents expressed as minimum–maximum.
[c] Values are expressed as minimum–maximum of Refs. [30,31,71], based on ~42.5 g (or 1.5 ounces) chestnut serving.
* Values are expressed as AI.

consuming 42.5 g of edible chestnut are summarized in Table 11.3. Consuming the recommended daily amount of 42.5 g chestnut [29] supplies 16.67%–26.67%, 6.09%–37.78%, and 2.97%–4.38% of copper, manganese, and potassium, respectively, for RDA or AI for adults (Table 11.3) [30,31,71].

McCarthy and Meredith [23] reported that the potassium content of American, Chinese, and European chestnuts was 504, 447, and 378 mg/100 g, respectively. Chestnut also contains phosphorus, magnesium, calcium, sodium, manganese, iron, zinc, and copper. In American and Chinese chestnuts, phosphorus content was 96 mg/100 g. European chestnut had a lower phosphorus content (50 mg/100 g) than the other two varieties. Calcium content in American, Chinese, and European chestnuts varied between 18 and 24 mg/100 g. The range of magnesium content in American, Chinese, and European chestnuts was 32–84 mg/100 g. The sodium content for these varieties was 3 mg/100 g [23]. Pereira-Lorenzo et al. [25] investigated the mineral content of several chestnut cultivars from Spain and reported that potassium, phosphorus, calcium, and magnesium contents were present at 900, 188, 42, and 68 mg/100 g of dry weight, respectively.

Korel and Balaban [27] investigated the potassium and calcium contents of raw and candied chestnuts using inductively coupled plasma-atomic emission spectrometry (ICP-AES). The potassium content was 659 mg/100 g for raw chestnut and 180–288 mg/100 g for candied chestnut. The calcium contents of raw and candied chestnuts were 39 and 19.08 mg/100 g, respectively. Bellini et al. [26] found that Marrone del Mugello PGI chestnut had 381.1 mg/100 g of potassium and 25 mg/100 g of calcium on a fresh weight basis.

Essential trace elements such as copper, iron, zinc, and manganese are also present in chestnut. The copper content of raw and candied chestnuts was 0.48 and 0.19 mg/100 g, respectively. The iron and zinc contents were 1.24 and 1.98 mg/100 g for raw chestnut and 0.95 and 2.55 mg/100 g for candied chestnut, respectively. The manganese content was different in raw and candied chestnuts, with a value of 1.69 mg/100 g in raw chestnut and 0.76 mg/100 g in candied chestnut [27]. Bellini et al. [26] found that Marrone del Mugello PGI chestnut had 0.29 mg/100 g of copper, 0.42 mg/100 g of iron, 0.67 mg/100 g of zinc, 24.4 mg/100 g of magnesium, and 2.3 mg/100 g of sodium on a fresh weight basis. Pereira-Lorenzo et al. [25] reported the contents of microelements in Spanish chestnuts. The average manganese content was 38.7 mg/100 g on a dry weight basis, followed by 18.1 mg of iron, 12.3 mg of zinc, and 7.1 mg of copper.

11.4.3 Vitamins

The vitamin content of Chinese, European, and Japanese chestnuts is given in Table 11.2 [22]. Chestnut has considerable amounts of vitamin C, folate, and vitamin A. Vitamin C content is higher in European chestnut (40.2 mg/100 g) compared to Chinese and Japanese chestnuts (36.0 and 26.3 mg/100 g, respectively). On the other hand, folate and vitamin A contents are higher in Chinese chestnut compared to others. Chestnut also contains thiamin, riboflavin, niacin, pantothenic acid, and pyridoxine. Salvini et al. [32] reported that chestnut flour had appreciable amounts of vitamin E and B group vitamins. Bellini et al. [26] found that Marrone del Mugello PGI chestnut had 0.6 mg/100 g of δ-tocopherol and 7.9 mg/100 g of γ-tocopherol on a fresh weight basis.

11.4.4 Amino Acids

Amino acid concentrations of American, Chinese, and European chestnuts were investigated by Meredith et al. [33]. The amounts (g amino acid/100 g chestnut) present were as follows: alanine 0.14–0.25, arginine 0.14–0.44, aspartic acid 0.33–0.85, cystine 0.06–0.11, glutamic acid 0.25–0.62, glycine 0.11–0.23, histidine 0.05–0.13, isoleucine 0.08–0.19, leucine 0.11–0.30, lysine 0.12–0.27, methionine 0.05–0.10, phenylalanine 0.08–0.22, proline 0.10–0.21, serine 0.10–0.21, threonine 0.07–0.19, tryptophan 0.02–0.05, tyrosine 0.06–0.16, and valine 0.11–0.26. Amino acid contents of Chinese, European, and Japanese chestnuts are also given in Table 11.2 [22]. Aspartic acid was the highest and tyrptophan was the lowest among three chestnut varieties.

Even though chestnut has lower protein content than cereal grains, its proteins have unique nutritional characteristics. Globulins are the main storage proteins and its albumin content is relatively high. Amino acid composition of chestnut globulins presents high similarity to the 11S globulins from leguminous seeds, and has a relatively high content of albumins as those encountered in some legumes. Chestnut proteins are high in lysine and threonine, but methionine is the limiting amino acid [34]. Chestnut also contains considerable amounts of γ-amino butyric acid (GABA) [35]. GABA functions as a neurotransmitter in the central nervous system by decreasing neuron activity. It can be taken to calm the body and when it is used with niacinamide and inositol it prevents anxiety and stress-related messages from reaching the motor centers of the brain by filling its receptor site [36]. Deficiency of GABA may cause poor sleep in the elderly since this is a substance found in the body as it prepares for sleep [37].

11.4.5 Sugars

Sugars are responsible for the sweetness of foods. The carbohydrates, mainly starch and sucrose, are the major components of chestnut. Sucrose, representing one-third of the total sugar present, is an important quality parameter because consumers prefer sweeter fruits [38]. One of the first studies on compositional characteristics of Galician chestnut was carried out by Charro and Barreiro [39], who reported a sucrose content of 5.7 g/100 g in fresh nuts. Senter et al. [40] reported the sucrose content of different varieties of European chestnut as 9.2 g/100 g, on a dry weight basis, with only trace amounts of glucose and fructose present. They also reported the sucrose content of Chinese chestnut, chinkapin [*Castanea pumila* (L) Mill], and American chestnut as 8.0, 8.3, and 10.3 g/100 g on a dry weight basis, respectively. Bernárdez et al. [38] assessed sugar content in different varieties of European chestnuts and found that the contents of sucrose, glucose, and fructose ranged from 6.5 to 19.5 g/100 g, 0 to 0.30 g/100 g, and 0.04 to 0.30 g/100 g, respectively. The fruits of the Calva cultivar had the lowest content of sucrose, whereas Monfortiña and Casarella were the sweetest. A small amount of glucose and fructose was present, possibly due to partial hydrolysis of sucrose, and may not be the natural components of the fruit. Míguelez et al. [5] investigated the starch and sugar contents of 15 chestnut varieties from the region of Verín-Monterrei (Spain) and reported that starch content of these varieties ranged from 56.74 to 81.7 g/100 g; the sucrose content was high (6.5–19.5 g/100 g) and glucose and fructose contents were very low. Pereira-Lorenzo et al. [25] reported the starch content of Spanish chestnut cultivars as 57 g/100 g on an average basis. They stated that the high starch content should be taken into account in selecting the cultivars for consumption and for animal feed since starch is partially hydrolyzed into total sugar, which gives sweetness to chestnuts.

Several factors such as variety, storage conditions, and time, among others, could influence the sugar content of chestnuts [41]. Nomura et al. [41] investigated the changes of sugar content and activities of enzyme on starch degradation in five Japanese cultivars (Ganne, Ginyose, Kunimi, Riheiguri, and Tanzawa) during storage at 1°C. Sucrose was the main sugar detected in chestnuts and other sugars were present in trace amounts regardless of cultivars. Tanzawa and Kunimi are the early season, Ginyose is the midseason, and Ganne and Riheiguri are the late season cultivars. The initial sucrose content in the early season cultivars (Tanzawa: 1.64 g/100 g and Kunimi: 2.22 g/100 g) was lower compared to that of the midseason cultivar (Ginyose: 2.99 g/100 g). The late season cultivars (Ganne: 3.48 g/100 g and Riheiguri: 5.74 g/100 g) had the highest initial sucrose content. Sucrose content increased in each cultivar while the starch content decreased during storage; increase in sucrose content in the early cultivars was 8- to 10-fold after 10 weeks; however, those in medium and late cultivars was 3- to 4-fold. In Riheiguri, sucrose content increased approximately 2-fold. The sucrose content in each cultivar was almost the same after 1 month of storage. Chestnut amylase and β-amylase increased gradually during storage, but no significant ($P > 0.05$) relationship was observed regarding starch and sucrose contents [41]. These researchers concluded that storage of chestnut at chill temperatures for more than one month could be an effective method to improve chestnut quality.

11.4.6 Organic Acids

Important factors affecting the flavor of vegetables and fruits as well as chestnut include the nature and concentration of their taste-active components such as organic acids [42]. The antioxidant activity of organic acids (such as ascorbic acid) may also have a protective role against various diseases [43]. The organic acids present in two European chestnut varieties (Judia and Longal) are oxalic, *cis*-aconitic, citric, ascorbic, malic, quinic, and fumaric acids [44]. The organic acid content is higher in the Longal variety (~25.1 mg/100 g) than Judia variety (~20.5 mg/100 g). Ascorbic acid is the highest in both varieties representing ~42.6% of total acids in Judia and ~37.0% of total acids in Longal. The composition of organic acids is affected by processing (roasting, boiling, and frying). These processes cause a significant decrease in the total organic acid content of these varieties and frying renders the highest loss of these compounds (~12.7 and 15.5 mg/100 g in Julia and Longal varieties, respectively). Fumaric acid, a minor compound in both chestnut samples, disappeared totally or was reduced to trace amounts after each treatment [44].

11.4.7 Phenolics

Phenolic compounds in foods have health benefits since they are considered as powerful antioxidants. The total phenolic acid contents of American chestnut, a hybrid American, and a Chinese chestnut were investigated by Senter et al. [45]. They found that American, hybrid, and Chinese chestnuts contained 4.97, 1.65, and 4.33 μg of phenolic acids/g of sample, respectively. Gallic acid was the predominant phenolic acid present in these samples (4.21, 0.65, and 3.87 μg of gallic acid/g of sample extracted from American, hybrid, and Chinese chestnuts, respectively). Protocatechuic, caffeic, *p*-hydroxybenzoic, and syringic acids were found in low concentrations in chestnut samples [45]. The highest concentration of the total phenolics was in the seed coat of fresh chestnut (*C. henryi*) (as catechin ~36 mg/g of dry weight), followed by the pericarp (~10 mg/g of dry weight), and the lowest content in the endosperm (~4 mg/g of dry weight). The total phenolics content decreased during the 6 months of storage at 4°C and −20°C [46].

Flavonoids are also an important subgroup of phenolic compounds and one of their functions in foods includes their antioxidant properties. They may also contribute to flavor such as bitterness [47]. In raw European chestnut, flavonoids were found in trace amounts (0.01 mg catechin and 0.01 mg gallocatechin in 100 g of edible portion) [48].

The pellicle (skin) of chestnut is rich in tannins, a special group of phenolics, which contributes to the astringency in chestnut. The tannin of chestnut flesh is mainly composed of gallic acid occurring as 3,6-digalloyl glucose, as well as pyrogallol and resorcinol. The tannins of commercial chestnut bark are mainly ellagitannins [49]. Hwang et al. [50] investigated the tannin contents of 14 varieties of Korean chestnuts and found that their contents in the inner and outer shells of chestnuts were 7.83%–71.42% and 0.31%–2.04%, respectively.

11.4.8 Fatty Acids and Lipid Classes

American chestnut has the highest lipid content (9.5 g/100 g), followed by chinkapin (4.0 g/100 g), European (2.9 g/100 g), and Chinese chestnuts (2.1 g/100 g) [40]. The high lipid content in American chestnut may enhance its sensory preference since it is well known that mouthfeel and flavor of many food products are influenced by lipids [51].

Fatty acid contents of Chinese, European, and Japanese chestnuts are given in Table 11.2 [22]. European chestnut has the highest percentage of saturated fatty acids (SFA) and polyunsaturated fatty acids (PUFA) (19.9% and 42.8%, respectively). On the other hand, Japanese and Chinese chestnuts contain higher percentages of monounsaturated fatty acids (MUFA) (56.5% and 56.3%, respectively) than European chestnut (37.3%).

Fatty acids present in chestnut varieties (American, European, Chinese, and chinkapin) are palmitic (16:0), stearic (18:0), oleic (18:1ω9), linoleic (18:2ω6), and α-linolenic (18:3ω3) acids. Chinese and American chestnuts have the highest percentage of unsaturated fatty acids (89% and 87%, respectively). Palmitic acid is the predominant SFA in these chestnut samples. Stearic acid is found in measurable quantities only in American and European chestnuts, and trace amounts are present in Chinese chestnut and chinkapin. Oleic acid is the major fatty acid in American and Chinese chestnuts and chinkapin, whereas linoleic acid has been found to be the major fatty acid in European chestnut [40]. Chinkapin has the highest linolenic acid content followed by European and American chestnuts. The lipid content of 15 varieties of chestnuts from Verín-Monterrei region (Spain) ranged from 1.3% to 3.0% [5] and the lipid content of 17 cultivars of chestnuts from three Protected Designations of Origin (PDO) areas from the Trás-os-Montes e Alto Douro region (Portugal) were between 1.67% and 3.50% and had low SFA (17%) and high unsaturated fatty acids (83%). The major fatty acids of these chestnuts were linoleic, oleic, and palmitic, which constituted more than 85% to the total fatty acids present [52].

The lipid content of 23 shelled chestnut samples, corresponding to 7 different cultivars of European chestnut, from 3 geographical origins in Portugal, ranged from 1.02% to 1.76% (dry weight basis) [53]. The average values for nonpolar (essentially triacylglycerols, TAG) and polar (phospholipids) fractions in the lipid composition of total fat extracted were 66.98% and 33.02%, respectively. Unsaturated fatty acids of the polar and nonpolar fractions for all samples examined were 72.2% and 50.7%, respectively. Linoleic (51.8%), palmitic (26.7%), and oleic (14.6%) acids were the main fatty acids present in polar fractions, whereas lauric (34.3%), linoleic (26.4%), oleic (19.3%), and palmitic (14.6%) acids were the main fatty acids present in nonpolar fractions [53].

11.5 CHESTNUT ALLERGENS

There is limited information available on the allergenicity of chestnuts. The only information found is the research conducted by Lee [54] who reported allergens of chestnut, and noted that chestnut is the third prevalent food allergen in both adult and pediatric allergy patients. Allergens identified in tree nuts including chestnut are given in Chapter 4.

11.6 USE OF CHESTNUT BY-PRODUCTS

Tannin from bark and wood of trees is the prime source for leather processing. Romans discovered the potential of chestnut tree and used the rind, leaves, and flowers of chestnut in pharmacopoeia.

Dioscoride Pedanio, who was one of the founders of the pharmacopoeia and worked at Nero's court, described astringent, antitoxic, stomatic, and tonic properties of chestnut [4].

A flavorful honey, produced from the flowers of the European chestnut, is a popular product in specialty food shops in Italy and other European countries [55]. The amber-colored honey is aromatic and at the same time is considered as a healthy and energetic food [56].

Reddening of sensitive skin was avoided or reduced by the bark-infusion, and chestnut leaves were used to make after-shave lotions. Famous Italian stylists use chestnut leaves for dyeing of fabrics [56].

In Corsica, Spain, and southern Italy, swine are fed with chestnut to obtain high quality salami. Nuts discarded during processing can be used as livestock feed. A very good soil is obtained by burying the debris of chestnut trees (leaves, branches, and husks). Leaves are a good bedstead for livestock and a good fertilizer. Chestnut forests, due to their nourishing nuts, also maintain a good variety of wildlife [57].

11.7 HEALTH ASPECTS OF CHESTNUT

Owing to its nutritional qualities and potential beneficial health effects, chestnut plays an important role in the human diet. In recent years, consumers have been paying attention to the composition of foods they consume. Chestnut provides a good source of essential fatty acids [40,53,58], which play an important role in preventing cardiovascular disease (CVD) in adults and in promoting the development of brain and retina of infants [59]. The significant presence of PUFA, especially linoleic acid, may help in reducing cholesterol levels and prevent coronary heart disease (CHD) [60]. Consumption of 60 g nuts/day is recommended for people suffering from or having a high risk of CHD and this amount can have a positive effect on the reduction of blood cholesterol levels. On the other hand, consumption of this amount of nuts could lead to a modest increase in oxalate intake, increasing the risk of kidney stone formation. However, chestnut contains very low levels of oxalates (<85 mg/100 g on fresh weight basis) [61].

Nuts are generally recommended along with vegetables, fruits, and cereals to increase the fiber intake of consumers [62]. Chestnut has similar crude fiber content (3 g/100 g) to other nuts such as walnut (2.1 g/100 g), pecan (2.3 g/100 g), and pistachio (1.9%), but less than hazelnut (6.1 g/100 g) [62]. It is also a valuable food because of its vitamins (in particular vitamin C), minerals, amino acids, and antioxidant phenolic compounds. Arginine, potassium, copper, and magnesium contribute to the positive nutritional value of chestnut [63]. In addition, the unique and distinctive flavor of chestnut adds value to it as an ingredient in a variety of food products.

11.8 CONSUMPTION OF CHESTNUT

Consumers' interest in new and healthy foods is increasing. Chestnut is gaining popularity in many European countries, Australia, New Zealand, and the United States [64]. There are efforts to revitalize chestnut production and consumption in the United States [65]. The U.S. and Korean consumption of chestnut are 0.05 and 1.8 kg per capita, respectively [66]. An increase in the U.S. consumption up to the European levels (0.5 kg per capita) would boost domestic production and replace existing imports by providing a locally produced, fresh product, and would help meet the increased demand [65,66].

Marketing opportunities for chestnut have been identified by a study conducted by Food Processing Center of the University of Nebraska. The study investigated the upscale restaurant chefs' interest in value-added chestnut products, but also looked at the ingredient and retail markets. Results showed that freshness and quality of the product are very important. This creates a market niche for locally produced chestnut delivering a fresh and high-quality product. The study also showed that producers in Midwestern United States have an excellent market opportunity with value-added chestnut products, including shelled and frozen vacuum-packed chestnuts [67].

Midwest Nut Producers Council and Michigan State University have conducted another study on chestnut marketing and identified market opportunities in upscale restaurants in Michigan for peeled and unpeeled chestnuts [68]. Peeled chestnut was preferred by the chefs participating in this study who used them in many dishes [64].

The University of Missouri held Missouri Chestnut Roast Festival in 2003 and 2004, with the primary objective of increasing domestic demand and consumption of chestnut. During this festival, information and samples of fresh, roasted, and prepared chestnuts were offered. Survey questionnaires were distributed to the participants. It was concluded that as of 2004, the U.S. consumers remained unfamiliar with chestnut, they did not know about its health aspects, in what form and where to buy chestnut, or how to prepare it. Participants of these surveys preferred to buy roasted or fresh chestnut from grocery stores or farmers' markets. Quality and nutrition-diet-health were perceived as the most important factors influencing the decision to buy chestnut [65,69,70].

11.9 CONCLUSION

Chestnut is a good source of food and has a steady consumption in the main producing countries. Chestnut has started to come back into the daily eating habits in countries where it was rarely consumed in the past due to its nutritional value and dietary characteristics. Chestnut can be included in several different preparations and unique dishes such as polenta, soups, flans, stuffings, and sweets. It can also be consumed either boiled or roasted. Due to its essential minerals and trace elements, chestnut can be considered as a healthy food, and can easily be grown organically. Compared to other tree nuts, chestnut has a low fat content, and as expected, it is cholesterol-free. It is rich in essential fatty acids and has a moderate, but high-quality protein content. Chestnut contains high levels of vitamin C and is rich in potassium and low in sodium. Products containing chestnut flour are a good snack with high fiber content. The knowledge about nutritional properties of chestnut may create new opportunities to increase its demand, and it is a good method of marketing and promotion.

REFERENCES

1. Food and Agricultural Organization (FAO), Chestnut Production, Published online at: http://faostat.fao.org (accessed July 10, 2007).
2. Vossen, P., Chestnut Culture in California, University of California, Division of Agriculture and Natural Resources, Santa Rosa, CA, Publication 8010, 2000, 18pp.
3. Anon, *Eating Chestnuts—What is a Chestnut?* The Chestnut Growers Information Book, 2006, Published online at: http://www.chestnutgrowers.com.au (accessed July 21, 2006).
4. Adua, M., The sweet chestnut throughout history from Miocene to the third millennium, *Acta Hort.*, 494, 29–36, 1999.
5. Míguelez, J.D.L.M., Bernárdez, M.M., and Queijeiro, J.M.G., Composition of varieties of chestnuts from Galicia (Spain), *Food Chem.*, 84, 401–404, 2004.
6. McLaren, D., *Growing Chestnuts—Chestnut Production*, The Chestnut Growers Information Book, 1999, Published online at: http://www.chestnutgrowers.com.au (accessed August 14, 2006).
7. Rieger, M., Chestnuts—*Castanea* spp., Published online at: http://www.uga.edu/fruit/chestnut.html (accessed October 26, 2006).
8. Soylu, A., *Chestnut Growing and Specialties*, Atatürk Horticulture Research Institute, Yalova, Turkey, 1984.
9. Anon, Chestnuts—Postharvest Handling, Michigan State University Extension, 1999, Published online at: http://web1.msue.msu.edu/imp/mod01/01600673.html (accessed July 28, 2004).
10. Anon, Chestnuts, Clemson Extension Home and Garden Information Center, 2004, Published online at: http://dprod4.clemson.edu/factsheets/hgic3270.htm (accessed July 28, 2004).
11. Anon, Chestnuts—An Oregon Product, Oregon State University Extension Service, 1999, Published online at: http://extension.oregonstate.edu/lane/food-preservation/ docs/ SP%2050%20762.pdf (accessed October 26, 2006).
12. Sacchetti, G., Pinnavaia, G.G., Guidolin, E., and Dalla Rosa, M., Effects of extrusion temperature and feed composition on the functional, physical and sensory properties of chestnut and rice flour-based snack-like products, *Food Res. Int.*, 37, 527–534, 2004.

13. Bounous, G. and Giacalone, G., Chestnut storage, processing and usage in Italy, presented at the World Chestnut Industry Conference, Morgantown, WV, July 8–10, 1992.
14. Breisch, H., Harvesting, storage and processing of chestnuts in France and Italy, in *Proceedings of the 1st International Congress Chestnut*, Spoleto, Italy, October 20–23, 1993, pp. 429–436.
15. Sacchetti, G., Pittia, P., Mastrocola, D., and Pinnavaia, G.G., Stability and quality of traditional and innovative chestnut based products, *Acta Hort.*, 693, 63–69, 2005.
16. Beuchat, L.R., Relationship of water activity to moisture content in tree nuts, *J. Food Sci.*, 43, 754–755, 1978.
17. Beuchat, L.R. and Worthington, R.E., Technical note: Fatty acid composition of tree nut oil, *J. Food Technol.*, 13, 355–358, 1978.
18. Dudek, J.A. and Elkins, E.R., *Determination of the Nutrient Content of Selected Candies, Nuts, Condiments, Beverages and Vegetables*, National Food Processors Association Research Foundation, Washington, DC, 1984.
19. Kuzayli, M.V., Cowan, J.W., and Sabry, Z.I., Nutritive value of Middle Eastern foodstuffs. II. Composition of pulses, seeds, nuts and cereal products of Lebanon, *J. Sci. Food Agric.*, 17, 82–84, 1966.
20. Payne, J.A., Jaynes, R.A., and Kays, S.J., Chinese chestnuts production in the United States: Practice, problems and possible solutions, *Econ. Botany*, 37, 187–200, 1983.
21. Rhee, C.-O. and Kim, Z.-U., Analysis of the lipid components in chestnut (*Castanea crenata*). Part I. Composition of lipid fraction of inner and outer part of chestnut, *J. Korean Agric. Chem. Soc.*, 25, 239–247, 1982.
22. U.S. Department of Agriculture (USDA), Nutrient Data Lab, National Agricultural Library, Agricultural Research Service, 2006, Published online at: http://www.nal.usda.gov/fnic/foodcomp/cgi-bin/list_nut_edit.pl (accessed June 27, 2006).
23. McCarthy, M.A. and Meredith, F.I., Nutrient data on chestnuts consumed in the United States, *Econ. Botany*, 42, 29–36, 1988.
24. Gomes, A.L., Abreu, C.G., and Castro, L.T., COLUTAD Um Clone de Castanheiro com Resistência à doenÇa da Tinta, NATO/SFS Programme III PO-Chestnut, Vila Real, Universidade de Tras os Montes e Alto Douro, Vila Real, Portugal, 1997.
25. Pereira-Lorenzo, S., Ramos-Cabrer, A.M., Díaz-Hernández, M.B., Ciordia-Ara, M., and Ríos-Mesa, D., Chemical composition of chestnut cultivars from Spain, *Scientia Hort.*, 107, 306–314, 2006.
26. Bellini, E., Giordani, E., Marinelli, C., and Perucca, B., Marrone del Mugello PGI chestnut nutritional and organoleptical quality, *Acta Hort.*, 693, 97–101, 2005.
27. Korel, F. and Balaban, M.O., Composition, color and mechanical characteristics of pretreated candied chestnuts, *Int. J. Food Prop.*, 9, 559–572, 2006.
28. Belitz, H.-D. and Grosch, W., Minerals, in *Food Chemistry*, 2nd ed., Belitz, H.-D. and Grosch, W., Eds., Springer Verlag, Berlin, Germany, 1999, pp. 395–401.
29. Food and Drug Administration (FDA), Qualified Health Claims: Letter of Enforcement Discretion—Nuts and Coronary Heart Disease, Docket No 02P-0505, Food and Drug Administration, Washington, DC, 2003.
30. *Dietary Reference Intakes for Vitamin A, Vitamin K, Arsenic, Boron, Chromium, Copper, Iodine, Iron, Manganese, Molybdenum, Nickel, Silicon, Vanadium, and Zinc*, The National Academies Press, Washington, DC, 2001.
31. *Dietary Reference Intakes for Water, Potassium, Sodium, Chloride, and Sulfate*, The National Academies Press, Washington, DC, 2004.
32. Salvini, S., Parpinel, M., Gnagnarella, P., Maisonneuve, P., and Turrini, A., *Banca Dati di Composizione Degli Alimenti per Studi Epidemiologici in Italia*, Milano, Istituto Europeo di Oncologia, Milano, Italia, 1998, 370pp.
33. Meredith, F.I., McCarthy, M.A., and Leffler, R., Amino acid concentrations and comparison of different hydrolysis procedures for American and foreign chestnuts, *J. Agric. Food Chem.*, 36, 1172–1175, 1988.
34. Colloda, C., Casado, R., Barber, D., Caleya, R.F., and Aragoncillo, C., Extraccion y fraccionamento de las proteinas de castaña, Congresso Internacional sobre el Castaña, Pontevedra: Departamento de Investigação Florestal de Lourizan, Vila Real, Portugal, 417, 1984.
35. Tixier, M. and Desmaison, A.M., Relation metabolique entre l'arginine et l'acide gamma aminobutirique dans le fruit de *Castanea sativa*, *Phytochemistry*, 19, 1643–1646, 1980.
36. Anon, GABA-Kordial, 2007, Published online at: http://organicpharmacy.org/products/ GABA.-.Kordial (accessed April 15, 2007).
37. Kastenbaum, R., Wine and elderly person, *J. Nutr. Elderly*, 4, 15–24, 1985.

38. Bernárdez, M.M., Miguélez, J.D.M., and Queijeiro, J.G., HPLC determination of sugars in varieties of chestnut fruits from Galicia (Spain), *J. Food Comp. Anal.*, 17, 63–67, 2004.
39. Charro, A. and Barreiro, E., Estudio Bromatológico de la castaña, *Anales de Bromatológia*, 2, 453–490, 1957.
40. Senter, S.D., Payne, J.A., Miller, G., and Anagnostakis, S.L., Comparison of total lipids, fatty acids, sugars and nonvolatile organic acids in nuts from *Castanea* species, *J. Sci. Food Agric.*, 65, 223–227, 1994.
41. Nomura, K., Ogasawara, Y., Uemukoi, H., and Yoshida, M., Change of sugar content in chestnut during low temperature storage, *Acta Hort.*, 398, 265–276, 1995.
42. Vaughan, J.G. and Geissler, C.A., *The New Oxford Book of Food Plants*, Oxford University Press, Oxford, 1997.
43. Silva, B.M., Andrade, P.B., Valentão, P., Ferreres, F., Seabra, R.M., and Ferreira, M.A., Quince (*Cydonia oblonga* Miller) fruit (pulp, peel, and seed) and jam: Antioxidant activity, *J. Agric. Food Chem.*, 52, 4705–4712, 2004.
44. Ribeiro, B., Rangel, J., Valentão, P., Andrade, P.B., Pereira, J.A., Bölke, H., and Seabra, R.M., Organic acids in two Portuguese chestnut (*Castanea sativa* Miller) varieties, *Food Chem.*, 100, 504–508, 2007.
45. Senter, S.D., Horvat, R.J., and Forbus, W.R., Comparative GLC-MS analysis of phenolic acids of selected tree nuts, *J. Food Sci.*, 48, 798–799, 824, 1983.
46. Xu, J., The effect of low-temperature storage on the activity of polyphenol oxidase in *Castanea henryi* chestnuts, *Postharvest Biol. Technol.*, 38, 91–98, 2005.
47. von Elbe, J.H. and Schwartz, S.J., Colorants, in *Food Chemistry*, 3rd ed., Fennema, O.R., Ed., Marcel Dekker, New York, 1996, pp. 681–694.
48. U.S. Department of Agriculture (USDA), USDA Data Base for the Flavonoid Content of Selected Foods, Release 2, Nutrient Data Lab, Food Composition Lab, Beltsville Human Nutrition Research Center, Agricultural Research Service, 2006, Published online at: http://www.ars.usda.gov/nutrientdata (accessed January 10, 2007).
49. Vivas, N., Bourgeois, G., Vitry, C., Glories, Y., and de Freitas, V., Determination of the composition of commercial tannin extracts by liquid secondary ion mass spectrometry (LSIMS), *J. Sci. Food Agric.*, 72, 309–317, 1996.
50. Hwang, J.-Y., Hwang, I.-K., and Park, J.-B., Analysis of physicochemical factors related to the automatic pellicle removal in Korean chestnut (*Castanea crenata*), *J. Agric. Food Chem.*, 49, 6045–6049, 2001.
51. Pilgram, F.J. and Kamen, J.M., Predictors of human food consumption, *Science*, 139, 501–502, 1963.
52. Borges, O.P., Carvalho, J.S., Correia, P.R., and Silva, A.P., Lipid and fatty acid profiles of *Castanea sativa* Mill. chestnuts of 17 native Portuguese cultivars, *J. Food Comp. Anal.*, 20, 80–89, 2007.
53. Ferreira-Cardoso, J.V., Sequeira, C.A., Rodrigues, L., and Gomes, E.F., Lipid composition of *Castanea sativa* Mill. fruits of some native Portuguese cultivars, *Acta Hort.*, 494, 133–138, 1999.
54. Lee, S., Chestnut as a food allergen: Identification of the major allergens and effect of gastric and intestinal fluids, *J. Allergy Clin. Immunol.*, 113 (Suppl.1), 234S, 2004.
55. Anon, Flowers and Foliage (Honey-Castanea sativa), 2005, Published online at: http://www.fao.org/documents/show_cdr.asp?url_file=/DOCREP/005/Y4351E/y4351e08.htm (accessed February 6, 2005).
56. Bounous, G., The chestnut: A multipurpose resource for the new millennium, *Acta Hort.*, 693, 33–40, 2005.
57. Diamandis, S. and Perlerou, C., The mycoflora of the chestnut ecosystems in Greece, *For. Snow Landsc. Res.*, 76, 499–504, 2001.
58. Künsch, U., Schärer, H., Patrian, B., Hurter, J., Conedera, M., Sassella, A., Jermini, M., and Jelmini, G., Quality assessment of chestnut fruits, *Acta Hort.*, 494, 119–127, 1999.
59. Conner, W.E., The beneficial effects of omega-3 fatty acids: Cardiovascular disease and neuro-development, *Curr. Opin. Lipidol.*, 8, 1–3, 1997.
60. Garcia, J.M., Agar, I.T., and Streif, J., Fat content and fatty acid composition in individual seeds of pistachio varieties grown in Turkey, *Gartenbauwissenschaft*, 57, 130–133, 1992.
61. Ritter, M.M.C. and Savage, G.P., Soluble and insoluble oxalate content of nuts, *J. Food Comp. Anal.*, 20, 169–174, 2007.
62. Ensminger, A.H., Ensminger, M.E., Konlande, J.E., and Robson, J.R.K., *The Concise Encyclopedia of Foods and Nutrition*, 2nd ed., CRC Press, Boca Raton, FL, 1995.
63. Hu, F.B., Stampfer, M.J., Manson, J.E., Rim, J.E., Colditz, G.A., Rosner, B.A., Speizer, F.E., Hennekens, C.H., and Willett, W.C., Frequent nut consumption and risk of coronary heart disease in women: Prospective cohort study, *Br. Med. J.*, 317, 1341–1345, 1998.

64. Kelley, K.M. and Behe, B.K., Chefs' perceptions and uses of 'Colossal' chestnuts, *HortTechnol.*, 12, 172, 2002.
65. Gold, M., Cernusca, M.M., and Godsey, L., Consumer preferences for chestnuts, Eastern black walnuts, and pecans, *HortTechnol.*, 14, 583–589, 2004.
66. Bodet, L., Woods, T., and Ernst, M., The International Marketing Situation, University of Kentucky, Department of Agricultural Economics, Lexington, Kentucky, Staff Paper No. 411, 2001, 26pp.
67. Southeast Iowa Nut Growers Cooperative, Chestnut Market Opportunities: Assessing Upscale Restaurant Interest in Value Added Chestnut Products, University of Nebraska, Institute of Agricultural and Natural Resources, Food Processing Center, Lincoln, Nebraska, 2002, 39pp.
68. Smith, B., Behe, B.K., Fulbright, D.F., Walden, R.M., and Kelley, K.M., Marketing Chestnuts to Investigate Chefs and Consumer Interests and Develop Product Quality Criteria, A final report on the research conducted for Federal State Mktg. Improvement Program, Midwest Nut Producers Council and Michigan State University, East Lansing, MI, 2002.
69. Gold, M., Cernusca, M.M., and Godsey, L., Update on consumers' preferences for chestnuts, *HortTechnol.*, 15, 904–906, 2005.
70. Gold, M., Cernusca, M.M., and Godsey, L., Competitive market analysis: Chestnut producers, *HortTechnol.*, 16, 360–369, 2006.
71. *Dietary Reference Intakes for Calcium, Phosphorus, Magnesium, Vitamin D, and Fluoride*, The National Academies Press, Washington, DC, 1997.

12 Compositional Characteristics and Health Effects of Hazelnut (*Corylus avellana* L.): An Overview

Cesarettin Alasalvar, Fereidoon Shahidi, Joana S. Amaral, and Beatriz P.P. Oliveira

CONTENTS

12.1 INTRODUCTION

Hazelnut (*Corylus avellana* L.) belongs to the *Betulaceae* family and is a popular tree nut worldwide, mainly distributed in the coasts of the Black Sea region of Turkey, southern Europe (Italy, Spain, Portugal, and France), and in some areas of the United States (Oregon and Washington). Hazelnut is also cultivated in some other countries such as New Zealand, China, Azerbaijan, Chile, Iran, and Georgia, among others. Turkey is the world's largest producer of hazelnuts, contributing ~74% to the total global production, followed by Italy (~16%), the United States (~4%), and Spain (~3%). Other countries contribute ~3% to the total production [1]. Hazelnut is, therefore, the most popular tree nut in Europe. Several commercial hazelnut varieties are available [2–10]. Although each hazelnut variety has its own nutritional composition, distinctive taste, and aroma, for example, Tombul (round) hazelnut variety, grown throughout the Giresun province of Turkey, is classified as Giresun quality (also known as prime quality) among 17 commercial hazelnut varieties cultivated in Turkey [11,12]. The prime quality varieties differ for each production region as Tombul for Turkey, Tonda gentille for Italy, Negret for Spain, Fertille de Coutard (also known as Barcelona) and Segorbe in Portugal, and Barcelona, Ennis, Daviana, and Butler for the United States.

The detailed overview on the nutritional composition and health promoting components of hazelnut provided in this chapter summarizes the existing knowledge and appreciation for the use of hazelnut and its products in a variety of food and specialty products. Besides nutritional value and health aspects, the presence of taste- and aroma-active components contribute to the sensory characteristics of products. Thus, better taste and aroma/flavor of hazelnut may increase the consumption of this nutritionally important nut, as discussed in this chapter. In addition, characteristics of raw (natural) hazelnut as well as its health promotion and disease prevention aspects are given in detail. Aroma-active components of roasted hazelnut are also discussed.

12.2 COMPOSITIONAL CHARACTERISTICS OF HAZELNUT

12.2.1 Proximate Composition and Caloric Value

Hazelnut contains all major macronutrients: fat, carbohydrate, and protein. Table 12.1 shows the proximate composition and caloric value of various hazelnut varieties grown in different countries [6,10,11,13,14]. Based on the date given in Table 12.1, fat is the predominant component (58.40–64.10 g/100 g), followed by carbohydrate (15.50–17.61 g/100 g), protein (10.86–16.30 g/100 g), moisture (3.90–5.40 g/100 g), and ash (2.20–2.69 g/100 g). These values are comparable with those published in the literature [6,9,15–18]. Several factors have been reported to affect the proximate composition of hazelnut [3,6,9,13,16,17,19,20].

Hazelnut is characterized by high fat content, thus being considered as an excellent source of energy (604–690 kcal/100 g) (Table 12.1). The mean energy value of Portuguese hazelnuts has been reported as 690 kcal/100 g [10]. This high value is due to calculation of total carbohydrate content without deducting the total dietary fiber, which is not used for energy calculation. The energy requirement for adult men ranges from 2300 to 2900 kcal/day and is 1900–2200 kcal/day for adult women [21]. At the recommended consumption level of 42.5 g (1.5 ounces) of most tree nuts [22], hazelnut supplies approximately 10%–13% of the total energy requirement per day for adults.

12.2.2 Minerals

A total of 24 minerals (essentials and nonessentials) have been reported so far in hazelnut varieties, of which 13 essential mineral content are summarized in Table 12.2 by various studies [11,14,23,24,214]. In addition to the data in Table 12.2, mineral content of various hazelnut varieties has also been reported by several researchers [9,15,16,25–27,215]. In general, potassium is the most abundant mineral, followed by phosphorus, calcium, and/or magnesium. Several studies have indicated that mineral composition of hazelnut is affected by variety, geographical origin, harvest year, climate, composition of soil, irrigation, use of fertilizer, and method of cultivation [15,24,26,28].

TABLE 12.1
Proximate Composition and Caloric Value of Various Hazelnut Varieties Grown in Different Countries

Proximate Composition	Unit	Turkey[a] [11]	United States[b] [14]	Portugal[c] [10]	Italy[d] [13]	New Zealand[e] [6]
Protein	g/100g	15.35 ± 0.42	14.95 ± 0.16	10.86 (9.3–12.7)[f]	13.70 ± 0.5	16.30 (14.3–18.2)
Lipid (fat)	g/100g	61.21 ± 0.99	60.75 ± 0.39	63.97 (59.2–69.0)	64.10 ± 0.4	58.40 (54.6–63.2)
Carbohydrate, by difference	g/100g	17.30 ± 0.48	16.70	17.61 (12.1–21.1)	15.50	17.50 (15.6–20.5)
Moisture	g/100g	3.90 ± 0.20	5.31 ± 0.20	4.87 (3.5–6.4)	4.50 ± 0.2	5.40 (4.6–6.0)
Ash	g/100g	2.24 ± 0.03	2.29 ± 0.02	2.69 (2.4–3.4)	2.20 ± 0.1	2.40 (2.1–2.7)
Energy	kcal	631	628	690[g]	642	604

[a] Variety grown in Turkey (Tombul). Data are expressed as means ± the standard deviation.
[b] Variety grown in the United States (Unknown). Data are expressed as means ± the standard deviation.
[c] Varieties grown in Portugal (Butler, Campanica, Cosford, Couplat, Daviana, Ennis, Fertille de Coutard, Grossal, Gunslebert, Lansing, Longa d'Espanha, Merveille de Bollwiller, Morell, Negreta, Pauetet, Round du Piemont, Segorbe, Santa Maria de Jesus, and Tonda de Giffoni).
[d] Variety grown in Italy (Tonda Gentile Romana). Data are expressed as means ± the standard deviation.
[e] Varieties grown in New Zealand (Whiteheart, Barcelona, Butler, Ennis, Tonda di Giffoni, and Campanica).
[f] Data are expressed as means (minimum–maximum values) among varieties.
[g] Energy value contains total dietary.

TABLE 12.2
Essential Mineral Content of Hazelnuts and Percentage of RDA or AI

Mineral	RDA or AI for Adult Males[a]	RDA or AI for Adult Females[a]	Unit	Mineral Content: Ref. [11]	Ref. [14]	Ref. [23]	Ref. [24]	Percent of RDA or AI for Adult Males[b]	Percent of RDA or AI for Adult Females[b]
Calcium	1000 mg/day[*] [29]	1000 mg/day[*] [29]	mg/100 g	193.4	114	140	115–244[c]	4.8–10.4[*]	4.8–10.4[*]
Chromium	35 µg/day[*] [32]	25 µg/day[*] [32]	µg/100 g	10.0	—	—	—	12.1[*]	17.0[*]
Copper	0.9 mg/day [32]	0.9 mg/day [32]	mg/100 g	1.60	1.72	1.23	1.2–2.2	56.7–104	56.7–104
Iodine	150 µg/day [32]	150 µg/day [32]	µg/100 g	—	—	17	—	4.8	4.8
Iron	8 mg/day [32]	18 mg/day [32]	mg/100 g	4.97	4.70	3.2	3.4–4.4	17.0–26.4	7.6–11.7
Magnesium	400–420 mg/day [29]	310–320 mg/day [29]	mg/100 g	176.5	163	160	167–225	16.6–23.3	21.6–30.4
Manganese	2.3 mg/day[*] [32]	1.8 mg/day[*] [32]	mg/100 g	3.29	6.17	4.9	1.4–3.1	25.9–114[*]	33.1–146[*]
Molybdenum	45 µg/day[*] [32]	45 µg/day[*] [32]	µg/100 g	3.13[d]	—	—	—	2.9[*]	2.9[*]
Phosphorus	700 mg/day [29]	700 mg/day29	mg/100 g	355.7	290	300	—	17.6–18.2	17.6–18.2
Potassium	4700 mg/day [33]	4700 mg/day [33]	mg/100 g	761	680	730	415–530	3.8–6.9	3.8–6.9
Selenium	55 µg/day [31]	55 µg/day [31]	µg/100 g	60.0	2.4	tr[e]	—	1.9–46.4	1.9–46.4
Sodium	1500 mg/day [33]	1500 mg/day [33]	mg/100 g	3.13	0.05	6.0	1.8–3.9	0.0–0.2	0.0–0.2
Zinc	11 mg/day [32]	8 mg/day [32]	mg/100 g	1.94	2.45	2.1	2.3–2.9	7.5–11.2	10.3–15.4

[a] Recommended dietary allowances (RDA) or adequate intake (AI) for adults (aged 19–50 years).
[b] Values are expressed as minimum–maximum of Refs. [11,14,23,24], based on ~42.5 g (or 1.5 ounces) hazelnut serving.
[c] Contents expressed as minimum–maximum.
[d] Ref. [214].
[e] tr, trace.
* Values are expressed as AI.

With regard to nutritional aspects, percentage of recommended dietary allowances (RDA) or adequate intake (AI) for essential minerals for adult males and females (aged 19–50 years) are also given in Table 12.2. Hazelnut serves as an excellent source of copper, manganese, and selenium and good sources of chromium, iron, magnesium, phosphorus, and zinc. Consuming recommended daily amount of 42.5 g hazelnut [22] supplies 56.7%–104% of copper, 25.9%–146% of manganese, 1.9%–46.4% of selenium, 12.1%–17% of chromium, 7.6%–26.4% of iron, 16.6%–30.4% of magnesium, 17.6%–18.2% of phosphorus, and 7.5%–15.4% of zinc for RDA or AI for adults [29,31–33]. Hazelnut contains significant amounts of essential minerals that are associated with improved health status when consumed at doses beyond those necessary for preventing a deficiency state. Compared to other common foodstuffs, nuts including hazelnut, have an optimal nutritional density with respect to healthy minerals, such as calcium, magnesium, and potassium. A high intake of these minerals, together with a low sodium intake, is associated with protection against bone demineralization, atrial hypertension, insulin resistance, and overall cardiovascular risk [35].

Although each essential mineral has its own health benefits, minerals are vital to overall mental and physical well-being [36–39]. The essential trace mineral, selenium, is of fundamental importance to human health. As shown in Table 12.2, Tombul hazelnut [11] is an excellent source of selenium (60 μg/100 g), which plays a major antioxidant role, protecting cell membranes by neutralizing the deleterious effects of free radicals. The selenium content of (86.5 μg/100 g) Sicilian hazelnut has recently been reported [40]. Among tree nuts, only Brazil nut contains a much higher selenium content (1917 μg/100 g) [14] than that of Tombul and Sicilian hazelnut varieties. Increased selenium intake can reduce the risk of chronic diseases, such as coronary heart disease (CHD), certain types of cancer, and may preserve tissue elasticity of blood vessels [41–45].

12.2.3 Vitamins

Vitamins are essential nutrients that play many beneficial roles in the body [30–32,39,46–49]. Hazelnut contains both fat-soluble vitamins (A, E, and K) and water-soluble vitamins (thiamin, riboflavin, niacin, pantothenic acid, pyridoxine, biotin, folate, vitamin C, choline, and betaine) [11,12,14,23].

As can be seen in Table 12.3, hazelnut is an excellent source of vitamin E (15.03–25.66 mg/100 g), which is a lipid-soluble phenolic antioxidant. With regard to RDA of vitamins, 42.5 g of hazelnut provides up to ~42.6%–72.7% vitamin E of the daily 15 mg of vitamin E recommended for adult males and females [31]. Vitamin E is a popular and the most powerful lipid-soluble antioxidant in the body and serves as the primary defense against lipid peroxidation [50] by protecting the body's cells from free radical damage [49,51]. Hazelnut is also an excellent source of biotin (76–80 μg/100 g). Consuming only ~37–40 g hazelnut per day supplies 100% of the AI for biotin (Table 12.3) for adults. Moreover, hazelnut is a good source of thiamin (0.42–0.64 mg/100 g), pyridoxine (0.56–0.63 mg/100 g), pantothenic acid (0.92–1.51 mg/100 g), and folate (72–120 μg/100 g). Small and/or trace amounts of fat-soluble vitamins (A and K) and water-soluble vitamins (riboflavin, niacin, vitamin C, choline, and betaine) are also present in hazelnut. Significant differences in vitamin contents among different Turkish hazelnut varieties have been reported [9,26]. Among 11 tree nuts (hazelnut, almond, Brazil nut, cashew, macademia, pecan, pine nut, pistachio, walnut, chestnut, and coconut), hazelnut contains the highest amount of vitamin E, folate, and biotin [23].

12.2.4 Amino Acids

Table 12.4 presents the amino acid profile of various hazelnut varieties grown in different countries [6,9,11,14]. Hazelnut is a good source of both essential and nonessential amino acids. Glutamic acid is the most abundant (2.84–3.71 g/100 g) amino acid, followed by arginine (1.87–2.21 g/100 g) and aspartic acid (1.33–1.68 g/100 g). The high content of arginine in hazelnut, in part, accounts for its hypercholesterolemic effect in subjects receiving a diet supplemented with hazelnut [52]. The three

TABLE 12.3
Vitamin Content of Hazelnuts and Percentage of RDA or AI

Vitamin	RDA or AI for Adult Males[a]	RDA or AI for Adult Females[a]	Unit	Vitamin Content: Refs. [11,12]	Vitamin Content: Ref. [14]	Vitamin Content: Ref. [23]	Percent of RDA or AI for Adult Males[b]	Percent of RDA or AI for Adult Females[b]
Vitamin A (retinol activity equivalents)	900 μg/day [32]	700 μg/day [32]	μg/100 g	—	1.0	—	0.05	0.1
Vitamin C (total ascorbic acid)	90 mg/day [31]	75 mg/day [31]	mg/100 g	5.54	6.3	—	2.6–3.0	3.1–3.6
Vitamin D	5 μg/day* [29]	5 μg/day* [29]	μg/100 g	—	—	—	—	—
Vitamin E (α-tocopherol equivalents)	15 mg/day [31]	15 mg/day [31]	mg/100 g	25.66	15.03[c]	24.98	42.6–72.7	42.6–72.7
Vitamin K (phylloquinone)	120 μg/day* [32]	90 μg/day* [32]	μg/100 g	—	14.2	—	5.0*	6.7*
Thiamin (B_1)	1.2 mg/day [30]	1.1 mg/day [30]	mg/100 g	0.42	0.64	0.43	14.9–22.7	16.2–24.7
Riboflavin (B_2)	1.3 mg/day [30]	1.1 mg/day [30]	mg/100 g	0.10	0.11	0.16	3.3–5.2	3.9–6.2
Niacin (B_3)	16 mg/day [30]	14 mg/day [30]	mg/100 g	1.94	1.80	1.10	2.9–5.2	3.3–5.9
Pantothenic acid (B_5)	5 mg/day* [30]	5 mg/day* [30]	mg/100 g	1.12	0.92	1.51	7.8–12.8*	7.8–12.8*
Pyridoxine (B_6)	1.3 mg/day [30]	1.3 mg/day [30]	mg/100 g	0.63	0.56	0.59	18.3–20.6	18.3–20.6
Cobalamin (B_{12})	2.4 μg/day [30]	2.4 μg/day [30]	μg/100 g	—	—	—	—	—
Betaine	na	na	mg/100 g	—	0.4	—	—	—
Biotin	30 μg/day* [30]	30 μg/day* [30]	μg/100 g	80	—	76	107.7–113.3*	107.7–113.3*
Choline	550 mg/day* [30]	425 mg/day* [30]	mg/100 g	—	45.6	—	3.5*	4.6*
Folate	400 μg/day [30]	400 μg/day [30]	μg/100 g	120	113	72	7.7–12.8	7.7–12.8

Note: na, not available.

[a] Recommended dietary allowances (RDA) or adequate intake (AI) for adults (aged 19–50 years).

[b] Values are expressed as minimum–maximum of Refs. [11,12,14,23], based on ~42.5 g (or 1.5 ounces) hazelnut serving.

[c] α-Tocopherol.

* Values are expressed as AI.

TABLE 12.4
Amino Acid Profile (g/100 g) of Various Hazelnut Varieties Grown in Different Countries

Amino Acid	Turkey[a] [11]	Turkey[b] [9]	United States[c] [14]	New Zealand[d] [6]
Alanine	0.70	0.72 (0.631–0.825)[e]	0.73	0.54 (0.448–0.619)
Arginine	2.16	2.00 (1.187–2.322)	2.21	1.87 (1.532–2.205)
Aspartic acid	1.52	1.49 (0.489–1.697)	1.68	1.33 (1.064–1.778)
Cysteine	0.46	—	0.28	0.27 (0.231–0.301)
Glutamic acid	3.13	2.84 (2.196–3.475)	3.71	2.86 (2.330–3.310)
Glycine	0.71	0.64 (0.513–0.724)	0.72	0.54 (0.434–0.588)
Histidine[f]	0.45	0.42 (0.315–0.590)	0.43	0.32 (0.272–0.364)
Hydroxyproline	0.06	—	—	—
Isoleucine[f]	0.58	0.56 (0.318–0.689)	0.55	0.47 (0.429–0.525)
Leucine[f]	1.07	1.15 (0.924–1.271)	1.06	0.82 (0.696–0.922)
Lysine[f]	0.41	0.45 (0.378–0.519)	0.42	0.42 (0.365–0.477)
Methionine[f]	0.23	0.16 (0.124–0.189)	0.22	0.18 (0.149–0.201)
Phenylalanine[f]	0.66	0.64 (0.542–0.767)	0.66	0.56 (0.500–0.626)
Proline	0.56	0.59 (0.513–0.819)	0.56	0.49 (0.397–0.540)
Serine	0.65	0.72 (0.494–1.082)	0.74	0.56 (0.472–0.671)
Threonine[f]	0.53	0.46 (0.416–0.517)	0.50	0.39 (0.314–0.461)
Tryptophan[f]	0.04	—	0.19	—
Tyrosine	0.53	0.47 (0.414–0.597)	0.36	0.40 (0.337–0.454)
Valine[f]	0.71	0.66 (0.616–0.807)	0.70	0.60 (0.508–0.677)
Total essential amino acids	4.68	4.51	4.74	3.75
Total amino acids	15.16	13.99	15.72	12.60

[a] Variety grown in Turkey (Tombul).
[b] Varieties grown in Turkey (Acı, Cavcava, Çakıldak, Foşa, İncekara, Kalınkara, Kan, Karafındık, Kargalak, Kuş, Mincane, Palaz, Sivri, Tombul, Uzunmusa, Yassı Badem, and Yuvarlak Badem).
[c] Variety grown in the United States (Unknown).
[d] Varieties grown in New Zealand (Whiteheart, Barcelona, Butler, Ennis, Tonda di Giffoni, and Campanica).
[e] Data are expressed as means (minimum–maximum values) among varieties.
[f] Essential amino acids.

nonessential amino acids contribute 44.9%–48.3% to the total amino acids present. The quality of a protein is related mainly to its essential amino acid composition and digestibility. Hazelnut contains all nine essential amino acids (except tryptophan in some studies), which contribute 29.8%–32.2% to the total amino acids present (Table 12.4). Alasalvar et al. [11] found that except for lysine and tryptophan, which exist below the levels (mg/g protein) of reference protein [34,53], other essential amino acids are present above the reference values. Some of the nonessential amino acids present in hazelnut, such as arginine, cysteine, glycine, and tyrosine are now considered as conditionally essential. They are needed in the diet unless abundant amounts of their precursors are available for their synthesis [54]. Hazelnut has a low lysine-to-arginine ratio (~0.21), which is inversely associated with the risk of developing hypercholesterolemia and atherosclerosis [55]. Asparagine and glutamine are not found in any hazelnut varieties [6,9,11,14,25]. Additionally, the proportion of amino acids in hazelnuts varies according to genotype, variety, growing seasons, environmental factors, and maturity [9,25,56].

In general, animal proteins (meat, fish, poultry, milk, cheese, and eggs), which contain ample amounts of all essential amino acids are considered good sources of complete proteins. On the other hand, plant proteins (including nuts) are often considered incomplete because they generally do not have enough of one or more of the essential amino acids. Although hazelnut protein contained all essential amino acids, lysine and tryptophan were the limiting amino acids [11].

12.2.5 Dietary Fiber

Although dietary fiber (indigestible carbohydrate) is not a nutrient, it has a range of metabolic health benefits. High fiber intakes are associated with lower serum cholesterol concentrations, risk of CHD, and body mass index (BMI), improved laxation, enhanced weight control, better glycemic control, reduced blood pressure, risk of certain forms of cancer, and the postprandial glucose response, and improved gastrointestinal function, among others [34,57–61]. Dietary fiber can be categorized into water-soluble and water-insoluble components. Besides several other health benefits, water-soluble dietary fiber reduces total and low-density lipoprotein (LDL) cholesterols [59].

After cereals, nuts are the vegetable foods that are richest in fiber, which may partly explain their benefit on the lipid profile and cardiovascular health [61]. Alasalvar et al. [11] reported that total dietary fiber content of Tombul hazelnut was 12.88 g/100 g, of which 2.21 g/100 g was soluble fiber (fresh weight basis). Total dietary fiber content of Tombul hazelnut is slightly higher than those reported in the literature for different varieties of hazelnut, ranging from 6.5 to 9.7 (fresh weight basis) [14,23,62]. In addition, Savage and McNeil [6] compared the total dietary fiber content of six varieties of hazelnut grown in New Zealand. Their results range from 9.8 to 13.2 g/100 g (dry weight basis), being lowest in Tonda di Giffoni and highest in Campanica. These results show that differences in total dietary content exist among hazelnut varieties.

Most health/nutrition professionals agree on the benefit of increased consumption of dietary fiber up to 25–35 g/day [63,64]. The AI of dietary fiber has been recently set for adult women (25 g/day) and for adult men (38 g/day) [34]. Consuming recommended daily amount of 42.5 g hazelnut [22] per day is adequate for 21.9% and 15.6% of the recommended AI for adult women and men, respectively [11].

12.3 TASTE AND AROMA CHARACTERISTICS OF HAZELNUT

12.3.1 Taste-Active Compounds

The presence of taste-active components such as sugars, organic acids, free amino acids, and tannins can improve the sensory characteristics of different food products. The presence and composition of these components of hazelnut maybe affected by various factors such as variety, growing conditions, maturity, season, geographic origin, fertilization, soil type, storage conditions, amount of sunlight received, and time of harvest, among others [2,65–67]. Thus, better taste and flavor of hazelnut may increase the consumption of this nutritionally important nut. Sugars, organic acids, free amino acids, and tannins in hazelnut are discussed below in more detail.

12.3.1.1 Sugars

Sugars are responsible for the sweetness of foods. Individual sugars possess different relative sweetness scores; fructose is almost 1.8 times sweeter than sucrose, while glucose is as sweet as sucrose [68]. Six different sugars have been positively identified in Turkish Tombul hazelnut variety [11], including monosaccharides (fructose, glucose, and *myo*-inositol) as well as sucrose and its galactosides, namely raffinose and stachyose (Table 12.5). The total sugar content of hazelnut averages 3.58 g/100 g, and sucrose contributes 74.6% to the total amount, followed by stachyose at 13.4%. Other sugars (fructose, glucose, *myo*-inositol, and raffinose) are present in low amounts, 12% (3.9% fructose, 3.1% glucose, 1.1% *myo*-inositol, and 3.9% raffinose) of the total. The same six sugars have also been found by Botta et al. [2] in 12 different varieties of Oregon and Italian hazelnuts although at different levels with a mean total value of 4 g/100 g (dry weight basis) ranging from

TABLE 12.5
Taste-Active Compounds of Tombul Hazelnut

Taste-Active Component	g/100 g
Sugar	
Fructose	0.14 ± 0.05
Glucose (dextrose)	0.11 ± 0.03
Sucrose	2.67 ± 0.43
myo-Inositol	0.04 ± 0.01
Raffinose	0.14 ± 0.10
Stachyose	0.48 ± 0.08
Total	3.58 ± 0.56
Organic acid	
Oxalic	0.080 ± 0.002
Maleic	0.001 ± 0.000
Citric	0.412 ± 0.005
Malic[a]	1.050 ± 0.036
Lactic	0.032 ± 0.001
Acetic	0.049 ± 0.001
Total	1.624 ± 0.032
Free amino acid	
Alanine	68.13 ± 0.99
Arginine	204.92 ± 2.94
Aspartic acid	71.32 ± 1.31
Asparagine	53.65 ± 0.72
Cysteine	14.14 ± 1.52
Glutamic acid	128.68 ± 1.34
Glutamine	7.99 ± 0.11
Glycine	16.10 ± 0.27
Histidine	8.16 ± 0.13
Hydroxyproline	0.31 ± 0.01
Isoleucine	11.07 ± 1.17
Leucine	11.16 ± 0.54
Lysine	14.86 ± 0.46
Methionine	0.72 ± 0.01
Phenylalanine	10.18 ± 0.25
Proline	14.52 ± 0.85
Serine	14.23 ± 0.03
Threonine	19.73 ± 0.15
Tryptophan	13.50 ± 1.27
Tyrosine	12.52 ± 0.37
Valine	19.36 ± 0.30
Total	715.25 ± 3.40

Source: From Alasalvar, C., Shahidi, F., Liyanapathirana, C.M., and Ohshima, T., *J. Agric. Food Chem.*, 51, 3790, 2003. With permission.

Note: Data are expressed as means ± the standard deviation ($n = 3$) on a fresh weight basis.

[a] Malic acid may contain 5.54 mg/100 g ascorbic acid.

2.8 to 5.6 g/100 g with sucrose being predominant. These sugars, with the exception of *myo*-inositol, were also reported by Ruggeri et al. [13] in the Italian variety 'Tonda Gentile Romana,' with a total content of 4.1 g/100 g (fresh weight basis). Only three sugars in hazelnut (sucrose, glucose, and fructose) have recently been identified by the USDA [14], sucrose being the predominant sugar (4.20 g/100 g) followed by the same amount of glucose and fructose (0.07 g/100 g). Several factors such as variety, environmental factors, maturity, and growing seasons, among others, may influence the presence and composition of sugars in hazelnut.

12.3.1.2 Organic Acids

Organic acids have generally been reported to be responsible for sour, tart, acidic, and characteristic fruity taste of many foods [69–72]. They are important intermediate products of metabolism and have been reported as important factors influencing the sensory characteristics of fruits and vegetables [73]. Data in the literature referring organic acid composition of hazelnut are limited. Alasalvar et al. [11] reported the positive identification of six organic acids in Turkish cultivar "Tombul" (Table 12.5). Malic acid is the predominant compound (1.050 g/100 g), followed by lesser amounts of citric acid (0.412 g/100 g) and oxalic acid (0.080 g/100 g). Malic acid, which is the predominant organic acid in hazelnut, has a characteristic fruity, mellow, smooth tart, and sour taste in fresh fruits and vegetables. According to Alasalvar et al. [11], the contents of organic acids of Tombul hazelnut are somewhat higher than those reported for different varieties of Oregon and Italian hazelnuts [2]. Malic, galacturonic, levulinic, succinic, citric, oxalic, acetic, and butyric acids are present in 12 varieties of hazelnuts; malic acid being the most abundant (ranging from 42 to 209 mg/100 g, dried weight basis). Galacturonic, succinic, levulinic, and butyric acids are not detected in Tombul hazelnut [11]. The observed differences maybe due to either varietal or soil type.

12.3.1.3 Free Amino Acids

The literature data on free amino acid composition of hazelnut are scarce. Table 12.5 shows the free amino acids present in Tombul hazelnut. Hazelnut contains large amounts of arginine (204.9 mg/100 g) and glutamic acid (128.7 mg/100 g) along with lesser amounts of aspartic acid (71.3 mg/100 g), alanine (68.1 mg/100 g), and asparagine (53.6 mg/100 g). These five amino acids constitute 71.6% of the total free amino acids in Tombul hazelnut. Although individual free amino acids have been reported to impart distinct taste and flavor to foods [74–77], the dominant free amino acids found in highest amounts in Tombul hazelnut are noted as being responsible for bitter, sour, and sweet taste in foods [77,78]. Besides taste-active properties of free amino acids in foods, they can also play an important role in the formation of color, acrylamide, and aroma during roasting of nuts [79–81].

12.3.1.4 Tannins

Tannins, partly responsible for the bitter and astringent taste as well as the brown color, are water-soluble naturally occurring complex polyphenols that are present in many plant foods, including most nuts. In addition, tannins have a puckering effect in the mouth because of precipitating proteins. Tannins are divided into two groups, namely condensed tannins (also known as proanthocyanidins) and hydrolyzable tannins [82]. Detailed information about both types of tannins and their potential health benefits are given in Chapter 13. Hazelnut contains condensed tannins [18,83–85]. Pecan, walnut, almond, hazelnut, and pistachio have all been reported to contain high amounts of tannins in descending order, whereas trace amounts of tannins are present in Brazil nut, macadamia, and pine nut [18]. In addition, hazelnut contains the highest amount of condensed tannins among seven tree nuts (hazelnut, almond, cashew, chestnut, pecan, pistachio, and walnut) [84,86].

12.3.2 Aroma-Active Compounds

Hazelnut may be consumed in the natural (raw state) or preferably as roasted. The roasting process improves the flavor, color, crispiness, and crunchiness of hazelnut [87–89]. Improvements in the aroma of hazelnut in different ways may affect the aroma-active components of raw hazelnut and could lead to an increased consumption of hazelnuts, including their use as snack food and ingredient in baked goods, confectionery products, cereals, ice cream, coffee, yoghurt, various chocolate bars, sweet products, nougat, flavors, and fragrances, among others.

Volatile components of natural and roasted hazelnuts have been investigated by several researchers [7,88,90–98]. Among several volatile aroma-active compounds detected in roasted hazelnut, 5-methyl-(*E*)-2-hepten-4-one (filbertone) has been reported as the primary odorant (nutty-roasty and hazelnut-like) [88,93,94,96]. Alasalvar et al. [98] studied the comparison of natural and roasted Turkish Tombul hazelnuts and found a total of 39 compounds in natural hazelnut and 79 compounds in roasted hazelnut. These included ketones, aldehydes, alcohols, aromatic hydrocarbons, terpenes, furans, pyrroles, pyrazines, and acids. Pyrazines, pyrroles, terpenes, and acids are detected in roasted hazelnut only. The combination of several volatile aroma-active components that increases upon roasting may contribute to the distinctive and unique flavor of roasted hazelnut. Pyrazines together with ketones, aldehydes, furans, and pyrroles may contribute to the characteristic roasted aroma of hazelnut. Detail information about flavor and volatile compounds in major tree nuts are detailed in Chapter 7.

12.4 FUNCTIONAL LIPID CHARACTERISTICS OF HAZELNUT

12.4.1 Fatty Acid Composition

A total of 18 fatty acids have been reported in oils extracted from different varieties of hazelnuts from different growing areas and harvesting years; representative results are summarized in Table 12.6 [4,10,11,18,99]. Among the identified fatty acids, 18:1ω9 is by far the dominant fatty acid (ranging from 77.50% to 82.95%), followed by 18:2ω6 (ranging from 7.55% to 13.69%), 16:0 (ranging from 4.85% to 5.79%), 18:0 (ranging from 1.88% to 3.12%), and 18:1ω7 (mean 1.35%). The remaining 13 fatty acids are present in less than 1% (Table 12.6). Vaccenic acid (18:1ω7), an isomer of 18:1ω9, was only reported in 19 cultivars of hazelnuts grown in Portugal [10]. The major fatty acids presented in Table 12.6 are, in general, comparable to those reported in the literature on different hazelnut varieties [3–5,8,9,12,13,19,20,24,100–104]. However, the composition and amount of fatty acids both between and within the same hazelnut varieties maybe influenced by different factors [4,10,19,24,99,105,106].

As shown in Table 12.6, the total saturated fatty acids (SFA) make up a small proportion (6.90%–8.52%) of the total fatty acids of hazelnut oil, whereas total monounsaturated fatty acids (MUFA) is the highest (78.90%–83.16%). The heart-healthy fatty acids [MUFA + polyunsaturated fatty acids (PUFA)] account for ~92% of the total fatty acids present. Hazelnut oil, in terms of its high proportion of heart-healthy fatty acids, is much more desirable than other vegetable oils. The health benefits of these fatty acids are given in the upcoming sections. Compared to other tree nuts and vegetable oils, hazelnut oil is among the ones with highest contents of MUFA and lowest content of SFA [14,18]. Consumption of a high amount of SFA raises LDL cholesterol and lowers high-density lipoprotein (HDL) cholesterol concentrations, therefore increasing the risk of developing heart disease, stroke, and certain types of cancer, among others [107–110]. Hazelnut oil, due to its low content of SFA, promotes health benefits.

Amaral et al. [104] reported that oils extracted from raw and roasted hazelnuts, in addition to the 18 fatty acids mentioned, contain trace amounts (<0.1%) of *trans* fatty acid (18:1ω9t). Although negative health effects of *trans* fatty acids have been reported [111–114], the contents found in both raw and roasted hazelnuts are very low, thus having a negligible impact on the nutritional value of hazelnuts and are not expected to pose any adverse effects [115].

TABLE 12.6
Fatty Acid Composition (%) of Oils Extracted from Various Hazelnut Varieties Grown in Different Countries

Fatty Acid	Turkey[a] [117]	Portugal[b] [10]	Spain[c] [4]	United States[d] [18]	New Zealand[e] [99]
14:0	0.03 ± 0.00	0.03 (0.02–0.05)[f]	—	0.03 ± 0.00	tr[g]
15:0	0.02 ± 0.00	0.01 (nd-0.02)	—	—	—
16:0	4.85 ± 0.02	5.55 (4.84–6.75)	5.79 (5.40–6.48)	5.78 ± 0.01	4.86 (4.08–5.94)
16:1	0.16 ± 0.00	0.20 (0.14–0.26)	0.28 (0.22–0.40)	0.15 ± 0.00	0.16 (0.12–0.24)
17:0	0.04 ± 0.00	0.05 (0.04–0.06)	—	0.05 ± 0.00	0.07 (tr-0.24)
17:1	0.07 ± 0.00	0.08 (0.07–0.10)	—	—	0.08 (0.07–0.10)
18:0	2.73 ± 0.00	2.73 (2.08–3.70)	1.97 (1.64–2.28)	3.12 ± 0.05	1.88 (1.62–2.03)
18:1ω9	82.72 ± 0.04	80.36 (76.71–82.81)	79.10 (77.05–84.05)	82.95 ± 0.04	77.50 (73.80–80.07)
18:1ω7	—	1.35 (0.71–1.76)	—	—	—
18:1ω11	—	—	—	—	1.02 (0.87–1.18)
18:2ω6	8.89 ± 0.01	9.15 (7.20–11.37)	12.58 (6.62–14.99)	7.55 ± 0.02	13.69 (11.96–16.53)
18:3ω3	0.10 ± 0.00	0.10 (0.08–0.12)	0.12 (0.11–0.16)	0.24 ± 0.00	0.14 (0.12–0.16)
20:0	0.14 ± 0.00	0.13 (0.11–0.17)	—	0.14 ± 0.00	0.09 (0.08–0.11)
20:1ω9	0.16 ± 0.00	0.15 (0.12–0.20)	0.18 (0.14–0.22)	—	0.14 (0.12–0.15)
22:0	0.03 ± 0.00	0.03 (0.02–0.04)	—	—	tr
22:1ω9	0.03 ± 0.01	0.05 (0.04–0.07)	—	—	—
24:0	0.01 ± 0.00	—	—	—	—
24:1ω9	0.02 ± 0.00	—	—	—	—
Total SFA	7.85	8.52 (7.5–10.0)	7.76	9.11	6.90 (6.3–7.0)
Total MUFA	83.16	82.19 (78.7–84.6)	79.56	83.10	78.90 (75.1–81.4)
Total PUFA	8.99	9.25 (7.3–11.5)	12.70	7.79	13.82 (12.1–16.7)

[a] Variety grown in Turkey (Tombul). Data are expressed as means ± the standard deviation.
[b] Varieties grown in Portugal (Butler, Campanica, Cosford, Couplat, Daviana, Ennis, Fertille de Coutard, Grossal, Gunslebert, Lansing, Longa d'Espanha, Merveille de Bollwiller, Morell, Negreta, Pauetet, Round du Piemont, Segorbe, Santa Maria de Jesus, and Tonda de Giffoni).
[c] Varieties grown in Spain (Castanyera, Culpla, Gironell, Negret, Pauetet, Sant Giovanni, Tonda Gentille, Tonda di Giffoni, Tonda Italiana, and Tonda Romana).
[d] Variety grown in the United States (Unknown). Data are expressed as means ± the standard deviation.
[e] Varieties grown in New Zealand (Whiteheart, Barcelona, Butler, Ennis, Tonda di Giffoni, and Campanica).
[f] Data are expressed as means (minimum–maximum values) among varieties.
[g] tr, trace.

12.4.2 Lipid Classes

The lipid classes of hazelnut oil include triacylglycerols (TAG) as nonpolar lipids (98.4%) and glucolipids (1.4%) and phospholipids (<0.2%) [phosphatidylcholine (PC) and phosphatidylinositol (PI)] as polar lipids [116]. The 18:1ω9 is dominant in the nonpolar lipid class (TAG), whereas 16:0, 18:0, and 18:2ω6 are most predominant in the polar lipid class in hazelnut oil. Recently, Alasalvar et al. [117] examined lipid class composition of Tombul hazelnut oil and found that it contained 98.8% of nonpolar and 1.2% of polar constituents. The main nonpolar lipid class in hazelnut oil is TAG, contributing nearly 100% to the total amount. PC, phosphatidylethanolamine (PE), and PI are main polar lipids, contributing 56.4%, 30.8%, and 11.7% to the total polar lipids, respectively. Similar results were reported by Parcerisa et al. [105]. More recently, Miraliakbari and Shahidi [216] examined the lipid classes in tree nut oils that included TAG, sterols and sterol esters, phospholipids, and sphingolipids. Hazelnut oil contained TAG, sterols, sterol esters, phosphatidylserine (PS), PC, PI, phosphatidic acid, and sphingolipids.

12.4.3 Triacylglycerol Composition

TAG are increasingly used in the food industry as a tool to assess the quality and authenticity of vegetable oils [118,119], particularly adulteration of olive oil with hazelnut oil [120,121].

A total of 13 TAG (including one unknown) have been determined in crude hazelnut oils extracted from various hazelnut varieties grown in different countries (Table 12.7); these are LLL,

TABLE 12.7
Triacylglycerol Composition (%) of Oils Extracted from Various Hazelnut Varieties Grown in Different Countries

Triacylglycerol[a]	Turkey[b] [12]	Portugal[c] [122]	Spain[d] [4]
LLL	0.24 ± 0.02	0.22 (0.10–0.70)[e]	1.39 (0.63–1.99)
OLL	1.85 ± 0.10	1.76 (0.97–3.31)	5.20 (2.15–6.49)
PLL	0.05 ± 0.00	0.08 (0.04–0.24)	1.00 (0.52–1.22)
OOL	12.26 ± 0.38	14.51 (11.24–18.85)	18.10 (11.93–20.86)
POL	0.79 ± 0.07	1.46 (0.97–2.37)	4.31 (2.98–4.93)
PPL	0.01 ± 0.00	0.03 (0.01–0.05)	0.45 (0.32–0.59)
OOO	71.31 ± 1.11	66.61 (57.56–71.37)	48.32 (45.72–51.92)
POO	9.45 ± 0.76	11.34 (8.36–13.08)	13.73 (12.38–15.47)
PPO	0.08 ± 0.01	0.10 (0.03–0.20)	1.20 (0.72–1.68)
Unknown	0.07 ± 0.01	—	—
PPP	—	—	0.56 (0.19–0.89)
SOO	3.79 ± 0.24	3.74 (2.23–6.91)	4.69 (3.74–6.17)
PSO	0.10 ± 0.01	0.12 (0.03–0.36)	0.96 (0.51–1.46)

[a] Trilinoleoylglycerol (LLL), oleoyl-dilinoleoylglycerol (OLL), palmitoyl-dilinoleoylglycerol (PLL), dioleoyl-linoleoylglycerol (OOL), palmitoyl-oleoyl-linoleoylglycerol (POL), dipalmitoyl-linoleoylglycerol (PPL), trioleoylglycerol (OOO), palmitoyl-dioleoylglycerol (POO), dipalmitoyl-oleoylglycerol (PPO), tripalmitoylglycerol (PPP), stearoyl-dioleoylglycerol (SOO), and palmitoyl-stearoyl-oleoylglycerol (PSO).

[b] Variety grown in Turkey (Tombul). Data are expressed as means ± the standard deviation.

[c] Varieties grown in Portugal (Butler, Campanica, Cosford, Couplat, Daviana, Ennis, Fertille de Coutard, Grossal, Gunslebert, Lansing, Longa d'Espanha, Merveille de Bollwiller, Morell, Negreta, Pauetet, Round du Piemont, Segorbe, Santa Maria de Jesus, and Tonda de Giffoni). Year crop 2001.

[d] Varieties grown in Spain (Castanyera, Culpla, Gironell, Negret, Pauetet, Sant Giovanni, Tonda Gentille, Tonda di Giffoni, Tonda Italiana, and Tonda Romana).

[e] Data are expressed as means (minimum–maximum values) among varieties.

OLL, PLL, OOL, POL, PPL, OOO, POO, PPO, unknown, PPP, SOO, and PSO (where P, palmitoyl; S, stearoyl; O, oleoyl; and L, linoleoyl) [4,12,122]. In addition, Parcerisa et al. [4,105,123] detected trace amounts of PPP in different hazelnut varieties cultivated in Spain. In contrast, Ayorinde et al. [124] analyzed hazelnut oil by matrix-assisted laser desorption ionization–time-of-flight (MALDI-TOF) mass spectrometry and pointed out that minor quantities of SLL and SOL could be coeluted with OOL and OOO, respectively. Likewise, Holčapek et al. [125] also referred that minor amounts of SOL could exist in hazelnut oil.

The predominant TAG in hazelnut oils are OOO (48.32%–71.31%), followed by OOL (12.26%–18.10%), POO (9.45%–13.73%), SOO (3.74%–4.69%), OLL (1.76%–5.20%), and POL (0.79%–4.31%). As shown in Table 12.7, significant differences in TAG composition exist among hazelnut varieties, which can be influenced by genetic and environmental factors as well as detector used [4,120,122,123,126]. A good agreement between fatty acid profiles and the TAG composition has been found [4,10,12,122]. For example, oleic acid (18:1ω9) and OOO are the highest in all hazelnut oils.

12.4.4 Phytosterol and Phytostanol Compositions

Phytosterols and phytostanols (also called plant sterols and stanols) serve as important functional food ingredients due to their cholesterol-lowering effects. Thus, the Food and Drug Administration (FDA) has issued a rare "health claim" for sterol- and stanol-containing products [127]. Therefore, the presence and amount of phytosterols and phytostanols in hazelnuts is important.

Crews et al. [8] reported 13 phytosterols and phytostanols (cholestanol, campesterol, campestanol, stigmasterol, Δ^7-campesterol, $\Delta^{5,23}$-stigmastadienol, clerosterol, β-sitosterol, sitostanol, Δ^5-avenasterol, $\Delta^{5,24}$-stigmastadienol, Δ^7-stigmastenol, and Δ^7-avenasterol) and trace amounts of cholesterol in various raw hazelnut varieties grown in different countries (Table 12.8). The total contents among hazelnut oils range from 105.7 to 195.2 mg/100 g. It is obvious from the table that both the concentration and pattern of phytosterols and phytostanols vary widely among hazelnuts from different origins. β-Sitosterol, which is the most common and predominant phytosterol in hazelnut oils, accounts for 78.1%–83.6% of the total. Hazelnut oil also serves as a good source of campesterol, sitostanol, and Δ^5-avenasterol (Table 12.8). Recently, Alasalvar et al. [12] identified and quantified eight phytostrols (campesterol, stigmasterol, clerosterol, β-sitosterol, Δ^5-avenasterol + β-sitostanol, Δ^7-stigmastenol, and Δ^7-avenasterol) as well as trace amounts of cholesterol in Turkish Tombul hazelnut with a total content of 164.9 mg/100 g oil, respectively. Among these, β-sitosterol comprised 81.28% of the total, while Δ^5-avenasterol + β-sitostanol and campesterol were the second and third components of the group with values of 8.45% and 6.02%, respectively. The same sterols in 19 varieties of hazelnut oils have been identified by Amaral et al. [10]. Different patterns as well as individual and total sterols have been reported in the literature for several hazelnut oils grown in different countries [99,101–103,117,126]. Amaral [106] observed that phytsoterol and phytostanol compositions of hazelnut were more affected by genetic factors than geographical location or climate.

Phillips et al. [128] studied the sterol composition of nuts consumed in the United States and quantified six phytosterols and phytostanols present (β-sitosterol, campesterol, stigmasterol, Δ^5-avenasterol, sitostanol, and campestanol). Among 10 economically important nuts (hazelnut, almond, walnut, pistachio, pine nut, Brazil nut, cashew, pecan, macademia and peanut), the total phytosterol content ranged from 95 to 279 mg/100 g of nut, being lowest in Brazil nut and the highest in pistachio nut. Hazelnut had a phytosterol content of 121 mg/100 g of nut in their study.

Plant oils, such as corn oil and rapeseed oil, are major dietary source of phytosterols, whereas phytostanols occur mainly in cereals, such as wheat, rice, and rye [129–131]. With respect to health aspects, phytosterols, phytostanols, and their esters have been shown to decrease the risk of certain types of cancer [132–135] and cardiovascular disease (CVD) [136,137], inhibit cholesterol absorption [138], and enhance immune functions [139], among others. Furthermore, numerous

TABLE 12.8
Phytosterol and Phytostanol Composition (mg/100 g) of Oils Extracted from Various Hazelnut Varieties Grown in Different Countries

Sterol and Stanol	Turkey[a]	Italy[b]	France[c]	Spain[d]
Cholesterol	nd-0.7	nd-0.5	nd-0.7	nd-1.4
Cholestanol	nd	nd	nd-0.7	nd
Campesterol	5.8–10.7	6.7–8.1	7.3–10.1	5.9–9.7
Campestanol	nd-1.3	nd-0.8	0.4–0.9	nd-0.7
Stigmasterol	1.5–2.8	2.0–5.4	1.4–2.7	1.5–1.9
Δ^7-Campesterol	nd-0.6	nd-1.7	nd-1.1	nd-0.6
$\Delta^{5,23}$-Stigmastadienol	nd	nd-0.4	nd	nd
Clerosterol	1.8–3.8	1.8–2.8	1.5–2.4	1.9–4.0
β-Sitosterol	84.5–143.1	116.4–138.8	112.5–161.2	92.9–149.1
Sitostanol	3.3–11.7	3.2–7.6	3.2–6.7	2.9–6.2
Δ^5-Avenasterol	0.8–9.4	4.2–6.2	4.5–6.1	3.8–6.5
$\Delta^{5,24}$-Stigmastadienol	nd-2.0	nd-1.4	0.3–1.5	nd-0.8
Δ^7-Stigmastenol	nd-2.1	nd-2.4	nd-3.2	nd-1.5
Δ^7-Avenasterol	nd-2.2	nd-1.8	nd-0.8	nd-0.7
Total	105.7–183.2	141.3–177.2	135.3–195.2	114.2–178.4

Source: From Crews, C., Hough, P., Godward, J., Brereton, P., Lees, M., Guiet, S., and Winkelmann, W., *J. Agric. Food Chem.*, 53, 4843, 2005. With permission.

Note: nd, not detected (<0.1 mg/100 g oil).

[a] Six unknown samples from different regions of Turkey.

[b] Eight unknown samples from different regions of Italy.

[c] Nine unknown samples from different regions of France.

[d] Eight unknown samples from different regions of Spain.

well-designed studies are known on the cholesterol-lowering effect of phytosterols, phytostanols, and their esters added to various fat-containing functional foods, such as margarines, spreads, cooking and salad oils, mayonnaise, yogurt drinks, milks, soy drinks, cheese-type products, and salad dressing, among others [127,131,133,140–148]. In view of the cited references, daily doses of 1.5–3 g of phytosterols, phytostanols, or their esters have been recommended for lowering total and LDL cholesterol concentrations significantly. According to the Novel Food Regulations of the European Union, the use of spreads, salad dressing, milk, and cheese products as well as soy drinks enriched with phytosterols, phytostanols, and their esters is now permitted at a maximum dose of 3 g/day [149]. In addition, the safety and activity of phytosterols, phytostanols, and their esters have been thoroughly reviewed by the FDA and their use in foods is currently allowed. Nowadays, these compounds are being frequently used for lowering plasma cholesterol levels, either as part of a regular diet or through food fortification. For example, consumption of margarine containing 10% phytosterols resulted in a 13% reduction in total and LDL cholesterols in normal and mildly hypercholesterolemic patients [141]. Hence, the consumption of foods such as hazelnut or hazelnut oil, which contain phytosterols and phytostanols, may provide health benefits.

12.4.5 Tocols (Tocopherol and Tocotrienol) Composition

Vitamin E is a group of fat-soluble compounds including four tocopherols (designated as α, β, γ, and δ) and four tocotrienols (designated as α, β, γ, and δ). The food that we consume may contain different combinations and amounts of tocols. α-Tocopherol has become synonymous with vitamin E because it is the predominant form in human and animal tissues. Recent research, however, shows that the other tocopherols and tocotrienols have important and unique antioxidant and other biological effects in nutrition and health and are now receiving increased attention [150–155]. In addition, Qureshi et al. [156] reported that tocotrienols lower serum cholesterol in hypercholesterolemic humans.

Table 12.9 presents a compilation of tocol composition of hazelnuts published in the literature, regarding different varieties and growing locations. Oils extracted from various hazelnut varieties showed significant tocol differences and patterns. A total of seven tocol isoforms have been reported in hazelnut oils including four tocopherols (α, β, γ, and δ) and three tocotrienols (α, β, and γ), with α-tocopherol being most abundant. As shown in Table 12.9, the mean content of total tocols range from 31.46 mg/100 g in Portuguese hazelnut oil to 51.31 mg/100 g in Turkish Tombul hazelnut oil. High levels of tocols detected by Crews et al. [8] could be due to methodology differences such as the use of detector. Fluorescence detector has been used for quantification of hazelnuts grown in New Zealand [99], and a fluorescence detector in series with a diode array detector for hazelnuts grown in Turkey [12] and Portugal [104], whereas Crews at al. [8] used ultraviolet detector for hazelnuts collected from Italy, France, and Spain.

Several researchers have only detected and quantified tocopherols in crude, refined, virgin, mixed, and pressed hazelnut oils, among which α-tocopherol is the predominant tocopherol with concentrations ranging from 11.9 to 61.88 mg/100 g oil [3,99,101,103,120,157,158]. However, small amounts of α-, β-, and γ-tocotrienols together with four tocopherols have recently been reported in hazelnut oils [8,100,104,159,160]. Varietal differences in tocopherol and tocotrienol content of hazelnut cultivars have been reported in the cited studies. In addition, geographic origin, climate, harvesting year, storage conditions, culture conditions, and soil type, among others, might also affect the tocopherol and tocotrienol compositions and contents in hazelnut oils.

Kornsteiner et al. [158] studied the tocopherol content of oils extracted from different nuts and found that vitamin E content was highest in hazelnut oil (33.1 mg/100 g), followed by, in descending order, almond > peanut > pistachio > pine nut > walnut > Brazil nut > pecan > cashew > macademia. Hazelnut oil has been reported to contain the highest α-tocopherol level among nut oils [3,158]. Depending on the variety, hazelnut oil also contained-two to three times more α-tocopherol than olive oil [100,127,157].

12.4.6 Squalene

Squalene is a 30-carbon, straight-chain terpenoid hydrocarbon that serves as a precursor of steroids, both in plant cells and animals. In plant cells, squalene can be converted to phytosterols. It has been reported that squalene significantly decreases total cholesterol, LDL cholesterol, and TAG levels in hypercholesterolemic patients [161,162]. There has been growing interest in squalene as a potential chemopreventative agent [163–165]. In addition, the possible protective effect of squalene maybe attributed to its antioxidant function [166–168].

Natural sources rich in squalene include the shark and whale liver oils and several vegetable oils such as olive oil [169]. Data concerning squalene content in nuts are limited. Among 10 different types of frequently consumed nuts (hazelnut, walnut, almond, peanut, macadamia, Brazil nut, pecan, pine nut, pistachio, and cashew), the levels of squalene range from 9.4 to 1377.8 μg/g oil, being lowest in walnut oil and highest in Brazil nut oil. Hazelnut presents the second highest content of squalene (186.4 μg/g oil), only exceeded by Brazil nut [170,171]. The level found in hazelnut oil is comparable with soybean oil (220 μg/g) [172]. Kalogeropoulos et al. [213] measured the squalene content in fresh nuts (almond, chestnut, hazelnut, pistachio, and walnut) and seeds (pumpkin, sesame, and sunflower) and found that hazelnut contains 14.8 mg/100 g squalene. Its concentrations ranged from 0.6 mg/100 g in sesame seed to 52 mg/100 g in pumpkin seed. The average intake of squalene in the United States is 30 mg/day. Consuming approximately 160 g hazelnut oil per day is adequate for squalene intake.

TABLE 12.9
Tocol Composition (mg/100 g) of Oils Extracted from Various Hazelnut Varieties Grown in Different Countries

Tocol	Turkey[a] [12]	Portugal[b] [106]	Italy[c] [8]	France[d] [8]	Spain[e] [8]	New Zealand[f] [99]
α-Tocopherol	40.40 ± 0.56	28.95 (15.87–38.81)[g]	11.8–44.0	34.3–54.5	31.4–59.0	33.99 (19.91–40.90)
β-Tocopherol	1.53 ± 0.02	1.15 (0.51–2.18)	nd-1.0	nd-1.0	nd-1.5	1.12 (0.62–1.67)
γ-Tocopherol	8.33 ± 0.16	0.88 (0.19–3.29)	nd-1.0	nd-1.0	nd-4.2	4.50 (1.86–14.89)
δ-Tocopherol	0.53 ± 0.01	0.16 (0.02–0.77)	nd-1.0	nd-1.0	nd-1.0	0.27 (0.10–0.66)
α-Tocotrienol	0.20 ± 0.02	0.20 (nd-0.46)	nd-20.9	nd	nd	nd
β-Tocotrienol	0.09 ± 0.00	0.04 (0.01–0.71)	nd	nd	nd-2.2	nd
γ-Tocotrienol	0.23 ± 0.01	0.08 (nd-0.57)	nd	nd	nd-3.4	nd
δ-Tocotrienol	nd	nd	nd	nd	nd	nd
Total	51.31 ± 0.51	31.46 (16.97–42.82)	35.4– ~50.0	34.3–54.5	31.4–61.3	39.88 (22.58–55.20)

Note: nd, not detected.

[a] Variety grown in Turkey (Tombul). Data are expressed as means ± the standard deviation.
[b] Varieties grown in Portugal (a mixture of two or three unknown varieties).
[c] Eight unknown samples from different regions of Italy.
[d] Nine unknown samples from different regions of France.
[e] Eight unknown samples from different regions of Spain.
[f] Varieties grown in New Zealand (Whiteheart, Barcelona, Butler, Ennis, Tonda di Giffoni, and Campanica).
[g] Data are expressed as means (minimum–maximum values) among varieties.

12.5 HEALTH EFFECTS OF HAZELNUT

Traditionally, nuts were perceived by the general public as being unhealthy due to their high fat content. However, recent epidemiologic and clinical studies have provided evidence that frequent nut consumption is associated with favorable plasma lipid profiles [52,173–178] and reduced risk of CHD, CVD, myocardial infarction, atherosclerosis, and other chronic disorders [179–183]. Epidemiologic studies [179,184,185] have estimated that the percentage decrease in CHD risk with frequent consumption (more than four times per week) of nuts is up to 30%–50%. The most recent recognition of nuts as "heart-healthy" foods by the FDA has provided a major boost to the image of nuts, including hazelnut [22]. According to the new Healthy Eating Pyramid [186], one to three serving of nuts and legumes should be incorporated into the diet each day for lifelong health.

There is an increasing interest in the lipid characteristics of nut oils as they seem to be an interesting source of bioactive constituents and functional ingredients [170,174,187]. Among tree nuts, hazelnut has many beneficial health attributes and is among three most popular and commonly consumed tree nuts in Europe [188] and other Western populations [179]. The benefits of hazelnut inclusion into the human diet is partly related to its fat components (around 60%), most of which are highly rich in MUFA (primarily oleic acid), tocols (α-tocopherol), phytosterols (β-sitosterol), polyphenols, and squalene [10,12,99,104,117,160,170,189].

12.5.1 Plasma Cholesterol and Lipoprotein Profiles

Although current recommended National Cholesterol Education Program/American Heart Association Step I or Step II diets have beneficial effects in lowering total and LDL cholesterol concentrations [190], they tend to decrease HDL cholesterol and increase TAG concentrations, thereby potentially adversely affecting coronary risk factors [191,192]. It is, therefore, imperative to identify alternative diets that can effectively modify the plasma lipid profiles and thus reduce CHD risk.

In contrast with a Step I or Step II diet, a high-MUFA-diet tends to raise HDL cholesterol and lower TAG concentrations [176,177,193,194]. In agreement with this evidence, hazelnut, an excellent source of MUFA and minor components [12,117] may prove to be beneficial in this respect. In addition to MUFA, other components in hazelnut oil have been reported to reduce plasma total and LDL cholesterol concentrations; these include PUFA [140,141,183,195], phytosterols and phytostanols [127,142,146], and tocotrienol [156]. Besides favorable changes in plasma lipid profiles, a MUFA-rich diet instead of carbohydrates for SFA calories may favorably affect CVD risk [196] and has positive effects on atherosclerosis [197]. Furthermore, phytosterols and phytostanols have been shown to decrease the risk of certain types of cancer [132–134] and CVD [136,137], and enhance immune functions [139]. Hazelnut oil, which is an excellent source of tocopherols [12,160], has also been shown to reduce the risk of CHD [46]. As stated above, health effects of vitamin E isoforms (α-, β-, γ-, and δ-tocopherols and tocotrienols) have been well documented [49,150,198,199].

Although a considerable number of clinical studies on different tree nuts have been performed in the last decade, the ones conducted specifically on hazelnut- or hazelnut oil-enriched diet are limited [52,200–202]. Recently, the effect of a high cholesterol diet with and without hazelnut oil supplementation has been investigated in rabbits [201]. The supplementation with hazelnut oil reduced lipid peroxide levels in plasma and apolipoprotein B 100-containing lipoproteins as well as aortic atherosclerotic lesions in rabbits fed high cholesterol diet without any decreasing effect on lipid levels. In addition, the results obtained also suggest that hazelnut can contribute to reduced hemolytic anemia together with a significant decrease in diene conjugate and H_2O_2-induced malondialdehyde (MDA) levels [201].

More recently, Mercanlıgil et al. [202] investigated the effects of a hazelnut-enriched diet (40 g/day) on plasma cholesterol and lipoprotein profiles in hypercholesterolemic adult males (aged 33–59 years) compared with baseline and control diet (Table 12.10). They found that compared with baseline, the hazelnut-enriched diet favorably decreased ($P < 0.05$) the concentrations of very-low density

TABLE 12.10
Planned Composition of the Control and Hazelnut-Enriched Diets

	Diets	
Constituent	**Control**	**Hazelnut-Enriched**
Carbohydrate (% of energy)	55–60	50–55
Protein (% of energy)	12–15	12–15
Fat (% of energy)	25–30	35–40
SFA (% of energy)	<7	<7
MUFA (% of energy)	13–15	17–20
PUFA (% of energy)	7–8	7–8
Cholesterol (mg/day)	≤300	≤300
Dietary fiber (g/day)	25–30	25–30

Source: From Mercanlıgil, S.M., Arslan, P., Alasalvar, C., Okut, E., Akgül, E., Pınar, A., Geyik, P. Ö., Tokgözoğlu, L., and Shahidi, F., *Eur. J. Clin. Nutr.*, 61, 212, 2007. With permission.

Note: 40 g hazelnut provides 254 kcal energy, 6.7 g carbohydrate, 5.04 g protein, 25 g fat (of which 1.84 g SFA, 19.64 g MUFA, and 2.4 g PUFA).

lipoprotein (VLDL) cholesterol, TAG, and apolipoprotein B by 29.5%, 31.8%, and 9.2%, respectively, while increasing HDL cholesterol levels by 12.6% (Figure 12.1). Although insignificant ($P > 0.05$), there was a decreasing trend for the rest of the parameters, particularly for total (5.2%) and LDL cholesterol (3.3%) in subjects consuming a hazelnut-enriched diet compared to that of the baseline. No changes were found in fasting levels of glucose, apolipoprotein A-1, and homocysteine

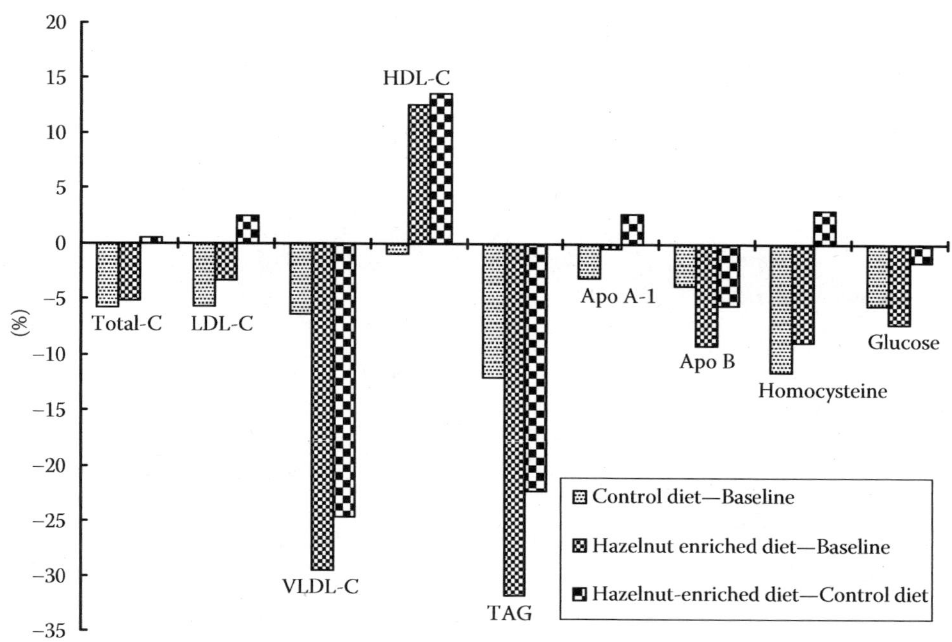

FIGURE 12.1 Comparison (%) of plasma cholesterol and lipoprotein profiles among baseline, control, and hazelnut-enriched diet. *Abbreviations*: TAG, triacylglycerol; Apo A-1, apolipoprotein A-1; Apo B, apolipoprotein B. (From Mercanlıgil, S.M., Arslan, P., Alasalvar, C., Okut, E., Akgül, E., Pınar, A., Geyik, P. Ö., Tokgözoğlu, L., and Shahidi, F., *Eur. J. Clin. Nutr.*, 61, 212, 2007. With permission.)

between the control and hazelnut-enriched diet. Therefore, a high-fat and high-MUFA-rich hazelnut diet was found to be superior to a low-fat control diet because of favorable changes in plasma lipid profiles of hypercholesterolemic adult males and, thereby positively affecting the CHD risk profile.

Hazelnut, which contains lipid-lowering bioactives such as MUFA, PUFA, phytosterols, phytostanols, polyphenols, and sphingolipids [11,12,107,117,140,141,147,175,183,189,203–205], offers an opportunity as a potential hypercholesterolemic "heart-healthy" diet component. In addition to these bioactives and minor components, there are a number of nonfat constituents (such as essential minerals, essential amino acids, antioxidant phenolics, soluble dietary fiber, and phytochemicals, among others) in hazelnut that may elicit additional cholesterol-lowering and cardioprotective effects [11,57,59,85,117,174,206].

12.5.2 Antiobesity Effects of Hazelnut

Traditionally, nuts including hazelnut have been considered a staple food, but because of their high energy and fat content were not considered good for body weight control or insulin sensitivity [207]. Although whether or not frequent consumption of nuts can cause weight gain and impair insulin sensitivity is not fully understood, review of the available data to date suggests that adding nuts to habitual diets of free-living individuals does not lead to any weight gain [208,209]. In fact, nuts have a tendency to lower body weight and fat mass [175,210].

Hazelnut oil contains certain polyphenols, which are different from olive oil polyphenols [189]. Hazelnut polyphenols show stronger antioxidant activity against autoxidation of soybean and fish oils than α-tocopherol at the same concentrations. It has been reported that tea catechins have anti-obesity activity by some biological pathways, such as by increasing fat oxidation and decreasing energy absorption, and by inducing apoptosis of adipocytes [211,212]. When hazelnut polyphenols were added to the fat cell and 3T3-L1 cell, the addition of polyphenols reduced the fat accumulation in the fat cell compared with the control. This effect is dose-dependent, suggesting a possible anti-obesity effect of hazelnut polyphenols [189]. More recently, Mercanlıgil et al. [202] found no significant ($P > 0.05$) changes in the body weight and waist to hip ratio throughout the 8-week diet period, whereas BMI and the percentage of body fat were reduced ($P < 0.05$) when subjects consumed a hazelnut-enriched diet (40 g/day) compared with the baseline. Both *in vitro* and *in vivo* studies suggest that hazelnut has an antiobesity effect. Thus, hazelnut enrichment as part of a health diet does not alter body weight, but favorably affects the plasma cholesterol and lipoprotein profiles despite an increase in the dietary fat content. Therefore, hazelnut or hazelnut oil can be classified as a heart-healthy food.

12.6 CONCLUSION

The data presented in this chapter show that hazelnut can play a major role in human nutrition and health because of its high and special nutritional components. Thus, these nutritional attributes show that hazelnut can serve as an important healthy food in the human diet. The content and composition of free amino acids, sugars, organic acids, and tannins may play a significant role in the taste and aroma/flavor of hazelnut. With respect to functional lipid characteristics of hazelnut, hazelnut oil serves as a good source of natural antioxidants and bioactives, thus reflecting its nutraceutical potential in different food and specialty applications. Moreover, hazelnut oil, a rich source of MUFA (oleic acid) and phytosterols (β-sitosterol), maybe considered as a supplement for daily diet plannings to reduce the risk of CHD. Hazelnut enrichment as part of a healthy diet favorably affects the plasma cholesterol and lipoprotein profiles despite an increase in the dietary fat content. A high-fat and high-MUFA-rich hazelnut diet is preferred to a low-fat control diet because of more favorable effects on the CHD risk profile. Hazelnut enrichment as part of a healthy diet does not alter body weight. The presence of essential minerals, vitamins, and amino acids; the high content of heart-healthy fat and the presence of soluble dietary fiber; bioactives, minor components, and phytochemicals as well as their antioxidant activities make the addition of hazelnut to healthy diets a useful tool for the prevention of CHD, diabetes, and sudden death, among others.

Although data on the compositional characteristics of hazelnut are available, more studies are still needed, especially those concerning minor components and dietary phytochemicals that provide potential health benefits and disease risk reduction. A great deal of research is also needed on the compositional, taste, and functional lipid characteristics of different varieties of roasted hazelnuts.

REFERENCES

1. Turkish Hazelnut Exporter's Union, *The Miracle Nut: Hazelnut*, Hazelnut Promotion Group, Giresun, Turkey, 2007.
2. Botta, R., Gianotti, C., Richardson, D., Suwanagul, A., and Sanz, C.L., Hazelnut variety organic acids, sugars, and total lipid fatty acids, *Acta Hort.*, 351, 693–699, 1994.
3. Ebrahem, K.S., Richardson, D.G., Tetley, R.M., and Mehlenbacher, S.A., Oil content, fatty acid composition, and vitamin E concentration of 17 hazelnut varieties, compared to other types of nuts and oil seeds, *Acta Hortic.*, 351, 685–692, 1994.
4. Parcerisa, J., Boatella, J., Codony, R., Rafecas, M., Castellote, A.I., García, J., López, A., and Romero, A., Comparison of fatty acid and triacylglycerol compositions of different hazelnut varieties (*Corylus avellana* L.) cultivated in Catalonia (Spain), *J. Agric. Food Chem.*, 43, 13–16, 1995.
5. Parcerisa, J., Richardson, D.G., Rafecas, M., Codony, R., and Boatella, J., Fatty acid, tocopherol and sterol content of some hazelnut varieties (*Corylus avellana* L.) harvested in Oregon (USA), *J. Chromatogr. A*, 805, 259–268, 1998.
6. Savage, G.P. and McNeil, D.L., Chemical composition of hazelnuts (*Corylus avellana* L.) grown in New Zealand, *Int. J. Food Sci. Nutr.*, 49, 199–203, 1998.
7. Alasalvar, C., Odabasi, A.Z., Demir, N., Balaban, M.Ö., Shahidi, F., and Cadwallader, K.R., Volatiles and flavor of five Turkish hazelnut varieties as evaluated by descriptive sensory analysis, electronic nose, and dynamic headspace analysis/gas chromatography–mass spectrometry, *J. Food Sci.*, 69, SNQ99–SNQ106, 2004.
8. Crews, C., Hough, P., Godward, J., Brereton, P., Lees, M., Guiet, S., and Winkelmann, W., Study of the main constituents of some authentic hazelnut oils, *J. Agric. Food Chem.*, 53, 4843–4852, 2005.
9. Köksal, A.İ., Artik, N., Şimşek, A., and Güneş, N., Nutrient composition of hazelnut (*Corylus avellana* L.) varieties cultivated in Turkey, *Food Chem.*, 99, 509–515, 2006.
10. Amaral, J.S., Casal, S., Citová, I., Santos, A., Seabra, R.M., and Oliveira, B.P.P., Characterization of several hazelnut (*Corylus avellana* L.) cultivars based in chemical, fatty acid and sterol composition, *Eur. Food Res. Technol.*, 222, 274–280, 2006.
11. Alasalvar, C., Shahidi, F., Liyanapathirana, C.M., and Ohshima, T., Turkish Tombul hazelnut (*Corylus avellana* L.). 1. Compositional characteristics, *J. Agric. Food Chem.*, 51, 3790–3796, 2003.
12. Alasalvar, C., Amaral, J.S., and Shahidi, F., Functional lipid characteristics of Turkish Tombul hazelnut (*Corylus avellana* L.), *J. Agric. Food Chem.*, 54, 10177–10183, 2006.
13. Ruggeri, S., Cappelloni, L., Gambelli, S., and Carnovale, E., Chemical composition and nutritive value of nuts grown in Italy, *Ital. J. Food Sci.*, 3, 243–252, 1998.
14. U.S. Department of Agriculture (USDA), USDA National Nutrition Database for Standard Reference, Release 20, National Technical Information Service, USDA, Springfield, VA, 2007.
15. Bonvehí, J.S. and Coll, F.V., A chemical study of the mineral fraction of Tarragona hazelnuts (*Corylus avellana* L.), *Acta Alimentaria*, 26, 243–253, 1997.
16. Özdemir, M. and Devres, O., Turkish hazelnuts: Properties and effect of microbiological and chemical changes on quality, *Food Rev. Int.*, 15, 309–333, 1999.
17. Yurttas, H.C., Schafer, H.W., and Warthesen, J.J., Antioxidant activity of nontocopherol hazelnut (*Corylus* spp.) phenolics, *J. Food Sci.*, 65, 276–280, 2000.
18. Venkatachalam, M. and Sathe, S.K., Chemical composition of selected edible nut seeds, *J. Agric. Food Chem.*, 54, 4705–4714, 2006.
19. Bonvehí, J.S. and Coll, F.V., Oil content, stability and fatty acid composition of the main varieties of Catalonian hazelnuts (*Corylus avellana* L.), *Food Chem.*, 48, 237–241, 1993.

20. Parcerisa, J., Boatella, J., Codony, R., Farràn, A., Garcia, J., López, A., Rafecas, M., and Romero, A., Influence of variety and geographical origin on the lipid fraction of hazelnuts (*Corylus avellana* L.) from Spain: (I). Fatty acid composition, *Food Chem.*, 48, 411–414, 1993.
21. NRC., *Recommended Dietary Allowances*, 10th ed., National Academy Press, Washington, DC, 1989.
22. Food and Drug Administration (FDA), Qualified Health Claims: Letter of Enforcement Discretion—Nuts and Coronary Heart Disease, Docket No 02P-0505, Food and Drug Administration, Washington, DC, 2003.
23. Holland, B., Welch, A.A., Unwin, I.D., Buss, D.H., Paul, A.A., and Southgate, D.A.T., *McCance and Widdowson's The Composition of Foods*, 5th ed., The Royal Society of Chemistry and Ministry of Agriculture, Fisheries and Food, Cambridge, United Kingdom, 1995.
24. Özdemir, M., Açkurt, F., Kaplan, M., Yıldız, M., Löker, M., Gürcan, T., Birigen, G., Okay, A., and Seyhan, F.G., Evaluation of new Turkish hybrid hazelnut (*Corylus avellana* L.) varieties: Fatty acid composition, α-tocopherol content, mineral composition and stability, *Food Chem.*, 73, 411–415, 2001.
25. Pala, M., Açkurt, F., Löker, M., Yildiz, M., and Ömeroglu, S., Composition and nutritional evaluation of different hazelnut varieties, *Turk. J. Agric. For.*, 20, 43–48, 1996 (in Turkish).
26. Açkurt, F., Özdemir, M., Biringen, G., and Löker, M., Effects of geographical origin and variety on vitamin and mineral composition of hazelnut (*Corylus avellana* L.) varieties cultivated in Turkey, *Food Chem.*, 65, 309–313, 1999.
27. Çağlarırmak, N. and Batkan, C., Nutrients and biochemistry of nuts in different consumption types in Turkey, *J. Food Process. Preserv.*, 29, 407–423, 2005.
28. Parcerisa, J., Rafecas, M., Castellote, A.I., Codony, R., Farràn, A., Garcia, J., Gonzalez, C., López, A., Romero, A., and Boatella, J., Influence of variety and geographical origin on the lipid fraction of hazelnuts (*Corylus avellana* L.) from Spain: (III). Oil stability, tocopherol content and some mineral contents (Mn, Fe, Cu), *Food Chem.*, 53, 71–74, 1995.
29. *Dietary Reference Intakes for Calcium, Phosphorus, Magnesium, Vitamin D, and Fluoride*, The National Academies Press, Washington, DC, 1997.
30. *Dietary Reference Intakes for Thiamin, Riboflavin, Niacin, Vitamin B_6, Folate, Vitamin B_{12}, Pantothenic Acid, Biotin, and Choline*, The National Academies Press, Washington, DC, 1998.
31. *Dietary Reference Intakes for Vitamin C, Vitamin E, Selenium, and Carotenoids*, The National Academies Press, Washington, DC, 2000.
32. *Dietary Reference Intakes for Vitamin A, Vitamin K, Arsenic, Boron, Chromium, Copper, Iodine, Iron, Manganese, Molybdenum, Nickel, Silicon, Vanadium, and Zinc*, The National Academies Press, Washington, DC, 2001.
33. *Dietary Reference Intakes for Water, Potassium, Sodium, Chloride, and Sulfate*, The National Academies Press, Washington, DC, 2004.
34. *Dietary Reference Intakes for Energy, Carbohydrate, Fiber, Fat, Fatty Acids, Cholesterol, Protein, and Amino Acids*, The National Academies Press, Washington, DC, 2005.
35. Segura, R., Javierre, C., Lizarraga, M.A., and Ros, E., Other relevant components of nuts: Phytosterols, folate and minerals, *Br. J. Nutr.*, 96(Suppl. 2), 36S–44S, 2006.
36. MAFF, *Manual of Nutrition*, 10th ed., Reference Book 342, HMSO, London, 1995.
37. O'Dell, B.L. and Sunde, R.A., Introduction, in *Handbook of Nutritionally Essential Minerals*, O'Dell, B.L. and Sunde, R.A., Eds., Marcel Dekker, New York, 1997, pp. 1–12.
38. Sardesai, V.M., *Introduction to Clinical Nutrition*, Marcel Dekker, New York, 1998.
39. Berdanier, C.D., *Advanced Nutrition: Micronutrients*, CRC Press, Boca Raton, FL, 1998.
40. Dugo, G., Pera, L.L., Turco, V.L., Mavrogeni, E., and Alfa, M., Determination of selenium in nuts by cathodic stripping potentiometry (CSP), *J. Agric. Food Chem.*, 51, 3722–3725, 2003.
41. Jansson, B., Selenium and cancer, *Cancer Bull.*, 36, 164–166, 1984.
42. Combs Jr. G.F. and Combs, S.B., *The Role of Selenium in Nutrition*, Academic Press, Orlando, FL, 1986.
43. Oldfield, J.E., Some implications of selenium for human health, *Nutr. Today*, 26, 6–11, 1991.
44. Rayman, M.P., The importance of selenium to human health, *Lancet*, 356, 233–241, 2000.
45. Brenneisen, P., Steinbrenner, H., and Sies, H., Selenium, oxidative stress, and health aspects, *Mol. Aspects Med.*, 26, 256–267, 2005.
46. Rimm, E.B., Stampfer, M.J., Ascherio, A., Giovannucci, E., Colditz, G.A., and Willett, W.C., Vitamin E consumption and the risk of coronary heart disease in men, *N. Engl. J. Med.*, 328, 1450–1456, 1993.

47. Blusztajn, J.K., Choline, a vital amine, *Science*, 281, 794–795, 1998.
48. Nogala-Kalucka, M., Fat-soluble vitamins, in *Chemical and Functional Properties of Food Lipids.*, Sikorski, Z.E. and Kolakowska, A., Eds., CRC Press, Boca Raton, FL, 2002.
49. Eitenmiller, R. and Lee, J., *Vitamin E: Food Chemistry, Composition, and Analysis*, Taylor & Francis, New York, 2004.
50. Burton, G.W., Joyce, A., and Ingold, K.U., First proof that vitamin E is major lipid-soluble, chain-breaking antioxidant in human blood plasma, *Lancet*, 8292 ii, 327, 1982.
51. Azzi, A. and Stocker, A., Vitamin E: Non-antioxidant roles, *Prog. Lipid Res.*, 39, 231–255, 2000.
52. Durak, I., Köksal, I., Kaçmaz, M., Büyükkoçak, S., Çimen, B.M.Y., and Öztürk, H.S., Hazelnut supplementation enhances plasma antioxidant potential and lowers plasma cholesterol levels, *Clin. Chim. Acta.*, 284, 113–115, 1999.
53. FAO/WHO/UNU., Energy and Protein Requirement, Report of a Joint FAO/WHO/UNU Expert Consultation, World Health Organization Technical Report Series 724, WHO, Geneva, Switzerland, 1985.
54. Garrow, J.S., James, W.P.T., and Ralph, A., *Human Nutrition and Dietetics*, 10th ed., Churchill Livingstone, London, 2000.
55. Brufau, G., Boatella, J., and Rafecas, M., Nuts: Source of energy and macronutrients, *Br. J. Nutr.*, 96(Suppl. 2), 24S–28S, 2006.
56. Villarroel, M., Biolley, E., Schneeberger, R., Ballester, D., and Ramirez, S., Amino acid composition of Chilean hazelnuts, *Food Chem.*, 25, 155–158, 1987.
57. Anderson, J.W., Smith, B.M., and Gustafson, N.J., Health benefits and practical aspects of high-fiber diets, *Am. J. Clin. Nutr.*, 59, 1242S–1247S, 1994.
58. Kritchevsky, D. and Banfield, C., *Dietary Fiber in Health and Disease*, Eagan Press, St. Paul, MN, 1995.
59. Brown, L., Rosner, B., Willett, W.W., and Sacks, F.M., Cholesterol-lowering effects of dietary fiber: A meta-analysis, *Am. J. Clin. Nutr.*, 69, 30–42, 1999.
60. Tariq, N., Jenkins, D.J.A., Vidgen, E., Fleshner, N., Kendall, C.W.C., Story, J.A., Singer, W., D'Coosta, M., and Struthers, N., Effect of soluble and insoluble fiber diets on serum prostate specific antigen in men, *J. Urol.*, 163, 114–118, 2000.
61. Salas-Salvadó, J., Bulló M., Pérez-Heras, A., and Ros, E., Dietary fibre, nuts and cardiovascular diseases, *Br. J. Nutr.*, 96(Suppl. 2), 45S–51S, 2006.
62. Lintas, C. and Cappelloni, M., Dietary fiber content of Italian fruit and nuts, *J. Food Comp. Anal.*, 5, 146–151, 1992.
63. NACNE, Proposals for nutritional guidelines for health education in Britain, *Lancet*, 12, 835–837, 1983.
64. Dreher, M.L., *Handbook of Dietary Fiber*, Marcel Dekker, New York, 1987.
65. Botta, R., Gianotti, C., and Me, G., Kernel quality in hazelnut cultivars and selections analyzed for sugars, lipids and fatty acid composition, *Acta Hort.*, 445, 319–326, 1997.
66. Keme, T., Messerli, M., Shejbal, J., and Vitali, F., The storage of hazelnuts at room temperature under nitrogen (II), *Rev. Choc. Confect. Bakery*, 8, 15–20, 1983.
67. Keme, T., Messerli, M., Shejbal, J., and Vitali, F., The storage of hazelnuts at room temperature under nitrogen (I), *Rev. Choc. Confect. Bakery*, 8, 24–28, 1983.
68. Alexander, R.J., *Sweeteners: Nutritive*, Eagan Press, St. Paul, MN, 1998.
69. Johnston, F.B. and Hammill, M.M., The nonvolatile organic acids of some fresh fruits and vegetables, *Can. Inst. Food Technol. J.*, 1, 3–5, 1968.
70. Ulrich, R., Organic acids, in *The Biochemistry of Fruits and Their Products*, vol. 1, Hulme, A.C., Ed., Academic Press, New York, 1970, pp. 89–118.
71. Jadhav, S.J. and Andrew, W.T., Effects of cultivars and fertilizers on nonvolatile organic acids in potato tubers, *Can. Inst. Food Sci. Technol. J.*, 10, 13, 1977.
72. Siebert, K.J., Modeling the flavor thresholds of organic acids in beer as a function of their molecular properties, *Food Qual. Pref.*, 10, 129–137, 1999.
73. Vaughan, J.G. and Geissler, C.A., *The New Oxford Book of Food Plants*, Oxford University Press, Oxford, 1997.
74. Kuninaka, A., Taste and flavor enhancers, in *Flavor Research: Recent Advances*, Teranishi, R. and Flath, R.A., Eds., Marcel Dekker, New York, 1981, pp. 305–353.
75. Kato, H., Rhue, M.R., and Nishimura, T., Role of free amino acids and peptides in food taste, in *Flavor Chemistry: Trends and Developments*, Teranishi, R., Buttery, R.G., and Shahidi, F., Eds., ACS Symposium Series 388, American Chemical Society, Washington, DC, 1989, pp. 158–174.
76. Fuke, S. and Konosu, S., Taste-active components in some foods: A review of Japanese research, *Physiol. Behav.*, 49, 863–868, 1991.

77. Fuke, S., Taste-active components of seafoods with special reference to umami substances, in *Seafoods: Chemistry, Processing, Technology and Quality*, Shahidi, F. and Botta, J.R., Eds., Blackie Academic and Professional, Glasgow, United Kingdom, 1994, pp. 115–139.
78. Macleod, G., The flavour of beef, in *Flavor of Meat, Meat Products and Seafoods*, 2nd ed., Shahidi, F., Ed., Blackie Academic & Professional, London, 1998, pp. 27–60.
79. Basha, S.M. and Young, C.T., Changes in the polypeptide composition of peanut (*Arachis hypogaea* L.) seed during oil roasting, *J. Agric. Food Chem.*, 33, 350–354, 1985.
80. Chiou, R.Y.-Y., Chang, Y.-S., Tsai, T.-T., and Hot, S., Variation of flavor-related characteristics of peanuts during roasting as affected by initial moisture contents, *J. Agric. Food Chem.*, 39, 1155–1158, 1991.
81. Amrein, T.M., Lukac, H., Andres, L., Perren, R., Escher, F., and Amad, R., Acrylamide in roasted almonds and hazelnuts, *J. Agric. Food Chem.*, 53, 7819–7825, 2005.
82. Shahidi, F. and Naczk, M., *Phenolics in Food and Nutraceuticals*, CRC Press, Boca Raton, FL, 2004.
83. Gu, L., Kelm, M.A., Hammerstone, J.F., Beecher, G., Holden, J., Haytowitz, D., and Prior, R.L., Screening of foods containing proanthocyanidins and their structural characterization using LC-MS/MS and thiolytic degradation, *J. Agric. Food Chem.*, 51, 7513–7521, 2003.
84. Gu, L., Kelm, M.A., Hammerstone, J.F., Beecher, G., Holden, J., Haytowitz, D., Gebhardt, S., and Prior, R.L., Concentrations of proanthocyanidins in common foods and estimations of normal consumption, *J. Nutr.*, 134, 613–617, 2004.
85. Alasalvar, C., Karamać, M., Amarowicz, R., and Shahidi, F., Antioxidant and antiradical activities in extracts of hazelnut kernel (*Corylus avellana* L.) and hazelnut green leafy cover, *J. Agric. Food Chem.*, 54, 4826–4832, 2006.
86. U.S. Department of Agriculture (USDA), Database for the Proanthocyanidin Content of Selected Foods, August 2004, Published online at: http://www.nal.usda.gov/fnic/foodcomp (accessed September 28, 2006)
87. Saklar, S., Ungan, S., and Katnas, S., Instrumental crispness and crunchiness of roasted hazelnuts and correlations with sensory assessment, *J. Food Sci.*, 64, 1015–1019, 1999.
88. Langourieux, S., Perren, R., and Escher, F., Influence of processing parameters on the aroma of dry-roasted hazelnuts, in *Frontiers of Flavour Science*, Schieberle, P. and Engel, K.H., Eds., Deutsche Forschungsanstalt für Lebensmittelchemie, Garching, Germany, 2000, pp. 527–535.
89. Fallico, B., Arena, E., and Zappalà, M., Roasting of hazelnut. Role of oil in colour development and hydroxymethylfurfural formation, *Food Chem.*, 81, 569–573, 2003.
90. Kinlin, T.E., Muralidhara, R., Pittet, A.O., Sanderson, A., and Walradt, J.P., Volatile components of roasted filberts, *J. Agric. Food Chem.*, 20, 1021–1028, 1972.
91. Sheldon, R.M., Lindsay, R.C., and Libbey, L.M., Identification of volatile flavor compounds from roasted filberts, *J. Food Sci.*, 37, 313–316, 1972.
92. Silberzahn, W., Gas Chromatographic-Mass Spectrometric Investigations on the Flavor Compounds in Hazelnuts (*Corylus avellana*) (in German), PhD thesis, Technical University of Berlin, Berlin, Germany, 1988.
93. Jauch, J., Schmalzing, D., Schurig, V., Emberger, R., Hopp, R., Köpsel, M., Silberzahn, W., and Werkhoff, P., Isolation, synthesis, and absolute configuration of filbertone—The principal flavor component of the hazelnut, *Angew. Chem. Int. Ed. Engl.*, 28, 1022–1023, 1989.
94. Güntert, M., Emberger, R., Hopp, R., Köpsel, M., Silberzahn, W., and Werkhoff, P., Chirospecific analysis in flavor and essential oil chemistry. Part A. Filbertone—The character impact compound of hazelnuts, *Z. Lebensm. Unters. Forsch.*, 192, 108–110, 1991.
95. Matsui, T., Guth, H., and Grosch, W., A comparative study of potent odorants in peanut, hazelnut, and pumpkin seed oils on the basis of aroma extract dilution analysis (AEDA) and gas chromatography–olfactometry of headspace sampling (GCOH), *Fett (Lipid)*, 100, 51–56, 1998.
96. Pfnuer, P., Matsui, T., Grosch, W., Guth, H., Hofmann, T., and Schieberle, P., Development of a stable isotope dilution assay for the quantification of 5-methyl-(*E*)-2-hepten-4-one: Application to hazelnut oils and hazelnuts, *J. Agric. Food Chem.*, 47, 2044–2047, 1999.
97. Wickland, S.E., Johnston, J.J., and Stone, M.B., Evaluation of roasted and natural hazelnut volatiles by purge-and-trap/gas chromatography/mass spectrometry, presented at the Institute of Food Technologists Annual Meeting and Food Expo, New Orleans, LA, June 23–27, 2001, Abstract 88B-17.
98. Alasalvar, C., Shahidi, F., and Cadwallader, K.R., Comparison of natural and roasted Turkish Tombul hazelnut (*Corylus avellana* L.) volatiles and flavor by DHA/GC/MS and descriptive sensory analysis, *J. Agric. Food Chem.*, 51, 5067–5072, 2003.

99. Savage, G.P., McNeil, D.L., and Dutta, P.C., Lipid composition and oxidative stability of oils in hazelnuts (*Corylus avellana* L.) grown in New Zealand, *J. Am. Oil Chem. Soc.*, 74, 755–759, 1997.
100. Benitez-Sánchez, P.L., León-Camacho, M., and Aparicio, R., A comprehensive study of hazelnut oil composition with comparisons to other vegetable oils, particularly olive oil, *Eur. Food Res. Technol.*, 218, 13–19, 2003.
101. Bada, J.C., León-Camacho, M., Prieto, M., and Alonso, L., Characterization of oils of hazelnuts from Asturias, Spain, *Eur. J. Lipid Sci. Technol.*, 106, 294–300, 2004.
102. Seyhan, F., Ozay, G., Saklar, S., Ertaş, E., Satır, G., and Alasalvar, C., Chemical changes of three native Turkish hazelnut varieties (*Corylus avellana* L.) during fruit development, *Food Chem.*, 105, 590–596, 2007.
103. Karabulut, I., Topcu, A., Yorulmaz, A., Tekin, A., and Ozay, S.D., Effects of the industrial refining process on some properties of hazelnut oil, *Eur. J. Lipid Sci. Technol.*, 107, 476–480, 2005.
104. Amaral, J.S., Casal, S., Seabra, R.M., and Oliveira, B.P.P., Effects of roasting on hazelnut lipids, *J. Agric. Food Chem.*, 54, 1315–1321, 2006.
105. Parcerisa, J., Codony, R., Boatella, J., and Rafecas, M., Triacylglycerol and phospholipid composition of hazelnut (*Corylus avellana* L.) lipid fraction during fruit development, *J. Agric. Food Chem.*, 47, 1410–1415, 1999.
106. Amaral, J.S., Contribuição para a caracterização química das folhas e sementes de aveleira (*Corylus avellana* L.) e de nogueira (*Juglans regia* L.) (in Portuguese), PhD thesis, Universidade do Porto, Porto, Portugal, 2005.
107. Committee on Diet and Health, Food and Nutrition Board, Commission on Life Sciences, National Research Council, *Diet and Health: Implications for Reducing Chronic Disease Risk*, National Academy Press, Washington, DC, 1989.
108. Mensink, R.P., Effects of the individual saturated fatty acids on serum lipids and lipoprotein concentrations, *Am. J. Clin. Nutr.*, 57, 711S–714S, 1993.
109. Willet, W.C., Specific fatty acids and risks of breast and prostate cancer: Dietary intake, *Am. J. Clin. Nutr.*, 66, 1557S–1563S, 1997.
110. Hu, F.B., Stampfer, M.J., Manson, J.E., Ascherio, A., Colditz, G.A., Speizer, F.E., Hennekens, C.H., and Willett, W.C., Dietary saturated fats and their food sources in relation to the risk of coronary heart disease in women, *Am. J. Clin. Nutr.*, 70, 1001–1008, 1999.
111. Mensink, R.P. and Katan, M.B., Effect of dietary trans fatty acids on high-density and low-density lipoprotein cholesterol levels in healthy subjects, *N. Engl. J. Med.*, 323, 439–445, 1990.
112. Allison, D.B., Denke, M.A., Dietschy, J.M., Emken, E.A., Kris-Etherton, P.M., and Nicolosi, R.J., *Trans* fatty acids and coronary heart disease risk. Report of the expert panel on *trans* fatty acids and coronary heart disease, *Am. J. Clin. Nutr.*, 62, 655S–708S, 1995.
113. Oomen, C.M., Ocké, M.C., Feskens, E.J.M., van Erp-Baart, M.A.J., Kok, F.J., and Kromhout, D., Association between *trans* fatty acid intake and 10-year risk of coronary heart disease in the Zutphen Elderly Study: A prospective population-base study, *Lancet*, 357, 746–751, 2001.
114. Stender, S. and Dyerberg, J., *A Report from the Danish Nutrition Council. The Influence of Trans Fatty Acids on Health*, 4th ed., The Danish Nutrition Council, Søborg, Denmark, 2003.
115. Food and Drug Administration (FDA), Guidance for Industry: Food Labeling: Trans Fatty Acids in Nutrition Labeling, Nutrient Content Claims, and Health Claims, Published online at: http://vm.cfsan.fda.gov/~dms/transgui.html (accessed January 15, 2006)
116. Parcerisa, J., Richardson, D.G., Rafecas, M., Codony, R., and Boatella, J., Fatty acid distribution in polar and nonpolar lipid classes of hazelnut oil (*Corylus avellana* L.), *J. Agric. Food Chem.*, 45, 3887–3890, 1997.
117. Alasalvar, C., Shahidi, F., Ohshima, T., Wanasundara, U., Yurttas, H.C., Liyanapathirana, C.M., and Rodrigues, F.B., Turkish Tombul hazelnut (*Corylus avellana* L.). 2. Lipid characteristics and oxidative stability, *J. Agric. Food Chem.*, 51, 3797–3805, 2003.
118. Ulberth, F. and Buchgraber, M., Authenticity of fats and oils, *Eur. J. Lipid Sci. Technol.*, 102, 687–694, 2000.
119. Aparicio, R. and Aparicio-Ruíz, R., Authentication of vegetable oils by chromatographic techniques, *J. Chromatogr. A*, 881, 93–104, 2000.
120. Parcerisa, J., Casals, I., Boatella, J., Codony, R., and Rafecas, M., Analysis of olive and hazelnut oil mixtures by high-performance liquid chromatography-atmospheric pressure chemical ionisation mass spectrometry of triacylglycerols and gas-liquid chromatography of non-saponifiable compounds (tocopherols and sterols), *J. Chromatogr. A*, 881, 149–158, 2000.

121. Vichi, S., Pizzale, L., Toffano, E., Bortolomeazzi, R., and Conte, L., Detection of hazelnut oil in virgin olive oil by assessment of free sterol and triacylglycerols, *J. AOAC Int.*, 84, 1534–1541, 2001.
122. Amaral, J.S., Cunha, S.C., Santos, A., Alves, M.R., Seabra, R.M., and Oliveira, B.P.P., Influence of cultivar and environmental conditions on the triacylglycerol profile of hazelnut (*Corylus avellana* L.), *J. Agric. Food Chem.*, 54, 449–456, 2006.
123. Parcerisa, J., Rafecas, M., Castellote, A.I., Codony, R., Farràn, A., Garcia, J., López, A., Romero, A., and Boatella, J., Influence of variety and geographical origin on the lipid fraction of hazelnuts (*Corylus avellana* L.) from Spain: (II). Triglyceride composition, *Food Chem.*, 50, 245–249, 1994.
124. Ayorinde, F.O., Eribo, B.E., Balan, K.V., Johnson Jr, J.H., and Wan, L.W., Determination of major triacylglycerol components of polyunsaturated specialty oils using matrix-assisted laser desorption/ionization time-of-flight mass spectrometry, *Rapid Commun. Mass Spectrom.*, 13, 937–942, 1999.
125. Holčapek, M., Jandera, P., Zderadička, P., and Hrubá, L., Characterization of triacylglycerol and diacylglycerol composition of plant oils using high-performance liquid chromatography–atmospheric pressure chemical ionization mass spectrometry, *J. Chromatogr. A*, 1010, 195–215, 2003.
126. Bernardo-Gil, M.G., Grenha, J., Santos, J., and Cardoso, P., Supercritical fluid extraction and characterisation of oil from hazelnut, *Eur. J. Lipid Sci. Technol.*, 104, 402–409, 2002.
127. Hicks, K.B. and Moreau, R.A., Phytosterols and phytostanols: Functional food cholesterol busters, *Food Technol.*, 55, 63–67, 2001.
128. Phillips, K.M., Ruggio, D.M., and Ashraf-Khorassani, M., Phytosterol composition of nuts and seeds commonly consumed in the United States, *J. Agric. Food Chem.*, 53, 9436–9445, 2005.
129. Padley, F.B., Gunstone, F.D., and Harwood, J.L., Occurrence and characteristics of oils and fats, in *The Lipid Handbook*, 2nd ed., Gunstone, F.D., Harwood, J.L., and Padley, F.B., Eds., Chapman and Hall, London, 1994, pp. 47–223.
130. Wester, I., Cholesterol-lowering effect of plant sterols, *Eur. J. Lipid Sci. Technol.*, 102, 37–44, 2000.
131. Weber, N. and Mukherjee, K.D., Plant sterols and steryl esters in functional foods and nutracuticals, in *Nutraceutical and Specialty Lipids and Their Co-Products*, Shahidi, F., Ed., CRC Press, Taylor & Francis Group, Boca Raton, FL, 2006, pp. 483–508.
132. Awad, A.B. and Fink, C.S., Phytosterols as anticancer dietary components: Evidence and mechanism of action, *J. Nutr.*, 130, 2127–2130, 2000.
133. Law, M., Plant sterol and stanol margarines and health, *Br. Med. J.*, 320, 861–864, 2000.
134. Awad, A.B., Williams, H., and Fink, C.S., Phytosterols reduce in vitro metastic ability of MDA-MB-231 human breast cancer cells, *Nutr. Cancer*, 40, 157–164, 2001.
135. Bradford, P.C. and Awad, A.B., Phytosterols as anticancer compounds, *Mol. Nutr. Food Res.*, 51, 161–170, 2007.
136. Plat, J. and Mensink, R.P., Effects of plant sterols and stanols on lipid metabolism and cardiovascular risk, *Nutr. Metab. Cardiovasc. Dis.*, 11, 31–40, 2001.
137. Hooper, L., Summerbell, C.D., Higgins, J.P.T., Thompson, R.L., Capps, N.E., Smith, G.D., Riemersma, R.A., and Ebrahim, S., Dietary fat intake and prevention of cardiovascular disease: Systematic review, *Br. Med. J.*, 322, 757–763, 2001.
138. Mel'nikov, S.M., Seijen ten Hoorn, J.W.M., and Bertrand, B., Can cholesterol absorption be reduced by phytosterols and phytostanols via a cocrystallization mechanism? *Chem. Phys. Lipids*, 127, 15–33, 2004.
139. Bouic, P.J., The role of phytosterols and phytosterolins in immune modulation: A review of the past 10 years, *Curr. Opin. Clin. Nutr. Metab. Care.*, 4, 471–475, 2001.
140. Jones, P.J.H., MacDougall, D.E., Ntanios, F.Y., and Vanstone, C.A., Dietary phytosterols as cholesterol-lowering agents in humans, *Can. J. Physiol. Pharm.*, 75, 217–227, 1997.
141. Weststrate, J.A. and Meijer, G.W., Plant sterol-enriched margarines and reduction of plasma total- and LDL-cholesterol concentrations in normocholesterolaemic and mildly hypercholesterolaemic subjects, *Eur. J. Clin. Nutr.*, 52, 334–343, 1998.
142. Gylling, H. and Miettinen, T.A., Cholesterol reduction by different stanol mixtures and with variable fat intake, *Metabolism*, 48, 575–580, 1999.
143. Andersson, A., Karlström, B., Mohsen, R., and Vessby, B., Cholesterol-lowering effects of a stanol ester-containing low-fat margarine used in conjunction with a strict lipid-lowering diet, *Eur. Heart J.*, 1, 80S–90S, 1999.
144. Hendriks, H.F.J., Weststrate, J.A., van Vliet, T., and Meijer, G.W., Spreads enriched with three different levels of vegetable oil sterols and the degree of cholesterol lowering in normocholesterolaemic and mildly hypercholesterolaemic subjects, *Eur. J. Clin. Nutr.*, 53, 319–327, 1999.

145. Plat, J., van Onselen, E.N.M., van Heugten, M.M.A., and Mensink, R.P., Effects on serum lipids, lipoproteins and fat soluble antioxidant concentrations of consumption frequency of margarines and shortenings enriched with plant stanol esters, *Eur. J. Clin. Nutr.*, 54, 671–677, 2000.
146. Hallikainen, M.A., Sarkkinen, E.S., Gylling, H., Erkkilä, A.T., and Uusitupa, M.I.J., Comparison of the effects of plant sterol ester and plant stanol ester-enriched margarines in lowering serum cholesterol concentrations in hypercholesterolaemic subjects on a low-fat diet, *Eur. J. Clin. Nutr.*, 54, 715–725, 2000.
147. Volpe, R., Niittynen, L., Korpela, R., Sirtori, C., Bucci, A., Fraone, N., and Pazzucconi, F., Effect of yoghurt enriched with plant sterols on serum lipids in patients with moderate hypercholesterolaemia, *Br. J. Nutr.*, 86, 233–239, 2001.
148. Zawistowski, J., European functional foods landmark: Phytosterols and coronary heart diseases, presented at 2nd International Conference on Functional Foods and Nutraceuticals, Istanbul, Turkey, May 4–6, 2006, Abstract OP 06023.
149. 2004/333/EC: Commission decision of 31 March 2004 authorizing the placing on the market of yellow fat spreads, salad dressings, milk type products, fermented milk type products, soya drinks and cheese-type products with added phytosterols/phytostanols as novel food ingredients under Regulation (EC) No 258/97 of the European Parliament and of the Council [notified under document number C(2004) 1243]. Official Journal 2004, L 105, 14/04/2004, 0040–0042.
150. Khor, H.T. and Chieng, D.Y., Effect of dietary supplementation of tocotrienols and tocopherols on serum lipids in the hamster, *Nutr. Res.*, 16, 1393–1401, 1996.
151. Papas, A.M., *The Vitamin E Factor*, Haper Collins, New York, 1999.
152. Theriault, A., Chao, J.-T., Wang, Q., Gapor, A., and Adeli, K., Tocotrienol: A review of its therapeutic potential, *Clin. Biochem.*, 32, 309–319, 1999.
153. Packer, L., Weber, S.U., and Rimbach, G., Molecular aspects of α-tocotrienol antioxidant action and cell signaling, *J. Nutr.*, 131 Symposium, 369S–373S, 2001.
154. Mishima, K., Tanaka, T., Pu, F., Egashira, N., Iwasaki, K., Hidaka, R., Matsunaga, K., Takata, J., Karube, Y., and Fujiwara, M., Vitamin E isoforms α-tocotrienol and γ-tocopherol prevent cerebral infarction in mice, *Neurosci. Lett.*, 337, 56–60, 2003.
155. Nesaretnam, K., Ambra, R., Selvaduray, K.R., Radhakrishnan, A., Reimann, K., Razak, G., and Virgili, F., Tocotrienol-rich fraction from palm oil affects gene expression in tumors resulting from MCF-7 cell inoculation in athymic mice, *Lipids*, 39, 459–467, 2004.
156. Qureshi, A.A., Qureshi, N., Wright, J.J., Shen, Z., Kramer, G., Gapor, A., Chong, Y.H., DeWitt, G., Ong, A., and Peterson, D.M., Lowering of serum cholesterol in hypercholesterolemic humans by tocotrienols (palmvitee), *Am. J. Clin. Nutr.*, 53, 1021S–1026S, 1991.
157. Dionisi, F., Prodolliet, J., and Tagliaferri, E., Assessment of olive oil adulteration by reversed-phase high-performance liquid chromatography/amperometric detection of tocopherols and tocotrienols, *J. Am. Oil Chem. Soc.*, 72, 1505–1511, 1995.
158. Kornsteiner, M., Wagner, K-.H., and Elmadfa, I., Tocopherols and total phenolics in 10 different nut types, *Food Chem.*, 98, 381–387, 2006.
159. Amaral, J.S., Casal, S., Torres, D., Seabra, R.M., and Oliveira, B.P.P., Simultaneous determination of tocopherols and tocotrienols in hazelnuts by a normal phase liquid chromatographic method, *Anal. Sci.*, 21, 1545–1548, 2005.
160. Amaral, J.S., Casal, S., Alves, M.R., Seabra, R.M., and Oliveira, B.P.P., Tocopherol and tocotrienol content of hazelnut cultivars grown in Portugal, *J. Agric. Food Chem.*, 54, 1329–1336, 2006.
161. Miettinen, T.A. and Vanhanen, H., Serum concentration and metabolism of cholesterol during rapeseed oil and squalene feeding, *Am. J. Clin. Nutr.*, 59, 356–363, 1994.
162. Chan, P., Tomlinson, B., Lee, C.B., and Lee, Y.S., Effectiveness and safety of low-dose pravastatin and squalene, alone and in combination, in elderly patients with hypercholesterolemia, *J. Clin. Pharmacol.*, 36, 422–427, 1996.
163. Newmark, H.L., Squalene, olive oil and cancer risk: A review and hypothesis, *Cancer Epidem. Biomar. Prevent.*, 6, 1101–1103, 1997.
164. Rao, C.V., Newmark, H.L., and Reddy, B.S., Chemopreventive effect of squalene on colon cancer, *Carcinogenesis*, 19, 287–290, 1998.
165. Smith, T.J., Squalene: Potentaial chemopreventive agent, *Expert Opin. Investig. Drugs*, 9, 1841–1848, 2000.
166. Kohno, Y., Egawa, Y., Itoh, S., Nagaoka, S., Takahashi, M., and Mukai, K., Kinetic study of quenching reaction of singlet oxygen and scavenging reaction of free radical by squalene in *n*-butanol, *Biochim. Biophys. Acta*, 1256, 52–56, 1995.

167. Owen, R.W., Mier, W., Giacosa, A., Hull, W.E., Spiegelhalder, B., and Bartsch, H., Phenolic compounds and squalene in olive oils: The concentration and antioxidant potential of total phenols, simple phenols, secoiridoids, lignans and squalene, *Food Chem. Toxicol.*, 38, 647–659, 2000.
168. O'Sullivan, L., Woods, J.A., and O'Brien, N.M., Squalene but not *n*–3 fatty acids protect against hydrogen peroxide-induced sister chromatid exchanges in Chinese hamster V79 cells, *Nutr. Res.*, 22, 847–857, 2002.
169. Ko, T.-F., Weng, Y.-M., and Chiou, R.Y.-Y., Squalene content and antioxidant activity of *Terminalia catappa* leaves and seeds, *J. Agric. Food Chem.*, 50, 5343–5348, 2002.
170. Maguire, L.S., O'Sullivan, S.M., Galvin, K., O'Connor, T.P., and O'Brien, N.M., Fatty acid profile, tocopherol, squalane and phytosterol content of walnuts, almonds, peanuts, hazelnuts and the macadamia nut, *Int. J. Food Sci Nutr.*, 5, 171–178, 2004.
171. Ryan, E., Galvin, K., O'Connor, T.P., Maguire, A.R., and O'Brien, N.M., Fatty acid profile, tocopherol, squalene and phytosterol content of brazil, pecan, pine, pistachio and cashew nuts, *Int. J. Food Sci. Nutr.*, 57, 219–228, 2006.
172. Mendes, M.F., Pessoa, F.L.P., and Uller, A.M.C., An economic evaluation based on an experimental study of the vitamin E concentration present in deodorizer distillate of soybean oil using supercritical CO_2, *J. Supercrit. Fluids*, 23, 257–265, 2002.
173. Edwards, K., Kwaw, I., Matud, J., and Kurtz, I., Effect of pistachio nuts on serum lipid levels in patients with moderate hypercholesterolemia, *J. Am. Coll. Nutr.*, 18, 229–232, 1999.
174. Kris-Etherton, P.M., You-Poth, S., Sabaté, J., Ratcliffe, H.E., Zhao, G., and Etherton, T.D., Nuts and their bioactive constituents: Effects on serum lipids and other factors that affect disease risk, *Am. J. Clin. Nutr.*, 70, 504S–511S, 1999.
175. Zambón, D., Sabaté, J., Muñoz, S., Campero, B., Casals, E., Merlos, M., Laguna, J.C., and Ros, E., Substituting walnuts for monounsaturated fat improves the serum lipid profile of hypercholesterolemic men and women: A randomized crossover trial, *Ann. Intern. Med.*, 132, 538–546, 2000.
176. Rajaram, S., Burke, K., Connell, B., Myint, T., and Sabaté, J., A monounsaturated fatty acid-rich pecan-enriched diet favorably alters the serum lipid profile of healthy men and women, *J. Nutr.*, 131, 2275–2279, 2001.
177. Hyson, D.A., Schneeman, B.O., and Davis, P.A., Almonds and almond oil have similar effects on plasma lipids and LDL oxidation in healthy men and women, *J. Nutr.*, 132, 703–707, 2002.
178. Garg, M.L., Blake, R.J., and Wills, R.B.H., Macadamia nut consumption lowers plasma total and LDL cholesterol levels in hypercholesterolemic men, *J. Nutr.*, 133, 1060–1063, 2003.
179. Fraser, G.E., Sabaté, J., Beeson, W.L., and Strahan, T.M., A possible protective effect of nut consumption on risk of coronary heart disease. The Adventist Health Study, *Arch. Intern. Med.*, 152, 1416–1424, 1992.
180. Hu, F.B. and Stampfer, M.J., Nut consumption and risk of coronary heart disease: A review of epidemiologic evidence, *Curr. Ather. Rep.*, 1, 204–209, 1999.
181. Sabaté, J., Radak, T., and Brown, J.Jr., The role of nuts in cardiovascular disease prevention, in *Handbook of Nutraceuticals and Functional Foods*, Wildman, R.E.C., Ed., CRC Press, Boca Raton, FL, 2000, pp. 477–496.
182. Albert, C.M., Gaziano, J.M., Willett, W.C., and Manson, J.E., Nut consumption and decreased risk of sudden cardiac death in the Physicians' Health Study, *Arch. Intern. Med.*, 162, 1382–1387, 2002.
183. Feldman, E.B., The scientific evidence for a beneficial health relationship between walnuts and coronary heart disease, *J. Nutr.*, 132, 1062S–1101S, 2002.
184. Hu, F.B., Stampfer, M.J., Manson, J.E., Rimm, E.B., Colditz, G.A., Rosner, B.A., Speizer, F.E., Hennekens, C.H., and Willett, W.C., Frequent nut consumption and risk of coronary heart disease in women: Prospective cohort study, *Br. Med. J.*, 317, 1341–1345, 1998.
185. Kelly, J.H. and Sabaté, J., Nuts and coronary heart disease: An epidemiological perspective, *Br. J. Nutr.*, 96(Suppl. 2), 61S–67S, 2006.
186. Willett, W.C., *Eat, Drink, and be Healthy. The Harvard Medical School Guide to Healthy Eating*, Simon and Schuster, New York, 2004.
187. Shahidi, F. and Miraliakbari, H., Tree nut oils, in *Bailey's Industrial Oil and Fat Products, Edible Oil and Fat Products: Specialty Oils and Oil Products*, vol. 3, 6th ed., Shahidi F., Ed., Wiley Interscience, Hoboken, NJ, 2005, pp. 175–193.
188. Jenab, M. et al., Consumption of portion sizes of tree nuts, peanuts and seeds in the European Prospective Investigation into Cancer and Nutrition (EPIC) cohorts from 10 European Coutries, *Br. J. Nutr.*, 96(Suppl. 2), 12S–23S, 2006.

189. Miyashita, K., Tsubakihara, Y., and Alasalvar, C., Antioxidant and anti-obesity effects of hazelnut polyphenols, presented at 6th International Conference and Exhibition on Nutraceuticals and Functional Foods, Anaheim, CA, October 16–19, 2005, Abstract 28.
190. NCEP Expert Panel, Summary of the second report of the National Cholesterol Education Program (NCEP) Expert Panel on Detection, Evaluation, and Treatment of High Blood Cholesterol in Adults (Adult Treatment Panel II), *J. Am. Med. Assoc.*, 269, 3015–3023, 1993.
191. Connor, W.E., Connor, S.L., Katan, M.B., Grundy, S.M., and Willett, W.C., Should a low-fat, high-carbohydrate diet be recommended for everyone? *N. Engl. J. Med.*, 337, 562–567, 1997.
192. Krauss, R.M., Eckel, R.H., Howard, B., Appel, L.J., Daniels, S.R., Deckelbaum, R.J., Erdman, J.W. Jr., Kris-Etherton, P., Goldberg, I.J., et al., AHA dietary guidelines. Revision 2000: A statement for healthcare professionals from the nutrition committee of the American Hearth Association, *Circulation*, 102, 2284–2299, 2000.
193. Grundy, S.M., Comparison of monounsaturated fatty acids and carbohydrates for lowering plasma cholesterol, *N. Engl. J. Med.*, 314, 745–748, 1986.
194. Mensink, R.P. and Katan, M.B., Effect of monounsaturated fatty acids versus complex carbohydrates on high-density lipoproteins in healthy men and women, *Lancet*, 1, 122–125, 1987.
195. Horrobin, D.F. and Manku, M.S., How do polyunsaturated fatty acids lower plasma cholesterol levels? *Lipids*, 18, 558–562, 1983.
196. Kris-Etherton, P.M., Monounsaturated fatty acids and risk of cardiovascular disease, *Circulation*, 100, 1253–1258, 1999.
197. Pérez-Jiménez, F., Castro, P., López-Miranda, J., Paz-Rojas, E., Blanco, A., López-Segura, F., Velasco, F., Marín, C., Fuentes, F., and Ordovás, J.M., Circulating levels of endothelial function are modulated by dietary monounsaturated fat, *Atherosclerosis*, 145, 351–358, 1999.
198. Esterbauer, H., Dieber-Rotheneder, M., Striegl, G., and Waeg, G., Role of vitamin E in preventing the oxidation of low-density lipoprotein, *Am. J. Clin. Nutr.*, 53, 314S–321S, 1991.
199. Papas, A.M., Beyond α-tocopherol: The role of the other tocopherols and tocotrienols, in *Phytochemicals in Nutrition and Health*, Meskin, M.S., Bidlack, W.R., Davies, A.J., and Omaye, S.T., Eds., CRC Press, Boca Raton, FL, 2002, pp. 61–78.
200. Alphan, E., Pala, M., Açkurt, F., and Yilmaz, T., Nutritional composition of hazelnuts and its effects on glucose and lipid metabolism, *Acta Hort.*, 445, 305–310, 1997.
201. Balkan, J., Hatipoğlu, A., Aykaç-Toker, G., and Uysal, M., Influence on hazelnut oil administration on peroxidation status of erythrocytes and apolipoprotein B 100-containing lipoproteins in rabbits fed on a high cholesterol diet, *J. Agric. Food Chem.*, 51, 3905–3909, 2003.
202. Mercanlıgil, S.M., Arslan, P., Alasalvar, C., Okut, E., Akgül, E., Pınar, A., Geyik, P. Ö., Tokgözoğlu, L., and Shahidi, F., Effects of hazelnut-enriched diet on plasma cholesterol and lipoprotein profiles in hypercholesterolemic adult men, *Eur. J. Clin. Nutr.*, 61, 212–220, 2007.
203. Fang, F., Ho, C.-T., Sang, S., and Rosen, R.T., Determination of sphingolipids in nuts and seeds by a single quadrupole liquid chromatography–mass spectrometry method, *J. Food Lipids*, 12, 327–343, 2005.
204. Imaizumi, K., Tominaga, A., Sato, M., and Sugano, M., Effects of dietary sphingolipids on levels of serum and liver lipids in rats, *Nutr. Res.*, 12, 543–548, 1992.
205. Kobayashi, T., Shimizugawa, T., Osakabe, T., Watanabe, S., and Okuyama, H., A long-term feeding of sphingolipids affected the levels of plasma cholesterol and hepatic triacylglycerol but not tissue phospholipids and sphingolipids, *Nutr. Res.*, 17, 111–114, 1997.
206. Shahidi, F., Alasalvar, C., and Liyana-Pathirana, C.M., Antioxidant phytochemicals in hazelnut kernel (*Corylus avellana* L.) and hazelnut byproducts, *J. Agric. Food Chem.*, 55, 1212–1220, 2007.
207. Rajaram, S. and Sabaté, J., Nuts, body weight and insulin resistance, *Br. J. Nutr.*, 96(Suppl. 2), 79–86, 2006.
208. Sabaté, J., Nut consumption and body weight, *Am. J. Clin. Nutr.*, 78, 647S–650S, 2003.
209. St-Onge, M.P., Dietary fats, teas, dairy, and nuts: Potential functional foods for weight control? *Am. J. Clin. Nutr.*, 81, 7–15, 2005.
210. Chisholm, A., Mann, J., Skeaff, M., Frampton, C., Sutherland, W., Duncan A., and Tiszavari S., A diet rich in walnuts favourably influences plasma fatty acid profile in moderately hyperlipidaemic subjects, *Eur. J. Clin. Nutr.*, 52, 12–16, 1998.
211. Dulloo, A.G., Duret, C., Rohrer, D., Girardier, L., Mensi, N., Fathi, M., Chantre, P., and Vandermander, J., Efficacy of a green tea extract rich in catechin polyphenols and caffeine in increasing 24-h energy expenditure and fat oxidation in humans, *Am. J. Clin. Nutr.*, 70, 1040–1045, 1999.

212. Klaus, S., Pültz, S., Thöne-Reineke, C., and Wolfram, S., Epigallocatechin gallate attenuates diet-induced obesity in mice by decreasing energy absorption and increasing fat oxidation, *Int. J. Obes.*, 29, 615–623, 2005.
213. Kalogeropoulos, N., Chiou, A., Nomikos, T., Antonopoulou, S., Andrikopoulos, N., and Karathanos, V., Phytosterols, tocopherols, total polyphenols, squalene and fatty acids in fresh nuts and seeds usually consumed in Greece, presented at the Institute of Food Technologists Annual Meeting and Food Expo, Chicago, IL, July 27–August 1, 2007, Abstract 095-33.
214. Alasalvar, C., Amaral, J.S., Satir, G., and Shahidi, F., Lipid characteristics and essential minerals of native Turkish hazelnut varieties (*Coryluls avellana* L.), *Food Chem.*, 2009 (in press).
215. Simsek, A. and Aykut, O., Evaluation of the microelement profile of Turkish hazelnut (*Corylus avellana* L.) varieties for human nutrition and health, *Int. J. Food Sci. Nutr.*, 58, 677–688, 2007.
216. Miraliakbari, H. and Shahidi, F. Lipid class compositions, tocopherols and sterols of tree nut oils extracted with different solvents, *J. Food Lipids*, 15, 81–96, 2008.

13 Antioxidant Activities and Phytochemicals in Hazelnut (*Corylus avellana* L.) and Hazelnut By-Products

Cesarettin Alasalvar, Angela M. Hoffman, and Fereidoon Shahidi

CONTENTS

13.1 INTRODUCTION

Phytochemicals are defined as nonnutritive, naturally occurring, and biologically active plant-derived chemicals that have health supporting, protective, and disease preventive properties. Several thousand phytochemicals (such as phenolics, carotenoids, allylic compounds, sterols, chlorophylls, indoles, saponins, gingerols, and terpenes, among others) have been reported in fruits, vegetables, nuts, medicinal plants, herbal remedies, flowers, and red wines, among others [1–7]. Currently, there

FIGURE 13.1 Hazelnut kernel and hazelnut by-products.

is much interest in phytochemicals because of potential health benefits related to their substantial antioxidant and antiradical activities [3,8], anti-inflammatory and antiaggregatory properties [9,10], anticarcinogenic and antimutagenic effects [1], and antiproliferative potential [11]. Phytochemicals provide protection against harmful effects of free radicals and are known to reduce the risk of developing certain types of cancer, coronary heart disease (CHD), stroke, atherosclerosis, cataracts, osteoporosis, type-2 diabetes, inflammation, ulcers, diabetes, endothelial function, insulin resistance, sudden death, and hypertension, among others [1,3,12–24].

Nuts are regarded as nutritious foods with a high content of healthful lipid. Recent recognition of nuts as "heart-healthy" foods by the U.S. Food and Drug Administration (FDA) has provided a major boost to their image. Hazelnut (*Corylus avellana* L.), which belongs to the family Betulaceae, is one of the most popular tree nuts on a worldwide basis and ranks second in tree nut production after almond. Turkey, specifically its Black Sea region, is the major hazelnut producing area, which contributes around 75% to the total global production [25], followed by Italy, United States, and Spain. Hazelnut is typically consumed as the whole nut (raw or roasted) or used as an ingredient in a variety of processed foods, especially in bakery and confectionery products.

Figure 13.1 shows the ready-to-harvest hazelnut fruit. The hazelnut green leafy covers, occasionally together with hazelnut tree leaves, are mechanically removed from hazelnut hard shells soon after harvesting. The hazelnut hard shell, containing a kernel, is the nut of commerce. After cracking the hazelnut hard shell, hazelnut kernel maybe consumed as raw or roasted, and with or without skin. Hazelnut by-products include the hard shell, green leafy cover, and leaves. Among these by-products, only the hard shell has commercial value for fuel. The use of hazelnut by-products as potential sources of phytochemicals and functional food ingredients is of great interest to the hazelnut industry.

Although antioxidant activity and phenolic constituents of hazelnut (hazelnut kernel) [4,26–28] and some of its by-products [29–31,100] have been reported, little is known about phenolic phytochemicals and their potential health benefits in hazelnut and hazelnut by-products. Therefore, this chapter provides detailed information on nutraceuticals, phytochemicals, and health aspects of hazelnut and hazelnut by-products and evaluates their antioxidant activities by a number of indicators. Taxanes present in hazelnut by-products are discussed in detail. Where possible, the health benefits of relevant phytochemicals are described.

13.2 PHENOLICS IN HAZELNUT AND HAZELNUT BY-PRODUCTS

13.2.1 Total Phenolics

Phenolic compounds, a major group of phytochemicals, are naturally present in both edible (fruits, vegetables, cereal grains, and nuts, among others) and nonedible plant parts. They play important roles in human health due to their antioxidant activity, which is based on their ability to donate a hydrogen atom to free radicals [3]. Nuts and their by-products are able to reduce risk of cancer, diabetes, and other chronic diseases [32].

Wu et al. [4] measured the total phenolic content of 10 different tree nuts, expressed as milligrams of gallic acid equivalents (GAE/g) on a weight basis (see Chapter 2). A large variation in phenolics was observed among nuts ranging from 1.56 in macadamia to 20.16 in pecan. Hazelnut contained the fourth largest content of phenolics overall at 8.35 mg GAE/g. Recently, Shahidi et al. [31] studied the total phenolic content of hazelnut kernel and hazelnut by-products. Total phenolic content of hazelnut skin was the highest (577.7 mg catechin equivalents CE/g ethanolic extract) whereas that of hazelnut kernel was the lowest (13.7 mg CE/g ethanolic extract), with the following trend: hazelnut skin > hazelnut hard shell > hazelnut tree leaf > hazelnut green leafy cover > hazelnut kernel (Table 13.1). Significant differences ($P < 0.05$) in the phenolic contents existed among hazelnut extracts (kernel, skin, hard shell, green leafy cover, and tree leaf). Hazelnut skin afforded ~7.4-fold higher total phenolics (426.7–502.3 mg GAE/g extract depending on the solvent) than that of hazelnut hard shell (56.6–72.2 mg GAE/g extract depending on the solvent) [100]. In comparison, Siriwardhana and Shahidi [33] found that the whole almond seed (kernel), brown skin, and green shell (leafy) cover had a total phenolic content of 8.1, 87.8, and 71.1 mg CE/g ethanolic extract, respectively. Alasalvar et al. [30] reported that hazelnut extracts with 80% (v/v) ethanol had a significantly lower ($P < 0.05$) total phenolic content compared to those of extracts obtained using 80% (v/v) acetone.

13.2.2 Phenolic Acids

Phenolic acids in foods occur in the free, esterified, glycosidic, and insoluble-bound forms [3]. Free phenolic acids are known to contribute to the taste of foods.

TABLE 13.1
Contents of Phenolics and Total Antioxidant Activity in Extracts of Hazelnut Kernel and Hazelnut By-Products

Extract	Phenolics[a]	Total Antioxidant Activity[b]
Hazelnut kernel (with skin)	13.7 ± 0.5[c]	29.0 ± 3.5[c]
Hazelnut skin	577.7 ± 1.1[d]	132.0 ± 4.0[d]
Hazelnut hard shell	214.1 ± 0.3[e]	120.0 ± 3.0[e]
Hazelnut green leafy cover	127.3 ± 0.7[f]	117.0 ± 2.5[e]
Hazelnut tree leaf	134.7 ± 1.0[g]	148.0 ± 2.1[f]

Source: From Shahidi, F., Alasalvar, F., and Liyana-Pathirana, C.M., *J. Agric. Food Chem.*, 55, 1212, 2007. With permission.

Note: Data are expressed as means ± SD ($n = 3$) on an extract.

[a] Content of phenolics, expressed as milligrams of catechin equivalents per gram of extract (mg of CE/g extract).

[b] Total antioxidant activity, expressed as micromoles of Trolox equivalents per gram of extract (μmol of TE/g extract).

[c–g] Means ± SD followed by the same letter, within a column, are not significantly different ($P > 0.05$).

Five soluble phenolic acids (free and esterified), one of which is a hydroxylated derivative of benzoic acid (gallic acid) and four are cinnamic acid derivatives (caffeic, *p*-coumaric, ferulic, and sinapic acids), have been studied and tentatively identified in ethanolic extracts of hazelnut kernel and hazelnut by-products (Table 13.2) [31]. The order of total phenolic acid concentration was as follows: hazelnut hard shell > hazelnut green leafy cover > hazelnut tree leaf > hazelnut skin > hazelnut kernel. Different phenolic acids predominate in each plant part examined. Among the identified phenolic acids, *p*-coumaric acid was most abundant in hazelnut kernel, hazelnut green leafy cover, and hazelnut tree leaf, whereas gallic acid was most abundant in hazelnut skin and hazelnut hard shell, possibly implying the presence and perhaps the dominance of tannins in the latter samples (Table 13.2). The same number, but different concentration, of phenolic acids have also been reported in hazelnut kernel and hazelnut green leafy cover [30].

Senter et al. [34] compared phenolic acids of nine edible tree nuts produced in the United States. The extracts of nut samples exhibited great diversity in the phenolic acids present. Qualitative and quantitative differences existed in the phenolic acids present in the nut samples, with gallic acid being predominant except in pine nut, almond, and hazelnut (filbert). A total of eight phenolic acids (*p*-hydroxybenzoic, *p*-hydroxyphenylacetic, vanillic, protocatechuic, syringic, gallic, caffeic, and ferulic) were isolated and identified in nine different nuts. Among these, protocatechuic acid predominated the phenolic acids in testa (skin) of hazelnut with a concentration of 0.36 μg/g. The levels of the remaining phenolic acids did not exceed 10 ng/g of skin. Protocatechuic acid has not been detected in other studies [30,31]. Caffeic, sinapic, ferulic, and *p*-coumaric acids have been reported as being more antioxidative than syringic, vanillic, and protocatechuic acids [35].

Yurttas et al. [26] isolated and tentatively identified six phenolic aglycones in Turkish and American hazelnut extracts; these were gallic acid, *p*-hydroxybenzoic acid, epicatechin and/or caffeic acid, sinapic acid, and quercetin. However, the variety of hazelnut and extraction solvents used by Shahidi et al. [31] was different from that used by Yurttas et al. [26]. Variety and extraction exert a great influence on the concentration and variability of phenolic acids present. Recently, Amaral et al. [29] identified and quantified four phenolic acids, namely 3-caffeoylquinic, 5-caffeoylquinic, caffeoyltartaric, and *p*-coumaroyltartaric acids, in hazelnut leaves from 10 different cultivars grown in Portugal. Like hazelnut, some of the other tree nuts and their processing by-products have been reported to contain different patterns and levels of phenolic acids [3].

TABLE 13.2
Contents of Phenolic Acids (Free and Esterified) in Extracts of Hazelnut Kernel and Hazelnut By-Products

Extract	Gallic	Caffeic	*p*-Coumaric	Ferulic	Sinapic
Hazelnut kernel (with skin)	127 ± 5[a]	81 ± 2[a]	208 ± 15[a]	105 ± 5[a]	93 ± 5[a]
Hazelnut skin	387 ± 9[b]	trace[b]	231 ± 17[a]	124 ± 8[b]	124 ± 4[b]
Hazelnut hard shell	3261 ± 79[c]	212 ± 13[c]	757 ± 31[b]	333 ± 25[c]	235 ± 17[c]
Hazelnut green leafy cover	892 ± 43[d]	158 ± 6[d]	1662 ± 43[c]	327 ± 15[c]	64 ± 3[d]
Hazelnut tree leaf	157 ± 8[e]	362 ± 10[e]	884 ± 19[d]	237 ± 12[d]	241 ± 11[c]

Source: From Shahidi, F., Alasalvar, F., and Liyana-Pathirana, C.M., *J. Agric. Food Chem.*, 55, 1212, 2007. With permission.

Note: Data are expressed as means ± SD ($n = 3$) on an extract; Phenolic acids, expressed as micrograms per gram of extract (μg/g extract).

[a–e] Means ± SD followed by the same letter, within a column, are not significantly different ($P > 0.05$).

13.2.3 Flavonoids

Flavonoids are another group of phenolic compounds that can be classified into seven groups: flavanones, flavones, isoflavones, anthocyanidins, flavonols, flavononols, and flavanols (flavan-3-ols or catechins) [8]. Flavonoids, which are the most common and widely distributed group of plant phenolics, are increasingly appreciated as being an important component of the human diet. Humans consume ~1 g of flavonoids/day [36]. The intake of flavonoids has been associated with a lower incidence of various diseases such as cancer, stroke, cardiovascular disease (CVD), and other chronic disorders [12,37,38]. The positive health effects of flavonoids are probably related to their strong antioxidant properties, among other mechanisms and effects [39,40]. Many of the flavonoids and their related phenolic acids have shown marked antioxidant characteristics [41]. These compounds are potent antioxidants *in vitro*, scavenging superoxide radical ($O_2^{\bullet -}$), hydroxyl radical ($HO^{\bullet}$), and peroxyl radical ($ROO^{\bullet}$) [42], thus controlling lipid peroxidation [43], and protecting low-density lipoproteins (LDL) against oxidation [44]. They can also inhibit platelet aggregation [45] and enhance vasodilatation [46].

Different classes (anthocyanidins, flavan-3-ols, flavonones, and flavonols) and amounts of flavonoids have been reported for different tree nuts [7]. The total content of flavonoids in tree nuts varies between 0.02 mg/100 g in chestnut and 34.01 mg/ 100 g in pecan (Table 13.3). Brazil nut and macadamia contain undetectable amounts of flavonoids. Hazelnut contains one anthocyanidin (cyanidin) and four flavan-3-ols [(−)-epicatechin, (−)-epigallocatechin, (−)-epigallocatechin 3-gallate, and (+)-catechin], with a total amount of 11.96 mg/100 g (fourth highest among tree nuts). Cyanidin is the most abundant flavonoid (6.71 mg/100 g) and no flavonones and flavonols were found in hazelnut.

13.2.4 Proanthocyanidins (Condensed Tannins)

Depending on their structures, tannins are defined as hydrolyzable (gallotannins and ellagitannins) or condensed (monomers, dimers, oligomers, and polymers of flavan-3-ols). Condensed tannins are also known as proanthocyanidins [3,47]. Proanthocyanidins can be divided into propelargonidins, based on the hydroxylation pattern of the A- and B-rings [3]. Of these, procyanidins constitute the most common subclass of flavonoids in foods, and prodelphinidins and propelargonidins are also present [48,49].

Besides their astringent flavor in foods, proanthocyanidins are of great interest in nutrition and medicine because of their potent antioxidant activity and possible protective effects on human health [48,50–52]. Recently, it has been hypothesized that free radical scavenging properties of proanthocyanidins may reduce the risk of CVD [53,54], cancer [38,55], blood clotting [56], and that certain types of trimetric proanthocyanidins may protect against urinary tract infections [57].

The data on hazelnut proanthocyanidins are limited. Recently, Gu et al. [49] found that some tree nuts are good sources of proanthocyanidins with contents ranging from 0.05 in chestnut to 500.7 mg/100 g in hazelnut (Table 13.4). The order of total proanthocyanidin concentration content in tree nuts was as follows: hazelnut > pecan > pistachio > almond > walnut > cashew > chestnut. No proanthocyanidins have been detected in Brazil nut, macadamia, and pine nut [49]. Among the proanthocyanidins, polymers are most abundant in hazelnut and some other tree nuts such as almond, pecan, and pistachio. Average intake of proanthocyanidins is estimated at 58 mg/100 g in the United States [49].

The content of condensed tannins in hazelnut (kernel), expressed as milligrams of CE per gram of extract, varies quite markedly, from a low of 40.5 for 80% ethanolic extract to a high of 320 for 80% acetone extract. Acetone is a more effective solvent for the extraction of condensed tannins as tannins are relatively high-molecular-weight compounds and ethanol is not necessarily suitable for their extraction [30]. The reason for this is that tannins are relatively high-molecular-weight compounds and the polarity of ethanol is too low for total extraction of these polar compounds from plant sources.

TABLE 13.3
Comparison of Flavonoids (mg/100 g) in Hazelnut with Other Tree Nuts

Flavonoid		Hazelnut	Almond	Cashew	Chestnut	Pecan	Pine Nut	Pistachio	Walnut[a]
Anthocyanidins	Cyanidin	6.71 ± 1.18	2.46 ± 0.58	nd	nd	10.74 ± 1.50	nd	6.06 ± 0.79	2.71 ± 0.25
	Delphinidin	nd	nd	nd	nd	7.28 ± 0.92	nd	nd	nd
Flavan-3-ols	(−)-Epicatechin	0.22 ± 0.09	0.60 ± 0.10	0.93 ± 0.22	nd	0.82 ± 0.08	nd	0.83 ± 0.46	nd
	(−)-Epicatechin 3-gallate	nd	nd	0.15 ± 0.10	nd	nd	nd	nd	nd
	(−)-Epigallocatechin	2.78 ± 1.21	2.59 ± 0.31	nd	nd	5.63 ± 1.47	0.49 ± 0.25	2.05 ± 0.82	nd
	(−)-Epigallocatechin 3-gallate	1.06 ± 0.46	nd	nd	nd	2.30 ± 0.46	nd	0.40 ± 0.40	nd
	(+)-Catechin	1.19 ± 0.49	1.28 ± 0.33	0.90 ± 0.28	0.01 ± 0.01	7.24 ± 0.51	nd	3.57 ± 1.00	nd
	(+)-Gallocatechin	nd	nd	nd	0.01 ± 0.01	nd	nd	nd	nd
Flavanones	Eriodictyol	nd	0.25 ± 0.06	nd	nd	nd	nd	nd	nd
	Naringenin	nd	0.13 ± 0.03	nd	nd	nd	nd	nd	nd
Flavonols	Isorhamnetin	nd	7.05 ± 1.03	nd	nd	nd	nd	nd	nd
	Kaempferol	nd	0.52 ± 0.05	nd	nd	nd	nd	nd	nd
	Quercetin	nd	0.36 ± 0.11	nd	nd	nd	nd	1.46 ± 0.64	nd
Total		11.96	15.24	1.98	0.02	34.01	0.49	14.37	2.71

Source: From U.S. Department of Agriculture (USDA), Database for the Flavonoid Content of Selected Foods (Release 2), August 2006, Published online at: http://www.ars.usda.gov/nutrientdata (accessed September 28, 2006).

Note: Data are expressed as means ± SE ($n = 3-16$) on a fresh weight basis; Brazil nut and macadamia contain undetectable amounts of flavonoids; nd, not detected.

[a] English walnut.

TABLE 13.4
Comparison of Proanthocyanidins (mg/100 g) in Hazelnut with Other Tree Nuts

Proanthocyanidin	Hazelnut	Almond	Cashew	Chestnut	Pecan	Pistachio	Walnut[a]
Monomers	9.8 ± 1.6	7.78 ± 0.9	6.7 ± 2.9	0.02 ± 0.00	17.7 ± 2.5	10.9 ± 4.3	6.9 ± 3.4
Dimers	12.5 ± 3.8	9.5 ± 1.6	2.0 ± 0.4	0.01 ± 0.00	42.1 ± 5.4	13.3 ±1.8	5.6 ± 0.9
Trimers	13.6 ± 3.9	8.8 ± 1.7	nd	0.02 ± 0.00	26.0 ± 2.0	10.5 ± 1.2	7.2 ± 1.2
4–6mers	67.7 ± 20.3	40.0 ± 8.5	nd	nd	101.4 ± 10.4	42.2 ± 5.2	22.1 ± 3.3
7–10mers	74.6 ± 21.9	37.7 ± 8.4	nd	nd	84.2 ± 12.9	37.9 ± 4.9	5.4 ± 0.8
Polymers	322.4 ± 102.5	80.3 ± 28.1	nd	nd	223.0 ± 59.1	122.5 ± 37.1	20.0 ± 9.3
Total	500.7 ± 152.0	184.0 ± 48.2	8.7 ± 3.2	0.05 ± 0.00	494.1 ± 86.2	237.3 ± 52.0	67.3 ± 14.7
Types	PC, PD	PP, PC	PC	—	PC, PD	PC, PD	PC

Source: From U.S. Department of Agriculture (USDA), Database for the Proanthocyanidin Content of Selected Foods, August 2004, Published online at: http://www.nal.usda.gov/fnic/foodcomp (accessed September 28, 2006); Gu, L., Kelm, M.A., Hammerstone, J.F., Beecher, G., Holden, J., Haytowitz, D., Gebhardt, S., and Prior, R.L., *J. Nutr.*, 134, 613, 2004.

Note: Data are expressed as means ± SD ($n = 4-8$) on a fresh weight basis; Brazil nut, macadamia, and pine nut contain undetectable amounts of proanthocyanidins; nd, not detected; PC, Procyanidins; PD, Prodelphinidins; PP, Propelargonidins.

[a] English walnut.

Recently, Venkatachalam and Sathe [58] reported the appropriate composition of commercially important edible nuts (hazelnut, almond, Brazil nut, cashew, macadamia, pecan, pine nut, pistachio, walnut, and Virginia peanut). The authors used both absolute methanol and acidified methanol (1% HCl, v/v) to extract nonpolar and polar tannins, respectively. The content of nonpolar tannin ranged from 0.01 g/100 g in Brazil nut and pine nut to 0.84 g/100 g edible portion of pecan, whereas polar tannin ranged from 0.01 g/100 g in Brazil nut and pine nut to 0.88 g/100 g edible portion of pecan. The content of nonpolar and polar tannins in hazelnut was 0.04 and 0.23 g/100 g of edible portion, respectively. More tannins were extracted by acidified methanol from hazelnut, almond, cashew, pecan, pistachio, and Virginia peanuts, suggesting the presence of these nuts. On the contrary, the same amount of tannins was extracted from Brazil nut, macadamia, and pine nut, indicating that tannins in these nuts were more nonpolar [58].

High tannin content in hazelnut by-product extract (skin and hard shell) was reported by Contini et al. [100]. Total tannins represented the principal fraction of the phenolic substances of the extracts, containing nearly 60%–65% of the total phenol [100]. As a general rule, the highest amount of total tannins was detected in the extracts with the highest amount of total phenols. Hence, the high antioxidant activity in hazelnut by-product extracts maybe due to these polyphenolic compounds since they are expected to have powerful antioxidant activity [100].

13.2.5 Phytoestrogens

Phytoesterogens have gained much interest in recent years because of their potential protective effects against many diseases and conditions including several forms of cancer, cardiovascular and neurodegenerative diseases, osteoporosis, and menopausal symptoms [59–68]. They impart antioxidant, phytoestrogenic, antiproliferative, and enzyme modulating activities within the human metabolic system [68].

The major phytoestrogen groups are isoflavones, lignans, and coumestans. Thompson et al. [69] analyzed phytoestrogen content of 121 food samples including seven major tree nuts (almond, cashew, chestnut, hazelnut, pecan, pistachio, and walnut). Tree nuts analyzed (Table 13.5) contained four isoflavones (formononetin, daidzein, genistein, and glycitein), four lignans (matairesinol, lariciresinol, pinoresinol, and secoisolariciresinol), and one coumestan (coumestrol). Among seven tree nuts, pistachio was the richest source of total isoflavones (176.9 μg/100 g on an as is basis), total lignans (198.9 μg/100 g), and total phytoestrogens (382.5 μg/100 g). Hazelnut contained third highest total isoflavones (30.2 μg/100 g), primarily genistein, after pistachio and walnut and sixth highest total lignans (77.1 μg/100 g), primarily secoisolariciresinol, and total phyoestrogens (107.5 μg/100 g).

13.3 ANTIOXIDANT ACTIVITIES OF HAZELNUT AND HAZELNUT BY-PRODUCTS

Antioxidants, including food phenolics, are able to retard the process of oxidation by neutralizing free radicals and chelating metal ions, thereby protecting the body or the food from deleterious effects of oxidation [70,71]. In this section, total antioxidant activity, free radical scavenging activity tests (hydrogen peroxide, superoxide, and 2,2-diphenyl-1-picrylhydrazyl [DPPH]), together with antioxidant activity in the β-carotene–linoleate model system, inhibition of oxidation of human LDL cholesterol, inhibition of strand breaking of supercoiled DNA, and reducing power of the extracts of hazelnut and its by-products are discussed in more detail below.

13.3.1 Total Antioxidant Activity (Capacity)

Wu et al. [4] evaluated the total antioxidant capacity (activity) of more than 100 different kinds of foods, including 10 different types of nuts, using the oxygen radical absorbance capacity with

TABLE 13.5
Comparison of Phytoestrogen (μg/100 g) in Hazelnut with Other Tree Nuts

Phytoestrogen		Hazelnut	Almond	Cashew	Chestnut	Pecan	Pistachio	Walnut
Isoflavones	Formononetin	1.2	0.8	10.0	1.7	0.7	0.2	0.9
	Daidzein	3.6	2.1	1.4	2.8	1.6	73.1	35.2
	Genistein	24.8	14.4	10.3	16.4	0.9	103.3	16.4
	Glycitein	0.5	0.6	0.4	0.3	0.3	0.4	0.8
Lignans	Matairesinol	1.2	0.3	0.3	0.5	0.6	0.1	0.2
	Lariciresinol	14.3	32.2	60.5	7.8	8.4	123.0	7.2
	Pinoresinol	1.1	9.0	1.1	5.6	1.2	31.2	0.2
	Secoisolari-ciresinol	60.5	70.3	37.5	172.7	14.8	44.6	78.0
Coumestan	Coumestrol	0.3	1.5	0.4	2.4	0.3	6.7	0.6
Total isoflavones		30.2	18.0	22.1	21.2	3.5	176.9	53.3
Total lignans		77.1	111.7	99.4	186.6	25.0	198.9	85.7
Total phytoestrogens[a]		107.5	131.1	121.9	210.2	28.8	382.5	139.5

Source: From Thompson, L.U., Boucher, B.A., Liu, Z., Cotterchio, M., and Kreiger, N., *Nutr. Cancer*, 54, 184, 2006. With permission.

Note: Data are expressed as means ± SD ($n = 2$) on a fresh weight basis.

[a] Total phytoestrogens is the sum up of isoflavones, lignans, and coumestan.

fluorescein probe assay ($ORAC_{FL}$) and combining both lipophilic (L-$ORAC_{FL}$) and hydrophilic (H-$ORAC_{FL}$) components. The authors observed quite a large variation of H-$ORAC_{FL}$ (4.43–175.24 μmol of Trolox equivalents (TE)/g) among 10 nuts whereas the range of L-$ORAC_{FL}$ (1.72–5.57 μmol of TE/g) was much lower. Considering all the studied nuts, hazelnut had the third highest value (96.45 μmol of TE/g on an as is basis) of total antioxidant activity (by combining L-$ORAC_{FL}$ and H-$ORAC_{FL}$), with pecan and walnut having the highest. To perform an overall evaluation of the total antioxidant activity of the consumed foods, the authors considered serving size (28.4 g/day = 1 ounce) as well as concentrations of serving size, and divided foods among four groups based upon the range of their H-$ORAC_{FL}$ and L-$ORAC_{FL}$ per serving rather than per unit weight of food. In both cases (H-$ORAC_{FL}$ and L-$ORAC_{FL}$), hazelnut was positioned in the group with the highest ORAC value per serving size (2739 μmol of TE) (see Chapter 2).

Recently, Shahidi et al. [31] found that total antioxidant activity of hazelnut extracts ranged from 29 to 148 μmol TE/g of ethanolic extract, with lowest activity in hazelnut kernel and highest in hazelnut tree leaf (Table 13.1). The total antioxidant activity of hazelnut by-product extracts were approximately 4- to 5-fold higher than that of hazelnut kernel at the same extract concentration. At a given concentration, hazelnut by-product extracts would serve as a more effective antioxidant than hazelnut kernel extract. Consideration of defatted hazelnut (on an extract basis) and hazelnut (as is weight basis) make the quantitative comparison between the studies of Wu et al. [4] and Shahidi et al. [31] impossible.

Siriwardhana and Shahidi [33] evaluated the Trolox equivalents antioxidant capacity (TEAC) of almond and its by-product extracts and found that ethanolic extracts, at the same concentration the TEAC values, followed the order of brown skin > green shell cover > whole seed (kernel). Values of brown skin and green shell cover extracts were 12.6- and 9.8-fold higher than that of whole seed extract, respectively. Alasalvar et al. [30] observed that hazelnut kernel and its green leafy cover extracts from 80% (v/v) ethanol were characterized as having significantly lower ($P < 0.05$) total antioxidant activity compared to those of extracts obtained from 80% (v/v) acetone.

Yurttas et al. [26] measured the antioxidant activity of both hydrolyzed and nonhydrolyzed Turkish and American hazelnut kernel extracts in a linoleic acid-buffer system. Results showed that nonhydrolyzed extracts of hazelnut exhibited a greater antioxidant activity than the corresponding hydrolyzed extracts. They suggested that the antioxidant activity could be attributed to some phenolic compounds tentatively identified.

More recently, Contini et al. [100] measured natural antioxidants from hazelnut hard shell and hazelnut skin and found that the crude extracts from the skin manifested the strongest antioxidant activity, similar or superior to butylated hydroxyanisole (BHA), butylated hydroxytoluene (BHT), Trolox, and α-tocopherol, at equivalent concentrations. Table 13.6 represents the amount (μg) of antioxidant standards (BHA, BHT, Trolox, and α-tocopherol) that cause the same (50%) DPPH-scavenging effect as 1 μg of analyzed sample. One microgram of crude phenolic contained in skin extract exhibited radical scavenging ability equivalent to 2 μg of BHA (or Trolox), about 3 μg of BHT, and nearly 5 μg of α-tocopherol [100]. The finding indicates that the DPPH-scavenging activity of phenolic fractions contained in hazelnut by-products (particularly skin) is superior, when compared with the most largely employed commercial synthetic (BHA and BHT) or natural (α-tocopherol) antioxidative compounds. These considerations enable us to hypothesize that an eventual purification of hazelnut by-product extracts could yield extremely potent natural biophenolic antioxidants.

Contini et al. [100] also observed that different solvents [80% (v/v) methanol, 80% (v/v) ethanol, and 80% (v/v) acetone] used for the extraction had different capacities in extracting substances and phenols from hazelnut by-products. The most suitable phenolic solvents for hazelnut by-products were either 80% acetone or 80% ethanol [100]. These results are in agreement with that of Alasalvar et al. [30].

13.3.2 Free Radical Scavenging Activity Tests

Several free radical scavenging activity tests (hydrogen peroxide, superoxide, and DPPH) have been employed to evaluate the extracts of hazelnut kernel and hazelnut by-products. This information

TABLE 13.6
BHA, BHT, Trolox, and α-Tocopherol Equivalent Antioxidant Capacity of Hazelnut By-Product Extracts

		BHA		BHT		Trolox		α-Tocopherol	
Extract	Solvent[a]	μg/μg of extract	μg/μg of GAE[c]	μg/μg of extract	μg/μg of GAE	μg/μg of extract	μg/μg of GAE	μg/μg of extract	μg/μg of GAE
Hazelnut skin	M	0.06	1.03	0.08	1.50	0.06	1.15	0.13	2.35
Hazelnut skin	E	0.06	0.99	0.09	1.43	0.07	1.10	0.13	2.25
Hazelnut skin	A	0.08	1.14	0.12	1.66	0.09	1.28	0.19	2.61
Hazelnut hard shell	M	0.95	1.98	1.38	2.88	1.07	2.22	2.18	4.54
Hazelnut hard shell	E	1.09	2.17	1.58	3.15	1.22	2.43	2.49	4.96
Hazelnut hard shell	A	0.94	2.01	1.36	2.92	1.05	2.26	2.15	4.60

Source: From Contini, M., Baccelloni, S., Massantini, R., and Anelli, G., *Food Chem.*, 110, 659–669, 2008. With permission.

[a] Efficient concentration (EC_{50}) (μg of extract/mg DPPH): butylated hydroxyanisole (BHA), 134.3 ± 2.7; butylated hydroxytoluene (BHT): 195.2 ± 3.5; Trolox, 150.6 ± 2.4; α-Tocopherol, 307.3 ± 5.8.

[b] M, 80% (v/v) methanol; E, 80% (v/v) ethanol; A, 80% (v/v) acetone.

[c] Gallic acid equivalents.

is important for using hazelnut ingredients during the preparation of medical or functional foods as potential curative agents for specific disease conditions. Phenolic compounds display multiple biological effects, including radical scavenging activity. Free radical scavenging is the accepted mechanism for antioxidant action in inhibiting lipid oxidation. The types of free radical reactions that can be neutralized by an antioxidant are therefore important in determining their usefulness. Furthermore, evaluation of antioxidant efficacy in plant products cannot be carried out accurately by any single universal method due to the complex nature of phytochemicals present [72,73].

13.3.2.1 Hydrogen Peroxide Scavenging Activity

Hydrogen peroxide may cause damage to the structure of the cell through the formation of highly reactive oxygen species (ROS), such as hydroxyl radical. Effective scavenging of hydrogen peroxide can therefore prevent oxidative damage to lipids and other biomolecules. The scavenging activity of hydrogen peroxide by hazelnut extracts has recently been reported by Shahidi et al. [31] (Table 13.7). At 200 ppm level, all extracts exhibited 97%–99% scavenging of hydrogen peroxide with the exception of hazelnut kernel that scavenged only 77% of hydrogen peroxide. Thus, scavengers were mainly present in the outer portions of the hazelnut kernel. Scavenging activity varied between 60% and 95% at 100 ppm level. Catechin (reference antioxidant) exhibited 91% and 96% hydrogen peroxide scavenging activity at 100 and 200 ppm levels, respectively. Compared to catechin, extracts of hazelnut by-products showed stronger ($P < 0.05$) hydrogen peroxide scavenging activity than that of catechin at 200 ppm, except hazelnut green leafy cover ($P > 0.05$). Hence, hazelnut by-products may serve as effective scavengers and thereby protect cells from oxidative damage.

Siriwardhana and Shahidi [33] reported that hydrogen peroxide scavenging activities at 100 ppm were 59%, 63%, and 66% and at 200 ppm were 86%, 91%, and 91% for whole almond seed, brown skin, and green shell (leafy) cover extracts, respectively. The results obtained from both hazelnut and almond clearly show that extracts from by-products scavenged organic free radicals more effectively than kernel or seed extracts.

TABLE 13.7
Radical Scavenging Activities in Extracts of Hazelnut Kernel and Hazelnut By-Products

	Hydrogen Peroxide Scavenging Activity (%)		Superoxide Radical Scavenging Activity (%)		DPPH Radical Scavenging Activity (%)	
Extract	**100 ppm**	**200 ppm**	**100 ppm**	**200 ppm**	**50 ppm**	**100 ppm**
Hazelnut kernel (with skin)	60 ± 3^{a}	77 ± 2^{a}	82 ± 1^{a}	94 ± 1^{a}	86.1 ± 0.1^{a}	92.2 ± 0.1^{a}
Hazelnut skin	95 ± 1^{b}	99 ± 1^{b}	88 ± 1^{b}	99 ± 1^{b}	93.4 ± 0.2^{b}	99.5 ± 0.2^{b}
Hazelnut hard shell	$94 \pm 3^{b,c}$	99 ± 1^{b}	88 ± 2^{b}	99 ± 1^{b}	93.5 ± 0.1^{b}	99.4 ± 0.2^{b}
Hazelnut green leafy cover	85 ± 2^{d}	$97 \pm 2^{b,c}$	86 ± 2^{b}	99 ± 1^{b}	97.3 ± 0.1^{c}	99.5 ± 0.1^{b}
Hazelnut tree leaf	$93 \pm 2^{b,e}$	99 ± 1^{b}	87 ± 2^{b}	99 ± 1^{b}	94.8 ± 0.2^{d}	99.4 ± 0.2^{b}
Catechin	$91 \pm 1^{c,e}$	96 ± 1^{c}	90 ± 2^{b}	91 ± 1^{c}	100.0 ± 0.0^{e}	100.0 ± 0.0^{c}

Source: From Shahidi, F., Alasalvar, F., and Liyana-Pathirana, C.M., *J. Agric. Food Chem.*, 55, 1212, 2007. With permission.
Note: Data are expressed as means ± SD ($n = 3$) on an extract.
[a–e] Means ± SD followed by the same letter, within a column, are not significantly different ($P > 0.05$).

13.3.2.2 Superoxide Radical Scavenging Activity

Superoxide radical is a powerful oxidizing agent that can react with biological membranes and induce tissue damage [74]. It may also decompose to singlet oxygen, hydroxyl radical, or hydrogen peroxide [75]. The data on the efficacy of hazelnut extracts to scavenge superoxide radical are presented in Table 13.7 [31]. The activity of both hazelnut kernel and hazelnut by-products increase with increasing concentration. Extracts from by-products demonstrated superior activity (99%) compared to hazelnut kernel (94%) and catechin (91%) at 200ppm. A close scrutiny of the results assembled indicates that all extracts were more effective ($P < 0.05$) than catechin at 200ppm. Therefore, superoxide radical scavenging activity of hazelnut extracts, especially of by-products, would be one of the major mechanisms contributing to their antioxidant activities [31]. Superoxide radical scavenging activities at 100ppm are 76%, 89%, and 97% and at 200ppm were 85%, 95%, and 99% for whole almond seed, brown skin, and green shell (leafy) cover extracts, respectively [33].

13.3.2.3 DPPH Radical Scavenging Activity

The use of DPPH radical scavenging assay is advantageous in evaluating antioxidant effectiveness because it is more stable than hydroxyl and superoxide radicals [33]. The antioxidant potential of hazelnut extracts was evaluated using the stable DPPH radical. This method has been used extensively to predict antioxidant activity because of the relatively short time required for analysis [76]. The DPPH scavenging activities of all extracts as well as catechin at 50 and 100ppm concentrations are shown in Table 13.7 [31]. All hazelnut extracts exhibited fairly effective DPPH radical scavenging activity at both concentrations tested. On the other hand, catechin scavenged DPPH radical nearly completely at both concentrations. Thus, phenolic compounds present may have acted as free radical scavengers by virtue of their hydrogen-donating ability [77]. At 100ppm concentration, hazelnut kernel and hazelnut green leafy cover extracts exhibited stronger DPPH radical scavenging (92.2% and 99.5%, respectively) than that of almond seed and almond green shell (leafy) cover extracts (21% and 35%, respectively), whereas the skin extracts for both nuts were the same [31,33]. Similar to hazelnut and almond by-products, white and black and sesame hulls exhibited significantly ($P < 0.05$) higher DPPH radical activity than their corresponding whole seeds [78].

13.3.3 Retention of β-Carotene–Linoleate Model System

The antioxidant activity of hazelnut extracts, as measured by the β-carotene–linoleate model system, is presented in Figure 13.2. The reference compound, catechin, exhibited a more powerful antioxidant activity than any of the hazelnut extracts. The antioxidant activity in this model system was in the order of catechin > hazelnut skin > hazelnut hard shell > hazelnut tree leaf > hazelnut green leafy cover > hazelnut kernel [31]. Although the same order was obtained at 50ppm level, at 100ppm, hazelnut skin extract exhibited the highest antioxidative activity by retaining β-carotene in the medium, followed by hazelnut hard shell, catechin, hazelnut tree leaf, hazelnut green leafy cover, and hazelnut kernel (Table 13.8). The differences in the activity order between the two different sets of experiments at 100 and 200ppm concentrations (50 and 100ppm) could be explained. Alasalvar et al. [30] found a similar order of antioxidant efficacy when using two solvent systems (80%, v/v, ethanol and 80%, v/v, acetone) for comparing defatted hazelnut kernel and hazelnut green leafy cover. In that study, acetone is found to be a more effective solvent for the extraction process compared to ethanol.

13.3.4 Inhibition of Oxidation of Human LDL Cholesterol

It has now been recognized that oxidation of human LDL cholesterol by free radicals arising from lipid oxidation products maybe involved in the pathogenesis of atherosclerosis, and transition metal

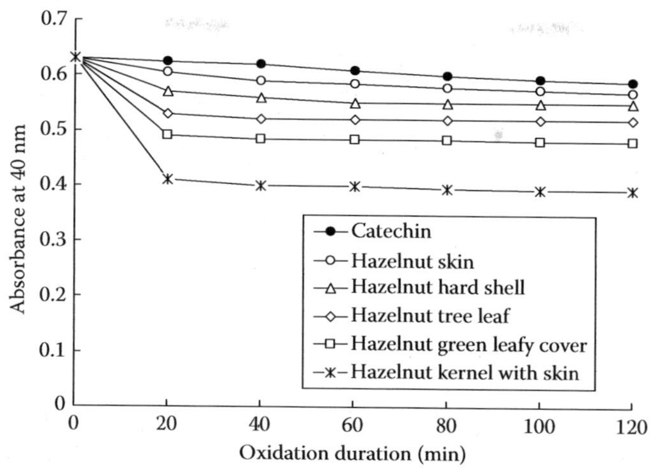

FIGURE 13.2 Antioxidant activity in extracts of hazelnut kernel and hazelnut by-products in the β-carotene–linoleate model system. (From Shahidi, F., Alasalvar, F., and Liyana-Pathirana, C.M., *J. Agric. Food Chem.*, 55, 1212, 2007. With permission.)

ions could promote oxidative modification through interaction with hydroperoxides [79]. It has been suggested that certain phenolic compounds may act as peroxyl radical scavengers, protecting LDL from oxidation. These phenolic compounds may function as chain-breaking antioxidants in inhibiting LDL oxidation [80].

There has been an increased interest in natural antioxidants from plant-derived foods and their by-products for their role in preventing oxidation of human LDL cholesterol. Table 13.8 shows the inhibition of copper-induced human LDL oxidation by hazelnut extracts. Hazelnut skin and hazelnut green leafy cover extracts at 50 ppm concentration effectively inhibited copper-induced oxidation of

TABLE 13.8
Retention of β-Carotene and Inhibition of Oxidation of Human LDL Cholesterol in Extracts of Hazelnut Kernel and Hazelnut By-Products

	Retention of β-Carotene–Linoleate Model System (%)		Inhibition of Oxidation of Human LDL Cholesterol (%)	
Extract	**50 ppm**	**100 ppm**	**50 ppm**	**100 ppm**
Hazelnut kernel (with skin)	62.5 ± 1.1[a]	63.5 ± 1.5[a]	42 ± 2[a]	99 ± 1[a]
Hazelnut skin	83.3 ± 1.2[b]	93.3 ± 2.0[b]	99 ± 1[b]	99 ± 1[a]
Hazelnut hard shell	83.1 ± 0.9[b]	89.1 ± 1.1[c]	56 ± 3[c,d]	99 ± 1[a]
Hazelnut green leafy cover	76.4 ± 1.9[c,e]	76.5 ± 1.8[e]	93 ± 1[e]	99 ± 1[a]
Hazelnut tree leaf	78.5 ± 1.6[e]	83.3 ± 1.0[d]	61 ± 2[c]	99 ± 1[a]
Catechin	83.6 ± 1.1[b]	83.6 ± 1.2[d]	53 ± 3[d]	99 ± 1[a]

Source: From Shahidi, F., Alasalvar, F., and Liyana-Pathirana, C.M., *J. Agric. Food Chem.*, 55, 1212, 2007. With permission.

Note: Data are expressed as means ± SD ($n = 3$) on an extract.

[a–e] Means ± SD followed by the same letter, within a column, are not significantly different ($P > 0.05$).

human LDL cholesterol (99% and 93%, respectively) compared to hazelnut kernel (42%), hazelnut hard shell (56%), and hazelnut tree leaf (61%) extracts, which reached the same level of efficacy (99%) at 100ppm. At 50ppm level, all hazelnut extracts, except hazelnut kernel, were far more effective in inhibiting human LDL oxidation than the standard catechin (53%). At 100ppm, catechin exhibited 99% inhibition, the same as that shown by each of the hazelnut extracts. Differences in the solubility and partitioning between aqueous and lipid phases in the LDL system are among factors responsible for the observed trends. Similar to this study, Wijeratne et al. [81] found that brown skin of almond exerted the highest preventive effect against LDL oxidation at 10, 50, and 100ppm levels compared to those of whole almond and its green shell (leafy) cover. At 200ppm, all extracts exerted the same effects. Kinsella et al. [82] reported the importance of dietary antioxidants in the inhibition of LDL cholesterol oxidation, thereby reducing the risk of atherosclerosis and CHD. Dietary antioxidants, including those from hazelnut extracts, may therefore moderate risk factors involved in CHD.

13.3.5 Inhibition of Supercoiled DNA Strand Breakage by Hydroxyl Radical

Hydroxyl radical-induced DNA single-strand breaks maybe a better index in the evaluation of the effects of phenolic compounds against hydroxyl radical [83]. Oxidative damage of DNA results in strand breakage and sister chromatid exchange, DNA–DNA and DNA–protein cross-linking, and base modification [84]. Damage to DNA can lead to mutation and cancer [85]. The effect of hazelnut extracts on DNA single-strand breaks, induced by Fenton reagent, has been examined by Shahidi et al. [31] and summarized in Table 13.9. Hazelnut skin extract exhibited the highest inhibition while hazelnut kernel extract showed the lowest effect at the four concentrations tested (5, 10, 25, and 50ppm). Extracts from by-products (skin, hard shell, green leafy cover, and tree leaf) displayed a stronger inhibition ($P < 0.05$) than hazelnut kernel in most cases. As the concentrations of antioxidative hazelnut extracts were increased, the protective effects against nicking of supercoiled DNA also increased (Table 13.9). The inhibitory effects of hazelnut extracts maybe attributed to their ability to scavenge hydroxyl radical. Hence, hazelnut products may participate in cancer prevention. Wijeratne et al. [81] investigated the inhibition of peroxyl- and hydroxyl radical-induced DNA scission of almond whole seed, brown skin, and green shell cover extracts between 2 and 100ppm levels. The authors found that green shell cover extract at 50ppm level completely arrested peroxyl radical-induced DNA scission, whereas 100ppm of brown skin and whole seed extracts was required for a similar effect. On the other hand, for hydroxyl radical-induced DNA strand scission, all three

TABLE 13.9
Retention (%) of Supercoiled DNA in Extracts of Hazelnut Kernel and Hazelnut By-Products in Free Radical Induced Strand Scission

Extract	5ppm	10ppm	25ppm	50ppm
Hazelnut kernel (with skin)	33.3 ± 1.9^{a}	39.6 ± 1.2^{a}	$53.4 \pm 1.7^{a,e}$	59.2 ± 2.1^{a}
Hazelnut skin	64.7 ± 2.7^{b}	73.2 ± 3.3^{b}	90.7 ± 0.9^{b}	95.4 ± 2.5^{b}
Hazelnut hard shell	48.1 ± 2.2^{c}	68.7 ± 3.7^{b}	86.3 ± 2.4^{c}	$94.7 \pm 1.9^{b,c}$
Hazelnut green leafy cover	$44.2 \pm 1.9^{c,d}$	54.4 ± 1.4^{c}	$83.0 \pm 2.5^{c,d}$	89.9 ± 1.3^{d}
Hazelnut tree leaf	38.9 ± 1.3^{e}	$45.2 \pm 0.9^{d,e}$	56.1 ± 1.4^{e}	65.7 ± 2.4^{e}
Catechin	33.1 ± 2.0^{a}	44.2 ± 2.1^{e}	51.8 ± 1.2^{a}	60.3 ± 1.7^{a}

Source: From Shahidi, F., Alasalvar, F., and Liyana-Pathirana, C.M., *J. Agric. Food Chem.*, 55, 1212, 2007. With permission.

Note: Data are expressed as means ± SD ($n = 3$) on an extract.

[a–e] Means ± SD followed by the same letter, within a column, are not significantly different ($P > 0.05$).

almond extracts exerted a total protection at 50 ppm against both site-specific and nonsite-specific strand scissions.

In the absence of any antioxidant, it maybe expected that a peroxyl radical abstracts a hydrogen atom from a nearby DNA to generate new radicals, which in turn evokes a free radical chain reaction resulting in the destruction of DNA molecules. However, in the presence of antioxidants, this chain reaction is terminated by abstracting a hydrogen atom from the antioxidant molecule [86].

13.3.6 Reducing Power

The main outcome of the reducing reaction is termination of radical chain reactions that may otherwise be very damaging [87]. Figure 13.3 depicts the reducing power of the extracts of hazelnut kernel and hazelnut green leafy cover [30]. At the same dose, the reducing power of green leafy cover was superior to that of hazelnut kernel. The lower reducing power of hazelnut kernel extract is due to a lower content of phenolic acids than those present in the extract of hazelnut green leafy cover extract (Table 13.2). Thus, phenolics present in the extracts of green leafy cover displayed a considerable reducing power, primarily due to their effect as electron donors and thereby suppressing radical chain reactions by converting free radicals to more stable products. Alasalvar et al. [30] reported that the reducing power of hazelnut kernel was higher than that of almond seed extract [88].

13.4 TAXANES IN HAZELNUT BY-PRODUCTS

Paclitaxel, an active ingredient in Taxol (marketed by Bristol Myers Squibb), has been approved by the U.S. FDA for the treatment of ovarian cancer, breast cancer, nonsmall cell lung cancer (NSCLC), and acquired immunodeficiency syndrome (AIDS)-related Kaposi's sarcoma [89–91]. Taxol is currently one of the largest selling cancer drugs worldwide. It is an effective inhibitor of cell division by interfering with cellular microtubule dissolution. Figure 13.4 shows the structure of tree common taxanes (A, Paclitaxel; B, Cephalomannine; C, Baccatin III). These compounds are known as "secondary metabolites" since they are not necessary for the plant growth, but aid in its defense against environmental and other external factors.

Pacific yew (*Taxus brevifolia*) bark, the original source for this drug [92], has been supplemented with other yew sources such as bark, needles, and roots of *T. media* x *Hicksii, T. canadensis*, and *T. cuspidata*. Semisynthesis of paclitaxel maybe accomplished using baccatin III or 10-deacetyl baccatin III from *T. baccata*, but full laboratory synthesis is too expensive for commercial uses.

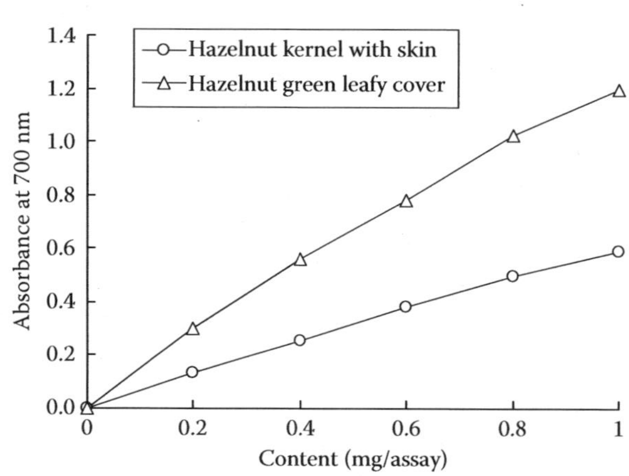

FIGURE 13.3 Reducing power of hazelnut kernel and hazelnut green leafy cover. (From Alasalvar, C., Karamać, M., Amarowicz, R., and Shahidi, F., *J. Agric. Food Chem.*, 54, 4826, 2006. With permission.)

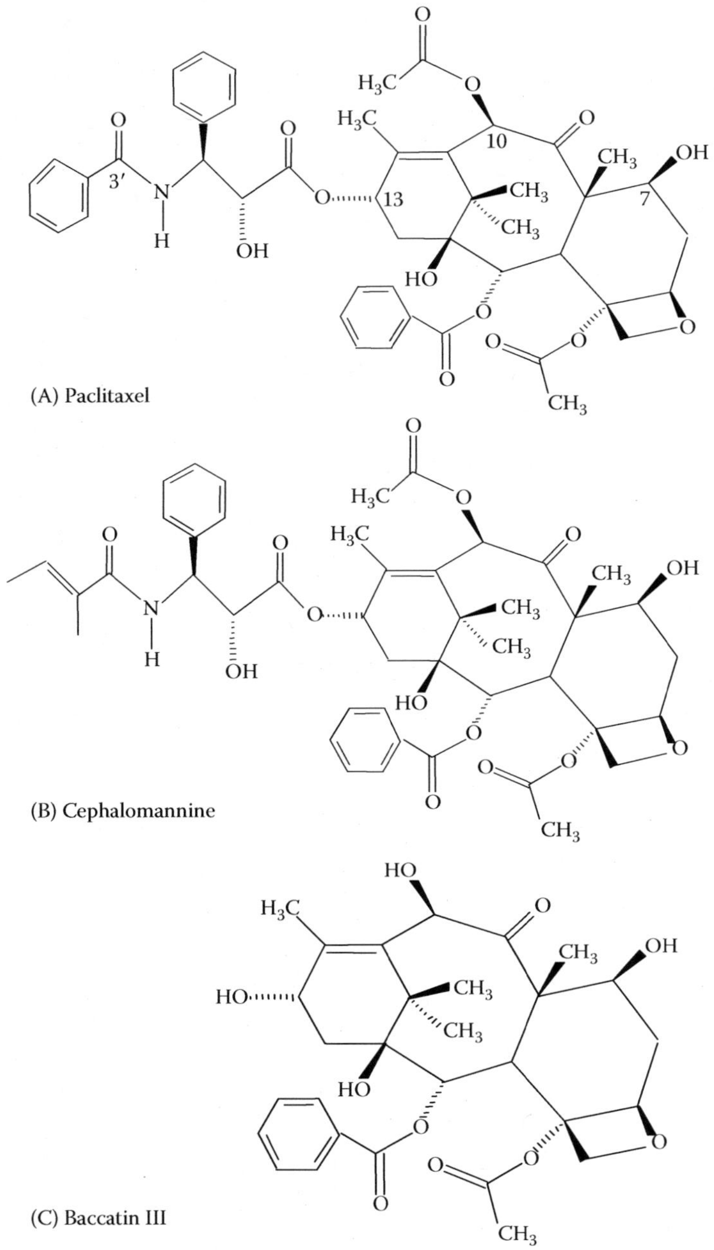

FIGURE 13.4 Structure of three common taxanes: A, Paclitaxel; B, Cephalomannine; C, Baccatin III.

While the supply of Taxol is generally meeting demand for currently approved cancer treatments and clinical trials, researchers are discovering an increasing number of other medical applications that will increase demand. Paclitaxel has shown promise for the treatment of psoriasis, polycystic kidney disease, multiple sclerosis, and Alzheimer's, among others [93].

An alternative source could stimulate competition among drug manufacturers, which may also decrease drug prices. Hoffman et al. [93,94] unexpectedly discovered paclitaxel in all parts of hazelnut trees (in the bark, leaves, limbs, shells, and kernel). The hazelnut trees also produced a host of related taxanes that chemists can convert into the pharmaceuticals. The function of paclitaxel in plants is not known, but there is evidence that it has some antifungal properties [93] and maybe a protective agent against stress [95]. Both the Pacific yew and the hazelnut thrive in humid climates, and this suggests that paclitaxel maybe expected in plants from humid areas.

Certain species of hazelnut trees produce as much as one-tenth of the paclitaxel found in the Pacific yew trees. For every dry weight gram of bark of the Pacific yew, about 50 to 70 μg of paclitaxel can be extracted, while in branches and leaves of the hazelnuts, about 5 to 7 μg of paclitaxel can be extracted per gram dry weight. Hazelnut kernel (raw) and shells have also been reported to contain small amounts of paclitaxel (<0.05 μg/g extract) [93].

Hoffman et al. [96] found that microbe-free cultures of these plants as well as fungi associated with branches of these plants are able to produce paclitaxel and several other taxanes responsible for the production of paclitaxel and related taxanes in hazelnut trees. Similar arrangements of endophytic fungi have been noted in the Pacific yew [97,98]. All the fungi that make paclitaxel are associated with healthy plants.

Small amounts of paclitaxel were also isolated from Turkish Tombul hazelnut tree leaf (0.05 μg/g extract) and hard shell (0.08 μg/g extract). In addition to paclitaxel, 10 deacetyl baccatin III and cephalomannine (leaf and hard shell) and baccatin III (green leafy cover) were found in the extracts (Table 13.10). Taxanes were not detected in the extracts of hazelnut kernel and hazelnut skin from Turkish Tombul cultivar.

Despite small amounts of taxanes present, hazelnut by-products could have an industrial value for the production of paclitaxel [99]. The reason behind this is that the hazelnut trees grow much faster than the Pacific yew (a slow-growing plant in limited quantities in the Pacific Northwest). This could reduce the cost of the commercial drug and make it more readily available. Commercial supplies of Taxol are now manufactured by a semisynthetic method that relies on extracts from leaves of other yew species. Although paclitaxel has been synthesized artificially in the laboratory without using any yew parts, this method is currently too complex and expensive to implement commercially.

13.5 CONCLUSION

Hazelnut compared with other tree nuts is an excellent source of proanthocyanidins and a good source of total phenolics, total antioxidant activity, and flavonoids. Hazelnut by-products, which are rich sources of phytochemicals, could provide potential health benefits. They exhibit stronger antioxidant activities compared to that of hazelnut (kernel) and could potentially be considered as an excellent and readily available source of natural antioxidants and used as functional food ingredients and nutraceuticals in a variety of food products. Among hazelnut by-products, hazelnut skin, in general, shows superior antioxidant efficacy and higher total phenolic content as compared to other hazelnut tissues. Therefore, hazelnut skin could be considered as possible industrial source for antioxidants. Some hazelnut by-products, which contain paclitaxel (an active cancer ingredient in Taxol drug), can be considered as an alternative source of this cancer drug, although the concentrations are low.

TABLE 13.10
Taxanes (μg/g Extract) in Extracts of Hazelnut By-Products

Extract	10d B III	B III	10d T	Ceph	7,10d T	Paclitaxel
Hazelnut tree leaf	0.47 ± 0.13	nd	nd	0.01 ± 0.00	nd	0.05 ± 0.01
Hazelnut hard shell	0.85 ± 0.14	nd	nd	1.73 ± 0.47	nd	0.08 ± 0.01
Hazelnut green leafy cover	nd	1.89 ± 0.04	nd	nd	nd	nd

Note: **10d B III**, 10 deacetyl baccatin III; **B III**, baccatin III; **10d T**, 10 deacetyl taxol; **Ceph**, cephalomannine; **7, 10d T**, 7 epi 10 deacetyl taxol; Data are expressed as means ± SD (n = 2) on an extract; nd, not detected.

REFERENCES

1. Surh, Y.-J., Cancer chemoprevention with dietary phytochemicals, *Nat. Rev. Cancer*, 3, 768–780, 2003.
2. Gu, L., Kelm, M.A., Hammerstone, J.F., Beecher, G., Holden, J., Haytowitz, D., and Prior, R.L., Screening of foods containing proanthocyanidins and their structural characterization using LC-MS/MS and thiolytic degradation, *J. Agric. Food Chem.*, 51, 7513–7521, 2003.
3. Shahidi, F. and Naczk, M., *Phenolics in Food and Nutraceuticals*, CRC Press, Boca Raton, FL, 2004.
4. Wu, X., Beecher, G.R., Holden, J.M., Haytowitz, D.B., Gebhardt, S.E., and Prior, R.L., Lipophilic and hydrophilic antioxidant capacities of common foods in the United States, *J. Agric. Food Chem.*, 52, 4026–4037, 2004.
5. Wu, X., Beecher, G.R., Holden, J.M., Haytowitz, D.B., Gebhardt, S.E., and Prior, R.L., Concentrations of anthocyanins in common foods in the United States and estimation of normal consumption, *J. Agric. Food Chem.*, 54, 4069–4075, 2006.
6. U.S. Department of Agriculture (USDA), Database for the Proanthocyanidin Content of Selected Foods, August 2004, Published online at: http://www.nal.usda.gov/fnic/foodcomp (accessed September 28, 2006).
7. U.S. Department of Agriculture (USDA), Database for the Flavonoid Content of Selected Foods (Release 2), August 2006, Published online at: http://www.ars.usda.gov/nutrientdata (accessed September 28, 2006).
8. Shahidi, F. and Ho, C.-T., Phenolics in food and natural health products: An overview, in *Phenolic Compounds in Foods and Natural Products*, Shahidi, F. and Ho, C.-T., Eds., ACS Symposium Series 909, American Chemical Society, Washington, DC, 2005, pp. 1–8.
9. Rice-Evans, C.A., Miller, N.J., Bolwell, P.G., Bramley, P.M., and Pridham, J.B., The relative antioxidant activities of plant-derived polyphenolic flavonoids, *Free Radical Res.*, 22, 375–383, 1995.
10. Salah, N., Miller, N.J., Paganga, G., Tijburg, L., Bolwell, G.P., and Rice-Evans, C.A., Polyphenolic flavanols as scavengers of aqueous phase radicals and as chain-breaking antioxidants, *Arch. Biochem. Biophys.*, 322, 339–346, 1995.
11. Yang, J., Halim, L., and Liu, R.H., Antioxidant and antiproliferative activities of common nuts, presented at the Institute of Food Technologists Annual Meeting and Food Expo, New Orleans, LA, July 16–20, 2005, Abstract 035-05.
12. Hertog, M.G.L., Feskens, E.J.M., Hollman, P.C.H., Katan, M.B., and Kromhout, D., Dietary antioxidant flavonoids and risk of coronary heart disease: The Zutphen elderly study, *Lancet*, 342, 1007–1011, 1993.
13. Bloch, A. and Thomson, C.A., Position of the American Dietetic Association: Phytochemicals and functional foods, *J. Am. Dent. Assoc.*, 95, 493–496, 1995.
14. Ness, A.R. and Powles, J.W., Fruit and vegetables, and cardiovascular disease: A review, *Int. J. Epidemiol.*, 26, 1–13, 1997.
15. Joseph, J.A., Shukitt-Hale, B., Denisova, N.A., Bielinski, D., Martin, A., McEwen, J.J., and Bickford, P.C., Reversals of age-related declines in neuronal signal transduction, cognitive, and motor behavioral deficits with blueberry, spinach, or strawberry dietary supplementation, *J. Neurosci.*, 19, 8114–8121, 1999.
16. Mazza, G., Fukumoto, L., Delaquis, P., Girard, B., and Ewert, B., Anthocyanins, phenolics, and color of Cabernet Franc, Merlot, and Pinot Noir wines from British Colombia, *J. Agric. Food Chem.*, 47, 4009–4017, 1999.
17. Dillard, C.J. and German, J.B., Phytochemicals: Nutraceuticals and human health, *J. Sci. Food Agric.*, 80, 1744–1756, 2000.
18. Watson, R.R., *Functional Foods and Nutraceuticals in Cancer Prevention*, Blackwell Publishing, Oxford, 2003.
19. Wien, M.A., Sabaté, J.M., Iklé, D.N., Coles, S.E., Kandeel, F.R., Almonds vs complex carbohydrates in a weight reduction program, *Int. J. Obes. Relat. Metab. Disord.*, 27, 1365–1372, 2003.
20. Tapsell, L.C., Gillen, L.J., Patch, C.S., Batterham, M., Owen, A., Bare, M., and Kennedy, M., Including walnuts in a low-fat/modified-fat diet improves HDL cholesterol-to-total cholesterol ratios in patients with type 2 diabetes, *Diabetes Care*, 27, 2777–2783, 2004.
21. Cortés, B., Núñez, I., Cofán, M., Gilabert, R., Pérez-Heras, A., Casals, E., Deulofeu, R., and Ros, E., Acute effects of high-fat meals enriched with walnuts or olive oil on postprandial endothelial function, *J. Am. Coll. Cardiol.*, 48, 1666–1671, 2006.
22. Clifford, M.N. and Brown, J.E., Dietary flavonoids and health-broadening the perspective, in *Flavonoids: Chemistry, Biochemistry and Applications*, Andersen, Ø.M. and Markham, K.R., Eds., CRC Press, Taylor & Francis Group, Boca Raton, FL, 2005, pp. 319–370.

23. Kelly, J.H. and Sabaté, J., Nuts and coronary heart disease: An epidemiological perspective, *Br. J. Nutr.*, 96 (Suppl. 2), 61S–67S, 2006.
24. Salas-Salvadó, J., Casas-Agustench, P., Murphy, M.M., López-Uriarte, P., and Bulló, M., The effect of nuts on inflammation, *Asia Pac. J. Clin. Nutr.*, 17 (Suppl. 1), 333S–336S, 2008.
25. Turkish Hazelnut Exporter's Union, *The Miracle Nut: Hazelnut*, Hazelnut Promotion Group, Giresun, Turkey, 2007.
26. Yurttas, H.C., Schafer, H.W., and Warthesen, J.J., Antioxidant activity of nontocopherol hazelnut (*Corylus spp.*) phenolics, *J. Food Sci.*, 65, 276–280, 2000.
27. Moure, A., Franco, D., Sineiro, J., Domínguez, H., Núñez, M.J., and Lema, J.M., Antioxidant activity of extracts from *Gevuina avellana* and *Rosa rubiginosa* defatted seeds, *Food Res. Int.*, 34, 103–109, 2001.
28. Parry, J.W., Su, L., Moore, J., Luther, M., and Yu, L.L., Chemical composition, antioxidant properties, and cancer cell growth inhibitory effects of Turkish and Oregon roasted hazelnuts, presented at the Institute of Food Technologists Annual Meeting, Orlando, FL, June 24–28, 2006, Abstract 003E-30.
29. Amaral, J.S., Ferreres, F., Andrade, P.B., Valentão, P., Pinheiro, C., Santos, A., and Seabra, R., Phenolic profile of hazelnut (*Corylus avellana* L.) leaves cultivars grown in Portugal, *Nat. Prod. Res.*, 19, 157–163, 2005.
30. Alasalvar, C., Karamać, M., Amarowicz, R., and Shahidi, F., Antioxidant and antiradical activities in extracts of hazelnut kernel (*Corylus avellana* L.) and hazelnut green leafy cover, *J. Agric. Food Chem.*, 54, 4826–4832, 2006.
31. Shahidi, F., Alasalvar, F., and Liyana-Pathirana, C.M., Antioxidant phytochemicals in hazelnut kernel (*Corylus avellana* L.) and hazelnut byproducts, *J. Agric. Food Chem.*, 55, 1212–1220, 2007.
32. Pszczola, D., Health and functionality in a nutshell, *Food Technol.*, 54, 54–59, 2000.
33. Siriwardhana, S.S.K.W. and Shahidi, F., Antiradical activity of extracts of almond and its by-products, *J. Am. Oil Chem. Soc.*, 79, 903–908, 2002.
34. Senter, S.D., Horvat, R.J., and Forbus, W.R., Comparative GLC-MS analysis of phenolic acids of selected tree nuts, *J. Food Sci.*, 48, 798–799, 824, 1983.
35. Cuvelier, M.E., Richard, H., and Berset, C., Comparison of the antioxidant activity of some acid-phenols: Structure–activity relationship, *Biosci. Biotech. Biochem.*, 56, 324–325, 1992.
36. Kuhnau, J., The flavonoids: A class of semi-essential food components: Their role in human nutrition, *World Rev. Nutr. Diet.*, 24, 117–191, 1976.
37. Keli, S.O., Hertog, M.G., Feskens, E.J., and Kromhout, D., Dietary flavonoids, antioxidant vitamins, and incidence of stroke: The Zutphen Study, *Arch. Intern. Med.*, 156, 637–642, 1996.
38. Neuhouser, M.L., Flavonoids and cancer prevention: What is the evidence in humans? *Pharm. Biol.*, 42, 36–45, 2004.
39. Wang, H., Cao, G., and Prior, R.L., Total antioxidant capacity of fruits, *J. Agric. Food Chem.*, 44, 701–705, 1996.
40. Pietta, P.-G., Flavonoids as antioxidants, *J. Nat. Prod.*, 63, 1035–1042, 2000.
41. Pratt, D.E. and Hudson, B.J.F., Natural antioxidants not exploited commercially, in *Food Antioxidant*, Hudson, B.J.F., Ed., Elsevier Applied Science, London, 1990, pp. 171–192.
42. van Acker, S.A.B.E., de Groot, M.J., van der Berg, D.-J., Tromp, M.N.J.L., den Kelder, G.D.-O., van der Vijgh, W.J.F., and Bast, A., A quantum chemicals explanation of the antioxidant activity of flavonoids, *Chem. Res. Toxicol.*, 9, 1305–1312, 1996.
43. Vinson, J.A. and Dabbagh, Y.A., Effect of green and black tea supplementation on lipids, lipid oxidation and fibrinogen in the hamster: Mechanisms for the epidemiological benefits of tea drinking, *FEBS Lett.*, 433, 44–46, 1998.
44. Fuhrman, B., Lavy, A., and Aviram, M., Consumption of red wine with meals reduces the susceptibility of human plasma and low-density lipoprotein to lipid oxidation, *Am. J. Clin. Nutr.*, 61, 549–554, 1995.
45. Renaud, S. and de Lorgeril, M., Wine, alcohol, platelets, and the French paradox for coronary hearth disease, *Lancet*, 339, 1523–1526, 1992.
46. Fitzpatrick, D.F., Hirschfield, S.L., and Coffey, R.G., Endothelium-dependent vasorelaxing activity of wine and other grape products, *Am. J. Physiol. Heart Circ. Physiol.*, 265, H774-H778, 1993.
47. Khanbabaee, K. and van Ree, T., Tannins: Classification and definition, *Nat. Prod. Rep.*, 18, 641–649, 2001.
48. Santos-Buelga, C. and Scalbert, A., Proanthocyanidins and tannin-like compounds—Nature, occurrence, dietary intake and effects on nutrition and health, *J. Sci. Food Agric.*, 80, 1094–1117, 2000.
49. Gu, L., Kelm, M.A., Hammerstone, J.F., Beecher, G., Holden, J., Haytowitz, D., Gebhardt, S., and Prior, R.L., Concentrations of proanthocyanidins in common foods and estimations of normal consumption, *J. Nutr.*, 134, 613–617, 2004.

50. Chung, K.-T., Wong, T.Y., Wei, C.-I.Y., Huang, Y.-W.Y., and Lin, Y.Y., Tannins and human health: A review, *Crit. Rev. Food Sci. Nutr.*, 38, 421–464, 1998.
51. Hagerman, A.E., Riedl, K.M., Alexander Jones, G., Sovik, K.N., Ritchard, N.T., Hartzfeld, P.W., and Riechel, T.L., High molecular weight plant polyphenolics (tannins) as biological antioxidants, *J. Agric. Food Chem.*, 46, 1887–1892, 1998.
52. Beecher, G.R., Proanthocyanidins: Biological activities associated with human health, *Pharm. Biol.*, 42, 2–20, 2004.
53. Reed, J., Cranberry flavonoids, atherosclerosis and cardiovascular health, *Crit. Rev. Food Sci. Nutr.*, 42 (Suppl. 3), 301S–316S, 2002.
54. Steinberg, F.M., Bearden, M.M., and Keen, C.L., Cacao and chocolate flavonoids: Implications for cardiovascular health, *J. Am. Diet. Assoc.*, 103, 215–223, 2003.
55. Bagchi, D., Bagchi, M., Stohs, S.J., Das, D.K., Ray, S.D., Kuszynski, C.A., Joshi, S.S., and Pruess, H.H., Free radicals and grape seed proanthocyanidin extract: Importance in human health and disease prevention, *Toxicology*, 148, 187–197, 2000.
56. Murphy, K.J., Chronopoulos, A.K., Singh, I., Francis, M.A., Moriarty, H., Pike, M.J., Turner, A.H., Mann, N.J., and Sinclair, A., Dietary flavanols and procyanidin oligomers from cocoa (*Theobroma cacao*) inhibit platelet function, *Am. J. Clin. Nutr.*, 77, 1466–1473, 2003.
57. Foo, L., Lu, Y., Howell, A., and Vorsa, N., A-type proanthocyanidin trimers from cranberry that inhibit adherence of uropathogenic p-fimbriated *Escherichia coli*, *J. Nat. Pro.*, 63, 1225–1228, 2000.
58. Venkatachalam, M. and Sathe, S.K., Chemical composition of selected edible nut seeds, *J. Agric. Food Chem.*, 54, 4705–4714, 2006.
59. Adlercreutz, H., Phytoestrogens and breast cancer, *J. Steroid Biochem. Mol. Biol.*, 83, 113–118, 2002.
60. Anthony, M.S., Phytoestrogens and cardiovascular disease: Where's the meat? *Arterioscler. Thromb. Vasc. Biol.*, 22, 1245–1247, 2002.
61. Wood, C.E., Barnes, S., and Cline, J.M., Phytoestrogen actions in the breast and uterus, in *Phytoestrogens and Health*, Gilani, G.S. and Anderson, J.J.B., Eds., AOCS Press, Champaign, IL, 2002, pp. 440–469.
62. Wu, A.H., Phytoestrogens and cancer: Epidemiologic evidence, in *Phytoestrogens and Health*, Gilani, G.S. and Anderson, J.J.B., Eds., AOCS Press, Champaign, IL, 2002, pp. 371–404.
63. Adlercreutz, H., Heinonen, S.M., and Penalvo-Garcia, J., Phytoestrogens, cancer and coronary heart disease, *BioFactors*, 22, 229–236, 2004.
64. Cornwell, T., Cohick, W., and Raskin, I., Dietary phytoestrogens and health, *Phytochemistry*, 65, 995–1016, 2004.
65. Magee, P.J. and Rowland, I.R., Phyto-etsrogens, their mechanism of action: Current evidence for a role in breast and prostate cancer, *Br. J. Nutr.*, 91, 513–531, 2004.
66. Webb, A.L. and McCullough, M.L., Dietary lignans: Potential role in cancer prevention, *Nutr. Cancer*, 51, 117–131, 2005.
67. Thompson, L.U., Flaxseed, lignans and cancer, in *Flaxseed in Human Nutrition*, 2nd ed., Thompson, L.U. and Cunnane, S.C., Eds., AOCS Press, Champaign, IL, 2003, pp. 194–222.
68. Gültekin, E. and Yıldız, F., Introduction to phytoestrogens, in *Phytoestrogens in Functional Foods*, Yıldız, F., Ed., CRC Press, Taylor and Francis Group, Boca Raton, FL, 2006, pp. 3–18.
69. Thompson, L.U., Boucher, B.A., Liu, Z., Cotterchio, M., and Kreiger, N., Phytoestrogen content of foods consumption in Canada, including isoflavones, lignans, and coumestan, *Nutr. Cancer*, 54, 184–201, 2006.
70. Decker, E.A., Warner, K., Richards, M.P., and Shahidi, F., Measuring antioxidant effectiveness in food, *J. Agric. Food Chem.*, 53, 4303–4310, 2005.
71. Shahidi, F., Phytochemicals, bioactives and natural antioxidants, presented at 2nd International Congress on Functional Foods and Nutraceuticals, Istanbul, Turkey, May 4–6, 2006, Abstract PL 06005.
72. Chu, Y.-H., Chang, C.-L., and Hsu, H.-F., Flavonoid content of several vegetables and their antioxidant activity, *J. Sci. Food Agric.*, 80, 561–566, 2000.
73. Prior, R.L., Wu, X., and Schaich, K., Standardized methods for the determination of antioxidant capacity and phenolics in foods and dietary supplements, *J. Agric. Food Chem.*, 53, 4290–4302, 2005.
74. Yoshikawa, T., Naito, Y., and Kondo, M., Free radicals and diseases, in *Food and Free Radicals*, Hiramastu, M., Yoshikawa, T., and Inoue, M., Eds., Plenum Press, New York, 1997, pp. 11–19.
75. Niki, E., Free radicals in chemistry and biochemistry, in *Food and Free Radicals*, Hiramastu, M., Yoshikawa, T., and Inoue, M., Eds., Plenum Press, New York, 1997, pp. 1–10.

76. Chen, Y., Wang, M., Rosen, R.T., and Ho, C.-T., 2,2-Diphenyl-1-picrylhydrazyl radical-scavenging active components from *Polygonum multiflorum* Thunb, *J. Agric. Food Chem.*, 47, 2226–2228, 1999.
77. Castelluccio, C., Bolwell, G.P., Gerrish, C., and Rice-Evans, C., Differential distribution of ferulic acid to the major plasma constituents in relation to its potential as an antioxidant, *Biochem. J.*, 316, 691–694, 1996.
78. Shahidi, F. and Liyana-Pathirana, C.M., Antioxidant activity of sesame fractions, in *Phenolic Compounds in Foods and Natural Products*, Shahidi, F. and Ho, C.-T., Eds., ACS Symposium Series 909, American Chemical Society, Washington, DC, 2005, pp. 33–45.
79. Decker, E.A., Ivanov, V., Zhu, B.-Z., and Frei, B., Inhibition of low density lipoprotein oxidation by carnosine and histidine, *J. Agric. Food Chem.*, 49, 511–516, 2001.
80. Castelluccio, C., Paganga, G., Melikian, N., Bolwell, G.P., Pridham, J., Sampson, J., and Rice-Evans, C., Antioxidant potential of intermediates in phenylpropanoid metabolism in higher plants, *FEBS Lett.*, 368, 188–192, 1995.
81. Wijeratne, S.S.K., Abou-Zaid, M.M., and Shahidi, F., Antioxidant polyphenols in almond and its coproducts, *J. Agric. Food Chem.*, 54, 312–318, 2006.
82. Kinsella, J.E., Frankel, E., German, B., and Kanner, J., Possible mechanisms for the protective role of antioxidants in wine and plant foods, *Food Technol.*, 47, 85–89, 1993.
83. Hiramoto, K., Kido, K., and Kikugawa, K., DNA breaking by maillard products of glucose–amino acid mixtures formed in an aqueous system, *J. Agric. Food Chem.*, 42, 689–694, 1994.
84. Dizdaroglu, M., Chemistry and free radical damage to DNA and nucleoproteins, in *DNA and Free Radicals*, Halliwell, B. and Aruoma, O.I., Eds., Ellis Harwood, Chichester, United Kingdom, 1993, pp. 19–39.
85. Ames, B.N. and Shigenaga, M.K., Oxidants are a major contributor to cancer and aging, in *DNA and Free Radicals*, Halliwell, B. and Aruoma, O.I., Eds., Ellis Horwood Ltd., Chichester, United Kingdom, 1993, pp. 1–15.
86. Hu, C. and Kitts, D.D., Studies on the antioxidant activity of *Echinacea* root extract, *J. Agric. Food Chem.*, 48, 1466–1472, 2000.
87. Yen, W.J. and Chen, B.H., Isolation of xanthophylls from Taiwanese orange peels and their effects on the oxidation stability of soybean oil, *Food Chem.*, 53, 417–425, 1995.
88. Amarowicz, R., Troszyñska, A., and Shahidi, F., Antioxidant activity of almond seed extract and its fractions, *J. Food Lipids*, 12, 344–358, 2005.
89. Chesnoff, S., The use of Taxol as a trademark, *Nature*, 374, 208, 1995.
90. Kingston, D.G.I., Taxol, a molecule for all seasons, *Chem. Commun.*, 867–880, 2001.
91. Mann, J., Natural products in cancer chemotheraphy: Past, present and future, *Nat. Rev. Cancer*, 2, 143–148, 2002.
92. Wani, M.C., Taylor, H.L., Wall, M.E., Coggon, P., and McPhail, A.T., Plant antitumor agents. VI. The isolation and structure of taxol, a novel antileukemic and antitumor agent from *Taxus brevifolia*, *J. Am. Chem. Soc.*, 93, 2325–2327, 1971.
93. Hoffman, A.M., Ross, M., Weigandt, H., Gulati, H., and Khan, W., Hazelnut (filbert), a new source for taxanes, presented at 219th ACS National Meeting, San Francisco, CA, March 26–30, 2000, Abstract BIOL 127.
94. Hoffman, A.M., Khan, W., Worapong, J., Strobel, G., Griffin, D., Arbogast, B., Barofsky, D., Boone, R.B., Ning, L., Zheng, P., and Daley, L., Bioprospecting for Taxol in angiosperm plant extracts: Using high performance liquid chromatography–thermospray mass spectrometry to detect the anticancer agent and its related metabolites in filbert trees, *Spectroscopy*, 13, 22–32, 1998.
95. Hoffman, A., Shock, C., and Feibert, E., Taxane and ABA production in yew under different soil water regimes, *HortScience*, 34, 882–885, 1999.
96. Hoffman, A.M., Ross, K., Weidandt, H., Gulati, W., and Khan, E., Feibert, hazelnut (Filbert): A new source for taxanes, NNGA 92nd Annual Report, 29–34, 2002.
97. Stierle, A., Strobel, G., and Stierle, D., Taxol and taxane production by *Taxomyces andreanae*, an endophytic fungus of Pacific yew, *Science*, 260, 214–216, 1993.
98. Stierle, A., Strobel, G., Stierle, D., Grothaus, P., and Bignami, G., The search for a taxol-producing microorganism among the endophytic fungi of the Pacific yew, *Taxus brevifolia*, *J. Nat. Prod.*, 58, 1315–1324, 1995.
99. Hoffman, A.M. and Shahidi, F., Paclitaxal and other taxanes in hazelnut, *J. Func. Foods*, 2009 (in press).
100. Contini, M., Baccelloni, S., Massantini, R., and Anelli, G., Extraction of natural antioxidants from hazelnut (*Corylus avellana* L.) shell and skin wastes by long maceration at room temperature, *Food Chem.*, 110, 659–669, 2008.

14 Phytochemical Profiles and Potential Health Benefits of Heartnut (*Juglans ailanthifolia* var. *cordiformis*): A Comparison with the Common Walnut (*Juglans regia* L.)

Rong Tsao and Li Li

CONTENTS

14.1 INTRODUCTION

While most consumers are not aware of its existence, heartnut (*Juglans ailanthifolia* var. *cordiformis)* has become a recent favor of the North American nut growers and nut enthusiasts in the Great Lakes Region of Canada and the United States for its sweeter taste and natural heart-shape. Heartnut is a naturally occurring genetic oddity of the more common Japanese walnut (*J. ailanthifolia*). Its hardy characteristic and the rapid increase in commercial production may bring to the region a great economic benefit [1].

Unlike the common walnut, there is virtually no information available in peer-reviewed literature about the nutritional composition and value of heartnut. The polyphenolic and lipid compositions of the heartnut and how they contribute to the total antioxidant activities have been recently reported [2,3]. This chapter provides a more comprehensive overview about this special tree nut and its potential as a healthy food for consumers and a value-added new crop for growers. Due to lack of information about heartnut, literature reviews on the nutritional composition will mainly be on the common walnut. Furthermore, we will focus our discussions on the phenolic, tocopherol, and fatty acid contents and their contribution to human health through cholesterol reduction and antioxidative properties.

14.2 HEARTNUT TREE

Heartnut (*J. ailanthifolia* var. *cordiformis*) tree is not native to North America, but originally came from Japan. It is well suited, however, to the Great Lakes fruit growing regions (zone 6), where the climate is similar to Japan. The Japanese walnut (*J. ailanthifolia*) is said to have been introduced to North America from Japan around 1870 by a nurseryman in San Jose, California [4]. From this and other subsequent introductions, a considerable number of heartnut trees have been grown and distributed in the United States and Canada, particularly in the states of Pennsylvania, New York, Connecticut, New Jersey, and Delaware, and in many provinces of Canada. However, the Great Lakes region of Canada, especially southern Ontario in the Niagara region and along the north shores of Lake Erie and Lake Ontario, is perhaps one of the only places where commercial production takes place. Ontario produces the most of the heartnut grown in Canada [5].

The heartnut tree is best suited to well-drained fertile sandy and clay loamy soils with a pH of 6 to 7, growing 50 to 100 cm or more in a year, reaching a height of 15 m and a spread of 20–30 m, and is largely unaffected by most insect pests [6]. Although heartnut tree will grow in colder regions than zone 6, it can be affected by the late frosts. Frost-injured flowers will abort and so reduce the crop. The nut, unlike the oval or egg-shaped Japanese walnut, is heart-shaped. Typical sizes of heartnut range from 2 to 5 cm in length. The Japanese walnut usually cracks out very poorly, making it unsuitable for commercial production, therefore, selection/breeding programs for heartnut have been focused on improving the cracking property, hardiness (grow better in colder regions), and the shape, color, and taste of the kernel [6]. Commercial orchards normally grow grafted trees in order to ensure the production of heartnut with consistent heart shape and good quality. Grafted trees will begin to bear fruits in 1–3 years, with commercial production expected in 6–8 years [6]. A single mature tree can produce up to 118 kg nut per season. Good heartnut can yield 773 kg/acre at 10 years old. It has been estimated that at maturity, good heartnut trees are capable of producing an average of 1000–3000 kg/acre, equivalent to California walnut production [1]. Heartnut is a low-cost management crop, readily machine harvested, husked, and cracked [1]. Several heartnut varieties, including Imshu, Campbell CW1, Campbell CW3, Campbell CWW, and Fodermaier, have been successfully selected and are in commercial production [6].

14.3 PHYTOCHEMICAL PROFILES AND HEALTH BENEFITS OF HEARTNUT AND WALNUT

14.3.1 NUT CONSUMPTION AND HEALTH BENEFITS

Due to the limited production and consumption of heartnut, no actual research studies have been conducted on its health benefits, however, from the similarity of its phytochemical composition with Persian walnut [2,3] (Tables 14.1 through 14.4) and the similarly strong antioxidant activities found in our recent research [2,3], one can expect that consumption of heartnut would result in similar physiological effect.

The seed of Persian walnut or English walnut (*Juglans regia* L.) is a rich source of essential fatty acids and tocopherols [7–9], the hormone and strong antioxidant melatonin [10], and polyphenolics [11]. The total phenolic content of walnut, determined by the Folin-Ciocalteu assay, was ~16 mg gallic acid equivalents (GAE)/g nut kernel [11]. Typical phytochemicals in walnut and heartnut are shown in Figure 14.1. The total oil content of walnut cultivars varied from 62.6% to 70.3% while the crude protein was typically 13.6%–18.1%. Dietary fiber in walnut kernels ranged from 4.2% to 5.2% and the starch content made up no more than 2.8% of the remaining portion of the walnut kernel [12,13]. The amino acid content of the walnut was similar between cultivars and the patterns of essential amino acids were characteristic of a high-quality protein [12]. In a recent study by Venkatachalam and Sathe [13], walnut was found to contain ~2% sugar and 0.34% total tannins, which were extracted using 1% HCl in methanol.

TABLE 14.1
Total Phenolic Contents in Different Fractions of Heartnut and Persian Walnut

	Extractable Phenolic Acids		
Sample	**FPA[a]**	**AHPA[b]**	**BPA[c]**
Heartnut[d]	195.6 ± 50	148.3 ± 15	244.2 ± 44
Walnut[e]	1001.9 ± 58	366.5 ± 17	324.1 ± 70

Source: From Li, L., Tsao, R., Yang, R., Liu, C., Zhu, H., and Young, J.C., *J. Agric. Food Chem.*, 54, 8033, 2006. With permission.

Note: Total phenolic, expressed as milligrams of gallic acid equivalents per 100 g nut (mg of GAE/100 g nut).

[a] Free phenolic acids.

[b] Acid-hydrolyzable phenolic acids.

[c] Bound phenolic acids.

[d] Heartnut (average value of three varieties: Campbell CW1, Campbell CW3, and Imshu).

[e] Walnut (average value of two varieties: Combe Persian and Lake Persian).

Consumption of nuts has been associated with a lower risk of several diseases, such as coronary heart diseases (CHD) and type 2 diabetes. Jiang et al. [14] recently examined associations between nut consumption and C-reactive protein, interleukin-6, and fibrinogen in the Multi-Ethnic Study of Atherosclerosis involving 6080 participants aged 45–84 years. They found that frequent nut and seed consumption was associated with lower levels of inflammatory markers, which may partially explain the inverse association of nut consumption with CHD and diabetes risk. They also found that the associations of nut and seed consumption with these biomarkers were not modified by body mass index (BMI), waist to hip ratio, or race/ethnicity, or by physiological status such as hypertension, diabetes, medication use, and lipid levels of the participants [14]. Consumption of whole walnut has been related to lower total and low-density lipoprotein (LDL) cholesterol concentrations, not only in normal young men [15], but also in men and women with polygenic hypercholesterolemia [16]. It also has favorable effects on human serum lipid profiles [15,17,18]. The positive health benefit of walnut was reported in increased levels of high-density lipoprotein (HDL) cholesterol concentration and apolipoprotein A-1 [18]. In another study, consumption of walnut was found

TABLE 14.2
Major Phenolic Acids in Different Fractions of Heartnut (Imshu) and Walnut (Lake Persian)

	Ellagic Acid (mg/g nut)			Valonic Acid Dilactone (Ellagic Acid Equivalents) (mg/g nut)		
Sample	**FPA[a]**	**AHPA[b]**	**BPA[c]**	**FPA**	**AHPA**	**BPA**
Heartnut	0.24 ± 0.01	0.58 ± 0.01	0.85 ± 0.01	nd	0.10 ± 0.00	0.41 ± 0.01
Walnut	0.25 ± 0.00	1.33 ± 0.01	0.64 ± 0.00	nd	0.71 ± 0.01	0.44 ± 0.00

Source: From Li, L., Tsao, R., Yang, R., Liu, C., Zhu, H., and Young, J.C., *J. Agric. Food Chem.*, 54, 8033, 2006. With permission.

Note: nd, not detected.

[a] Free phenolic acids.

[b] Acid-hydrolyzable phenolic acids.

[c] Bound phenolic acids.

TABLE 14.3
Fatty Acid Composition (Relative %) of Oils Extracted from Heartnut (Imshu) and Walnut (Lake Persian)

Fatty Acid	Heartnut	Walnut
16:0	2.71	5.59
18:0	1.03	2.83
18:1ω9	13.91	16.39
18:2ω6	72.08	60.96
18:3ω3	7.97	12.11
Total SFA	3.87	8.61
Total MUFA	15.74	17.89
Total PUFA	80.10	73.10

Source: From Li, L., Tsao, R., Yang, R., Kramer, J.K.G., and Hernandez, M., *J. Agric. Food Chem.*, 55, 1164, 2007. With permission.

to improve the lipid profile of patients with type 2 diabetes; there was a significant increase in HDL and HDL/total cholesterols ratio, and a 10% reduction in LDL cholesterol concentrations [19]. The cholesterol-lowering effect was considered to arise from the polyunsaturated fatty acids (PUFA) rather than the antioxidant phytochemicals in walnut [19].

Supplementation of walnut in the diet of 13 postmenopausal women and five men also showed increased serum concentrations of linoleic acid (18:2ω6) and α-linolenic acid (18:3ω3), and decreased plasma total and LDL cholesterol concentrations [20]. Measurements of lipoprotein subclasses and particle size suggested that walnut supplementation lowered cholesterol preferentially in small LDL. However, the researchers of the same study also observed a decrease in HDL cholesterol concentrations, and they attributed it primarily to decreases of the large HDL particles [20].

14.3.2 Phenolic Compounds and Health Benefits

Phenolic compounds are major contributors to the delicate and slightly astringent flavor of walnut fruits [21]. They are mostly found in a special protective tan-brown skin known as the pellicle that surrounds the kernel. Despite the small amount of this thin cover (5% of the fruit weight), it is rich

TABLE 14.4
Tocopherol Contents (mg/ 100 g oil) of Oils Extracted from Heartnut (Imshu) and Walnut (Lake Persian)

Sample	Oil (%)	α-Tocopherol	β-Tocopherol	γ-Tocopherol	δ-Tocopherol	Total
Heartnut	51 ± 1	0.13 ± 0.01	nd	18.73 ± 0.23	1.17 ± 0.16	20.03 ± 0.40
Walnut	60 ± 4	0.51 ± 0.09	nd	20.55 ± 0.27	1.16 ± 0.03	22.22 ± 0.40

Source: From Li, L., Tsao, R., Yang, R., Kramer, J.K.G., and Hernandez, M., *J. Agric. Food Chem.*, 55, 1164, 2007. With permission.

Note: nd, not detected.

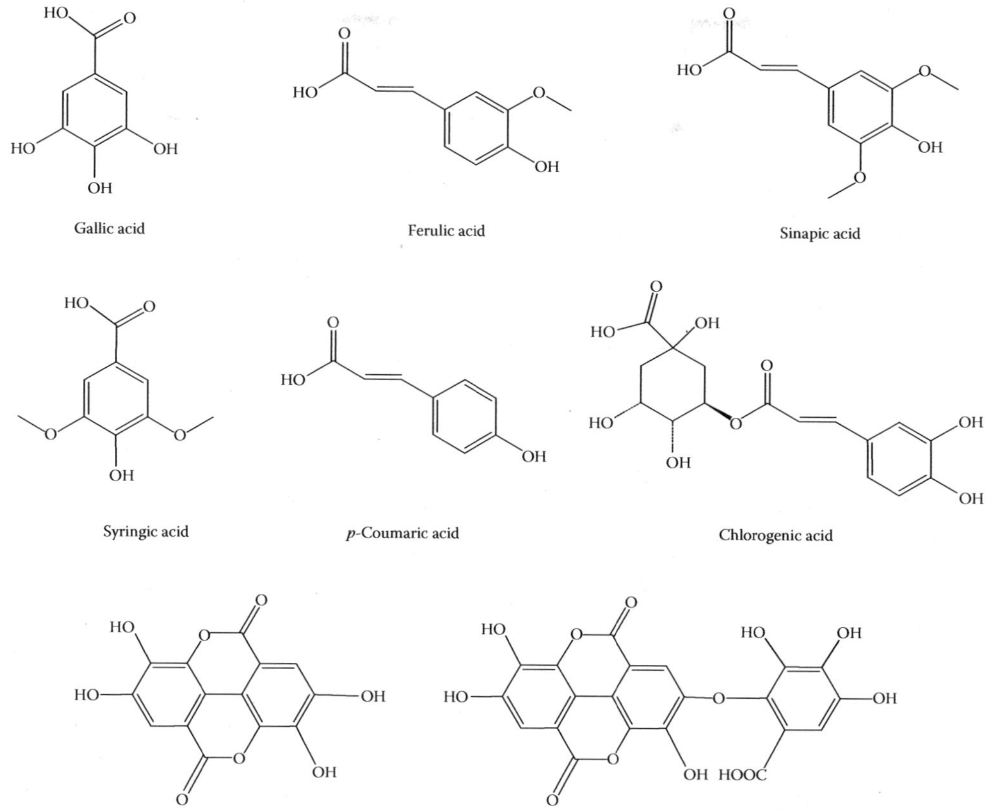

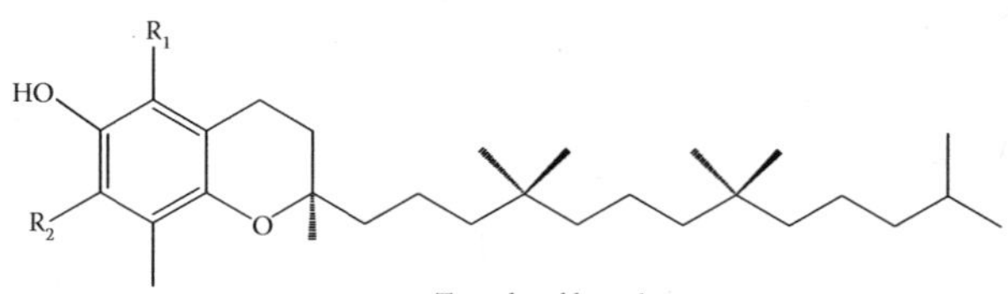

Tocopherols structure	R_1	R_2
α-Tocopherol	CH_3	CH_3
β-Tocopherol	CH_3	H
γ-Tocopherol	H	CH_3
δ-Tocopherol	H	H

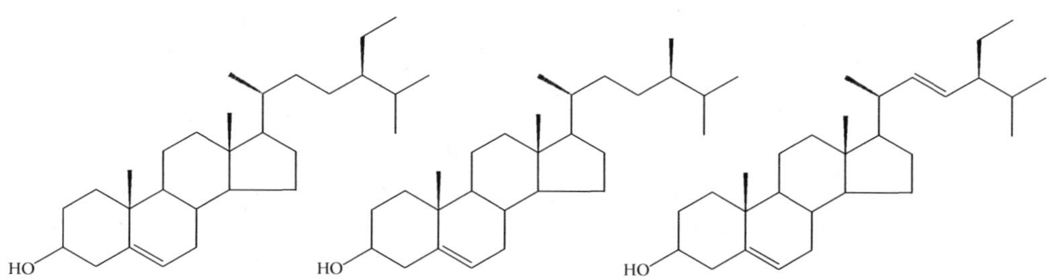

FIGURE 14.1 Typical phytochemicals in heartnut and walnut.

in antioxidant polyphenolic compounds that help protect the oil-rich kernel against oxidation [22]. The total phenolic content was estimated between 10 and 20 mg of GAE/g walnut [7]. Only a few studies have examined and obtained quantitative data of specific polyphenolic phytochemicals in walnut [2,7,23–26] (Figure 14.1). Several phenolic acids, ellagic acid, and its derivatives have been identified and found to contribute significantly to the *in vitro* and *in vivo* antioxidant activities of walnut [7,24–26]. Significant differences in the contents of identified phenolics have been observed among different walnut cultivars. The kernel contained an average of 34, 12, and 6 mg/100 g of syringic acid, juglone, and ellagic acid, respectively; these compounds were mostly found in the pellicle with an average concentration of 1003, 318, and 129 mg/100 g of pellicle, respectively [26]. Other phenolics such as *p*-coumaric acid, chlorogenic acid, ferulic acid, and sinapic acid were also found in walnut, but at lower concentrations [26]. In another study, three hydrolyzable tannins, glansrins A-C, were isolated from the *n*-butanol extract of walnut, and characterized as ellagitannins with a tergalloyl group or related polyphenolic acyl group [24]. Thirteen other tannins (gallic acid and ellagic acid derivatives) were also identified [24]. Most recently, we studied the polyphenolic content in two varieties of Persian walnut and found that the free forms of total phenolic content of the 80% methanol-extractable fraction of walnut was 10 mg of GAE/g nut, however significant amount of phenolics was found in the bound forms, e.g., they were only released upon acid hydrolysis. Ellagic acid, although not necessarily the dominant phenolic compound in the crude extract of walnut, was the predominant bound phenolic present [2]. Persian walnut has been reported to contain 0.29, 1.31, and 0.93 mg of free, hydrolyzable, and bound ellagic acid/g nut, respectively [2]. Using high-performance liquid chromatography (HPLC) combined with diode array detector (DAD) and electrospray ionization–mass spectrometry (ESI-MS), an ellagic acid derivative and valoneic acid dilactone were also identified, but it was only found in the bound form [2]. In spite of these studies on individual phenolics in edible walnut fruits of different cultivars [2,7,21–24], the complete profiles of walnut polyphenolics are often missing, particularly quantitative data [25].

Others have also reported that, ellagitannins, which are often referred to as hydrolyzable tannins, dominate the phenolic profile of the seed of *J. regia* L. [23–25]. Phenolic compounds from walnut fruits have a positive influence on human health such as risk reduction of CHD, prevention of several kinds of cancer, and anti-inflammatory and antimutagenic activities. The longer shelf life and lower tocopherol content of walnut as compared to other nuts imply that polyphenolic compounds such as ellagic acid, gallic acid, and flavonoids may be more important contributors to the inhibition of lipid autoxidation in walnut and in humans [11,24]. Using the ferric reducing antioxidant power (FRAP) assay, Halvorsen et al. [27] evaluated a few dozens of dietary plants and found that walnut ranked second just after dog rose and before pomegranates in terms of the FRAP value. FRAP assay also showed that the antioxidant activity of the different fractions, including free and bound (acid hydrolyzable) phenolic fractions, positively correlated with the total phenolic content; ellagic acid and valoneic acid dilactone, as the major individual compounds, were considered to contribute significantly to the total antioxidant activity of the bound phenolic fractions [2]. The strong antioxidant activity of the phenolic fractions of walnut and the contribution of ellagic acid and valoneic acid dilactone were also confirmed with another *in vitro* model, the photochemiluminescence (PCL) method, which measures the scavenging activity against the superoxide radical [2]. Walnut was found to have the highest antioxidant activity among commonly consumed foods and drinks in Turkey [28]. Walnut extracts, containing ellagic acid monomers, polymeric tannins, and other phenolics, effectively inhibited human plasma and LDL oxidation *in vitro* [11]. Ellagic acid and flavonoids in walnut have also been indicated as potential serum cholesterol modulators [29]. Juglone, a naphthaquinone found mostly in walnut hulls also considered a phenolic due to the presence of a hydroxyl group on the aromatic ring, has many biological effects, particularly its ability to lower the incidence of tumors of the small intestine in rats [30]. Several studies have demonstrated that dietary ellagitannins from different plant sources including walnut are not directly absorbed in humans, but they are converted to ellagic acid *in vivo* as a result of digestion, and it is the ellagic acid that is absorbed and further metabolized by the human colonic microflora to yield the bioavailable 3,8-dihydroxy-6*H*-dibenzo [*b,d*]pyran-6-one (durolithin AT) derivatives [25,31]. Larrosa et al. [32] suggested that the anticarcinogenic effect of dietary ellagitannins could be mainly due to their hydrolysis product, ellagic acid, which induces apoptosis via the mitochondrial pathway in colon cancer Caco-2 cells, but not in normal colon cells. Ellagitannins and ellagic acid from different plant sources have also been studied for their antioxidant properties *in vitro* [16,17,26–28]. Fukuda et al.

[24] isolated and identified 15 ellagitannins from the aqueous ethanol extract of walnut seeds and found that these compounds had a superoxide dismutase (SOD)-like activity with an EC_{50} of 21.4–190 mM and a remarkable radical scavenging effect against 1,1-diphenyl-2-picrylhydrazyl (DPPH) radical (EC_{50} 0.34–4.72 mM). Their study indicated that ellagitannins with a galloyl group plus a hexahydroxydiphenoyl (HHDP) group, or valoneoyl and its isomeric group, tend to exhibit a more potent radical scavenging effect against DPPH than those with only an HHDP group. Anderson et al. [11] also found that ellagic acid (1.0 μmol/L) and walnut (*J. regia* L.) extract significantly inhibited 2,2′-azobis(2-amidino propane) hydrochloride (AAPH)-induced LDL oxidation. Similarly, copper-mediated LDL oxidation was also inhibited by ellagic acid and walnut extract, indicating that ellagic acid may be completely or partly responsible for the observed *in vitro* effect on maintaining LDL α-tocopherol [11].

In terms of the phenolic content of heartnut and their implications in maintaining good human health, there is essentially no information available other than recent report from our laboratory [2]. We studied three varieties of heartnut from the Niagara region, Ontario, and Canada, and found totally different profiles of phenolics as compared to the common walnut, particularly the free phenolic acids in the 80% methanol-extractable fraction, which contained 195.6 mg of GAE/100 g heartnut, a mere one-fifth of what was found in the same fraction of walnut [2] (Tables 14.1 and 14.2). This may explain well why heartnut is "sweeter" than the common Persian walnut; they contain less polyphenols, particularly those in the free form, which contribute to the astringency of walnut. At the individual phenolics level, ellagic acid and valoneic acid dilactone were the predominant bound phenolics of the acid hydrolyzable 80% methanol extractable fraction and the residue [2]. The antioxidant activity of the heartnut extracts and their different fractions was also evaluated using the FRAP assay and the PCL method [2]. In general, heartnut had slightly weaker antioxidant activities, mostly owing to their lower total and individual phenolic contents, but among the different fractions, the strongest antioxidant activity was found, in both assays, in the bound phenolic acid fraction of the residue that contained mainly ellagic acid and valoneic acid dilactone [2].

14.3.3 Lipid Contents and Health Benefits

Walnut is a good source of lipids including essential fatty acids, tocopherols, and phytosterols that contribute to the lowering of serum LDL cholesterol concentration and the risk reduction of CHD [8,33]. The lipid content of walnut kernel is in the range of 57%–68% of the total nut [3,8,34]. Walnut lipids are mostly consisted of different fatty acids: saturated fatty acids (SFA), monounsaturated fatty acids (MUFA), and PUFA. The positive effects of walnut consumption on human health via the reduction of total plasma total cholesterol and LDL cholesterol concentrations, and the improvement of serum lipid profiles [14–19] maybe attributed to the fatty acid profile of the oil component, in particular, omega-3 and omega-6 PUFA, of which walnut oil is a particularly good source [35]. Among walnut fatty acids, PUFA contributes the most predominant group, ranging from 71% to 74%, followed by MUFA and SFA ranging from18% to 20% and from 9% to 10%, respectively [3,8]. MUFA is mostly oleic acid (18:1ω9) with a concentration at 13%–22%, whereas PUFA is mostly 18:2ω6 and 18:3ω3 [3,8,33,36–39]. 18:2ω6 is the single most prevalent fatty acid (57%–64%) of walnut lipid, while 18:3ω3 ranges from 10% to 16%. SFA are mostly palmitic acid (16:0, 5%–8%) and stearic acid (18:0, 1%–3%) [3,8,35]. It is widely recognized that the type of fat in the diet influences plasma cholesterol levels to a greater extent than total fat intake. Therefore, replacing saturated fat with unsaturated fat maybe more effective in lowering the risk of CHD than reducing the total fat intake [40,41]. Diets high in MUFA were shown to have a favorable effect on the ratio of total/HDL cholesterols, which is a more accurate indicator of risk for CHD than total cholesterol level alone. Diets high in MUFA reduce levels of LDL without adversely affecting the HDL fraction, thereby reducing the risk of CHD [19,41].

The average total lipid content of the three heartnut varieties was significantly lower (51%) than that of the walnut (60%), but the major fatty acid profiles were generally similar, e.g., 18:2ω6,

18:1ω9, 18:3ω3, 16:0, and 18:0, in a decreasing order [3] (Tables 14.3 and 14.4). However, what was interesting to note is that heartnut contained significantly higher PUFA (80.1%) than walnut (73.1%), and lower MUFA and SFA contents (15.74% vs. 17.89%, and 3.87% vs. 8.61%, respectively) [3]. The higher PUFA and lower SFA contents suggest that heartnut may potentially contribute more significantly to health promotion [35]. In terms of individual fatty acids, heartnut had significantly higher 18:2ω6 content, but lower 18:3ω3, 16:0, 18:0, and 18:1ω9 compared to the Persian walnut [3] (Table 14.3).

Vitamin E is a group of fat-soluble phenolic compounds consisting four tocopherols (α-, β-, γ-, and δ) (Figure 14.1) and four tocotrienols (α-, β-, γ-, and δ). These compounds are believed to be essential to a diverse physiological and biochemical functions, mainly due to their action as antioxidants but also by acting as membrane stabilizers [42]. Different homologues of tocopherols have been demonstrated to have different antioxidant activities. The relative antioxidant activity of tocopherols *in vivo* is in the order $\alpha > \beta > \gamma > \delta$ [43]. For this reason, many studies have focused on the health benefit of α-tocopherol in humans, whereas little attention has been paid to the other homologues other than the evaluation of their relative antioxidant activity. However, in recent years other vitamin E homologues have been shown to possess important bioactivities [44,45], indicating that all homologues of vitamin E may contribute to the total bioactivity [43–45]. In fact, several reports have indicated that γ-tocopherol was a stronger antioxidant in some model systems than α-tocopherol [46,47]. In our recent study, we indicated significant differences among the different tocopherols in their radical scavenging capacity (RSC) against the DPPH [3]. γ-Tocopherol showed the highest RSC followed by α-, β-, and δ-tocopherols; however, using the PCL method, the antioxidant activities of the tocopherols were in the order of $\delta > \alpha > \beta \approx \gamma$, against the superoxide anion [3].

The total tocopherol concentration of walnut ranges from 22.2 to 36.0 mg/100 g nut kernel [3,7,9,33,48]. γ-Tocopherol has been identified as the major (up to 89% of the total) vitamin E homologue in walnut followed by δ- and α-tocopherols, whereas β-tocopherol was only at trace level [3] (Figure 14.1). The actual concentration may vary significantly due to many different factors [3,8,33,39,48,49]. Tocopherols may provide another important defense mechanism in walnut, in addition to the polyphenols, that protect the high lipid content from being oxidized. The antioxidant activity of walnut oil has been attributed to the presence of tocopherols therein [39,50] (e.g., the higher the total tocopherol content, the stronger the antioxidant activity [3]).

Heartnut is also a rich source of tocopherols with a profile similar to the common walnut [3] (Table 14.4). The total tocopherol concentration (average of three varieties) in heartnut oil ranged from 12.84 to 20.03 mg/100 g, which was significantly lower than in walnut oil (from 22.22 to 30.80 mg/100 g) [3]. However, the percentage of γ-tocopherol to the total vitamin E homologues was higher (up to 96% of the total).

Espín et al. [51] suggested that the RSC of lipid fraction of different plant oils, including walnut oil was mainly due to their different concentrations and types of tocopherols. Our recent study showed that the RSC (EC_{50} values) of the lipid fractions of nuts (both walnut and heartnut) correlated well with the total tocopherol contents ($r^2 = 0.97$), indicating that tocopherols are perhaps the components responsible for the antioxidant effect [3]. The EC_{50} values also correlated well with γ-tocopherol concentration ($r^2 = 0.97$). Considering that γ-tocopherol was the predominant vitamin E homologue in walnut and heartnut oils, γ-tocopherol, therefore, represents the single major contributor to the total antioxidant activity of walnut and heartnut oils. Tocopherols contributed 75%–99% to the total antioxidant activity of all nut oils in the DPPH assay, and γ-tocopherol alone contributed 65%–96% [3]. In another *in vitro* system using the PCL method, we found that the antioxidant activities of Persian walnut and heartnut were similar ($P = 0.48$), despite the significant differences in total and individual tocopherol contents [3]. However, the antioxidant activity in the PCL experiment showed good correlation with both total tocopherol and γ-tocopherol contents with coefficients of determinations ($r^2 = 0.81$ and 0.77, respectively). Tocopherols contributed 41%–58% to the total antioxidant activity of all nut oils in the PCL assay, and γ-tocopherol alone contributed 35%–52%, both less than in the DPPH assay, indicating other compounds may have also played a role in the PCL assay against the superoxide anion radical [3].

The health benefits of the lipids of walnut and heartnut may largely arise from the fatty acids and tocopherols as discussed above; however, other minor components found in walnut, particularly phytosterols may also be important. Phytosterols are found in plant-based foods and are structurally and functionally analogous to cholesterol in humans and animals. β-Sitosterol, campesterol, and stigmasterol (Figure 14.1) are the most commonly occurring phytosterols and constitute 95% of total phytosterols in the diet [33]. The effects of dietary supplementation with phytosterols on serum cholesterol levels in humans have been reviewed comprehensively [52–55], and the general conclusion is that phytosterol supplementation tends to decrease total and LDL cholesterol concentrations and has little effect on serum levels of HDL cholesterol and triacylglycerols (TAG) [33]. Walnut oil, as many other nut oils, contains mainly β-sitosterol (109–175 mg/100 g oil) and some minor phytosterols such as campesterol (5.1–11 mg/100 g oil), stigmasterol (0.8–5.6 mg/100 g oil) [8,33,56], and trace amount of phytostanols [56]. Consumption of walnut and heartnut may receive added health benefits from these phytochemicals.

14.4 CONCLUSION

Heartnut, as a special variety of the Japanese walnut, contains similar polyphenolic, fatty acid, and tocopherol profiles to those present in the common Persian walnut, despite the different concentrations of individual constituents, particularly the total phenolic content as found in recent studies from our research groups. Although the relatively lower phenolic contents and the tocopherol concentrations generally led to weaker antioxidant activities, other unknown contributing factors could not be excluded. The health benefits of heartnut in human can only be deduced from the existing data on walnut; however, the similar phytochemical profiles to walnut suggest that this special tree nut has a good potential as a reliable source of phytochemical antioxidants and bioactives. The lower phenolic content may also explain the "sweeter" taste of the heartnut that many consumers prefer. The high concentration of PUFA (particularly 18:2ω6 and 18:3ω3) also adds to the value of heartnut. Although the ratio of 18:3ω3/18:2ω6 was favorable for the Persian walnut, heartnut in general had lower SFA, which is generally considered better. In addition, the varietal differences in phytochemical compositions found in walnut and heartnut indicate that improvement can be achieved through breeding. Considering heartnut breeding and growing has only been a recent trend, better quality heartnut with good nutritional value are just a matter of time.

REFERENCES

1. Grimo, E., Personal Communication, Published online at: http://www.grimonut.com/section1.htm (accessed July 10, 2007).
2. Li, L., Tsao, R., Yang, R., Liu, C., Zhu, H., and Young, J.C., Polyphenolic profiles and antioxidant activities of heartnut (*Juglans ailanthifolia* var. *cordiformis*) and Persian walnut (*Juglans regia* L.), *J. Agric. Food Chem.*, 54, 8033–8040, 2006.
3. Li, L., Tsao, R., Yang, R., Kramer, J.K.G., and Hernandez, M., Fatty acid profiles, tocopherol contents and antioxidant activities of heartnut (*Juglans ailanthifolia* var. *cordiformis*) and Persian walnut (*Juglans regia* L.), *J. Agric. Food Chem.*, 55, 1164–1169, 2007.
4. Neilson, J.A., Some Notes on the Japanese Walnut in North America, Published online at: http://www.songonline.ca/library/articles/japanesewalnut_1930.htm (accessed July 10, 2007).
5. Society of Ontario Nut Growers, Published online at: http://www.songonline.ca/nuts/heartnut.htm (accessed July 10, 2007).
6. Mohammed, G.H., Non-Timber Forest Products in Ontario: An Overview, Forest Research Information Paper No. 145, Ontario Forest Research Institute, Ontario, Ministry of Natural Resources, Published online at: http://www.mnr.gov.on.ca/MNR/forests/t&t_research/publications/145.pdf (accessed January 10, 2007).
7. Kornsteiner, M., Wagner, K.H., and Elmadfa, I., Tocopherol and total phenolics in 10 different nut types, *Food Chem.*, 98, 381–387, 2006.
8. Amaral, J.S., Casal, S., Pereira, J.A., Seabra, R.M., and Oliveira, B.P.P., Determination of sterol and fatty acid compositions, oxidative stability, and nutritional value of six walnut (*Juglans regia* L.) cultivars grown in Portugal, *J. Agric. Food Chem.*, 51, 7698–7702, 2003.

9. Amaral, J.S., Alves, M.R., Seabra, R.M., and Oliveira, B.P.P., Vitamin E composition of walnuts (*Juglans regia* L.): A 3-year comparative study of different cultivars, *J. Agric. Food Chem.*, 53, 5467–5472, 2005.
10. Reiter, R.J., Manchester, L.C. and Tan, D.X., Melatonin in walnuts: Influence on levels of melatonin and total antioxidant capacity of blood, *J. Nutr.,* 21, 920–924, 2005.
11. Anderson, K.J., Teuber, S.S., Gobeille, A., Cremin, P., Waterhouse, A.L. and Steinberg, F.M., Walnut polyphenolics inhibit in vitro human plasma and LDL oxidation, *J. Nutr.,* 131, 2837–2842, 2001.
12. Savage, G.P., Chemical composition of walnuts (*Juglans regia* L.) grown in New Zealand, *Plant Foods Human Nutr.,* 56, 75–82, 2001.
13. Venkatachalam, M. and Sathe, S.K., Chemical composition of selected edible nut seeds, *J. Agric. Food Chem.,* 54, 4705–4714, 2006.
14. Jiang, R., Jacobs, D.R. Jr., Mayer-Davis, E., Szklo, M., Herrington, D., Jenny, N.S., Kronmal, R., and Barr, R.G., Nut and seed consumption and inflammatory markers in the multi-ethnic study of atherosclerosis, *Am. J. Epidemiol.*, 163, 222–231, 2006.
15. Sabate, J., Fraser, G.E., Burke, K., Knutsen, S.F., Bennett, H., and Lindsted, K.D., Effects of walnuts on serum lipid levels and blood pressure in normal men, *N. Engl. J. Med.*, 328, 603–607, 1993.
16. Zambon, D., Sabate, J., Munoz, S., Campero, B., Casals, E., Merlos, M., Laguna, J., and Ros, E., Substituting walnuts for monounsaturated fat improves the serum lipid profile of hypercholesterolemic men and women: A randomised crossover trial, *Ann. Intern. Med.*, 132, 538–546, 2000.
17. Chisholm, A., Mann, J., Skeaff, M., Frampton, C., Sutherland, W., Duncan, A., and Tiszavari, S., A diet rich in walnuts favourably influences plasma fatty acid profile in moderately hyperlipidaemic subjects, *Eur. J. Clin. Nutr.*, 52, 12–16, 1998.
18. Lavedrine, F., Zmirou, D., Ravel, A., Balducci, F., and Alary, J., Blood cholesterol and walnut consumption: A cross-sectional survey in France, *Prev. Med.*, 28, 333–339, 1999.
19. Tapsell, L.C., Gillen, L.J., Patch, C.S., Batterham, M., Owen, A., Bare, M., and Kennedy, M., Including walnuts in a low-fat/modified-fat diet improves HDL cholesterol-to-total cholesterol ratios in patients with type 2 diabetes, *Diabetes Care*, 27, 2777–2783, 2004.
20. Almario, R.U., Vonghavaravat, V., Wong, R., and Kasim-Karakas, S.E., Effects of walnut consumption on plasma fatty acids and lipoproteins in combined hyperlipidemia, *Am. J. Clin. Nutr.*, 74, 72–79, 2001.
21. Prasad, R.B.N., Walnuts and pecans, in *Encyclopaedia of Food Sciences and Nutrition*, 2nd ed., Caballero, B., Trugo, L.C., and Finglas, P.M., Eds., Academic Press, London, 2003, pp. 6071–6079.
22. Jurd, L., Plant polyphenols. I. The polyphenolic constituents of the pellicle of the walnut (*Juglans regia*), *J. Am. Chem. Soc.*, 78, 3445–3448, 1956.
23. Jurd, L., Plant polyphenols II. The benzylation of ellagic acid, *J. Am. Chem. Soc.*, 79, 6043–6047, 1957.
24. Fukuda, T., Ito, H., and Yoshida, T., Antioxidative polyphenols from walnuts (*Juglans regia* L.), *Phytochemistry,* 63, 795–801, 2003.
25. Cerdá, B., Tomás-Barberán, F.A., and Espín, J.C., Metabolism of chemopreventive and antioxidant ellagitannins from strawberries, raspberries, walnuts and oak-aged wines in humans: Identification of biomarkers and individual variability, *J. Agric. Food Chem.*, 53, 227–235, 2005.
26. Colaric, M., Veberic, R., Solar, A., Hudina, M., and Stampar, F., Phenolic acids, syringaldehyde, and juglone in fruits of different cultivars of *Juglans regia* L., *J. Agric. Food Chem.*, 53, 6390–6396, 2005.
27. Halvorsen, B.L., Holte, K., Myhrstad, M.C.W., Barikmo, I., Hvattum, E., Remberg, S.F., Wold, A.-B., Haffner, K., Baugerod, H., Andersen, L.F., Moskaug, J.O., Jacobs, D.R., and Blomhoff, R., A systematic screening of total antioxidants in dietary plants, *J. Nutr.*, 132, 461–471, 2002.
28. Gunduc, N. and El, S.N., Assessing antioxidant activities of phenolic compounds of common Turkish food and drinks on *in vitro* low density lipoprotein oxidation, *J. Food Sci.*, 68, 2591–2595, 2003.
29. Kris-Etherton, P.M., You-Poth, S., Sabaté, J., Ratcliffe, H.E., Zhao, G., and Etherton, T. D., Nuts and their bioactive constituents: Effects on serum lipids and other factors that affect disease risk, *Am. J. Clin. Nutr.*, 70, 504S–511S, 1999.
30. Sugie, S., Okamoto, K., Rahman, K.M.W., Tanaka, T., Kawai, K., Yamahara, J., and Mori, H., Inhibitory effects of plumbagin and juglone on azoxymethane-induced intestinal carcinogenesis in rats, *Cancer Lett.*, 127, 177–183, 1998.
31. Cerdá, B., Espín, J.C., Parra, A., Martínez, P., and Tomás-Barberán, F.A., The potent in vitro antioxidant ellagitannins from pomegranate juice are metabolized into bioavailable but poor antioxidant hydroxy-6*H*-dibenzopyran-6-one derivatives by the colonic microflora of healthy humans, *Eur. J. Nutr.*, 43, 205–220, 2004.

32. Larrosa, M., Francisco, A., Tomás-Barberán, F.A., and Espín, J.C., The dietary hydrolysable tannin punicalagin releases ellagic acid that induces apoptosis in human colon adenocarcinoma Caco-2 cells by using the mitochondrial pathway, *J. Nutr Biochem.*, 17, 611–625, 2006.
33. Maguire, L.S., O'Sullivan, S.M., Galvin, K., O'Connor, T.P., and O'Brien, N.M., Fatty acid profile, tocopherol, squalene and phytosterol content of walnuts, almonds, peanuts, hazelnuts and the macadamia nut, *Int. J. Food Sci. Nutr.*, 55, 171–178, 2004.
34. Zeneli, G., Kola, H., and Dida, M., Phenotypic variation in native walnut populations of Northern Albania, *Sci. Hort.*, 105, 91–100, 2005.
35. Crews, C., Hough, P., Godward, J., Brereton, P., Lees, M., Guiet, S., and Winkelmann, W., Study of the main constituents of some authentic walnut oils, *J. Agric. Food Chem.*, 53, 4853–4860, 2005.
36. Zwarts, L., Savage, G.P., and McNeil, D., Fatty acid content of New Zealand-grown walnuts, *Int. J. Food Sci. Nutr.,* 50, 189–194, 1999.
37. Greve, C., McGranahan, G., Hasey, J., Snyder, R., Kelly, K., Goldhamerer, D., and Labavitch, J., Variation in polyunsaturated fatty acid composition of Persian walnut, *J. Soc. Hort. Sci.*, 117, 518–522, 1992.
38. Ruggeri, S., Cappelloni, M., Gambelli, L., Nicoli, S., and Carnovale, E., Chemical composition and nutritive value of nuts grown in Italy, *Ital. J. Food Sci.*, 3, 243–252, 1996.
39. Savage, G.P., Dutta, P.C., and Mcneil, D.L., Fatty acid and tocopherol contents and oxidative stability of walnut oils, *J. Am. Oil Chem. Soc.*, 76, 1059–1063, 1999.
40. O'Byrne, D.J., Knauft, D.A., and Shireman, R.B., Low fat-monounsaturated rich diets containing high-oleic peanuts improve serum lipoprotein profiles, *Lipids*, 32, 687–695, 1997.
41. Kris-Etherton, P.M., Pearson, T.A., Wan, Y., Hargrove, R.L., Moriarty, K., Fishell, V., and Etherton, T.D., High monounsaturated fatty acid diets lower both plasma cholesterol and triacylglycerol concentrations, *Am. J. Clin. Nutr.*, 70, 1009–1015, 1999.
42. Azzi, A. and Stocker, A., Vitamin E: Non-antioxidant roles, *Prog. Lipid Res.*, 39, 231–255, 2000.
43. Dayong, W., Meydani, M., Beharka, A.A., Serafini, M., Martin, K.R., and Meydani, S.N., In vitro supplementation with different tocopherol homologues can affect the function of immune cells in old mice, *Free Radical Biol. Med.*, 28, 643–651, 2000.
44. Mishima, K., Tanaka, T., Pu, F., Egashira, N., Iwasaki, K., Hidaka, R., Matsunaga, K., Takata, J., Karube, Y., and Fujiwara, M., Vitamin E isoforms α-tocotrienol and γ-tocopherol prevent cerebral infarction in mice, *Neurosci. Lett.*, 337, 56–60, 2003.
45. Campbell, S., Stone, W., Whaley, S., and Krishnan, K., Development of gamma (γ)-tocopherol as a colorectal cancer chemopreventive agent, *Crit. Rev. Oncol. Hematol.*, 47, 249–259, 2003.
46. Saldeen, T., Li, D., and Mehta, J.L., Differential effects of alpha and gamma-tocopherol on low-density lipoprotein oxidation, superoxide activity, platelet aggregation and arterial thrombogeneis, *J. Am. Coll. Cardiol.*, 34, 1208–1215, 1999.
47. Wolf, G., Gamma-tocopherol: An efficient protector of lipids against nitric oxide-initiated peroxidative damage, *Nutr. Rev.*, 55, 376–378, 1997.
48. Lavedrine, F., Ravel, A., Poupard, A., and Alary, J., Effect of geographic origin, variety and storage on tocopherol concentrations in walnuts by HPLC, *Food Chem.*, 58, 135–140, 1997.
49. Amaral, J.S., Casal, S., Oliveira, B.P., and Seabra, R.M., Development and evaluation of a normal phase liquid chromatographic method for the determination of tocopherols and tocotrienols in walnuts, *J. Liq. Chromatogr. Relat. Technol.*, 28, 785–795, 2005.
50. Baldioli, M., Servili, M., Perretti, G., and Montedoro, G.F., Antioxidant activity of tocopherols and phenolic compounds of virgin olive oil, *J. Am. Oil Chem. Soc.*, 73, 1589–1593, 1996.
51. Espín, J.C., Soler-Rivas, C., and Wichers, H.J., Characterization of the total free radical scavenger capacity of vegetable oils and oil fractions using 2,2-diphenyl-1-picrylhydrazyl radical, *J. Agric. Food. Chem.*, 48, 648–656, 2000.
52. Gylling, H. and Miettinen, T.A., Plant sterols in nutrition, *Scand. J. Nutr.*, 44, 155–157, 2000.
53. Law, M., Plant sterol and stanol margarines and health, *Br. Med. J.*, 320, 861–864, 2000.
54. Moreau, R.A., Whitaker, B.D., and Hicks, K.B., Phytosterols, phytostanols, and their conjugates in foods: Structural diversity, quantitative analysis, and health-promoting uses, *Prog. Lipid Res.*, 41, 457–500, 2002.
55. Neil, H.A.W., and Huxley, R.R., Efficacy and therapeutic potential of plant sterols, *Atheroscler. Suppl.*, 3, 11–15, 2002.
56. Phillips, K.M., Ruggio, D.M., and Ashraf-Khorassani, M., Phytosterol composition of nuts and seeds commonly consumed in the United States, *J. Agric. Food Chem.*, 53, 9436–9445, 2005.

15 Nutrient Composition and Health Beneficial Effects of Macadamia Nuts

Irene A. Munro and Manohar L. Garg

CONTENTS

15.1 INTRODUCTION

Macadamia is an evergreen native rainforest tree originating from the Australian east coast. The survival of these trees in their natural habitat was threatened when the land was cleared for agricultural use in the mid-1800s. However, commercial cultivation of the macadamia nut through the establishment of plantations in the 1870s ensured its survival, and today macadamia is Australia's only indigenous, commercially grown food crop. By the 1890s, the macadamia nut had been introduced to some regions of the United States, such as Hawaii, where hybridization and cloning developed different varieties, such as Cates, Beaumont's, and James, turning the nut into a valuable commercial crop. In addition to Australia and Hawaii, the tree is now also cultivated as a food crop in California, Florida, New Zealand, southern Africa, and parts of South and Central America.

In Australia, there are at least five species of macadamia trees, but only two produce edible nuts and are grown as a food crop. They are *Macadamia integrifolia* (smooth shelled), which contains 80% oil and 4% sugar, and *Macadamia tetraphylla* (rough shelled) with an oil content which ranges from 65% to 75% and a sugar content of 6% to 8%. *M. integrifolia* is the species most commonly grown for its crops of nuts.

This chapter focuses on the nutrient composition and health benefits of consuming macadamia nuts as part of a healthy diet.

15.2 GROWING AND HARVESTING OF MACADAMIA NUTS

Macadamia trees are slow growing and can reach a mature height of between 12 and 15 m in fertile, well drained soils. They require high annual rainfall and high levels of sunshine to ensure maximum growth. The dark green, bushy trees start to produce nuts after 4 or 5 years and they are fully productive after about 6 years. From a slow beginning, they have a long life and will continue to fruit indefinitely.

When the trees reach maturity, sweet smelling white or pink flowers develop in long, narrow sprays of between 40 and 50 flowers from which 4 to 15 nutlets form. These will eventually ripen into mature nuts encased in a woody shell surrounded by a green-brown fibrous husk. The shell surrounding the nut is extremely hard and difficult to crack open without damaging the nuts. For this reason, the commercial processing of the nuts was initially slow and only took off when a mechanized processing plant, to safely break the shells, was established in 1954.

In its natural environment, there will be flowers, nutlets, and mature nuts all growing simultaneously on the macadamia tree for most of the year. Under cultivation, however, harvesting is seasonal. Depending on the variety, the nuts usually drop from the tree when they are ready for harvesting, though they may also need to be picked from the trees when ripe. Nuts on the ground need to be harvested at regular intervals, because nuts left on the ground for longer than 2 to 3 weeks begin to germinate, which gives the nut an unpleasant taste. The fibrous, outer husks should be removed within 24 h of harvesting in order to reduce heat respiration, facilitate drying, and reduce the risk of developing mold. It is also easier to remove the husk while it is green and relatively soft.

At harvest, the nut has a moisture content of up to 30% and it completely fills the shell. As the shell requires a pressure of up to 135 kg (300 lb) to crack, the nut could very easily be damaged in the process. Therefore, while still in their shells, the nuts are air-dried in the shade or in specially built, fan-forced silos. When the moisture content of the nut reaches 15% or lower, the nuts in their shells can be moved to drying bins with circulating warm air to dry down to a moisture content of 1.5%. At this level of moisture, the nut kernel shrinks away from the shell so that it is not damaged when the shell is cracked for removal. The nut is then ready for consumption or processing. However, if preferred, the nuts can be left in their shells where they keep well for up to 12 months.

15.3 BY-PRODUCTS OF MACADAMIA NUTS

Only 31% of the macadamia nut, that is the kernel, is edible. The remaining 69% is waste, the disposal of which can be a problem for processors. In recent years, however, uses have been found for macadamia shells, which contain lignin and cellulose, two components that make them dense and strong. These properties have led to the production of charcoal-like substances, called activated carbons, from the macadamia nut shells to be used largely in water purification and the identification of pollutants [1–3]. The shells are highly flammable and can be used as a renewable fuel source for energy production [4] and the fibrous husks can be composted and used as garden mulch.

15.4 MARKETING MACADAMIA PRODUCTS

To maximize the profitability of macadamia nuts as a food crop, they are processed to varying degrees for value-added products and are in high demand from the snack, bakery, and confectionary sectors of the food industry.

15.4.1 Macadamia Oil

Macadamia nuts have the highest fat content of any nut, which is advantageous for the production of macadamia oil [5]. Macadamia oil is cold pressed and is 100% pure. It has a lower flash point than other vegetable oils so it is limited as cooking oil and is better used as flavoring in cooking in a similar way to sesame oil. It has an appealing flavor used in salad dressings and with pasta.

15.4.2 Macadamia Nut Snacks

Raw macadamia nuts can be bought whole or, because they are quite large, in halves as individual snacks. To enhance the flavor, they can be dry roasted and are also available roasted and salted, roasted and smoked, and roasted with barbeque flavor. Because of the high fat content the nuts can quickly become rancid so they are usually sold in vacuum-packed containers. Once opened, they should be stored in the fridge or freezer.

15.4.3 Macadamia Nuts in Confectionary, Cakes, and Desserts

Whole nuts can be sold as confectionary, coated with honey or caramel or chocolate. They are also halved or chopped and added as an ingredient to confectionary bars, biscuits, and ice cream. Ground macadamia nuts can be added to cakes, pastry, and biscuits as a shortening to enhance the texture and to impart flavor.

15.4.4 Savory Dishes

Chopped macadamia nuts can be added to salads, rice dishes, curries, and stir-fries. Ground macadamia nuts can be used to thicken sauces, as a coating for chicken and fish prior to baking, or blended into a creamy butter for use as a spread.

15.5 COMPOSITIONAL AND FUNCTIONAL LIPID CHARACTERISTICS OF MACADAMIA NUTS

The detailed information on chemical composition of 14 edible nuts is given in Chapter 2. In this section, some important components of macadamia nuts are discussed.

15.5.1 Proximate Composition

Macadamia nuts are a rich source of nutrients, high in unsaturated fats, vegetable protein, and important micronutrients. The content of the various chemicals and nutrients in macadamia nuts can vary considerably depending on the cultivar, seed maturity, growing locations, and growing conditions. Thus, the results of proximate analysis of macadamia nuts carried out in different places and at different times, can produce different results. For example, of all the nuts, macadamia nuts have the highest fat content at approximately 75% of the edible nut, making them energy dense with 733 kcal/100 g of nuts [6]. However, analyses of the nuts in different studies determined different levels of fat, from 78.4% [7], 75.8% [8], and 72.5% [9] to as low as 59.2% [10]. In New Zealand alone, the lipid content of locally grown macadamia nuts ranges from 69.1% to 78.4% [7].

The sugar content of the macadamia nut varies from 1.36% [11] to 4.57% [5], and is composed of fructose, glucose, maltose, and sucrose. Nuts, in general, contain 5%–10% fiber [12]; in the United States, the fiber content of macadamia nuts is approximately 8.6% [5], while the fiber content of macadamia nuts grown in Australia is 6.4% [6]. Macadamia nuts have a low moisture content, between 1.2% [6] and 2.1% [11], which is desirable, as a low moisture content decreases the probability of microbial growth and improves the keeping quality and shelf life of the nuts.

15.5.2 Amino Acids

Macadamia nuts are a good source of plant protein, but compared to the other major tree nuts (almonds, Brazil nuts, cashews, hazelnuts, pecans, pine nuts, pistachios, and walnuts), macadamia nuts have the lowest protein content at 7.9% to 8.4% [5,13]. The amino acids found in the highest levels in nuts in general are the hydrophobic ones with macadamia nuts containing the lowest combined total of these amino acids of all the nuts (37.2%). Individually, these are proline (6.77 ± 0.81 g/100 g of protein), leucine (6.55 ± 0.04 g/100 g of protein), glycine (4.87 ± 0.33 g/100 g of protein), alanine (4.51 ± 0.09 g/100 g of protein), valine (4.31 ± 0.03 g/100 g of protein), phenylalanine (3.34 ± 0.04 g/100 g of protein), isoleucine (3.26 ± 0.03 g/100 g of protein), methionine (2.15 ± 0.05 g/100 g of protein), cysteine (0.84 ± 0.23 g/100 g of protein), and tryptophan (0.59 ± 0.04 g/100 g of protein) [11]. At a slightly lower total concentration are the acidic amino acids, glutamic acid (23.65 ± 1.04 g/100 g of protein) and aspartic acid (8.69 ± 0.51 g/100 g of protein), giving a total of 32.3% in macadamia nuts [11].

With the essential amino acids, which cannot be synthesized by the body, tryptophan is the first limiting amino acid in macadamia nuts. Apart from this, macadamia nuts appear to contain adequate amounts of all the essential amino acids [11].

15.5.3 Minerals

Macadamia nuts contain significant amounts of the essential micronutrients [5]. Of all the nuts, macadamias contain the lowest level of potassium at 368 mg/100 g, with pistachio nuts containing the highest level at 1025 mg/100 g of edible portion. The magnesium content of macadamias is 130 mg/100 g, slightly more than is contained in pecans and pistachios, which are the lowest at 121 mg/100 g each. Brazil nuts have the highest level of magnesium at 376 mg/100 g. The amount of calcium in macadamia nuts is 85 mg/100 g, directly above cashews, which have the least at 37 mg/100 g. The nut with the most calcium is almonds at 248 mg/100 g. Macadamia nuts contain 188 mg/100 g of phosphorous and small amounts of iron, zinc, and copper [5].

15.5.4 Vitamins

Macadamia nuts contain B complex vitamins, niacin (2.5 mg/100 g), thiamin (1.2 mg/100 g), riboflavin (0.2 mg/100 g), and pyridoxine (0.3 mg/100 g). They also contain vitamin C (1.2 mg/100 g) and pantothenic acid (0.8 mg/100 g). Of all the nuts, macadamias have the lowest folate content at 11 μg/100 g edible portion, with peanuts containing the highest level of folate at 240 μg/100 g edible portion [5].

15.5.5 Fatty Acids

Although nuts are high in fat, over 75% of the fat is unsaturated, predominantly monounsaturated fatty acids (MUFA). For every 100 g of fat in macadamia nuts, 77.4% is MUFA, of which 58.5% is oleic acid (18:1ω9) and 18.7% is palmitoleic acid (16:1). While the hazelnut has the highest total MUFA content (83.1%), macadamias contain the highest levels of 16:1. Only 4.4% of the lipid content of macadamia nuts is composed of polyunsaturated fatty acids (PUFA), of which 2.6% is α-linolenic acid (18:3ω3), second only to walnuts (13.2%); 1.8% is linoleic acid (18:2ω6). Macadamia nuts contain 18.2% saturated fatty acids (SFA), made up largely of 8.9% palmitic acid (16:0), 4.3% stearic acid (18:0), and 2.95% eicosanoic acid (20:0) [11]. Brazil, pine, and cashew nuts all contain higher amounts of SFA (25.4%, 24.1%, and 21.1%, respectively) [10,11].

15.5.6 Tocols

Vitamin E is an important lipid-soluble phenolic antioxidant that exists in eight different forms, four tocopherol and four tocotrienols, both existing in α-, β-, γ-, and δ-forms. Analysis of macadamia nut oil by Kornsteiner et al. [14] revealed that α-, β- and γ-tocopherols were nondetectable. In their analysis of macadamia nut oil from nuts grown in New Zealand, Kaijser et al. [7] detected 0.8–1.1 μg of α-tocopherol/g of oil, and 3.5–4.8 μg of δ-tocopherol/g of oil, but the other tocopherols were nondetectable. Maguire et al. [10] detected a much higher level of α-tocopherol in macadamia nut oil at 122.3 μg/g of oil, and a trace of γ-tocopherol was also detected. An explanation for the different levels of tocopherols found in three studies could be attributed to different cultivars, growing locations, and conditions. Unfortunately, the growing location for the nuts for two of the studies is unknown.

15.5.7 Phytosterols and Phytostanols

The total phytosterol level in macadamia nut oil is 1618.3 μg/g of oil, present as 73.3 μg campesterol/g of oil, 38.3 μg stigmasterol/g of oil, and 1506.7 μg β-sitosterol/g of oil [10]. In addition, macadamia nuts contain 185 μg squalene/g of oil, which is converted to phytosterols in plant cells [10]. The range of phytosterols in all nuts is from 2178.4 μg/g of oil, found in almonds, to 1096 μg/g of oil, found in hazelnuts.

The total phytosterol level in the macadamia nut itself is 116 to 187 mg/100 g of edible nut, present as 8–10 mg campesterol/100 g of edible nut and 108–144 mg β-sitosterol/100 g of edible nut. There is no stigmasterol present [15].

15.6 PHENOLICS IN MACADAMIA NUTS

Phenolic compounds have been shown to possess antioxidant, anti-inflammatory, and antithrombotic activities in cell culture and *in vitro* studies [16–18]. It has been speculated that phenolic components of the plant foods, particularly those present in virgin olive oil, may contribute to the lower incidence of coronary heart disease (CHD) and certain types of cancer in Mediterranean countries. Phenolic compounds have been shown to improve ischemic reactive hyperemia during the postprandial state, a phenomenon thought to be mediated via reduction in oxidative stress and the increase of nitric oxide metabolites [19]. Tree nuts are a good source of phenolic compounds ranging from 32 mg gallic acid equivalents (GAE)/100 g in pine nuts to 1625 mg GAE/100 g in walnuts. The total phenolic content of macadamia nuts is approximately 46 mg GAE/100 g (fresh weight) [14].

Phenolic compounds in macadamia nuts and shells have been identified and their antioxidant activities have been assessed in refined macadamia nut oil [20]. The phenolic compounds in macadamia oil are 2,6-dihydroxybenzoic acid (24 μg/g), 2′-hydroxy-4′-methoxyacetophenone (6.9 μg/g), 3′5′-dimethoxy-4′-hydroxyacetophenone (10.5 μg/g), and 3,5-dimethoxy-4-hydroxycinnamic acid (7.3 μg/g), with a total concentration of phenolic compounds of 48.7 μg/g of oil [20] (Table 15.1). Using gas chromatography (GC) to analyze the macadamia nut shells, the same as they had used to analyze the oil, Quinn and Tang [20] found the concentrations of phenolic compounds in the shell very much higher, with 285.6 μg/g of 2,6-dihydroxybenzoic acid (~12 times higher), 83.7 μg/g of 2′-hydroxy-4′-methoxyacetophenone (~12 times higher), 202.3 μg/g of 3′5′-dimethoxy-4′-hydroxyacetophenone (~19 times higher), and 266.3 μg/g of 3,5-dimethoxy-4-hydroxycinnamic acid (~36.5 times higher). The total concentration of phenolic compounds in the macadamia nut shells is 837.9 μg/g of shell, over 17 times higher than the concentrations in the oil [20] (Table 15.1). It is also suggested that other phenolic compounds might be present in the oil and shells, such as catechols, phrogallol, and 3,4,5-trihydroxy phenolic compounds.

In addition to the health benefits of phenolic compounds in oils, the antioxidants contribute to the oxidative stability of the nuts and oils and extend their shelf life. When macadamia oil is obtained from the crushed nuts and shells by cold processing, some of the antioxidants leech from the shells into the oil and increase the antioxidant content. Unfortunately, during the refining of macadamia nut oil products, many of the antioxidants are removed [20]. Phenolic compounds also have organoleptic properties, which color and flavor the oils [20,21].

15.7 HEALTH ASPECTS OF MACADAMIA NUTS

As mentioned above, macadamia nuts contain a range of health-promoting constituents, which may form a part of healthy human diets for promotion of good health and prevention of degenerative diseases. Despite being rich in fat, dietary trials and population studies have demonstrated the health benefits of regular consumption of macadamia nuts. Macadamia nuts are packed with MUFA, proteins, dietary fiber, vitamins, minerals, and phytochemicals that are thought to be responsible for their positive effects on health. Although the research conducted directly on macadamia nuts is limited, there is a plethora of information on health aspects of tree nuts in general; the results may be applicable to macadamia nuts because of the composition similarities. Health effects of macadamia nuts include benefits for metabolic health, in terms of weight management and prevention of insulin resistance, dyslipidemia, and platelet function (e.g., characteristics of metabolic syndrome).

15.7.1 Macadamia Nuts and Plasma Lipids

Intervention trials have demonstrated that consumption of a small handful of macadamia nuts lower plasma cholesterol (Table 15.2). Macadamia nuts are an excellent source of MUFA (such as 18:1ω9 and 16:1) that are known for their beneficial effects on circulating lipid levels. Four human clinical trials have evaluated the lipid lowering potential of macadamia nuts [22–25]; another one involved macadamia oil [26].

The clinical trials were conducted using hypercholesterolemic subjects with the exception of one study [25] in which the participants were normocholesterolemic. The study by Garg et al. [24] included all male participants; the study by Hiraoka-Yamamoto et al. [25] had all female participants while the other two studies [22,23] involved equal numbers of male and female participants. The duration of

TABLE 15.1
Phenolic Compounds in Macadamia Oil and Nut Shells

Compound	μg/g Oil [20]	μg/g Nut Shells [20]	mg GAE/100 g [19]
2,6-Dihydroxybenzoic acid	24 ± 3.61	285.6 ± 56.9	—
2′-Hydroxy-4′methoxyacetophenone	6.9 ± 0.9	83.7 ± 27.1	—
3′5′-Dimethoxy-4′-hydroxyacetophenone	10.5 ± 0.7	202.3 ± 47.0	—
3,5-Dimethoxy-4-hydroxycinnamic acid	7.3 ± 0.3	266.3 ± 22.6	—
Total concentration	48.7	837.9	46.0

TABLE 15.2
Cholesterol Lowering Effects of Macadamia Nuts

Participants (Male/Females)	Duration	Lipid Changes	Ref.
n = 14 (7/7); 25–59 years	21 days	Reduced serum total and LDL cholesterol	[22]
n = 30 (15/15); 18–53 years	30 days	Reduced serum total and LDL cholesterol	[23]
n = 17 (17/0); mean 54 years	28 days	Reduced serum total and LDL cholesterol	[24]
n = 24 (0/24); 19–23 years	21 days	Reduced serum total and LDL cholesterol	[25]

Note: LDL, low-density lipoprotein.

macadamia nut supplementation ranged from 21 to 30 days. The amount of macadamia nuts varied from 45 to 100 g/day, replacing 15%–20% of daily energy intake.

All these trials were consistent in reporting a reduction in serum cholesterol. Colquhoun et al. [22] reported a 7.9% reduction in serum total cholesterol and a 10.7% reduction in low-density lipoprotein (LDL) cholesterol following macadamia nut intervention compared to baseline values. In this study, plasma triacylglycerols (TAG) levels were reduced by 20.9% with no change in high-density lipoprotein (HDL) cholesterol after 3 weeks of macadamia nut consumption. Curb et al. [23] reported a 6.8% reduction in serum total cholesterol and a 7.2% reduction in LDL cholesterol following macadamia nut intervention compared to baseline values. In this study, plasma TAG levels were reduced by 13.2% with a small, but significant reduction (3.5%) in HDL cholesterol after 30 days of macadamia nut consumption. In the trial by Garg et al. [24], a 3.0% reduction in serum total cholesterol and a 5.3% reduction in LDL cholesterol following macadamia nut intervention compared to baseline values were observed. In this study, plasma TAG levels remained unchanged, but a larger increase (7.9%) in HDL cholesterol after 28 days of macadamia nut consumption was noted. Finally, Hiraoka-Yamamoto et al. [25] reported a 6.1% reduction in serum total cholesterol and a 7.2% reduction in LDL cholesterol following macadamia nut intervention compared to baseline values. In this study, plasma TAG levels remained unchanged; however, a significant reduction (8.0%) in HDL cholesterol after 21 days of macadamia nut consumption was observed.

The lipid lowering potential of macadamia nuts has been attributed to the presence of high levels of MUFA. However, macadamia nuts contain considerably high levels (up to 20%) of 16:1 fatty acid in addition to be being a rich source of 18:1ω9 fatty acid. The effects of 16:1 fatty acid on human health have not been studied in great detail. Nestel et al. [26] concluded that 16:1 fatty acid present in macadamia oil may behave like a SFA, not a MUFA, in its effects on plasma total and LDL cholesterol. Using young growing swans as a model, Smith et al. [27] demonstrated that a diet enriched with purified 14:1 and 16:1 fatty acids resulted in the highest increase in LDL cholesterol, however, in this study even 18:1ω9 fatty acid–enriched diet also increased plasma cholesterol concentration. When consumed in the form of whole nuts as part of a healthy diet, the role of 16:1 fatty acid in human health merits further investigation. The plant sterols and dietary fiber present in macadamia nuts may also contribute to their lipid-lowering effects. It is apparent that unknown factors, other than the fat, present in tree nuts are responsible for their cholesterol lowering potential as quantitatively, the fat alone could not account for the extent of lipid reductions in the blood circulation [28].

15.7.2 Macadamia Nuts and Body Weight

Although all tree nuts in general are high in fats and kcal, macadamia nuts contain particularly high levels of fat (75% by weight). However, despite containing the highest fat content of all the tree nuts, macadamia nut consumption has been shown to have no adverse effects on body weight. There was no change in body weight following macadamia nut intervention in two of the trials [22,23] while the other two, including the one from our research group, demonstrated a small but significant reduction in body weight [24,25] after 28 days of consuming a handful of macadamia nuts daily. Dietary supplementation with 45 to 90 g of macadamia nuts had no effect on overall energy intake of study participants despite an increase in the fat content of the diet [24]. Apparently, macadamia nuts with high fat content may provide greater satiety resulting in automatic adjustments of the amount of other foods consumed. Weight gain was similar in the study in young growing swine when raised on diets enriched with various fatty acids including the group fed a diet supplemented with 14:1 and 16:1 fatty acids [27].

15.7.3 Macadamia Nuts and Platelet Aggregation

Generally speaking, research on tree nuts has been limited to their ability to modulate plasma lipid levels and information on their potential to modulate other health risk factors, including oxidative damage, inflammation, and clotting tendency, is lacking in the literature. Recently Garg et al. [29] have demonstrated that consumption of macadamia nuts for 4 weeks results in a nonsignificant reduction in circulating levels of thromboxane (26.2%) and prostacyclin (7.8%). The mean thromboxane TXB2 to prostacyclin ratio was reduced by 23.6%, following macadamia nut consumption; however, this change was not significant [29]. Thromboxane is a potent proaggregatory substance while prostacyclin is antiaggregatory. An overall reduction in the ratio of thromboxane and prostacyclin signify a tendency to reduce platelet aggregability following an injury. The study was of short duration and it is likely that longer intervention with macadamia nuts may result in a significant reduction in thromboxane to prostacyclin ratio resulting in reduced clotting tendency, a major manifestation of atherosclerosis. This will provide an additional mechanism by which macadamia nuts and possibly other tree nuts reduce the risk of cardiovascular disease (CVD). Further studies are warranted on this particular aspect of macadamia nuts.

15.7.4 Macadamia Nuts, Oxidative Damage, and Inflammation

Garg et al. [29] have recently demonstrated that macadamia nut consumption for 28 days was accompanied by a significant reduction in circulating levels of leukotrienes and 8-isoprostane [29]. 8-Isoprostane produced from arachidonic acid (20:4ω6) by a free radical catabolized reaction is a reliable *in vivo* marker of oxidative stress [30,31]. Plasma levels of 8-isoprostane were reduced by 18.9% following macadamia nut intervention. Previous studies have demonstrated that diets rich in MUFA reduce oxidizability of LDL and lower plasma 8-isoprostanes [32] and, therefore, maybe cardioprotective. It is, therefore, likely that MUFA present in the macadamia nuts are responsible for the reduction in 8-isoprostane levels in hypercholesterolemic subjects. It is also possible that the phenolic compounds present in macadamia nuts contribute to the reduction in plasma isoprostane levels [33]. Leukotrienes, derived from 20:4ω6 fatty acid by the action of lipoxygenase, are highly chemotactic substances, and elevated levels are suggestive of increased inflammation. Replacing 15% of daily energy intake with macadamia nuts for 28 days resulted in a 22.5% reduction of leukotriene concentration in the plasma. Anti-inflammatory effects of tree nuts, including macadamia nuts, have not previously been reported. A recent study using walnuts as an intervention in hypercholesterolemic subjects reported 64% increase in endothelium-dependent vasodialation (EDV) with a concomitant reduction of the vascular cell adhesion molecule (VCAM- 1) by 20% [34]. The authors attributed the change in VCAM-1 and EDV to changes in the plasma lipid profile and 18:3ω3 fatty acid present in walnuts. However, it is also likely that components of nuts, yet unidentified, may be responsible for the observed anti-inflammatory effects.

15.8 CONCLUSION

In summary, it would appear that consumption of macadamia nuts as part of a healthy diet is associated with a reduction in biomarkers of CHD. Combined cholesterol lowering, antiaggregatory, antioxidant, and anti-inflammatory effects of macadamia nuts, despite increasing the fat content of the diet, may be used alone or as an adjunct to drug therapy to reduce risk of CHD. Long-term health benefits of macadamia nut consumption remain to be elucidated. The constituents of macadamia nuts responsible for the reduction in biomarkers remain to be identified. In this respect, 16:1 fatty acid present in high amounts (~20% of total fatty acids) in macadamia nuts may appear promising. Effects of 16:1 fatty acid on eicosanoid metabolism are not known at present and are worthy of further investigations.

ACKNOWLEDGMENT

The research by Garg et al. [24,29] on macadamia nuts was funded by a grant from Horticultural Research and Development Corporation of Australia in association with Australian Macadamia Society.

REFERENCES

1. Conesa, J.A., Sakurai, M., and Antal, M.J., Synthesis of a high-yield activated carbon by oxygen gasification of macadamia nut shell charcoal in hot, liquid water, *Carbon*, 38, 839–848, 2000.
2. Dai, X. and Antal Jr, M.J., Synthesis of a high-yield activated carbon by air gasification of macadamia nut shell charcoal, *Ind. Eng. Chem. Res.*, 38, 3386–3395, 1999.
3. Suszkiw, J., Activated carbons in a nutshell, *Agr. Res.*, 47, 14–16, 1999.
4. ABC, Landline, Macadamia Nut Power, Published online at: www.abc.net.au/landline/stories/s962653.htm2003 (accessed July 25, 2007).
5. U.S. Department of Agriculture (USDA), National Nutrient Database for Standard Reference, Release 19, National Technical Information Service, USDA, Springfield, VA, 2006.
6. Australian Macadamia Society, Australian Macadamias—The Healthy Nut, Published on line at: http://macadamias.org/index.php?p=72007 (accessed July 25, 2007).
7. Kaijser, A., Duttah, P., and Savage, G., Oxidative stability and lipid composition of macadamia nuts grown in New Zealand, *Food Chem.*, 71, 67–70, 2000.
8. Ros, E. and Mataix, J., Fatty acid composition of nuts—Implications for cardiovascular health, *Br. J. Nutr.*, 96(Suppl. 2), 29S–35S, 2006.
9. Griel, A.E. and Kris-Etherton, P.M., Tree nuts and the lipid profile: A review of clinical studies, *Br. J. Nutr.*, 96(Suppl. 2), 68S–78S, 2006.
10. Maguire, L.S., O'Sullivan, S.M., Galvin, K., O'Connor, T.P., and O'Brien, N.M., Fatty acid profile, tocopherol, squalene and phytosterol content of walnuts, almonds, peanuts, hazelnuts and the macadamia nut, *Int. J. Food Sci. Nutr.*, 55, 171–178, 2004.
11. Venkatachalam, M. and Sathe, S.K., Chemical composition of selected edible nut seeds, *J. Agric. Food Chem.*, 54, 4705–4714, 2006.
12. Salas-Salvado, J., Bullo, M., Perez-Heras, A., and Ros, E., Dietary fibre, nuts and cardiovascular diseases, *Br. J. Nutr.*, 96(Suppl. 2), 45S–51S, 2006.
13. Brufau, G., Boatella, J., and Rafecas, M., Nuts: Source of energy and macronutrients, *Br. J. Nutr.*, 96(Suppl. 2), 24S–28S, 2006.
14. Kornsteiner, M., Wagner, K., and Elmadfa, I., Tocopherols and total phenolics in 10 different nut types, *Food Chem.*, 98, 381–387, 2006.
15. Segura, R., Javierre, C., Lizarraga, M.A., and Ros, E., Other relevant components of nuts: Phytosterols, folate and minerals, *Br. J. Nutr.*, 96(Suppl. 2), 36S–44S, 2006.
16. Moreno, J.A., Lopez-Miranda, J., Gomez, P., Benkhalti, F., El Boustani, E.S., and Perez-Jimenez, F., Effect of phenolic compounds of virgin olive oil on LDL oxidation resistance, *Med. Clin. (Barc).*, 120, 128–131, 2003.
17. Carluccio, M.A., Siculella, L., Ancora, M.A., Massaro, M., Scoditti, E., Storelli, C., Visioli, F., Distante, A., and De Caterina, R., Olive oil and red wine antioxidant polyphenols inhibit endothelial activation:

Antiatherogenic properties of Mediterranean diet phytochemicals, *Arterioscler. Thromb. Vasc. Biol.*, 23, 622–629, 2003.
18. Petroni, A., Blasevich, M., Salami, M., Papini, N., Montedoro, G.F., and Galli, C., Inhibition of platelet aggregation and eicosanoid production by phenolic components of olive oil, *Thromb. Res.*, 78, 151–160, 1995.
19. Ruano, J., Lopez-Miranda, J., Fuentes, F., Moreno, J.A., Bellido, C., Perez-Martinez, P., Lozano, A., Gomez, P., Jimenez, Y., and Perez Jimenez, F., Phenolic content of virgin olive oil improves ischemic reactive hyperemia in hypercholesterolemic patients, *J. Am. Coll. Cardiol.*, 46, 1864–1868, 2005.
20. Quinn, L.A. and Tang, H.H., Antioxidant properties of phenolic compounds in macadamia nuts, *J. Am. Oil Chem. Soc.*, 73, 1585–1588, 1996.
21. Kiritsakis, A., Flavor components of olive oil—A review, *J. Am. Oil Chem. Soc.*, 76, 673–681, 1998.
22. Colquhoun, D.M., Humphries, J.A., Moores, D., and Somerset, S.M., Effects of a macadamia nut enriched diet on serum lipids and lipoproteins compared to a low fat diet, *Food Australia*, 48, 216–221, 1996.
23. Curb, J.D., Wergowske, G., Dobbs, J.C., Abbott, R.D., and Huang, B., Serum lipid effects of a high-monounsaturated fat diet based on macadamia nuts, *Arch. Intern. Med.*, 160, 1154–1158, 2000.
24. Garg, M.L., Blake, R.J., and Wills, R.B., Macadamia nut consumption lowers plasma total and LDL cholesterol levels in hypercholesterolemic men, *J. Nutr.*, 133, 1060–1063, 2003.
25. Hiraoka-Yamamoto, J., Ikeda, K., Negishi, H., Mori, M., Hirose, A., Sawada, S., Onobayashi, Y., Kitamori, K., Kitano, S., Tashiro, M., Miki, T., and Yamori, Y., Serum lipid effects of a monounsaturated (palmitoleic) fatty acid-rich diet based on macadamia nuts in healthy, young Japanese women, *Clin. Exp. Pharmacol. Physiol.*, 31(Suppl. 2), 37S–38S, 2004.
26. Nestel, P., Clifton, P., and Noakes, M., Effects of increasing dietary palmitoleic acid compared with palmitic and oleic acids on plasma lipids of hypercholesterolemic men, *J. Lipid Res.*, 35, 656–662, 1994.
27. Smith, D.R., Knabe, D.A., Cross, H.R., and Smith, S.B., A diet containing myristoleic plus palmitoleic acids elevates plasma cholesterol in young growing swine, *Lipids*, 31, 849–858, 1996.
28. Kris-Etherton, P.M., Yu-Poth, S., Sabate, J., Ratcliffe, H.E., Zhao, G., and Etherton, T.D., Nuts and their bioactive constituents: Effects on serum lipids and other factors that affect disease risk, *Am. J. Clin. Nutr.*, 70, 504S–511S, 1999.
29. Garg, M.L., Blake, R.J., Wills, R.B., and Clayton, E.H., Macadamia nut consumption modulates favourably risk factors for coronary artery disease in hypercholesterolemic subjects, *Lipids*, 42, 583–587, 2007.
30. Mueller, T., Dieplinger, B., Gegenhuber, A., Haidinger, D., Schmid, N., Roth, N., Ebner, F., Landl, M., Poelz, W., and Haltmayer, M., Serum total 8-iso-prostaglandin F2alpha: A new and independent predictor of peripheral arterial disease, *J. Vasc. Surg.*, 40, 768–773, 2004.
31. Wood, L.G., Gibson, P.G., and Garg, M.L., Biomarkers of lipid peroxidation, airway inflammation and asthma, *Eur. Respir. J.*, 21, 177–186, 2003.
32. Nalsen, C., Vessby, B., Berglund, L., Uusitupa, M., Hermansen, K., Riccardi, G., Rivellese, A., Storlien, L., Erkkila, A., Yla-Herttuala, S., Tapsell, L., and Basu, S., Dietary (*n*–3) fatty acids reduce plasma F2-isoprostanes but not prostaglandin F2alpha in healthy humans, *J. Nutr.*, 136, 1222–1228, 2006.
33. Visioli, F., Caruso, D., Galli, C., Viappiani, S., Galli, G., and Sala, A., Olive oils rich in natural catecholic phenols decrease isoprostane excretion in humans, *Biochem. Biophys. Res. Commun.*, 278, 797–799, 2000.
34. Ros, E., Nunez, I., Perez-Heras, A., Serra, M., Gilabert, R., Casals, E., and Deulofeu, R., A walnut diet improves endothelial function in hypercholesterolemic subjects: A randomized crossover trial, *Circulation*, 109, 1609–1614, 2004.

16 Compositional Characteristics and Health Effects of Pecan [*Carya illinoinensis* (Wangenh.) K. Koch]

Ronald R. Eitenmiller and Ronald B. Pegg

CONTENTS

16.1 INTRODUCTION

Pecan [*Carya illinoinensis* (Wangenh.) K. Koch] is a monoecious, heterodichogamous, deciduous nut tree that is indigenous to the United States [1]. The native range follows the river bottoms of the Mississippi River and its many tributaries as well as the rivers of central and eastern Texas and their tributaries. The pecan is a member of the Juglandaceae family [2] and can grow 33 to 60 m in height, possess massive trunks up to 3.5 m in diameter, and live for a very long time. In fact, native trees are known to be 1000 years old, and some plantings are now ~120 years of age [3]. Although walnut, hickory nut, heartnut, and butternut also belong to the Juglandaceae family, the pecan is considered to be the traditional tree nut in the United States [4]. The word "pecan" comes from the Algonquin Indian term "pacaan," meaning "all nuts requiring a stone to crack" [5,6]. The pecan was an important food in the diet of Native American tribes of the central and southern regions of the United States long before the arrival of European settlers. These days they are distributed over an area extending from northern Illinois and southeastern Iowa to the gulf coast of the United States [7].

The industry has been active for ~120 years with the development of many cultivars. The pecan, however, does not come true from seeds, and is difficult to start from cuttings. Problems arose in early pecan husbandry with pecan trees being produced not true to type. Each cultivar, vegetatively propagated as clones, is genetically invariant. Beginning in the late 1800s, propagation was achieved mainly by budding and grafting selected cultivars onto seedling pecan rootstocks in order to perpetuate desirable varieties [8]. This has resulted in over 1000 named and documented pecan cultivars [9]. The pecan is an economically important agricultural crop for the United States, which produces >80% of the world's production [10]. During 1999–2005, the annual pecan production in the United States was 173–400 million pounds (in-shell) with an estimated market value of $201–$407 million [11]. Georgia, one of the major producing states, has 11 commercially viable cultivars including Cape Fear, Curtis, Desirable, Elliot, Gloria Grande, Kiowa, Oconee, Pawnee, Schley, Stuart, and Sumner. Of these cultivars, Stuart and Desirable pecans are commercially the most important ones. The pecan has a smooth shell and the kernels make up 40%–60% of the in-shell. Pecan is marketed as in-shell or shelled forms with the greatest sales occurring between the U.S. Thanksgiving and Christmas periods. The pecan is among the most preferred of all nuts in foods. The even, tender, crunchy texture, pleasing aroma, and appetizing flavor make them especially suitable for: (1) flavoring bakery goods, such as fruit cakes, cookies, pies, icings, and assorted items; (2) candies of many kinds; (3) flavoring dairy products, such as ice cream, milkshakes, and others; (4) toasting, salting, and use as snacks; and (5) use in salads, desserts, fowl stuffing, and with vegetables and meats [8,12]. The United States is the world's largest producer of pecan, but other major pecan-producing countries include Mexico, Australia, Israel, South Africa, Argentina, Chile, and Brazil [6]. The pecan ranks third behind almond and English walnut in total U.S. production [13], but is tied with the walnut as the second most frequently consumed tree nut in the United States after almonds.

This chapter provides a detailed overview on the nutritional composition and health-promoting components of the pecan, and summarizes the existing knowledge and appreciation for the use of pecan and its products in a variety of foods. Completion of the literature search required for the preparation of this chapter revealed that much of the material available on composition of the pecan resulted from studies that are now several decades old. This observation does not negate the quality of most of the earlier research; it, however, emphasizes the fact that a concentrated research effort is needed to provide data for the pecan that reflects current cultivars in production and modern agronomic practices. Though data on the compositional characteristics of the pecan is described below, more studies are needed particularly concerning minor components and dietary phytochemicals that may offer potential health benefits and disease risk reduction.

16.2 COMPOSITIONAL CHARACTERISTICS OF PECAN

16.2.1 Proximate Composition

The proximate compositions for both raw and dry roasted pecans are given in Table 16.1 [14]. The numbers for raw pecan are slightly different from previous reports of the macronutrients, as they reflect the most updated values from the U.S. Department of Agriculture (USDA) National Nutrient Database for Standard Reference, Release 20. Moreover, the table lists the values for the dry roasted pecan, which is not typically reported. Similar to other agricultural commodities, composition varies with cultivar, maturity, disease and insect damage, environmental conditions, geographic area of growth, soil type, and most agronomic practices, including irrigation and fertilization regimes. Several studies [15–31] have considered variables affecting the composition of the pecan. However, generalizations about effects of the many variables and their interrelationships on the major constituents of the pecan are

TABLE 16.1
Proximate Composition (g/100 g) of Raw and Dry Roasted Pecans

Proximate Composition	Raw	Dry Roasted
Water	3.52	1.12
Total lipid	71.97	74.27
Protein	9.17	9.50
Ash	1.49	1.56
Carbohydrate	13.86	13.55
Fiber, total dietary	9.6	9.4

Source: From U.S. Department of Agriculture (USDA), USDA National Nutrition Database for Standard Reference, Release 20, National Technical Information Service, USDA, Springfield, VA, 2007.

difficult to make. For example in 1994, Santerre [32] summarized compositional information available at the time, but only reported limited information on factors influencing specific compositional properties of the pecan. Since Santerre's review, several excellent studies with well-defined sampling plans have been published. These reports are discussed in the following sections of this chapter.

16.2.1.1 Moisture

The moisture content of harvested pecan fluctuates, depending on the time of harvest and climatic conditions (wet/dry) during growth. In order to reduce kernel breakage and to improve shelling efficiency, in-shell pecans are conditioned to ~8% moisture before cracking and shelling. After shelling, the conditioned pecan is dried back to <5% moisture to facilitate storage stability/extended shelf life. Santerre [33] provides an excellent discussion of pecan processing. The pecan nut is semiperishable; in-shell nuts maintained at 3.5% to 4.5% moisture have an acceptable storage life of 6 months at 21°C and 1 month at 38°C [34]. Erickson [35] also states that optimum storage conditions are 3.5% to 4.5% moisture. Higher moisture levels can support mold growth, kernel darkening, and various undesirable biochemical changes that often accompany these. On the other hand, lower moisture levels produce brittleness with increased susceptibility to mechanical damage during processing and marketing. The USDA National Nutrient Database [14] provides a value of 3.52 g/100 g of water in the raw pecan, which is less than the 5.5 g/100 g average that has previously been reported.

Wansri et al. [27] followed moisture changes throughout the kernel maturation sequence in Wichita and Western Schley pecans grown in Australia. Nut-in-shell and kernel moisture decreased with maturity. Additionally, as the harvest date was delayed, both nut-in-shell and kernel moisture contents declined. Singanusong et al. [30] followed moisture changes during maturation of the Western Schley pecan from green-closed to brown-open shuck. Nut-in-shell and kernel moisture levels of green-closed shuck were always higher than those of green-open shuck samples. The moisture content of the brown-open shuck pecan was always lower than green-open shuck counterpart. Observations by Singanusong et al. [30] generally agreed with earlier studies, indicating that kernel moisture content decreased as harvest was delayed [16,36–38].

The study by Heaton et al. [16] suggested that the optimum harvest date of the pecan is not necessarily associated with levels of natural kernel moisture. The authors generalized that sap moisture fluctuates with sequences of wet and dry weather and cannot be relied upon as a guide to harvest. Heaton et al. [16] concluded that an early harvest leads to improved color and flavor stability, which is related to reduced exposure of the nuts to severe weathering conditions of cycling drying and rewetting. Early studies by Woodroof and Heaton [12] showed that maximum flavor was reached about 4 weeks after maturity and then declined to rancidity development. Such work led to industry adoption of early harvest followed by drying to 4%–5% moisture and refrigeration [16].

16.2.1.2 Lipids

Pecan is rich in lipids. Santerre [32] reported that they contain 65%–75% lipid and that the lipid content depends upon growing conditions, horticultural practices, maturity, cultivar, and past productivity of the tree. The USDA National Nutrient Database [14] gives a value of 71.97% lipid for pecan; however, the value is based upon only three observations from "unknown" cultivars. Cultivar variations were clearly shown by early research conducted by Heaton et al. [16]. These investigators studied three cultivars—Schley, Stuart, and Wichita, which are commonly grown in the southeastern United States—over three growing seasons, and reported mean lipid contents of 76.7% for Wichita, 75.2% for Schley, and 75.1% for the Stuart cultivar. The investigators stated that the lipid content was unusually high as a result of exceptionally good kernel fill, indicating optimal production conditions over the 3-year period. Senter and Horvat [18] provided in-depth analytical data on the lipid components of six pecan cultivars (Table 16.2) including Stuart, Mahan, and four newly developed cultivars (Schley × Barton, Schley × McCulley, Cheyenne, and Shoshoni). The total lipid content ranged from 72% to 75% on a dry weight basis (dwb) and did not vary significantly ($P > 0.05$) among the cultivars at the 5% level. Triacylglycerols (TAG) comprised 97.6% of the total lipid fraction. Minor lipid fractions quantified in the study included complex lipids, monoacylglycerols, α,β-diacylglycerols, α,α′-diacylglycerols, and sterols.

Probably, the most in-depth report on pecan kernel lipid content is the study by Rudolph et al. [24]. The study encompassed 70 cultivars and breeding lines with nine commercial cultivars followed over four production years. The total lipid content of the 70 cultivars and lines ranged from 60.3% to 76.6%. The oil content of the commercial cultivars (three or four year mean) ranged from 75.1% in Stuart to 67.1% in Patrick on a fresh weight basis (fwb) (<5% moisture). Total lipid composition of the nine cultivars varied a great deal over the four production years, although statistical analyses were not provided. The percent oil in Schley varied only 1.9% over 3 years whereas Success varied 13.6% between the high and low years. The mean variation between years across cultivars was

TABLE 16.2
Major and Minor Lipid Constituents (g/100 g Nutmeat, dwb) from Six Pecan Cultivars

Cultivar	Lipid Content	Complex	Monoacylglycerol	α,β-Diacylglycerol	α,α′-Diacylglycerol	Sterol	Triacylglycerol
Stuart	72	0.48^{a}	0.79^{a}	0.79^{a}	0.33^{a}	0.21^{a}	69.39^{b}
Mahan	74	0.30^{a}	0.75^{a}	0.55^{b,c}	0.18^{a}	0.15^{a}	72.07^{a,b}
Cheyenne	75	0.71^{a}	0.05^{a}	0.27^{c}	0.02^{a}	0.07^{a}	74.41^{a}
Shoshoni	72	0.50^{a}	0.30^{a}	1.17^{a}	0.43^{a}	0.30^{a}	69.32^{b}
Schley × Barton	74	0.50^{a}	0.70^{a}	0.54^{b,c}	0.24^{a}	0.17^{a}	71.91^{a,b}
Schley × McCulley	72	0.35^{a}	0.59^{a}	0.52^{b,c}	0.09^{a}	0.08^{a}	70.38^{b}
Mean values	73	0.38	0.53	0.64	0.22	0.16	71.25

Source: From Senter, S.D. and Horvat, R.J., *J. Food Sci.*, 41, 1201, 1976. With permission.
Note: Each reported value is the mean of triplicate analyses.
[a–c] Values within a column followed by the same letter are not significantly different ($P > 0.05$).

8.4%. The authors attributed the variation by production year to environmental factors, although specific factors and the interrelationships to lipid content were not defined.

The Rudolph et al. [24] study was the first to dramatically show the impact of maturity on lipid content of the pecan kernel. For the Stuart cultivar, oil content was less than 2% at 9 weeks before maturity. Up to the sixth week before harvest, kernel oil content increased rapidly and did not substantially change over the final 4 weeks before harvest. In an earlier investigation, Wood and McMeans [23] found that lipid biosynthesis rapidly increases during early September in pecan grown in central Georgia, which is approximately 1 month prior to harvest. Lipid biosynthesis is rapid for ~2 to 3 weeks. Practically, all of the kernel dry weight accumulates during a 6-week period following the late water stage of development [23].

Maturation studies have been completed on pecan grown in Australia [27,30]. Wansri et al. [27] examined the oil content of Wichita and Western Schley cultivars at seven stages of maturity over the final 2 months of kernel development. Rapid lipid accumulation on a dwb was noted for both cultivars (Wichita, 42.3% to 62.8%; Western Schley, 44.1% to 68.7%). Wichita nuts were shown to reach maturity, and a stable lipid content, earlier than Western Schley nuts. Singanusong et al. [30] studied maturation of Western Schley over nine maturity stages characterized by shuck description (green-closed to brown-open) during the last 3 months of kernel development. In each of two production years (1999 and 2000), lipid content rapidly increased to final maturity. For example, in 1999, the initial lipid content rose from 38.05% to 70.58% and in 2000 from 50.23% to 72.58%. The authors concluded that lipid content was a good objective measure of maturity to ensure low moisture content and high sensory quality.

Toro-Vazquez and Pérez-Briceño [25] and Toro-Vazquez et al. [26] studied compositional properties of pecan grown in central Mexico. Mature nuts from 22 native trees not identified to cultivar were examined with their lipid content ranging from 70% to 75% (dwb).

Lipid content of the mature pecan is somewhat cultivar-dependent. However, Wells et al. [22] reported that orchard management practices can produce variations of a considerable extent within the same cultivar. They reported that lipid contents of irrigated pecan are generally higher than nonirrigated pecan of the same cultivar. Further, tree to tree variation of the same cultivar in the same orchard can be quite large on a year-to-year basis. In general, low yielding trees produce nuts higher in oil, and high yielding trees produce nuts lower in oil. Therefore, irregular bearing can influence lipid content and nut quality. Good management practices including irrigation as well as pest and disease control can, therefore, decrease variability of lipid content in a pecan cultivar on a year-to-year basis.

16.2.1.3 Protein

Protein is the third major constituent of the pecan. Literature values for protein in the raw pecan range from 5.00% to 16.9% [24–26,28,30]. Most recent USDA data [14] for the raw pecan sets the protein content at 9.17 g/100 g, which represents an upward revision from the 1984 data compilation [32,39]. Protein content varies by cultivar [24,40–42]. Rudolph et al. [24] noted a range in protein content of 7% to 17% in the 70 cultivars and selections included in their study. The frequency distribution of the percent protein for this large sample set is shown in Figure 16.1. Results indicated that pecan germplasm contains substantial genetic variation for protein content.

16.2.1.4 Carbohydrates

The USDA National Nutrient Database for Standard Reference, Release 20 [14] provides the following compositional data for carbohydrates in the raw pecan of unspecified cultivar origin: carbohydrate (by difference), 13.86 g/100 g; total dietary fiber, 9.6 g/100 g; total sugars, 3.97 g/100 g; sucrose, 3.90 g/100 g; fructose, 0.04 g/100 g; and starch, 0.46 g/100 g. This data is derived from only three data points, indicating that little definitive information exists on the carbohydrate components of pecan. Data on carbohydrate content presented by Santerre [32] was based on compositional

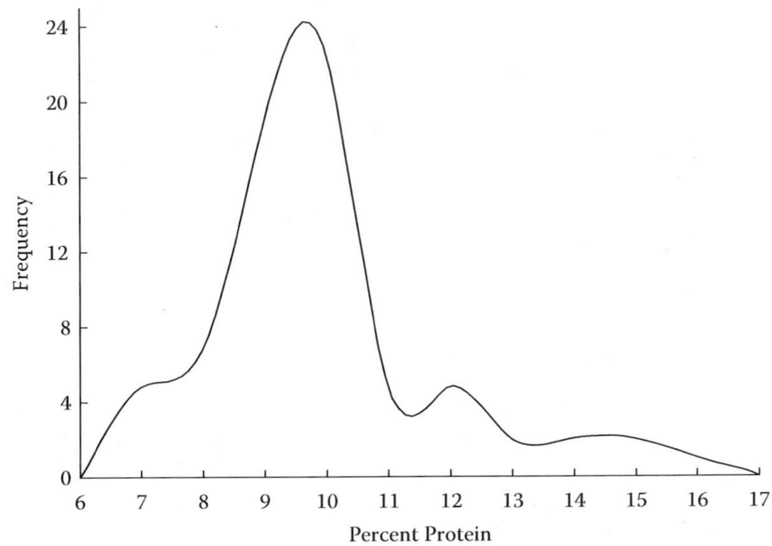

FIGURE 16.1 Frequency distribution of the percent protein at harvest for 70 pecan cultivars and unnamed selections. (From Rudolph, C.J., Odell, G.V., Hinrichs, H.A., Hopfer, D.A., and Kays, S.J., *J. Food Qual.*, 15, 263, 1992. With permission.)

information presented by USDA in the 1984 edition of *Agricultural Handbook* No. 8-12 [39] "Composition of Foods: Nuts and Seed Products," which reports total carbohydrate content of 18.2% and a dietary fiber value of 1.6%, indicating large discrepancies from the 2008 database values of 13.86 g/100 g for total carbohydrate and 9.6% for total dietary fiber.

At the present time, only five peer-reviewed research articles exist, which provide compositional information on the carbohydrates present in the pecan [23,27,28,30,43]. Four of the research studies [23,27,30,43] only deal with sugars, whereas research by Wakeling et al. [28] offers data on sucrose content and total dietary fiber. It is highly evident that a systematic study to define the carbohydrate composition of pecan does not exist to allow in-depth discussions and comparisons to other nuts or foods. Therefore, the following sections only review the available research reports and point out discrepancies in the data where they exist.

As cited by Santerre [32], USDA's 1984 edition of *Agricultural Handbook* No. 8–12 [39] provides a total carbohydrate value of 18.24 g/100 g. This value obtained by difference is based upon a moisture content of 4.82 g/100 g, protein content of 7.75 g/100 g, lipid content of 67.64 g/100 g, and an ash content of 1.56 g/100 g. The most recent USDA data in Standard Release 20 [14] gives a total carbohydrate value of 13.86 g/100 g. This value is based upon a moisture content of 3.52 g/100 g, protein content of 9.17g/100 g, lipid content of 71.97 g/100 g, and an ash value of 1.49 g/100 g. It is clear that the differences in the listed total carbohydrate content stem from the higher protein and lipid contents reported in the most recent USDA data release. Since the most recent USDA data reports a fat content (71.97 g/100 g) nearer to that cited in research works described in Section 16.2.1.2, the later USDA value is most likely a closer approximation to the "true" carbohydrate content of the pecan. Nevertheless, available literature sources do not provide proximate data obtained from comprehensive studies supported by a valid sampling plan to clearly define proximate composition, including total carbohydrate content.

16.2.2 Amino Acids

No recent studies have been conducted that report on amino acid content of pecan protein. Earlier work by Meredith [40], Wood and Reilly [42], and Elmore and Polles [44] showed that glutamic acid and arginine were the major amino acids found in pecan protein. Data from the Meredith [40] study for the Stuart, Sioux, and Schley cultivars is given in Table 16.3. For each pecan cultivar studied by Meredith [40], lysine was the first limiting amino acid, though tryptophan has been reported as the first limiting amino acid in other tree nuts. Calculated chemical scores using egg protein as the reference protein ranged from 37 to 46 [40].

Arginine content ranged from 0.691 to 1.226 g/100 g [40]. As noted for most tree nuts and peanut, the arginine content is higher than that reported in many other food categories. As arginine is the direct precursor of nitric oxide in mammalian metabolism, interest in the availability of arginine from dietary sources has increased. Nitric oxide is recognized as a key messenger compound that acts as a potent vasodilator. For this reason, it is responsible for maintenance of blood pressure and is known as the endothelium-derived relaxing factor (EDRF), which signals smooth muscles in the blood vessels to relax. Various roles in neurotransmission have been recognized for nitric oxide as well. Arginine is converted into citrulline with release of nitric oxide by nitric oxide synthase. It, therefore, functions as the primary circulating precursor necessary for nitric oxide synthesis.

TABLE 16.3
Amino Acid Composition (g/100 g) of Stuart, Sioux, and Schley Pecan Cultivars

Amino Acid	Stuart	Sioux	Schley
Alanine[a]	0.356	0.223	0.296
Arginine[b]	1.226	0.691	0.942
Aspartic acid[a]	0.771	0.400	0.620
Cysteine[b]	0.099	0.057	0.079
Glutamic acid[a]	1.677	0.963	1.580
Glycine[b]	0.397	0.240	0.314
Histidine[c]	0.250	0.151	0.194
Isoleucine[c]	0.353	0.212	0.285
Leucine[c]	0.553	0.338	0.445
Lysine[c]	0.332	0.202	0.265
Methionine[c]	0.191	0.136	0.145
Phenylalanine[c]	0.441	0.282	0.332
Proline[b]	0.408	0.278	0.319
Serine[a]	0.391	0.246	0.315
Threonine[c]	0.271	0.176	0.228
Tryptophan[c]	0.186	0.159	0.238
Tyrosine[b]	0.315	0.208	0.227
Valine[c]	0.408	0.249	0.334
Total indispensable amino acids	2.985	1.905	2.466
Total conditionally indispensable amino acids	2.445	1.474	1.881
Total dispensable amino acids	3.195	1.832	2.811

Source: From Meredith, F.I., Proceedings of the 87th Annual Meeting of the Florida State Horticultural Society, Miami Beach, FL, November 5–7, 1974, 362–365. With permission.

Note: Pecan samples tested were grown in Albany, Georgia in 1973.

[a] Dispensable amino acids.

[b] Conditionally indispensable amino acids.

[c] Indispensable amino acids.

16.2.3 Minerals

In the 1994 review of pecan composition, Santerre [32] reported data only from one study published in 1976 by Senter [17]. This study reported data for 16 elements from 10 pecan cultivars (Table 16.4). Earlier, Sparks [45] had reported concentrations of 14 elements in the pecan. Since these studies on pecan mineral composition, further information of the elements in the pecan were published by Furr et al. [21], Wakeling et al. [28], Singanusong et al. [30], and Moodley et al. [46]. One study indicates that the iron content in pecan kernels plays an important role with regard to coloration [47].

Wakeling et al. [28] followed elemental composition in Wichita and Western Schley over three growing seasons. These results showed no significant differences due to year of harvest or cultivar for calcium, potassium, sulfur, phosphorus, boron, copper, iron, or aluminum. Levels of magnesium, sodium, and zinc were significantly ($P < 0.05$) affected by harvest year, and the levels of manganese, sodium, and zinc changed significantly with cultivar. Except in very few instances, Wakeling et al. [28] results closely compared to earlier results [17,21,45]. Data for calcium presented by Senter [17] were ~10 times lower than calcium levels reported by other researchers, indicating a possible analytical error.

Singanusong et al. [30] followed elemental composition through maturation and found that calcium, magnesium, manganese, and zinc levels decreased significantly ($P < 0.05$) as maturity increased. The authors indicated that calcium levels in the pecan could be used as an index of

TABLE 16.4
Elemental Composition (mg/100 g Nutmeat) of Selected Pecan Cultivars

Cultivar	Cu	Fe	Co	Cr	Al	Mn	B	Zn
Cheyenne	1.44	1.93	tr	0.11	0	2.42	0.57	5.60
Western	1.22	2.52	tr	0.15	0	4.39	0.42	8.21
Tejas	1.22	2.65	tr	0.13	0	1.85	0.90	7.18
Cherokee	1.10	2.41	tr	0.20	0	1.73	0.74	8.03
Schley × Barton	1.09	2.15	tr	0.13	0	4.83	0.63	5.30
Shoshoni	1.06	2.28	tr	0.16	0	3.11	0.52	6.26
Stuart	1.08	2.02	tr	0.16	0	2.21	0.63	8.16
Schley × McCulley	0.90	2.00	tr	0.15	0	5.33	0.63	5.65
Mahan	0.87	2.11	tr	0	0	2.97	0.32	5.40
Desirable	0.82	1.93	tr	0	0	3.99	0.80	10.40
Mean	1.08	2.20	tr	0.12	0	3.28	0.62	7.02
Cultivar	**Mo**	**Sr**	**Ba**	**Na**	**P**	**K**	**Ca**	**Mg**
Cheyenne	0.07	0.52	0.67	0	390	330	0	140
Western	0.08	0.53	0.63	0.63	430	370	5.3	130
Tejas	0.05	0.74	0.90	0.21	470	440	5.3	160
Cherokee	0.07	0.64	0.32	0.84	430	540	0	150
Schley × Barton	0.07	0.52	0.47	0.21	440	390	5.3	150
Shoshoni	0.05	0.57	0.47	0.62	500	490	5.2	160
Stuart	0.06	0.58	0.48	0	470	470	0	120
Schley × McCulley	0.08	0.52	0.63	0.84	400	430	10.5	120
Mahan	0.07	0.53	0.27	0.63	340	440	5.3	170
Desirable	0.05	0.69	0.80	0.42	610	660	21.2	170
Mean	0.06	0.58	0.56	0.44	450	460	5.8	140

Source: From Senter, S.D., *J. Food Sci.*, 41, 963, 1976. With pemission.

Note: tr, trace.

TABLE 16.5
Percentage of RDA or AI Provided by the Pecan[a]

Mineral	RDA or AI for Adult Males[a]	RDA or AI for Adult Females[a]	Amount Provided in 42.5 g of Pecan[b]	Percent of RDA or AI for Adult Males[c]	Percent of RDA or AI for Adult Females[c]
Ca	1000 mg/day* [49]	1000 mg/day* [49]	30 mg	3*	3*
Cr	35 μg/day* [50]	25 μg/day* [50]	—	—	—
Cu	0.9 mg/day [50]	0.9 mg/day [50]	0.5 mg	56	56
Fe	8 mg/day [50]	18 mg/day [50]	1.1 mg	14	6
Mg	400–420 mg/day [49]	310–320 mg/day [49]	51 mg	12–13	15–16
Mn	2.3 mg/day* [50]	1.8 mg/day* [50]	1.9 mg	83*	106*
Mo	45 μg/day* [50]	45 μg/day* [50]	—	—	—
P	700 mg/day [49]	700 mg/day [49]	118 mg	17	17
K	4700 mg/day [51]	4700 mg/day [51]	174 mg	4	4
Na	1500 mg/day [51]	1500 mg/day [51]	0 mg	0	0
Zn	11 mg/day [51]	8 mg/day [51]	1.9 mg	17	24

[a] Recommended dietary allowances (RDA) or adequate intake (AI) for adults (aged 19–50 years).
[b] Quantity based upon USDA National Nutrient Database values [14].
[c] Values are based on a 42.5 g (or 1.5 ounces) pecan serving.
* Values are expressed as AI.

maturity. In a recent review, Griel and Kris-Etherton [48] stated that one serving of pecan provides ~10% of the daily value of both zinc and fiber.

Table 16.5 gives percentages of the recommended dietary allowances (RDA) or adequate intake (AI) levels for adult males and females of minerals provided by 1.5 ounces (42.5 g) of pecan using USDA Databank values [14]. A single serving of pecan provided significant amounts of manganese, copper, zinc, phosphorous, and magnesium. Reported selenium levels range from nondetectable [46] to 10 μg/100 g [52]. Data from the Food and Drug Administration's Total Diet Study show that pecan samples contain 2 μg/100 g [53], which closely compares to data reported in 1979 by Furr et al. [21]. The current USDA Databank value is 3.8 μg/100 g [14]. Data collected by Kannamkumarath et al. [52] was obtained with high-performance liquid chromatography/inductively coupled plasma/mass spectrometry (HPLC/ICP/MS), and showed pecan selenium levels to be low compared to Brazil nut, walnut, and cashew.

16.2.4 Vitamins

Published studies do not exist that define the water-soluble vitamin profile of the pecan. The U.S. Department of Agriculture Databank values [14] most likely were derived from analyses contracted by the USDA Nutrient Data Laboratory for inclusion to the databank. The following levels per 100 g were derived from three analyses: vitamin C, 1.1 mg; thiamin, 0.66 mg; riboflavin, 0.13 mg; niacin, 1.17 mg; pantothenic acid, 0.86 mg; and folate dietary folate equivalents (DFE), 22 μg. Clearly, the pecan can be considered as an excellent source of niacin and food folate. See Chapter 2 for detailed information.

16.2.5 Sugars

The literature agrees that sucrose is the primary sugar present in mature pecan. In an early study relying on gas chromatographic (GC) sugar analysis, Wood and McMeans [23] followed sugar concentrations throughout kernel maturation. During endosperm expansion, fructose and glucose accumulated whereas during embryo and cotyledon expansion, these sugars decreased to practically nondetectable levels and sucrose content increased to maturity. Throughout maturation, total sugars decreased. Sucrose levels reported by Wood and McMeans [23] were ~3.0 g/100 g.

Employing GC, Fourie and Basson [43] reported that mature pecan grown in South Africa contained 2.02 g/100 g of sucrose with small but detectable levels of inositol, glucose, and fructose (0.01, 0.01, and 0.02 g/100 g, respectively). Maturation studies completed by Wansri et al. [27] using a high-performance liquid chromatography (HPLC) technique for the analysis of sucrose levels in Wichita and Western Schley cultivars found that the disaccharide continuously increased during maturation reaching 2.37 g/100 g and 2.48 g/100 g in Wichita and Western Schley, respectively. Wakeling et al. [28] reported a mean sucrose content of 1.97 g/100 g for Wichita and Western Schley and that significant cultivar variations did not occur. The mean sucrose level of 1.97 g/100 g was derived from pecan collected in 1995–1997. It was skewed lower due to a low sucrose level in pecan collected in 1997. Individual crop year values for sucrose content were 2.65 g/100 g in 1995, 2.1 g/100 g in 1996, and 1.17 g/100 g in 1997. The pecan grown in 1997 was, however, from stressed trees that had experienced flooding conditions.

Singanusong et al. [30] also reported the sucrose content for Western Schley grown in Australia in 1999 and 2000. Final sucrose concentrations determined were 0.92 and 0.82 g/100 g, respectively, for the 1999 and 2000 crop years. These investigators also noted low concentrations of raffinose in the pecan collected in 2000.

16.2.6 Total Dietary Fiber

The USDA National Nutrient Database value for total dietary fiber in the pecan is 9.6 g/100 g [14], which is considerably higher than dietary fiber values reported in older USDA nutrient composition tables. Santerre [32] reported the dietary fiber value from the 1984 Handbook [39], which was 1.6 g/100 g. Prior to acceptance of Association of Official Analytical Chemists' (AOAC) Method 985.29 in 1985, most food fiber determinations were completed by crude fiber methods or modifications using acids or detergents [54]. Those methods normally gave variable fiber values compared to the current total dietary fiber procedure. Yet, the large discrepancy of 1.6 g/100 g compared to 9.6 g/100 g is difficult to understand. Only one report exists in the literature with application of a total dietary fiber methodology to pecan composition studies. Wakeling et al. [28] reported total dietary fiber values of 2.85 g/100 g for Wichita and 3.45 g/100 g for Western Schley. Production year and cultivar did not significantly ($P > 0.05$) affect the total dietary fiber concentrations.

16.3 LIPID CHARACTERISTICS OF PECAN

16.3.1 Fatty Acid Composition

One of the most complete studies of the fatty acid composition of pecan lipids was reported in 1966 by Heaton et al. [15] at the University of Georgia. The comprehensive study provided fatty acid profiles and total lipid values for 11 commercial cultivars (56 samples) and for 34 unidentified varieties (43 samples) collected over two production years. The fatty acid composition from the Heaton et al. [15] and more recent studies is reported in Table 16.6. Additionally, differences by growth location (Georgia, Texas, Louisiana, Florida, Mississippi, and Oklahoma), nitrogen fertilization rates, and insecticide applications (ethyl *p*-nitrophenyl thionobenzenephosphonate [EPN] and *O*,*O*-diethyl-*O*-(2-isopropyl-6-methylpyrimidine-4-yl)phosphorothioate [Diazinon]) were studied. Observations included the following:

- Lipid content of the commercial cultivars was greater than 71%. For all samples, lipid content ranged from 52.5% to 76.8%.

TABLE 16.6
Fatty Acid Composition (%) of Selected Pecan Cultivars

Cultivar	16:0	16:1ω7	17:0	17:1ω7	18:0	18:1ω9	18:2ω6	18:3ω3	20:0	20:1ω9
Cheyenne [18]	6.0	0.5	0.4	0.4	3.5	48.5	34.3	1.7	0.5	0.7
Curtis [15]	5.8	—	—	—	2.4	59.1	28.4	1.8	—	—
Desirable [15]	4.9	—	—	—	2.0	71.0	20.9	1.2	—	—
Elliot [15]	6.4	—	—	—	2.1	62.6	27.0	1.8	—	—
Farley [15]	5.4	—	—	—	2.2	66.5	24.6	1.3	—	—
Mahan [15]	5.9	—	—	—	2.4	57.3	32.4	1.9	—	—
Moneymaker [15]	6.5	—	—	—	2.0	64.2	25.9	1.5	—	—
Schley [15]	6.2	—	—	—	2.0	64.2	26.3	1.2	—	—
Schley [16]	5.9	—	—	—	2.1	68.7	21.8	1.1[a]	—	—
Shoshoni [18]	5.0	0.7	0.4	0.4	1.9	55.0	28.0	2.4	0.6	0.8
Stuart [15]	6.1	—	—	—	2.1	62.9	27.3	1.7	—	—
Stuart [16]	6.3	—	—	—	2.1	65.2	24.6	1.5[a]	—	—
Stuart [18]	5.7	0.7	0.4	0.3	1.7	50.1	32.6	2.6	0.6	0.6
Success [15]	5.5	—	—	—	2.3	68.0	23.2	1.0	—	—
Wichita [16]	6.2	—	—	—	2.5	73.1	16.9	1.1[a]	—	—
Wichita, Western Schley [28] (mean, 3 years)	6.6	0.2	—	—	2.5	55.3	32.9	1.7	—	—

[a] 18:3ω3 + 20:1ω9.

- Lipid components comprised 96% TAG with 4% other lipid constituents (e.g., phospholipids, sterols, mono- and diacylglycerols).
- Major fatty acids included 16:0 (4.6%–7.0%), 18:0 (1.5%–3.8%), 18:1ω9 (48.2%–73.9%), 18:2ω6 (16.0%–40.6%), and 18:3ω3 (0.7%–3.4%). A much tighter range was noted for fatty acids of the improved commercial cultivars: 16:0 (4.9%–6.4%), 18:0 (2.0–2.4%), 18:1ω9 (59.1%–71.0%), 18:2ω6 (20.9%–32.4%), and 18:3ω3 (1.0%–2.0%).
- Fatty acid profiles and lipid content were not affected by growing location.
- Increased nitrogen fertilization yielded nuts with a lower lipid content.
- 18:1ω9 content was lower in nuts from higher nitrogen treated trees.
- 18:2ω6 content was higher in nuts from higher nitrogen treated trees.
- Insecticide applications did not significantly ($P > 0.05$) affect oil content and fatty acid levels.

In a later study, Senter and Horvat [18] provided fatty acid data for six pecan cultivars. Statistical analysis showed no significant ($P > 0.05$) differences among replicate analyses within cultivars. Significant differences ($P < 0.05$) among cultivars in the quantities of each fatty acid were found, but with the exceptions of 18:3ω3, 20:0, and 20:1. Senter and Horvat [19] later indentified 23 fatty acids in four cultivars (2 years production). Fatty acids not previously identified included 10:0, 12:0, 12:1, 14:0, 14:1, 14:2, 15:0, 15:1, 15:2, 16:2, 17:2, 20:1, 21:0. U.S. Department of Agriculture data [14] shows that pecan fatty acids are 9.0% saturated, 59.5% monounsaturated, and 31.5% polyunsaturated.

Definitive research by Rudolph et al. [24] provided fatty acid composition data for nine cultivars analyzed over four production years. This work identified the relationships between 18:1ω9 and 18:2ω6 namely, that an increase or decrease in 18:1ω9 led to a commensurate but opposite change in 18:2ω6. The ratio of 18:1ω9 to 18:2ω6 (oleic to linoleic acid, O/L) varied among cultivars, ranging from a mean of 3.47 (Barton) to 1.66 (Western Schley). The O/L ratio within cultivars varied with production years and the susceptibility to variation was cultivar-dependent. Barton had the greatest differential in O/L ratio between years (3.89) and Schley the least (0.5). Thus, the fatty acid profile of Barton was the most susceptible to environmental influences. The authors indicated that O/L ratios provided a quality trait useful for breeding selection, but environmental factors influencing the O/L ratio required further study.

16.3.2 Tocopherol Composition

Tree nuts, in general, are recognized as excellent sources of vitamin E. Due to the highly unsaturated nature of pecan lipids, tocopherols are the primary, natural defense against oxidative events that can lead to both off-flavor development (oxidative rancidity) and darkening of the testa. Furthermore, their role as a natural antioxidant in the body against reactive oxygen species (ROS) along with endogenous phytochemicals (i.e., phenolic acids, flavan-3-ols, and proanthocyanidins) found in the pecan may offer health benefits against the development of chronic disease states. Various studies have reported tocopherol levels in the pecan [24,29,55–59]. Rudolph et al. [58] reported tocopherol levels as total tocopherol in the oil based on GC analysis. This study, while not providing information on the tocopherol profile, showed that tocopherol levels varied between cultivars and that the total vitamin E level declined during the final weeks of maturation. Rudolph et al. [58] also showed that total tocopherol concentration in pecan oil decreased as oxidation progressed with concurrent darkening of the oil.

Application of HPLC for the quantitation of tocopherols in the pecan provided information on the individual tocopherol homologs (α-, β-, γ-, δ-tocopherol). Summation of the four tocopherols yields a measure of the total vitamin E level, as tocotrienols are not present.

Fourie and Basson [56] noted that changes in tocopherol contents in the pecan during storage were related to development of oxidative rancidity. Quantified α- and γ-tocopherol levels decreased over a 16-month storage at 30°C. Yao et al. [57] provided the complete tocopherol profile for Desirable, Stuart, and Schley cultivars showing that α-, β-, γ-, and δ-tocopherols were present in pecan. This study related tocopherol decrease during storage to darkening; however, decreases in tocopherol levels did not show a significant relationship to rancidity development as measured by conjugated diene formation. Changes in tocopherols in Schley pecan stored at 23.9°C and 60%–70% relative humidity are shown in Figure 16.2.

In a later study, Chun et al. [29] observed significant differences in tocopherol contents among seedling, Desirable, Stuart, and Schley cultivars. Crop year variations in total tocopherol were small and no statistical differences ($P > 0.05$) existed for γ-tocopherol between crop years for each cultivar. Since γ-tocopherol comprised over 90% of the total tocopherol fraction by weight in each cultivar studied, its presence at nonvariable levels constitutes an important antioxidant characteristic of the pecan. A summary of the Chun et al. [29] study is given in Table 16.7.

16.3.3 Phytosterol Composition

Nuts contain small quantities of several bioactive components that are relevant to health, phytosterols being one of them. The first report on the sterol content of the pecan was in a compilation provided by Weihrauch and Gardner [60]. Pecan was reported to contain a total sterol content of 108 mg/100 g of which over 80% was β-sitosterol. More recently, Phillips et al. [61] provided sterol content for the pecan derived from composites of three regional samples collected in the United States. Mean sterol levels in mg/100 g were β-sitosterol, 116.5; campesterol, 5.9; stigmasterol, 2.6; Δ^5-avenasterol, 14.6; sitostanol, <1.7; campestanol, 2.8; and other sterols, 14.1. Total sterol content in the composites ranged from 154 to 159 mg/100 g. The data reported by these authors gives greater sterol concentrations than those reported in existing databases, but this is likely due to the inclusion of steryl glycosides, which represent a significant portion of total sterols in nuts that are not usually measured.

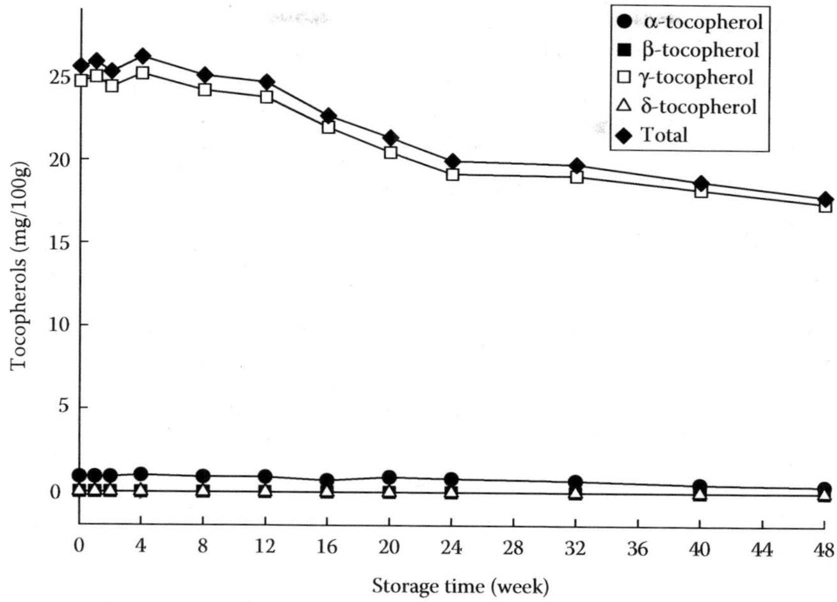

FIGURE 16.2 Tocopherols in Schley pecans stored at 23.9°C and 60%–70% relative humidity. (From Yao, F., Dull, G., and Eitenmiller, R., *J. Food Sci.*, 57, 1194, 1992. With permission.)

TABLE 16.7
Tocopherol Content (mg/100 g Nutmeat) of Selected Pecan Cultivars from the 1998 and 1999 U.S. Crop Year

Sample	Crop Year	α-Tocopherol	β-Tocopherol	γ-Tocopherol	δ-Tocopherol	Total	α -TE[a]
Seedling	1998 crop ($n = 6$)	1.2 ± 0.09	0.2 ± 0.03	22.7 ± 1.37	0.3 ± 0.10	24.4 ± 1.43	3.6 ± 0.16
	1999 crop ($n = 6$)	0.9 ± 0.08	0.2 ± 0.01	21.3 ± 0.96	0.2 ± 0.03	22.7 ± 1.00	3.2 ± 0.15
	Overall mean	1.0 ± 0.15	0.2 ± 0.03	22.0 ± 1.35	0.3 ± 0.08	23.6 ± 1.48	3.4 ± 0.25
Desirable	1998 crop ($n = 7$)	1.0 ± 0.11	0.3 ± 0.12	19.6 ± 1.59	0.4 ± 0.16	21.3 ± 1.41	3.1 ± 0.19
	1999 crop ($n = 6$)	1.3 ± 0.16	0.4 ± 0.05	20.5 ± 0.61	0.2 ± 0.03	22.4 ± 0.67	3.6 ± 0.19
	Overall mean	1.1 ± 0.22	0.4 ± 0.10	20.1 ± 1.27	0.3 ± 0.14	21.8 ± 1.25	3.3 ± 0.31
Stuart	1998 crop ($n = 7$)	1.4 ± 0.26	0.2 ± 0.03	23.8 ± 1.77	0.2 ± 0.04	25.6 ± 1.78	3.9 ± 0.30
	1999 crop ($n = 6$)	1.2 ± 0.08	0.3 ± 0.02	24.0 ± 0.91	0.2 ± 0.02	25.7 ± 0.94	3.7 ± 0.14
	Overall mean	1.3 ± 0.24	0.2 ± 0.05	23.9 ± 1.39	0.2 ± 0.03	25.6 ± 1.40	3.8 ± 0.25
Schley	1998 crop ($n = 7$)	1.2 ± 0.18	0.1 ± 0.14	30.5 ± 4.33	0.1 ± 0.04	32.0 ± 4.30	4.3 ± 0.51
	1999 crop ($n = 6$)	1.1 ± 0.15	0.1 ± 0.01	27.8 ± 1.98	0.1 ± 0.09	29.1 ± 1.89	4.0 ± 1.17
	Overall mean	1.2 ± 0.17	0.1 ± 0.10	29.3 ± 3.60	0.1 ± 0.07	30.7 ± 3.59	4.2 ± 0.43

Source: From Chun, J., Lee, J., Ye, L., and Eitenmiller, R.R., *J. Food Sci.*, 67, 1356, 2002. With permission.

Note: All samples were assayed in duplicate.

[a] α-TE (Tocopherol equivalent).

Studies have not been conducted to show cultivar, production year, environment, or agronomic practice effects on sterol content of the pecan.

16.4 PHYTOCHEMICALS

The phytochemicals in the pecan account for a portion of the nut's observed antioxidant and radical scavenging capacities. The health benefits derived from these secondary plant metabolites are discussed in the section below. The antioxidant activity originates mostly from the phenolic constituents (e.g., phenolic acids and tannins) and tocopherols. Early studies by Senter et al. [62] from the testa of Stuart pecan kernels revealed the presence of di- and trihydroxybenozic acid derivatives by gas chromatography–mass spectrometry (GC-MS). Eight phenolic acids were identified in pecan and include the following: gallic, gentisic, vanillic, protocatechuic, *p*-hydroxybenzoic, and *p*-hydroxyphenylacetic acid, with coumaric and syringic acids present in trace amounts. Acid quantities decreased significantly during 12 weeks of accelerated storage of the kernels at 21°C, 65% relative humidity. Strong correlations (r^2 = 0.95–0.97) were obtained between decreases in the hydroxybenzoic acid derivatives and declines in sensory quality of the kernels, thereby suggesting that these phenolic compounds may function antioxidatively and provide stability during storage. Gallic acid comprised 138 μg/g of the defatted kernel, which accounted for ~78% of the phenolic acid constituents. In a more recent study, Villarreal-Lozoya et al. [31] analyzed six pecan cultivars and found strong correlations in the kernels between total phenolic content and antioxidant activity. The total phenolic content ranged from 62 to 106 mg chlorogenic acid equivalents (CAE)/g defatted kernel and was significantly affected by pecan cultivar. After a base followed by an acid hydrolysis, HPLC chromatograms revealed the presence of gallic acid, ellagic acid, catechin, and epicatechin. Gallic acid was reported in the range of 651 to 1300 μg/g defatted kernel and ellagic acid in the range of 2505 to 4732 μg/g defatted kernel, with no significant differences among pecan cultivars for gallic acid ($P > 0.05$) and ellagic acid ($P > 0.05$). Furthermore, the findings confirm the presence of hydrolyzable tannins in the pecan kernels: both gallo- and ellagitannin types. The 2007 study reports a 10-fold higher content in gallic acid and identifies ellagic acid for the first time in pecan kernels. The authors postulated that the increase in gallic acid content may stem from the hydrolysis conditions employed and the release of this phenolic acid from hydrolyzable tannins and 3-*O*-gallate derivatives.

Polles et al. [63] assayed 31 pecan cultivars or seedling nutmeats for the presence of proanthocyanidins (also known as condensed tannins). Depending upon the cultivar, the percent condensed tannins of pecan nut kernels ranged from 0.70% to 1.71%. Tannin characteristics of pecan species are partly responsible for the coloration of the kernel; they are found in high quantities in the shuck and corky middle portion of the nut and to a lesser extent in the hull and kernel [63,64]. The color of the kernel is considered to be a primary factor in ascertaining general nut quality [3], and as the pecan ages during storage, it tends to darken. The importance of kernel color as a quality measure is due less to its effect on esthetic appeal than to the general association between dark kernel color and rancidity [3]. In 2002, Gu et al. [65] reported the presence of B-type proanthocyanidins (C4–C8 and C4–C6 bonds) in pecan kernels. These authors also identified prodelphinidines (e.g., 3-*O*-gallates) including epigallocatechin, epicatechin-3-*O*-gallate, catechin, and epicatechin. Gu et al. [66] characterized the degree of polymerization of the condensed tannins and reported a content of 494 ± 86 mg/100 g in pecan kernels, which represents almost 0.5% (w/w). The breakdown of polymerization for monomers, dimers, trimers, tetramers through hexamers, heptamers through decamers and polymers above 10 subunits was in the amounts of 17.2, 42.1, 26, 101, 84, and 223 mg/100 g, respectively. Villarreal-Lozoya et al. [31] found that condensed tannins, as evaluated by the vanillin assay, showed differences among cultivars and ranged from 23 to 47 mg catechin equivalents (CE)/g defatted kernel; this represented 0.7%–1.4% of the kernel weight, which was similar to the values reported by Polles et al. [63]. As tannins have been implicated in plant resistance to pathogens and insects in various plants [67], Polles et al. [63] postulated that the level of condensed tannins found in various pecan samples might be related to their insect and disease resistance.

In a report that screened common foods and vegetables across the United States, pecan kernels were shown to have the highest antioxidant capacity and total extractable phenolic content within the nut group, and the pecan ranked among the foods with the highest phenolic content [68]. The cultivar of the sample was never identified. Villarreal-Lozoya et al. [31], however, found that antioxidant capacity of pecan kernels is significantly affected by cultivar. Both oxygen radical absorbance capacity (ORAC) and 2,2-diphenyl-1-picryhydrazyl (DPPH) radical data on defatted kernels showed marked radical-scavenging capacity and strong correlations between these antioxidant assays with total phenolics; r^2 was 0.75 and 0.98, respectively. Mean ORAC values for the different pecan cultivars ranged from 373 to 817 μmol Trolox equivalents (TE)/g defatted meal; in their study, Wu et al. [68] reported an ORAC value of 583 μmol TE/g defatted kernel (179.40 μmol TE/g on an as-is basis). An important observation by Villarreal-Lozoya et al. [31] was that the proportions of condensed and hydrolyzable tannins differ for each cultivar, and this proportion determines the specific antioxidant activity of the phenolics present in each pecan cultivar. It was suggested that condensed tannins/total phenolics and ORAC/total phenolics ratios are important indices for antioxidant activity.

Finally, there is a growing body of evidence that polyphenols can facilitate circulatory function through increased production of the primary mediator of endothelial dilatation, viz., nitric oxide [69]. Maintaining the functional capacity of the endothelial cells lining blood vessels is vital to vascular health. Improvements in endothelial vasodilator function have been reported with high nut consumption [70]. To date, however, no such studies have involved the pecan.

16.5 HEALTH ASPECTS OF PECAN

According to the American Heart Association [71], cardiovascular disease (CVD) was responsible for 35.2% of all deaths nationwide in 2005. Heart disease remains the number one killer in the United States. Epidemiologic studies have consistently demonstrated an inverse association between nut consumption and risk markers of coronary heart disease (CHD) [72–74]. For instance, the Adventist Health Study [75] and the Nurses' Health Study [76] in the United States found a 30% reduction in overall relative risk of CVD with increased nut consumption. Development of CHD has been linked to many risk factors, some of which include gender, age, smoking habits, hypertension, obesity, lack of exercise, elevated levels of total blood cholesterol as well as low-density lipoprotein (LDL) cholesterol, and nutritional status [77]. In relation to individuals who ate nuts more than 1 time/week, those who ate them 1 to 4 times/week had a 25% reduced risk of dying from CHD; people who ate nuts ≥5 times/week experienced a ~50% reduction in risk [78]. The Food and Drug Administration (FDA) was petitioned and approved a qualified health claim with the following statements: "Scientific evidence suggests, but does not prove, that eating 1.5 ounces per day of some nuts, as part of a diet low in saturated fat and cholesterol, may reduce the risk of heart disease."

A number of mechanisms exist as to why nuts, such as the pecan, impart favorable effects on our cardiovascular system; the most important one being the lipid-lowering in blood serum. Yet, the lipid effects of nut intake only explain in part the CHD risk reduction observed in prospective studies, suggesting that nuts might have antiatherosclerotic effects beyond cholesterol lowering [70]. The pecan is low in saturated fatty acids (SFA) and rich in the monounsaturated fatty acids (MUFA), particularly oleic acid, which is known for its positive effects on blood lipids [79,80]. Reports and evaluations of the dietary habits in the Mediterranean region have linked diets, albeit rich in fat, to lower the incidences of CHD resulting from the consumption of olive oil [81–83]. In fact, MUFA levels in the pecan are similar to those of olive oil. The Scientific Advisory of the American Heart Association reported that high MUFA diets tend to raise high-density lipoprotein (HDL) cholesterol and lower TAG concentrations compared with low-fat, carbohydrate-rich, cholesterol-lowering diets [84,85]; this has the benefit of reducing the process of atherosclerosis and thus the risk of CHD. Evidence further suggests that other components in the pecan further reduce total cholesterol and LDL cholesterol concentrations beyond the effects predicted by equations based solely on fatty acid profiles [86,87]. The pecan is also rich in antioxidant vitamins, minerals, and numerous bioactives including flavonoids, stilbenes, and phytosterols that may have health benefits. Kris-Etherton et al. [73] point out that it is conceivable, although not proven, that many nutrients in nuts, like the pecan, may act synergistically to exert beneficial effects.

There have been four dietary studies on the effects of pecan consumption on serum (blood) lipid profiles. The first, a randomized control study from New Mexico State University compared the serum lipid profiles and dietary intakes of individuals with normal lipid levels (i.e., normolipidemic) who consumed pecan and those who did not consume the nuts at all [88]. The pecan treatment group consumed 68 g pecans per day for 8 weeks plus "self-selected" diets, whereas the control group avoided pecan as well as other nuts and also consumed "self-selected" diets. What is most interesting about this study is that though there is variability in the dietary habits of food choices made and calories consumed on a daily basis, this approach represents perhaps a realistic appraisal of the impact of the pecan set forth in the FDA-qualified health claim for tree nuts. Total, LDL, HDL, and TAG levels were measured at the beginning of the study to offer baseline data, at week 4 and then finally at week 8. Results showed that LDL cholesterol was lowered in the pecan treatment group from 2.61 ± 0.49 mmol/L at baseline to 2.35 ± 0.49 at week 4 ($P < 0.05$) and to 2.46 ± 0.59 at week 8 ($P < 0.05$). For the control group, LDL cholesterol levels increased from 2.74 ± 0.26 mmol/L at week 0 to 3.03 ± 0.57 at week 8. In terms of total cholesterol and HDL cholesterols, the numbers for the pecan treatment group at week 8 were significantly ($P < 0.05$) lower than in the control group (total cholesterol: 4.22 ± 0.83 vs. 5.02 ± 0.54 mmol/L; HDL cholesterol: 1.37 ± 0.23 vs. 1.47 ± 0.34 mmol/L). Additionally, dietary fat, MUFA, polyunsaturated fatty acids (PUFA), insoluble fiber, magnesium, and energy were significantly higher in the pecan treatment group than in the control group. Body mass indexes (BMI) and body weight were unchanged in both groups. Morgan and Clayshulte [88] concluded that the pecan can be included in a healthful diet when energy intake and potential weight gain are addressed.

The clinical study from New Mexico State University might be the first published study to specifically examine the effects of pecan ingestion on blood cholesterol and TAG levels. Though the study involved only 19 people, its findings are supportive of the FDA-qualified health claim for nuts and heart disease prevention. A study from Loma Linda University published in 2001 confirmed and extended the findings put forward by Morgan and Clayshulte [88]. Like previous investigations on the effects of almond and walnut on blood lipid levels [89–93], the study by Rajaram et al. [4] incorporated strict dietary regimens to control nutrient intake in addition to pecan supplementation. Their study examined the effects of pecan lipids as an alternative to the American Heart Association's Step I diet (e.g., a diet recommended by the National Cholesterol Education Program to lower cholesterol). Although the Step I diet is deemed favorable [94,95] due to its relatively high carbohydrate and low fat content, it has the disadvantage of tending to lower HDL cholesterol and raise TAG levels in the blood serum, which is an undesirable characteristic. Rajaram et al. [4] designed a single-blind, randomized, controlled, crossover feeding study for 23 subjects to follow two diets each of 4 weeks: a Step I diet, and a pecan-enriched diet (72 g/day), which proportionately reduced all food items of the Step I diet by one-fifth to provide a 20% isoenergetic replacement with pecan. Both diets improved lipid profiles of the subjects; however, the pecan-enriched diet decreased both total and LDL cholesterol concentrations by 0.32 mmol/L (e.g., 6.7% and 10.4%, respectively) and TAG by 0.14 mmol/L (~11.1%) beyond the Step I diet while increasing HDL cholesterol by 0.06 mmol/L. Furthermore, other serum lipoprotein markers decreased as a result of pecan supplementation to the diet. The authors concluded that the pecan, which is rich in MUFA, may be recommended as part of a prescribed cholesterol-lowering diets for patients or as part of the diet for healthy individuals. They also postulated that the unique nonfat component of the pecan may also have a role in favorably modifying the blood lipid profile and potentially other cardiovascular risk factors.

In 2005, a second clinical study on the potential health benefits of pecan consumption from New Mexico State University was published [96]. An 8-week, randomized, controlled study was conducted to assess the effect of pecan supplementation to the diets of hyperlipidemic adults (e.g., all 17 participants had total plasma cholesterol >200 mg/dL and LDL cholesterol >130 mg/dL). Again as in their previous study, individuals were assigned to a control group with no nut consumption or to a group where 68 g of pecan were consumed daily over the 8-week trial. All other aspects of the diet were self-selected. The researchers wanted to determine whether a significant effect on blood lipids would be found with a smaller (e.g., 68 g/day) and possibly a more dietarily acceptable level of nut supplementation than the 100 g/day of almond supplementation and 84 g/day of walnut supplementation from the Abbey et al. [93] and Sabaté et al. [92] studies, respectively. The results suggested that the inclusion of pecan in the diets of hyperlipidemic individuals only had a transient influence on the total and LDL cholesterol profiles, and did not yield sustainable lipid-lowering effect when total dietary fat intake was not limited. The researchers suggested that the 45% to 48% of energy as fat in the self-selected diets of the hypercholesterolemic participants consuming pecan might have been a mitigating factor in the transient lowering than rising of total and LDL cholesterol. What is most interesting is that if the study had been concluded after 4 weeks instead of 8, pecan consumption would have appeared to exert a lipid-lowering effect (Table 16.8). Despite the lack of a sustained lipid-lowering effect, the authors recommended that the inclusion of modest amounts of pecan, as a regular part of a healthy diet, is sound advice because of their overall nutritional profile.

In addition to MUFA, emerging evidence indicates there are other bioactive molecules in nuts, such as the pecan, that elicit cardioprotective effects. These include plant protein, dietary fiber, micronutrients such as copper and magnesium, plant sterols, and phytochemicals [72]. The pecan is also an excellent source of tocopherols, particularly γ-tocopherol, as has been described above. Recent studies, however, suggest that γ-tocopherol does not get the respect it deserves as a nutrient. γ-Tocopherol may have unique functions in detoxifying nitrogen dioxide and other reactive nitrogen species [97,98]. According to findings from Saldeen et al. [99] and Liu et al. [100], the effect of

TABLE 16.8
Effects of a Pecan-Enriched Diet on Plasma Cholesterol in Hyperlipidemic Adults

Serum Lipid Values, mg/dL (mmol/L)	Weeks of Supplementation		
	0	4	8
Total cholesterol			
Control group	259 ± 64 (6.71 ± 1.66)	257 ± 60 (6.66 ± 1.55)*	258 ± 67 (6.68 ± 1.74)
Pecan treatment group	233 ± 19 (6.03 ± 0.49)	221 ± 18 (5.72 ± 0.47)*	232 ± 35 (6.01 ± 0.91)
LDL cholesterol			
Control group	171 ± 60 (4.43 ± 1.55)	173 ± 60 (4.48 ± 1.55)**	169 ± 64 (4.38 ± 1.66)
Pecan treatment group	152 ± 21 (3.94 ± 0.54)‡	136 ± 22 (3.52 ± 0.57)**†	153 ± 33 (3.96 ± 0.85)†
HDL cholesterol			
Control group	52 ± 12 (1.35 ± 0.31)	55 ± 13 (1.42 ± 0.34)	56 ± 16 (1.45 ± 0.41)
Pecan treatment group	45 ± 11 (1.17 ± 0.29)	46 ± 12 (1.19 ± 0.31)	42 ± 18 (1.09 ± 0.47)
TAG			
Control group	183 ± 84 (2.07 ± 0.95)	143 ± 54 (1.62 ± 0.61)	160 ± 78 (1.81 ± 0.88)
Pecan treatment group	205 ± 122 (2.32 ± 1.38)	204 ± 123 (2.31 ± 1.39)	221 ± 173 (2.50 ± 1.95)

Source: From Eastman, W.A. and Clayshulte, B.J., *Fam. Consum. Sci. Res. J.*, 33, 197, 2005. With permission.

Note: Values reported as group means ± standard deviations; $n = 9$ for the control group and $n = 8$ for the pecan treatment group.

* $P < 0.05$, ** $P < 0.01$, significant difference between groups (ANOVA).

† $P < 0.05$, ‡ $P < 0.01$, significant difference over time within group (ANOVA).

α- and γ-tocopherol supplementation on platelet aggregation and thrombosis in rats revealed that γ-tocopherol leads to a greater decrease in platelet aggregation and delay of arterial thrombogenesis than α-tocopherol supplementation. Additionally, there was some evidence to suggest that γ-tocopherol may be protective against CVD because plasma γ-tocopherol levels were inversely associated with increased morbidity and mortality due to CVD in population studies [101,102]. Haddad et al. [103] examined the effect of plasma tocopherol concentrations on indices of antioxidant capacity and of oxidative stress as affected by pecan consumption. Despite the favorable effects of diets high in unsaturated fat on lipid profiles, concern exists that such diets could increase lipid peroxidation, thereby negating some of the cardioprotective effects [104,105]. A randomized, single-blind, crossover, controlled-feeding trial involving 24 healthy subjects were assigned to either a control or a pecan-enriched (20% of energy) diet for 4 weeks, after which the diets were reversed and the study was continued for an additional 4 weeks. Results showed that plasma γ-tocopherol (cholesterol-adjusted) increased by 10.1% ($P < 0.001$), α-tocopherol decreased by 4.6% ($P < 0.001$), and 2-thiobarbituric acid reactive substances (TBARS) decreased by 7.4% ($P < 0.05$) on the pecan diet. The decrease in plasma TBARS is noteworthy as it indicates that tocopherols and polyphenols in the pecan may be effective in inhibiting *in vivo* lipid peroxidation and degradation. Criticism exists, however, that TBARS may not be a reliable indicator of oxidative stress; nevertheless, these findings are in line with those of Actis-Goretta et al. [106], who reported an 11% decrease in plasma TBARS concentrations in healthy individuals after 30 days of supplementation with a low-dose mixture of lipid-soluble antioxidants. No changes were observed for the ferric-reducing ability of plasma (FRAP) assay, an *in vitro* antioxidant assay, or the Trolox equivalent antioxidant capacity (TEAC) assay, an *in vitro* antioxidant assay. The reduction in α-tocopherol was ascribed to adjustment in the basic diet upon incorporation of pecan at 20% of energy. The observed reduction in cholesterol-adjusted α-tocopherol may also be due to the vitamin being transported in the plasma mainly by LDL cholesterol; thus, a change in the lipoprotein levels impacts the vitamin's concentration [107]. This conclusion concurs with the results of Ros et al. [70], who found a decrease in α-tocopherol concentration in LDL when subjects were placed on a walnut-enriched diet.

16.5.1 Weight Gain, Diabetes, and Metabolic Syndrome

There have been no definitive studies involving pecan consumption and the metabolic benefits in relation to body weight regulation, preventing diabetes, and counteracting metabolic syndrome (e.g., central adiposity clustering with insulin resistance ± glucose intolerance, atherogenic dyslipidemia, hypertension, proinflammatory state, and prothrombotic state) [108]. Although metabolic syndrome is characterized by a constellation of cardiovascular risk factors, as evident from the definition given above, these factors are closely associated with insulin resistance and elevated insulin concentrations [109]. There is growing evidence to suggest that tree nut consumption may have a beneficial effect. That being said, it would not be surprising to find in the near future an intervention trial examining a cause and effect relationship between, say, the consumption of pecan in a diet on various antioxidant status markers of subjects diagnosed with metabolic syndrome. The pecan is tied with the walnut as being the second most frequently consumed tree nut in the United States after almond.

In light of the current obesity epidemic in North America, questions arise as to whether the reported beneficial effects of nuts on coronary risk reduction and lipid serum and lipoprotein profile outweigh the detrimental effects of possible weight gain [110]. There is in fact no epidemiological evidence to support the disquiet that the high energy density of nuts can cause weight gain. Sabaté [111] quite correctly points out that in Mediterranean countries where the per capita consumption of nuts is almost double than that in the United States, the rate of obesity is significantly lower. Data collected from over 12,000 participants in the U.S. Department of Agriculture's Continuing Survey of Food Intakes by Individuals [112] revealed that BMI was lower in nut consumers than in those who never ate nuts (23.8 ± 0.1 kg/m^2 vs. 25.0 ± 0.1 kg/m^2) despite their higher energy intakes

(2191 ± 20 kcal/day vs. 1997 ± 9 kcal/day). In the context of calorie-restricted diets, adding nuts generates a more lasting and greater magnitude of weight loss amongst obese subjects while improving insulin sensitivity. From limited epidemiological studies, there appears to be an inverse or no relationship between the frequency of nut consumption and body weight, but additional studies are needed to clarify the effect of long-term consumption of nuts on body weight and their role in altering insulin sensitivity both in normal and type-2 diabetics [112,113]. Reviews by García-Lorda et al. [114] and St-Onge [115] suggest that while the evidence for nut consumption facilitating weight loss is weak, there is support for a role of nuts in contributing to the maintenance of a healthy weight.

Several mechanisms can potentially explain why nuts do not result in weight gain: nuts are energy- and nutrient-dense foods with high fiber, protein, and low glycemic index (GI), all of which are dietary factors that have been shown to increase satiety [116]. The fatty acid profiles of nuts may also influence appetite suppression to some extent. Though lipids do not have the same satiating capacity as protein or carbohydrate, fatty acid chain length and degree of unsaturation have been hypothesized to influence appetite through different rates of oxidation [117]. Regarding the pecan study from Loma Linda University, isoenergetic diets (i.e., pecan-enriched diet and control diet) were tested for their effects on blood lipids and lipoproteins. According to the researchers, subjects were weighed daily during the 2-week run-in period and once a week thereafter throughout the trial. Those subjects on the pecan diet tended to lose weight [4,113]. Nuts are a complex matrix of nutrients; consequently, it is likely that fatty acid availability from them is decreased on account of incomplete digestion and or absorption. For example, in an intervention feeding trial with pecan, the lipid content in stools was greater when subjects were on the nut diet compared to the control diet (25.2 ± 3.8 compared with 6.3 ± 1.0 g/day, $P < 0.01$, respectively). This represented 8.3 ± 1.1 and 2.9% ± 0.5%, $P < 0.001$, of the dietary fat of the pecan and control diet, respectively [118].

Both chain length and degree of saturation of fat have been suggested to impact on glucose and insulin regulation [119]. An excellent review of this thesis is provided by Hu et al. [120]. Evidence to support a beneficial effect on nuts on insulin sensitivity comes from the Nurses Health Study [121]; an inverse association between nut consumption and the risk of developing type-2 diabetes was observed for women who consumed nuts ≥5 times per week. Although the benefit was ascribed mainly to the fatty acid profile of the nuts, other components within nuts such as dietary fiber or magnesium have also been reported to demonstrate an inverse relationship with the risk of developing type-2 diabetes [122].

Current research is pointing to a protective role for nuts in reducing the risk of type-2 diabetes and in improving insulin sensitivity in obese diabetics under a weight loss situation. Obviously more research—likely mechanistic studies—is needed to support such a position and to ultimately determine whether nuts do influence insulin sensitivity in both normal and type-2 diabetic individuals. There is also some evidence relating oxidative stress to the extent of insulin resistance [123]. Giugliano et al. [124], for example, suggested that high oxidative stress promotes an impaired insulin efficacy, which in turn might aggravate the degree of oxidation. Little, however, is known about the antioxidant status of individuals who suffer from metabolic syndrome. Because a low antioxidant status is associated with obesity, which is high amongst subjects with metabolic syndrome, this chronic disease state may be associated with a low antioxidant status [125]. Whether the nutritional composition of nuts can help one suffering from metabolic syndrome is highly speculative. For the time being, however, there is sufficient evidence to continue promoting the inclusion of nuts, like the pecan, as part of healthy diets [113].

16.6 CONCLUSION

The data presented in this chapter show that the pecan can play a significant role in human nutrition and health on account of its high and special nutritional components. Hence, these nutritional attributes indicate that the pecan can serve as an important healthy food in the human diet. With respect to functional lipid characteristics of the pecan, the nuts are good sources of natural antioxidants (e.g., γ-tocopherol) and bioactives, thus reflecting their nutraceutical potential in different food and specialty applications. Despite an increase in dietary fat content, pecan enrichment as part of a healthy diet favorably affects plasma LDL and HDL cholesterol levels as well as lipoprotein profiles, which are major risk factors of CVD. A high MUFA-rich pecan diet is preferred to a low-fat control diet in decreasing plasma LDL cholesterol concentrations. The presence of essential minerals, vitamins, and amino acids, the high content of heart-healthy fats, and the presence of soluble dietary fiber, bioactives, and phytochemicals, including their antioxidant and radical scavenging capacities, make the choice of pecan addition to healthy diets an important dietary consideration in assisting against the potential development of chronic disease states.

REFERENCES

1. Sparks, D., Adaptability of pecan as a species, *HortScience*, 40, 1175–1189.
2. Hall, G.D., Pecan food potential in prehistoric North America, *Econ. Bot.*, 54, 103–112.
3. Woodroof, J.G., *Tree Nuts: Production, Processing, Products*. 2nd ed., AVI Publishing Company, Inc., Westport, CT, 1979.
4. Rajaram, S., Burke, K., Connell, B., Myint, T., and Sabaté, J., A monounsaturated fatty acid-rich pecan-enriched diet favorably alters the serum lipid profile of healthy men and women, *J. Nutr.*, 131, 2275–2279, 2001.
5. McHatton, T.H., The history, distribution and naming of the pecan (*Hicoria pecan*), in Proceedings of the 50th Annual Convention of the Southeastern Pecan Growers Association, Biloxi, MS, March 3–5, 1957, pp. 10–34.
6. Wood, B.W., Payne, J.A., and Grauke, L.J., An overview of the evolution of the U.S. pecan industry, in *Pecan Technology*, Santerre, C.R., Ed., Chapman & Hall, New York, 1994, pp. 1–11.
7. Grauke, L.J., Appropriate name for pecan, *HortScience*, 26, 1358, 1991.
8. Duke, J.A., *Handbook of Nuts*, CRC Press, Boca Raton, FL, 2001.
9. Thompson, T.E. and Young, F., *Pecan Cultivars—Past and Present*, Texas Pecan Growers Association, College Station, TX, 1985.
10. Venkatachalam, M., Teuber, S.S., Peterson, W.R., Roux, K.H., and Sathe, S.K., Antigenic stability of pecan [*Carya illinoinensis* (Wangenh.) K. Koch] proteins: Effects of thermal treatments and in vitro digestion, *J. Agric. Food Chem.*, 54, 1449–1458, 2006.
11. U.S. Department of Agriculture (USDA), Fruit and Tree Nuts Situation and Outlook Yearbook/FTS2006/ October 2006, Economic Research Service, U.S. Department of Agriculture, 2006.
12. Woodroof, J.G. and Heaton, E.K., Pecans for Processing, University of Georgia College of Agriculture Bulletin N.S. 80, 1961.
13. Roux, K.H., Teuber, S.S., and Sathe, S.K., Tree nut allergens, *Int. Arch. Allergy Immunol.*, 131, 234–244, 2003.
14. U.S. Department of Agriculture (USDA), USDA National Nutrition Database for Standard Reference, Release 20, National Technical Information Service, USDA, Springfield, VA, 2007.
15. Heaton, E.K., Marion, J.E., and Woodroof, J.G., Pecan oil is highly unsaturated, *Peanut J. Nut World*, 45, 36–38, 1966.
16. Heaton, E.K., Worthington, R.E., and Shewfelt, A.L., Pecan nut quality. Effect of time of harvest on composition, sensory and quality characteristics, *J. Food Sci.*, 40, 1260–1263, 1975.
17. Senter, S.D., Mineral composition of pecan nutmeats, *J. Food Sci.*, 41, 963–964, 1976.
18. Senter, S.D. and Horvat, R.J., Lipids of pecan nutmeats, *J. Food Sci.*, 41, 1201–1203, 1976.
19. Senter, S.D. and Horvat, R.J., Minor fatty acids from pecan kernel lipids, *J. Food Sci.*, 43, 1614–1615, 1978.
20. Beuchat, L.R. and Worthington, R.E., Technical note: Fatty acid composition of tree nut oils, *J. Food Technol.*, 13, 355–358, 1978.
21. Furr, A.K., MacDaniels, L.H., St. John, L.E., Jr., Gutenmann, W.H., Pakkala, I.S., and Lisk, D.J., Elemental composition of tree nuts, *Bull. Environm. Toxicol.*, 21, 392–396, 1979.
22. Wells, J.M., McMeans, J.L., and Payne, J.A., Pecan oil content and composition as affected by variety and orchard conditions, in Proceedings of the 73rd Annual Convention of the Southeastern Pecan Growers Association, Panama City, FL, March 9–11, 1980, pp. 112–113.
23. Wood, B.W. and McMeans, J.L., Carbohydrates and fatty acids in developing pecan fruit, *J. Am. Soc. Hort. Sci.*, 107, 47–50, 1982.

24. Rudolph, C.J., Odell, G.V., Hinrichs, H.A., Hopfer, D.A., and Kays, S.J., Genetic, environmental, and maturity effects on pecan kernel lipid, fatty acid, tocopherol, and protein composition, *J. Food Qual.*, 15, 263–278, 1992.
25. Toro-Vazquez, J.F. and Pérez-Briceño, F., Chemical and physicochemical characteristics of pecan (*Carya illinoensis*) oil native of the central region of Mexico, *J. Food Lipids*, 5, 211–231, 1998.
26. Toro-Vazquez, J.F., Charó-Alonso, M.A., and Pérez-Briceño, F., Fatty acid composition and its relationship with physicochemical properties of pecan (*Carya illinoensis*) oil, *J.Am. Oil Chem. Soc.*, 76, 957–965, 1999.
27. Wansri, R., Mason, R.L., Wakeling, L.T., and Nottingham, S.M., Maturation studies of pecan nuts grown in Queensland, *Food Aust*, 53, 562–567, 2001.
28. Wakeling, L.T., Mason, R.L., D'Arcy, B.R., and Caffin, N.A., Composition of pecan cultivars Wichita and Western Schley [*Carya illinoinensis* (Wangenh.) K. Koch] grown in Australia, *J. Agric. Food Chem.*, 49, 1277–1281, 2001.
29. Chun, J., Lee, J., Ye, L., and Eitenmiller, R.R., Effects of variety and crop year on tocopherols in pecans, *J. Food Sci.*, 67, 1356–1359, 2002.
30. Singanusong, R., Mason, R.L., and D'Arcy, B.R., Compositional changes of Australia-grown Western Schley pecans [*Carya illinoinensis* (Wangenh.) K. Koch] during maturation, *J. Agric. Food Chem.*, 51, 406–412, 2003.
31. Villarreal-Lozoya, J.E., Lombardini, L., and Cisneros-Zevallos, L., Phytochemical constituents and antioxidant capacity of different pecan [*Carya illinoinensis* (Wangenh.) K. Koch] cultivars, *Food Chem.*, 102, 1241–1249, 2007.
32. Santerre, C.R., Pecan composition, in *Pecan Technology*, Santerre, C.R., Ed., Chapman & Hall, New York, 1994, pp. 98–110.
33. Santerre, C.R., Pecan processing, in *Pecan Technology*, Santerre, C.R., Ed., Chapman & Hall, New York, 1994, pp. 49–67.
34. Woodroof, J.G. and Heaton, E.K., Year round on pecans by refrigerated storage, *Food Eng.*, 25, 83–85, 141–142, 1953.
35. Erickson, M.C., Methods for measurement of pecan quality, in *Pecan Technology*, Santerre, C.R., Ed., Chapman & Hall, New York, 1994, pp. 111–133.
36. Herrera, E.A., Early harvest and oven drying temperatures influence pecan kernel flavor, *HortScience*, 29, 671–672, 1994.
37. Herrera, E.A., Cárdenas, M., and Meléndez, G., Effects of early harvesting on pecan moisture content, weight loss, and kernel percent, *Phyton. Int. J. Exp. Bot.*, 55, 89–96, 1994.
38. Silva, J.L., Matta, F.B., and Herrera, E.A., Quality indices and sensory scores support early harvest of pecans, *HortTechnology*, 5, 208–211, 1995.
39. United States Department of Agriculture, *Composition of Foods: Nuts and Seed Products, Agriculture Handbook,* No. 8-12, Human Nutrition Information Service, Washington, DC, 1984.
40. Meredith, F.I., Amino acid composition and quality in selected varieties of pecans, *Carya illinoensis*, in Proceedings of the 87th Annual Meeting of the Florida State Horticultural Society, Miami Beach, FL, November 5–7, 1974, pp. 362–365.
41. Hammar, H.E. and Hunter, J.H., Some physical and chemical changes in the composition of pecan nuts during kernel filling, *Plant Physiol.*, 21, 476–491, 1946.
42. Wood, B.W. and Reilly, C.C., Pecan kernel proteins and their changes with kernel development, *Hort-Science*, 19, 661–663, 1984.
43. Fourie, P.C. and Basson, D.S., Sugar content of almond, pecan, and macadamia nuts, *J. Agric. Food Chem.*, 38, 101–104, 1990.
44. Elmore, C.D. and Polles, S.G., Nitrogen fertilization effects on amino acid composition of pecan (*Carya illinoensis*) nutmeats, *J. Agric. Food Chem.*, 28, 902–904, 1980.
45. Sparks, D., Concentration and content of 14 elements in fruit of the pecan, *HortScience*, 10, 517–519, 1975.
46. Moodley, R., Kindness, A., and Jonnalagadda, S.B., Elemental composition and chemical characteristics of five edible nuts (almond, Brazil, pecan, macadamia and walnut) consumed in Southern Africa, *J. Environ. Sci. Health Part B*, 42, 585–591, 2007.
47. von Wandruszka, R.M.A., Smith, C.A., and Kays, S.J., The role of iron in pecan kernel color, *Lebensm. -Wiss. u. -Technol.*, 13, 38–39, 1980.
48. Griel, A.E. and Kris-Etherton, P.M., Tree nuts and the lipid profile: A review of clinical studies, *Brit. J. Nutr.*, 96 (Suppl. 2), S68-S78, 2006.
49. *Dietary Reference Intakes for Calcium, Phosphorus, Magnesium, Vitamin D, and Fluoride*, The National Academies Press, Washington, DC, 1997.

50. *Dietary Reference Intakes for Vitamin A, Vitamin K, Arsenic, Boron, Chromium, Copper, Iodine, Iron, Manganese, Molybdenum, Nickel, Silicon, Vanadium, and Zinc*, The National Academies Press, Washington, DC, 2001.
51. *Dietary Reference Intakes for Water, Potassium, Sodium, Chloride, and Sulfate*, The National Academies Press, Washington, DC, 2004.
52. Kannamkumarath, S.S., Wrobel, K., Wrobel, K., Vonderheide, and Caruso, J.A., HPLC-ICP-MS determination of selenium distribution and speciation in different types of nut, *Anal. Bioanal. Chem.*, 373, 454–460, 2002.
53. Pennington, J.A.T., Schoen, S.A., Salmon, G.D., Young, B., Johnson, R.D., and Marts, R.W., Composition of core foods of the U.S. food supply, 1981–1991: III. Copper, manganese, selenium, and iodine, *J. Food Comp. Anal.*, 8, 171–217, 1995.
54. AOAC International, *Official Methods of Analysis,* 18th ed., AOAC International, Arlington, VA, 2005.
55. Lambertsen, G., Myklestad, H., and Brækan, O.R., Tocopherols in nuts, *J. Sci. Food Agric.*, 13, 617–620, 1962.
56. Fourie, P.C. and Basson, D.S., Changes in the tocopherol content of almond, pecan and macadamia kernels during storage, *J.Am. Oil Chem. Soc.*, 66, 1113–1115, 1989.
57. Yao, F., Dull, G., and Eitenmiller, R., Tocopherol quantification by HPLC in pecans and relationship to kernel quality during storage, *J. Food Sci.*, 57, 1194–1197, 1992.
58. Rudolph, C.J., Odell, G.V., Hinrichs, H.A., Thomson, H.J., and Kays, S.J., Chemical changes in pecan oils during oxidation, *J. Food Qual.*, 15, 279–293, 1992.
59. Kornsteiner, M., Wagner, K.-H., and Elmadfa, I., Tocopherols and total phenolics in 10 different nut types, *Food Chem.*, 98, 381–387, 2006.
60. Weihrauch, J.L. and Gardner, J.M., Sterol content of foods of plant origin, *J. Am. Diet. Assoc.*, 73, 39–47, 1978.
61. Phillips, K.M., Ruggio, D.M., and Ashraf-Khorassani, M., Phytosterol composition of nuts and seeds commonly consumed in the United States, *J. Agric. Food Chem.*, 53, 9436–9445, 2005.
62. Senter, S.D., Horvat, R.J., and Forbus, W.R., Jr., Relation between phenolic acid content and stability of pecans in accelerated storage, *J. Food Sci.*, 45, 1380–1382, 1392, 1980.
63. Polles, S.G., Hanny, B.W., and Harvey, A.J., Condensed tannins in kernels of thirty-one pecan [*Carya illinoensis* (Wangenh.) K. Koch] cultivars, *J. Agric. Food Chem.*, 29, 196–197, 1981.
64. Brison, F.R. and Cain, R.F., Tannin content of kernels of the bitter pecan, in Proceedings of the Texas Pecan Growers Association, Brison, F.R. and Storey, J.B., Eds., Texas Pecan Growers Association, College Station, TX, 1957, pp. 32–33.
65. Gu, L., Kelm, M., Hammersonte, J.F., Beecher, G., Cunningham, D., Vannozzi, S., and Prior, R.L., Fractionation of polymeric procyanidins from lowbush blueberry and quantification of procyanidins in selected foods with an optimized normal-phase HPLC-MS fluorescent detection method, *J. Agric. Food Chem.*, 50, 4852–4860, 2002.
66. Gu, L., Kelm, M.A., Hammersonte, J.F., Beecher, G., Holden, J., Haytowitz, D., Gebhardt, S., and Prior, R.L., Concentrations of proanthocyanidins in common foods and estimations of normal consumption, *J. Nutr.*, 134, 613–617, 2004.
67. Feeny, P., Plant apparency and chemical defense, *Recent Adv. Phytochem.*, 10, 1–40, 1976.
68. Wu, X., Beecher, G.R., Holden, J.M., Haytowitz, D.B., Gebhardt, S.E., and Prior, R.L., Lipophilic and hydrophilic antioxidant capacities of common foods in the United States, *J. Agric. Food Chem.*, 52, 4026–4037, 2004.
69. Stoclet, J-C., Chataigneau, T., Ndiaye, M., Oak, M-H., El Bedoui, J., Chataigneau, M., and Schini-Kerth, V.B., Vascular protection by dietary polyphenols, *Eur. J. Pharmacol.*, 500, 299–313, 2004.
70. Ros, E., Núñez, I., Pérez-Heras, A., Serra, M., Gilabert, R., Casals, E., and Deulofeu, R., A walnut diet improves endothelial function in hypercholesterolemic subjects: A randomized crossover trial, *Circulation*, 109, 1609–1614, 2004.
71. Rosamond., W., Flegal, K., Furie, K., Go, A., Greenlund, K., Haase, N., Hailpern, S.M., Ho, M., Howard, V., Kissela, B., Kittner, S., Lloyd-Jones, D., McDermott, M., Meigs, J., Moy, C., Nichol, G., O'Donnell, C., Roger, V., Sorlie, P., Steinberger, J., Thom, T., Wilson, M., and Hong, Y., Heart disease and stroke statistics 2008 update: A report from the American Heart Association statistics committee and stroke statistics subcommittee, *Circulation*, 117, e25–e146, 2008.
72. Kris-Etherton, P.M., Yu-Poth, S., Sabaté, J., Ratcliffe, H.E., Zhao, G., and Etherton, T.D., Nuts and their bioactive constituents: Effects on serum lipids and other factors that affect disease risk, *Am. J. Clin. Nutr.*, 70, 504S–511S, 1999.
73. Kris-Etherton, P.M., Zhao, G., Binkoski, A.E., Coval, S.M., and Etherton, T.D., The effects of nuts on coronary heart disease risk, *Nutr. Rev.*, 59, 103–111, 2001.

74. Mukuddem-Petersen, J., Oosthuizen, W., and Jerling, J.C., A systematic review of the effects of nuts on blood lipid profiles in humans, *J. Nutr.*, 135, 2082–2089, 2005.
75. Fraser, G.E., Sabaté, J., Beeson, W.L., and Strahan, T.M., A possible protective effect of nut consumption on risk of coronary heart disease, *Arch. Intern. Med.*, 152, 1416–1424, 1992.
76. Hu, F.B., Stampfer, M.J., Manson, J.E., Rimm, E.B., Colditz, G.A., Rosner, B.A., Speizer, F.E., Hennekens, C.H., and Willett, W.C., Frequent nut consumption and risk of coronary heart disease in women: Prospective cohort study, *BMJ*, 317, 1341–1345, 1998.
77. Expert Panel, Summary of the second report of the National Cholesterol Education Program (NCEP) expert panel on detection, evaluation, and treatment of high blood cholesterol in adults (adult treatment panel II), *JAMA*, 269, 3015–3023, 1993.
78. Dreher, M.L., Maher, C.V., and Kearney, P., The traditional and emerging role of nuts in healthful diets, *Nutr. Rev.*, 54, 241–245, 1996.
79. Trevisan, M., Krogh, V., Freundenheim, J., Blake, A., Muti, P., Pancio, S., Farinaro, E., Mancini, M., Menotti, A., and Ricci, G., Consumption of olive oil, butter, and vegetable oils and coronary heart disease risk factors, *JAMA*, 263, 688–692, 1990.
80. Kris-Etherton, P.M., AHA science advisory: Monounsaturated fatty acids and risk of cardiovascular disease, *J. Nutr.*, 129, 2280–2284, 1999.
81. Baggio, G., Pagnan, A., Muraca, M., Martini, S., Opportuno, A., Bonanome, A., Ambrosio, G.B., Ferrari, S., Guarini, P., Piccolo, D., Manzato, E., Corrocher, R., and Crepaldi, G., Olive-oil-enriched diet: Effect on serum lipoprotein levels and bilary cholesterol saturation, *Am. J. Clin. Nutr.*, 47, 960–964, 1988.
82. Mattson, F.H., A changing role for dietary monounsaturated fatty acids, *J. Am. Diet. Assoc.*, 89, 387–391, 1989.
83. Grundy, S.M., What is the desirable ratio of saturated, polyunsaturated, and monounsaturated fatty acids in the diet? *Am. J. Clin. Nutr.*, 66, 988S–990S, 1997.
84. Grundy, S.M., Comparison of monounsaturated fatty acids and carbohydrates for lowering plasma cholesterol, *N. Engl. J. Med.*, 314, 745–748, 1986.
85. Mensink, R.P. and Katan, M.B., Effect of monounsaturated fatty acids versus complex carbohydrates on high-density lipoproteins in healthy men and women, *Lancet*, 329, 122–125, 1987.
86. Mensink, R.P. and Katan, M.B., Effect of dietary fatty acids on serum lipids and lipoproteins. A meta-analysis of 27 trials, *Arterioscler. Thromb. Vasc. Biol.*, 12, 911–919, 1992.
87. Hegsted, D.M., Ausman, L.M., Johnson, J.A., and Dallal, G.E., Dietary fat and serum lipids: An evaluation of the experimental data, *Am. J. Clin. Nutr.*, 57, 875–883, 1993.
88. Morgan, W.A. and Clayshulte, B.J., Pecans lower low-density lipoprotein cholesterol in people with normal lipid levels, *J. Am. Diet. Assoc.*, 100, 312–318, 2000.
89. Spiller, G.A., Jenkens, D.J.A., Cragen, L.N., Gates, J.E., Bosello, O., Berra, K., Rudd, C., Stevenson, M., and Superko, R., Effect of a diet high in monounsaturated fat from almonds on plasma cholesterol and lipoproteins, *J. Am. Coll. Nutr.*, 11, 126–130, 1992.
90. Spiller, G.A., Gates, J.E., Jenkins, D.J.A., Bosello, O., Nichols, S.F., and Cragen, L., Effect of two foods high in monounsaturated fat on plasma cholesterol and lipoproteins in adult humans [Abstract 90], *Am. J. Clin. Nutr.*, 51, 524, 1990.
91. Berry, E.M., Eisenberg, S., Friedlander, Y., Harats, D., Kaufmann, N.A., Norman, Y., and Stein, Y., Effects of diets rich in monounsaturated fatty acids on plasma lipoproteins—the Jerusalem nutrition study. II. Monounsaturated fatty acids vs carbohydrates, *Am. J. Clin. Nutr.*, 56, 394–403, 1992.
92. Sabaté, J., Fraser, G.E., Burke, K., Knutsen, S.F., Bennett, H., and Lindsted, K.D., Effects of walnuts on serum lipid levels and blood pressure in normal men, *N. Engl. J. Med.*, 328, 603–607, 1993.
93. Abbey, M., Noakes, M., Belling, G.B., and Nestel, P.J., Partial replacement of saturated fatty acids with almonds or walnuts lowers total plasma cholesterol and low-density-lipoprotein cholesterol, *Am. J. Clin. Nutr.*, 59, 995–999, 1994.
94. Stone, N.J., Nicolosi, R.J., Kris-Etherton, P., Ernst, N.D., Krauss, R.M., and Winston, M., Summary of the scientific conference on the efficacy of hypocholesterolemic dietary interventions, *Circulation*, 94, 3388–3391, 1996.
95. Ginsberg, H.N., Kris-Etherton, P., Dennis, B., Elmer, P.J., Ershow, A., Lefevre, M., Pearson, T., Roheim, P., Ramakrishnan, R., Reed, R., Stewart, K., Stewart, P., Phillips, K., and Anderson, N., Effects of reducing dietary saturated fatty acids on plasma lipids and lipoproteins in healthy subjects: The Delta study, protocol 1, *Arterioscler. Thromb. Vasc. Biol.*, 18, 441–449, 1998.
96. Eastman, W.A. and Clayshulte, B.J., Pecan effects on serum lipoproteins and dietary intakes of hyperlipidemic individuals consuming self-selected diets, *Fam. Consum. Sci. Res. J.*, 33, 197–207, 2005.

97. Christen, S., Woodall, A.A., Shigenaga, M.K., Southwell-Keely, P.T., Duncan, M.W., and Ames, B.N., γ-Tocopherol traps mutagenic electrophiles such as NO_x and complements α-tocopherol: Physiological implications, *Proc. Natl. Acad. Sci. USA*, 94, 3217–3222, 1997.
98. Jiang, Q. and Ames, B.N., γ-Tocopherol, but not α-tocopherol, decreases proinflammatory eicosanoids and inflammation damage in rats, *FASEB J.*, 17, 816–822, 2003.
99. Saldeen, T., Li, D., and Mehta, J.L., Differential effects of α- and γ-tocopherol on low-density lipoprotein oxidation, superoxide activity, platelet aggregation and arterial thrombogenesis, *J. Am. Coll. Cardiol.*, 34, 1208–1215, 1999.
100. Liu, M., Wallmon, A., Olsson-Mortlock, C., Wallin, R., and Saldeen, T., Mixed tocopherols inhibit platelet aggregation in humans: Potential mechanisms, *Am. J. Clin. Nutr.*, 77, 700–706, 2003.
101. Öhrvall, M., Sundlöf, G., and Vessby, B., Gamma, but not alpha, tocopherol levels in serum are reduced in coronary heart disease patients, *J. Intern. Med.*, 239, 111–117, 1996.
102. Kontush, A., Spranger, T., Reich, A., Baum, K., and Beisiegel, U., Lipophilic antioxidants in blood plasma as markers of atherosclerosis: The role of α-carotene and γ-tocopherol, *Atherosclerosis*, 144, 117–122, 1999.
103. Haddad, E., Jambazian, P., Karunia, M., Tanzman, J., and Sabaté, J., A pecan-enriched diet increases γ-tocopherol/cholesterol and decreases thiobarbituric acid reactive substances in plasma of adults, *Nutr. Res.*, 26, 397–402, 2006.
104. Porter, N.A., Caldwell, S.E., and Mills, K.A., Mechanisms of free radical oxidation of unsaturated lipids, *Lipids*, 30, 277–290, 1995.
105. Turpeinen, A.M., Basu, S., and Mutanen, M., A high linoleic acid diet increases oxidative stress in vivo and affects nitric oxide metabolism in humans, *Lipids*, 34 (Suppl. 1), 291S-292S, 1999.
106. Actis-Goretta, L., Carrasquedo, F., and Fraga, C.G., The regular supplementation with an antioxidant mixture decreases oxidative stress in healthy humans. Gender effect, *Clin. Chim. Acta*, 349, 97–103, 2004.
107. Kayden, H.J. and Traber, M.G., Absorption, lipoprotein transport, and regulation of plasma concentrations of vitamin E in humans, *J. Lipid Res.*, 34, 343–358, 1993.
108. Grundy, S.M., Brewer, H.B., Jr., Cleeman, J.I., Smith, S.C., Jr., and Lenfant, C., Definition of metabolic syndrome: Report of the National Heart, Lung, and Blood Institute/American Heat Association conference on scientific issues related to definition, *Arterioscler. Thromb. Vasc. Biol.*, 24, e13-e18, 2004.
109. Hansel, B., Giral, P., Nobecourt, E., Chantepie, S., Bruckert, E., Chapman, M.J., and Kontush, A., Metabolic syndrome is associated with elevated oxidative stress and dysfunctional dense high-density lipoprotein particles displaying impaired antioxidative activity, *J. Clin. Endocrinol. Metab.*, 89, 4963–4971, 2004.
110. Sabaté, J., Nut consumption and change in weight: The weight of the evidence, *Brit. J. Nutr.*, 98, 456–457, 2007.
111. Sabaté, J., Does nut consumption protect against ischaemic heart disease? *Eur. J. Clin. Nutr.*, 47 (Suppl. 1), 71S-75S, 1993.
112. Sabaté, J., Nut consumption and body weight, *Am. J. Clin. Nutr.*, 78, 647S–650S, 2003.
113. Rajaram, S. and Sabaté, J., Nuts, body weight and insulin resistance, *Brit. J. Nutr.*, 96 (Suppl. 2), 79S-86S, 2006.
114. García-Lorda, P., Megias Rangil, I., and Salas-Salvadó, J., Nut consumption, body weight and insulin resistance, *Eur. J. Clin. Nutr.*, 57 (Suppl. 1), 8S–11S, 2003.
115. St-Onge, M.-P., Dietary fats, teas, dairy, and nuts: Potential functional foods for weight control? *Am. J. Clin. Nutr.*, 81, 7–15, 2005.
116. Holt, S.H., Miller, J.C., Petocz, P., and Farmakalidis, E., A satiety index of common foods, *Eur. J. Clin. Nutr.*, 49, 675–690, 1995.
117. Kamphuis, M.M.J.W., Westerterp-Plantenga, M.S., and Saris, W.H.M., Fat-specific satiety in humans for fat high in linoleic acid vs fat high in oleic acid, *Eur. J. Clin. Nutr.*, 55, 499–508, 2001.
118. Haddad, E. and Sabaté, J., Effect of pecan consumption on stool fat, *FASEB J.*, 14, A294 [Abstract 227.7], 2000.
119. Coates, A.M. and Howe, P.R.C., Edible nuts and metabolic health, *Curr. Opin. Lipidol.*, 18, 25–30, 2007.
120. Hu, F.B., Manson, J.E., and Willett, W.C., Types of dietary fat and risk of coronary heart disease: A critical review, *J. Am. Coll. Nutr.*, 20, 5–19, 2001.
121. Jiang, R., Manson, J.E., Stampfer, M.J., Liu, S., Willett, W.C., and Hu, F.B., Nut and peanut butter consumption and risk of type 2 diabetes in women, *JAMA*, 288, 2554–2560, 2002.
122. Marlett, J.A., McBurney, M.I., and Slavin, J.L., Position of the American Dietetic Association: Health implications of dietary fiber, *J. Am. Diet. Assoc.*, 102, 993–1000, 2002.

123. Sentí, M., Tomás, M., Fitó, M., Weinbrenner, T., Covas, M.-I., Sala, J., Masiá, R., and Marrugat, J., Antioxidant paraoxonase 1 activity in the metabolic syndrome, *J. Clin. Endocrinol. Metab.*, 88, 5422–5426, 2003.
124. Giugliano, D., Ceriello, A., and Paolisso, G., Oxidative stress and diabetic vascular complications, *Diabetes Care*, 19, 257–267, 1996.
125. Davis, L., Stonehouse, W., Loots, D.T., Mukuddem-Petersen, J., van der Westhuizen, F.H., Hanekom, S.M., and Jerling, J.C., The effects of high walnut and cashew nut diets on the antioxidant status of subjects with metabolic syndrome, *Eur. J. Nutr.*, 46, 155–164, 2007.

17 Nutraceutical Potential of Pine Nut

Liangli Yu and Margaret Slavin

CONTENTS

17.1 INTRODUCTION

The pine nut, sometimes known as a pine kernel, is the ivory-colored, edible seed of a pine tree (genus *Pinus*) found in the pine cone under the scales. It is commonly consumed in a variety of cultures, particularly Mediterranean, Middle Eastern, Asian, and in the southwestern United States by Native Americans. It is a versatile food ingredient used in savory vegetable, meat, and poultry dishes, and also in sweet baked goods and confectionaries. Harvesting and shelling the pine nut is a labor-intensive process, which limits the number of species harvested in significant amounts, with the smaller kernel species not typically harvested. A list of nut-producing pine species in various geographical locations and their corresponding economic values are well summarized and can be found in Sharashkin and Gold [1].

Research on the possible health beneficial factors of various pine nut species has focused on seeds of the more commonly harvested species. However, there is potential for the smaller kernel species to have increased value if they are found to possess novel compounds/activities or larger amounts of known compounds. The available nutraceutical research on pine nut is quite broad, with promising research involving multiple chronic disease states, including hypertension, hypercholesterolemia, and obesity. There remains much potential to augment current knowledge of the health beneficial factors in the pine nut in each of these disease states and others.

17.2 COMPOSITIONAL CHARACTERISTICS OF PINE NUT

17.2.1 Proximate Composition

Detailed information about chemical composition of pine nut is given in Chapter 2. The nutritional composition of the pine nut varies according to species. Total lipid contents by weight of various species have been reported from 23% [2,3] to as high as 68% [1]. The pine nut is also a good source of protein, with protein contents by weight reported from approximately 12% [4,5] to 31% [6] and 34% [3], with the limiting essential amino acids of *Pinus pinea* to be lysine, threonine, and tryptophan [5]. Carbohydrate contents also vary, from as low as 2.4% to as high as 54% [3]. In addition, pine nut serves as a good source of a variety of important vitamins such as B_1 [6], vitamins K [7] and E [8] as well as minerals, including magnesium, zinc, iron, and phosphorus [6].

17.2.2 Antioxidants/Phenolics

Several studies have investigated the phenolic content of the pine nut. In 1983, Senter et al. [9] measured total phenolic acids in the defatted *Pinus edulis* pine nut meal. The meal was extracted with 0.1 M HCl in methanol, followed by acidic hydrolysis and preparation of methyl esters. The methyl esters of the phenolic acids were converted to their trimethylsilyl ethers, which were analyzed via gas–liquid chromatography–mass spectrometry (GLC-MS). This study found the total phenolics to be 0.137 mg gallic acid equivalents (GAE) per 100 g of defatted pine nut kernel with caffeic acid as the predominant phenolic acid [9]. In 2006, two additional studies evaluated phenolic contents of pine nut. One study reported the polyphenol content in the oil of *Pinus halepensis* to be 0.186 mg GAE/100 g oil [10]. The polyphenol content was determined using a colorimetric method using Folin-Ciocalteau reagent. Another group extracted phenolics from whole chopped nuts with a solvent consisting of 75% acetone and 526 mM sodium metabisulfate (3:1, v/v), then determined the phenolic content using the Folin-Ciocalteau reagent. It was reported that the total phenolics in pine nuts purchased in Austria and Greece (no species listed) to be low: 32 mg GAE/100 g fresh sample of pine nut [11]. Among most commonly consumed tree nuts and peanut, pine nut contained the lowest content of total phenolics (0.68 mg of GAE/g fresh sample) and total antioxidant activity (7.19 μmol of Trolox equivalents (TE)/g fresh sample) [46].

17.3 FUNCTIONAL LIPID CHARACTERISTICS OF PINE NUTS

17.3.1 Fatty Acid Composition

The complete fatty acid composition of pine nut is available in Chapter 2. Pine trees belong to the genus *Pinus* and contain an unusual group of polyunsaturated fatty acids (PUFA) present in considerable amounts in the seed oil of conifers. These particular PUFA contain 18 or 20 carbon atoms, with their first double bond at the Δ^5 location and a second double bond located at either $\Delta 9$ or $\Delta 11$. Thus, there is more than one methylene group separating double bonds on the fatty acid chain, and the term Δ^5 polymethylene-interrupted polyunsaturated fatty acid (PMI PUFA) is applied.

Six different Δ^5 PMI PUFAs have been identified in pine nut as shown in Figure 17.1. Pinolenic acid is the most prevalent Δ^5 PMI PUFA in most pine nut species, though contents vary greatly across species. Table 17.1 shows the Δ^5 PMI PUFA composition of selected pine nut species that are most commonly available in the food supply and/or are researched for their health-promoting qualities. A more extensive listing of pine species and their seed fatty acid compositions was done by Wolff et al. [12].

17.3.2 Phytosterols

Sterols are a group of cyclopentanoperhydrophenanthrene alcohols and are essential constituents of cellular membranes in animals and plants. Recently, the American Heart Association recommended

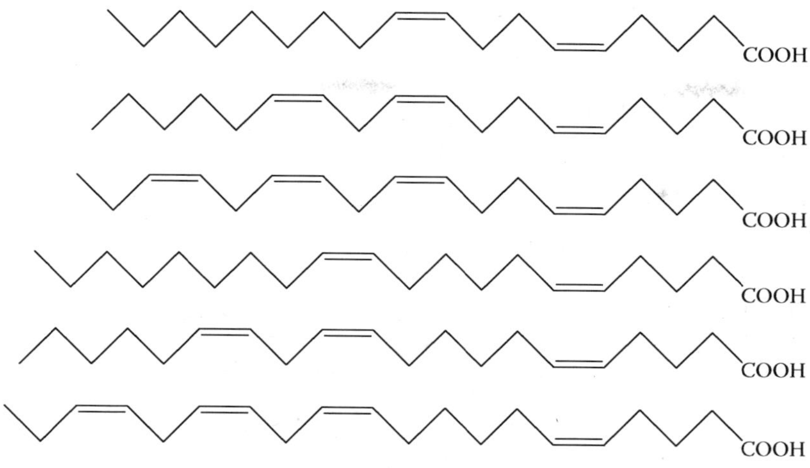

5,9 18:2 Taxoleic acid

5,9,12 18:3 Pinolenic acid

5,9,12,15 18:4 Coniferonic acid

5,11 20:2

5,11,14 20:3 Sciadonic acid

5,11,14,17 20:4 Juniperonic acid

FIGURE 17.1 Δ^5 Polymethylene-interrupted polyunsaturated fatty acids (PMI PUFA) occurring in *Pinus* seed oils.

TABLE 17.1
Δ^5 Polymethylene-Interrupted Polyunsaturated Fatty Acids (PMI PUFA) Composition of Selected Pine Nuts

Species[a]	Common Name	5,9 18:2 (Taxoleic)	5,9,12 18:3 (Pinolenic)	5,9,12,15 18:4 (Coniferonic)	5,11 20:2	5,11,14 20:3 (Sciadonic)	Ref.
P. edulis	Colorado pinyon	0.14	0.36	nd	nd	0.29	[13]
P. koraiensis	Korean pine	2.00	14.90	nd	0.10	0.80	[14]
		1.80	14.92	nd	0.10	0.90	[15]
		2.00	14.50	nd	0.10	0.90	[16]
		1.87	15.04	tr	0.08	0.79	[17]
P. halepensis	Aleppo pine	0.93–1.09	3.14–3.43	nd	<0.07	1.58	[18]
		0.20	0.10	0.0	0.50	4.50	[2]
		0.95	4.02	0.02	0.51	3.60	[19]
P. monophylla	Singleleaf pinyon	0.03	0.13	nd	0.07	0.34	[20]
P. pinaster	Maritime pine	0.74	7.13	nd	0.76	7.09	[15]
	Cluster pine	0.90	7.90	0.07	0.85	6.97	[12]
		0.10	8.10	0.10	1.00	7.20	[2]
P. pinea	Italian stone pine	0.14	0.35	nd	0.14	2.47	[15]
		0.10	0.00	0.0	0.50	2.20	[2]
		<0.05–1.19	0.29–0.42	nd	0.03–0.13	1.15–1.85	[18]
		1.00–3.10	2.50–3.30	nd	0.90–1.50	10.40–18.30	[21][b]
		1.00–5.30	1.60–3.70	nd	1.40–15.50	8.10–21.50	[21][c]
P. sibirica	Siberian pine	1.70	18.10	nd	0.10	0.90	[14]
	Russian cedar	1.98	18.49	nd	0.13	1.03	[19]

Note: Values expressed as weight % of total fatty acids; No data were available for 5,11,14,15, 20:4 (juniperonic) acid for any of the species; All double bonds are in the *cis* conformation; For a more extensive listing of pine species and their fatty acid compositions, see Ref. [12]; nd, not detected. tr, trace.

[a] All species are of the genus Pinus.

[b] Tunisian location.

[c] Mediterranean location.

supplementation of diet with plant sterol esters in order to reduce total and low-density lipoprotein (LDL) cholesterol concentrations in adults with hypercholesterolemia or high risk of atherosclerotic disease [22]. Growing evidence shows that consumption of approximately 2 g of plant sterols per day might reduce LDL cholesterol. Furthermore, current research appears to indicate that diets rich in naturally occurring plant sterols may reduce cholesterol absorption [23–25].

Pine nut has been studied for its phytosterol content. Segura et al. [26] summarized the phytosterol contents in pine nut from two previous studies and reported that pine nut (no species listed) may contain 141 to 236 mg total phytosterols/100 g edible portion of nut, with the highest individual component being β-sitosterol. Similarly, Nasri et al. [27] analyzed sterols in *P. pinea* populations from across Europe using GLC-MS. They reported β-sitosterol to be the most prominent phytosterol in the total extracted lipids, and an average total phytosterol amount of 429.8 mg/100 g oil. These data suggested that pine nut may serve as a dietary source of phytosterols.

Squalene is a precursor for steroid biosynthesis and may have cardioprotective and cancer-preventative properties. Though nuts are known to contain squalene, the limited literature available indicates that the pine nut contains a relatively low amount of squalene compared to other nuts [8]. Reverse-phase high-performance liquid chromatography (HPLC) analysis showed that pine nut oil had a squalene content of about 40 mg/100 g oil, which is approximately 23 mg/100 nuts. Thus, they would not be a likely choice for a squalene-based nutraceutical; however, this study did not report the species of pine nut measured, and variability between species of other compounds of interest may indicate that a different species could provide higher squalene content.

17.4 HEALTH EFFECTS OF PINE NUT

17.4.1 Plasma Cholesterol and Lipoprotein Profiles

It is known that dietary fatty acid composition may alter plasma lipoprotein concentrations, which are associated with the risk of cardiovascular disease (CVD). A number of studies have investigated the effects of pine nut oil intake on lipoprotein levels and the biological mechanisms involved, mainly using the French maritime pine (*Pinus pinaster*) [28–32]. In 1999, Asset et al. [29] investigated the effect of oils from *P. pinaster* and *Pinus koraiensis* seeds on plasma lipoprotein levels and apolipoprotein (Apo) gene expression in Wister rats. It was observed that 5% (w/w) *P. pinaster* seed oil in the diet significantly reduced serum triacylglycerol (TAG), very low-density lipoprotein (VLDL) TAG, and VLDL-cholesterol over the control diet with matching fatty acid composition except for the Δ^5 PMI PUFA, which were replaced with oleic acid [29]. Oil from *P. pinaster* resulted in a reduction of 30%, 40%, and 33% in serum TAG, VLDL TAG, and VLDL-cholesterol, respectively whereas feeding of *P. koraiensis* seed oil had a 16%, 21%, and 6% reduction in serum TAG, VLDL TAG, and VLDL-cholesterol, respectively. In addition, this study found out that *P. pinaster* seed oil did not alter the levels of circulating Apo, although it might slightly ($P < 0.05$) reduce liver Apo C-III mRNA level, but not that of Apo E, Apo A-I, or Apo A-II. These data suggested the potential health benefit of pine nut and its oil in improving the plasma lipid profile. These data also indicated that individual varieties of pine nut may significantly differ in their potential for improving human health. Also, in 1999, another study evaluated and compared the possible effects of *P. pinaster* seed oil on liprotein levels and atherosclerotic lesions to lard and sunflower oil using Apo E-deficient mice [28]. The Apo E-deficient mouse is an animal model of hyperlipidemia. It was observed that 10% (w/w) *P. pinaster* seed oil in the diet decreased plasma total and VLDL + intermediate-density lipoprotein (IDL)-cholesterol concentrations compared to control (lard), but not compared to sunflower oils [28]. It was also concluded from this study that *P. pinaster* seed oil might have a potential application in lowering VLDL and IDL cholesterol concentrations, but have no detectable preventive effect against atherosclerosis lesion formation in Apo E-deficient mice. Asset et al. [30] later compared the hypolipidemic effects of *P. pinaster* seed oil with fish oil in Apo E-deficient mice. Compared to the control group, both *P. pinaster* seed oil and fish oil significantly decreased VLDL and IDL cholesterol concentrations. Furthermore, *P. pinaster* seed oil had no difference to fish oil in lowering VLDL and IDL cholesterol concentrations. Interestingly, TAG levels in the Apo E-deficient mice increased in the mice fed pine and fish oils [28,30]. This observation was explained as a consequence of the Apo E deficiency and not an effect of the oils with normal metabolism.

In 2000, transgenic mice expressing human Apo A-I were used together with wild-type non-transgenic mice to study if *P. pinaster* nut oil may cause a change in high-density lipoprotein (HDL) and Apo A-I concentrations [31]. Animals were fed an isoenergenic diet containing 200 g/kg pine nut oil or lard for 2 weeks. The results indicated that a diet with 20% (w/w) pine nut oil might decrease HDL-cholesterol, along with total cholesterol and phospholipids, as compared to a diet with lard 20% (w/w) as the fat source. Further analysis of cholesterol efflux through an *in vitro* test with FU5AH rat hepatoma cells indicated a decrease in cholesterol efflux associated with the *P. pinaster* nut oil diet [31]. Cholesterol efflux is the first step in reverse cholesterol transport, the process where HDL delivers cholesterol back to the liver for excretion, thus a decrease in cholesterol efflux may promote development of atherosclerosis. It was suggested that these undesired effects maybe explained by the overexpression of human Apo A-I in the mice. In 2001, mice expressing human Apo B were used to study *P. pinaster* nut oil's effect on lipoprotein metabolism [32]. Animals were fed diets containing 20% pine nut, coconut, or sunflower oils [32]. The pine nut oil group experienced significant decreases in cholesterol, TAG, phospholipids, and Apo B as compared to the coconut oil group, but no differences when compared to the sunflower oil group. Another set of mice were fed sodium cholate in addition to a diet containing 20% pine nut, coconut, or sunflower oils. These mice were examined for atherosclerotic lesions, but no significant difference was seen in mean aortic lesion area, despite significant decreases in cholesterol, TAG, phospholipids, and Apo B in the pine nut and sunflower oil groups as compared to the lard group [32]. It was concluded that pine nut oil might not prevent cholesterol- and cholate feeding–induced aortic lesion formation.

In 2004, research was conducted to investigate the role of pinolenic acid on LDL-receptor activity [33]. A high-pinolenic acid-containing fatty acid extract was compared with a low-pinolenic acid-containing fatty acid extract for their effects on LDL receptor activity using HepG2 human liver carcinoma cells. The high-pinolenic acid-containing fatty acid extract treatment resulted in a significantly higher internalization of LDL in the HepG2 cells. These findings may help to explain some of the cholesterol-lowering effects seen in the mice studies discussed above.

In summary, pine nut shows promise in lowering cholesterol and therefore risk for CVD. The research is at times contradictory and sometimes shows detrimental effects. This is possibly due to the use of animal models that are inherently and sometimes unknowingly limited in their applicability to human diseases. Additional animal and human studies are required to understand the effects of pine nut or its individual components, such as pinolenic acid, on plasma lipids and ultimately CVD risk. It deserves mention that other chemical components in pine nut oils such as phytochemicals may contribute to its observed biological effects, and other components in the nut such as dietary fiber may also have potential cholesterol-lowering activity. Additional research is essential to investigate other components in pine nut and its effects on blood lipids.

17.4.2 Appetite Suppression

A nutraceutical product PinnoThin is marketed by Lipid Nutrition, The Netherlands, for its possible appetite-suppressing effects. According to the manufacturer, PinnoThin is a pine nut oil product derived from the Korean pine tree (*P. koraiensis*) [34]. The proposed mechanism of action of PinnoThin is that it causes an increase in release of satiety-inducing hormones cholecystokinin (CCK) and glucagon-like peptide1 (GLP1) [34]. The manufacturer cites an *in vitro* experiment using STC-1 mouse intestinal tumor cells, which release approximately 200% and 400% more CCK when treated with PinnoThin than Italian stone pine oil (*P. pinea*) and linolenic acid, respectively [34]. The only published information available on the efficacy of Pinnothin can be found in two abstracts [35,36], both referring to a single study of 18 overweight women, each of whom was fed 3 g PinnoThin with a light breakfast. Over the course of 4h, PinnoThin increased the release of CCK and GLP1 by 60% and 25% over the placebo (olive oil) response [35,36]. It also reduced "desire to eat" and "prospective food intake" by 29% and 36% of the placebo at 30min after intake [35,36]. However, it has to be kept in mind that little information is available from peer-reviewed literature on appetite-suppressing effects of pine nut oil, although a number of patents are owned surrounding the use of pinolenic acid as a weight loss adjuvant [37]. Also noted is that none of the animal studies using pine seed oil or pinolenic acid reported an effect on animal weight gain [28–32,38]. Notably, however, none of the animal models were specifically designed for weight studies. Further research to investigate active components is also needed.

17.4.3 Other Health Beneficial Activities

Pine nut oil has been evaluated for its effects in reducing blood pressure and modulating immune functions [38,39]. Early in 1994, pine nut oil containing approximately 18% pinolenic acid was compared with flaxseed, safflower, and evening primrose oils for their effects on PUFA metabolism, eicosanoid biosynthesis, and blood pressure in Sprague-Dawley rats [38]. After 5 weeks of receiving the treatment diet, which contained 10% of the tested oils by weight, the pine nut oil (*P. koraiensis*) appeared to halt the rise in systolic blood pressure of spontaneously hypertensive rats. This trend continued until the end of the study, at 8 weeks. A significant difference was seen between the diets supplemented with pine seed oil and evening primrose oil, but not with safflower oil [38]. Thus, it was concluded that pinolenic acid might not inhibit the conversion of linoleic acid to arachidonic acid. It was suggested that the antihypertensive effect of pine nut oil might be mediated by altering prostaglandin levels, but no direct evidence was provided. A Russian study suggests that taking 17.5g Siberian pine nut oil per day may improve systolic blood pressure in humans, but details of the experiment could not be found [40]. If valid, this would be the only *in vivo* study directly testing the effects of pine nut or its constituents on human health.

A few studies have examined possible effects of pine nut oil on immune functions because of its unique fatty acids [39,41]. In 2004, another mice feeding study evaluated the effects of pine nut oil on the intestinal immune system and macrophage from mice. Peyer's patch cells were prepared from the small intestine of C3H/He mice fed a diet with 5% pine seed oil (*P. koraiensis*) for 20 days and used in an *in vitro* immune response study [39]. These cells showed a 1.5-fold increase in immune system modulating activity over cells taken from the control mice, which were fed a corn oil diet. In addition, a diet with 5% pine seed oil fed to ICR mice over 20 days increased the stimulation of macrophages although this was statistically insignificant [39]. Another study that fed Brown-Norway rats a diet containing 10% pine seed oil showed an increase in the production of splenic $CD4^+$ T-lymphocytes and the splenic immunoglobulins IgG and IgE over the control of safflower oil [41]. These studies indicate that pine nut oil may modulate the immune system, but the exact mechanisms involved remain to be determined.

17.4.4 Toxicity

Little information is available on the toxicity of pine nut. In 2001, a study was conducted to investigate the effects of maritime pine seed oil (*P. pinaster*) on the developing brains of rat fetuses and pups [42]. It questioned whether the presence of Δ^5 PMI PUFA in *P. pinaster* seeds would prevent essential PUFA such as arachidonic acid from getting to the developing rat brain, thus inhibiting proper development of visual acuity and cognitive function in later life. Pregnant and lactating female Wistar rats were fed diets containing 10% oils (w/w). The treatment group received 14.1% (w/w) of total fatty acids as Δ^5 PMI PUFA, whereas the control group had a similar fatty acid profile in all other respects except that oleic acid replaced the Δ^5 PMI PUFA. The results from this study indicated that the pine nut oil in the diet of the mother did not have a detrimental effect on the ω-6 and ω-3 PUFA levels in the brains of her fetuses or pups [42].

17.5 BY-PRODUCTS FROM PINE NUT PROCESSING

The pine nut is generally used as an edible seed directly in the human diet and processed to obtain pine nut oil [1]. The by-products from pine nut processing include but are not limited to pine nut flakes, shell, and cones [1,43,44]. The flakes are used in a variety of processed food products including granolas, chocolates, and crunch bars [1]. Currently, pine nut shells are sometimes used as fuel for heat treating the seeds [45]. Chemical compositions of the flake, shell, and cone have been investigated in a few studies. A study isolated and identified six diterpenes with possible antitumor activities in the $CHCl_3$ extract of cones (*Pinus luchuensis*) [44]. These included a new *nor*-labsane-type diterpine compound, 15-*nor*-labda-8(17), 12*E*-dien-13,19-dienoic acid. The anticancer activity was measured as the inhibitory capacity on Epstein–Barr virus early antigen (EBV-EA) activation induced by 12-*O*-tetradecanoylphorbol 13-acetate (TPA) using β-carotene as the positive control. It was concluded that sandaracopimaric acid and other pimarane-type diterpenoids may serve as leading compounds for further development of potential cancer chemopreventive agents [44]. In 2005, another study investigated phenolic compounds in the shells (*P. sibirica*) [43]. Ground shells were extracted with 80% methanol, and the extract was further fractionated and isolated by chromatographic procedures to obtain pure phenolic compounds. Cedrusin, eriodyctiol, and 2-hydroxy-(4-octyloxy)benzophenone were isolated and from the pine nut shell for the first time, and their chemical structures were confirmed by ultraviolet-visible (UV-VIS) and nuclear magnetic resonance (NMR) analyses [43]. Further research to identify health beneficial factors in by-products of pine nut processing such as their shells may add additional monetary value to pine nut industries.

17.6 CONCLUSION

Research on the nutraceutical aspects of the pine nut is promising; however, current research remains scattered in its attempts to identify value in such a heterogeneous population. Future research in nutraceutical properties of the pine nut would benefit from defining specific goals aimed at specific species or nutritional components, i.e., researching the hypotensive properties of *P. pinea* or investigating the effects of pinolenic acid on blood sugar across various animal models and humans. Human consumption studies, whether epidemiological or clinical, have the potential to greatly enhance our understanding of the effects of the pine nut and its individual components on human health and metabolism. Additionally, safety studies on the effects of large doses of the Δ^5 PMI PUFA should be conducted, along with various trials to determine effective doses.

While there is a good amount of research on the health benefits of nuts in general, there has been relatively little research dedicated to the pine nut specifically. Perhaps, this is not without reason, as it is already one of the most expensive nuts: shelled pine nuts range from $20 to $35/kg and pine nut oils from $70 to $140/L [1]. The pine nut is largely harvested from natural forests, and good crops come an average of once every 3 to 5 years, causing demand to far exceed production for some years [1].

The main goal of nutraceutical research is to identify health beneficial factors, but often value is added to a crop as an intentional or unintentional consequence. Adding value to pine nut without the market being able to compensate has the potential to create more problems with supply and demand. This is not a valid reason to abstain from researching a botanical with exhibited potential in improving human health, but it is an important consideration that must be addressed in the development of any pine nut-based nutraceutical or functional food.

REFERENCES

1. Sharashkin, L. and Gold, M., Pine Nuts (Pignolia): Species, Products, Markets, and Potential for U.S. Production, Northern Nut Growers Association, 95th Annual Report, Columbia, MO, 2004.
2. Bagci, E. and Karaagacli, Y., Fatty acid and tocochromanol patterns of Turkish pines, *Acta Biol. Cracov. Bot.*, 46, 95–100, 2004.

3. Lopez-Mata, L., Proteins, amino acids, and fatty acid composition of nuts from the Mexican endemic rarity, *Pinus maximartinezii*, and its conservation implications, *Interciencia,* 26, 606–610, 2001.
4. Brufau, G., Boatella, J., and Rafecas, M., Nuts: Source of energy and macronutrients, *Br. J. Nutr.*, 96(Suppl. 2), 24S–28S, 2006.
5. Venkatachalam, M. and Sathe, S.K., Chemical composition of selected edible nut seeds, *J. Agric. Food Chem.*, 54, 4705–4714, 2006.
6. Nergiz, C. and Donmez, I., Chemical composition and nutritive value of *Pinus pinea* L. seeds, *Food Chem.*, 86, 365–368, 2004.
7. Dismore, M.L., Haytowitz, D.B., Gebhart, S.E., Peterson, J.W., and Booth, S.L., Vitamin K content of nuts and fruits in the US diet, *J. Am. Diet. Assoc.*, 103, 1650–1652, 2003.
8. Ryan, E., Galvin, K., O'Connor, T.P., Maguire, A.R., and O'Brien, N.M., Fatty acid profile, tocopherols, squalene and phytosterol content of brazil, pecan, pine, pistachio and cashew nuts, *Int. J. Food Sci. Nutr.*, 57, 219–228, 2006.
9. Senter, S.D., Horvat, R.J., and Forbus, W.R., Comparative GLC-MS analysis of phenolic acids of selected tree nuts, *J. Food Sci.*, 48, 798–799, 824, 1983.
10. Cheikh-Rouhou, S., Hentati, B., Besbes, S., Blecker, C., Deroanne, C., and Attia, H., Chemical composition and lipid fraction characteristics of Aleppo pine (*Pinus halepensis* Mill.) seeds cultivated in Tunisia, *Food Sci. Tech. Int.*, 15, 407–416, 2006.
11. Kornsteiner, M., Wagner, K.H., and Elmadfa, I., Tocopherols and total phenolics in 10 different nut types, *Food Chem.,* 98, 381–387, 2006.
12. Wolff, R.L., Pédrono, F., Pasquier, E., and Marpeau, A.M., General characteristics of *Pinus* spp. seed fatty acid compositions, and importance of Δ5-olefinic acids in the taxonomy and phylogeny of the genus, *Lipids*, 35, 1–22, 2000.
13. Wolff, R.L. and Marpeau, A.M., Δ5-olefinic acids in the edible seeds of nut pines (*Pinus cembroides edulis*) from the United States, *J. Am. Oil Chem. Soc.*, 74, 613–614, 1997.
14. Imbs, A.B. and Pham, L.Q., Fatty acids and triacylglycerols in seeds of *Pinaceae* species, *Phytochemistry*, 42, 1051–1053, 1996.
15. Wolff, R.L. and Bayard, C.C., Fatty acid composition of some pine seed oils, *J. Am. Oil Chem. Soc.*, 72, 1043–1046, 1995.
16. Wolff, R.L., A practical source of delta 5-olefinic acids for identification purposes, *J. Am. Oil Chem. Soc.*, 75, 891–892, 1998.
17. Takagi, T. and Itabashi, Y., *Cis*-5-olefinic unusual fatty-acids in seed lipids of gymnospermae and their distribution in triacylglycerols, *Lipids*, 17, 716–723, 1982.
18. Nasri, N., Khaldi, A., Hammami, M., and Triki, S., Fatty acid composition of two Tunisian pine seed oils, *Biotechnol. Prog.*, 21, 998–1001, 2005.
19. Wolff, R.L., Delue, L.G., Marpeau, A.M., and Comps, B., Chemotaxonomic differentiation of conifer families and genera based on the seed oil fatty acid compositions: Multivariate analyses, *Trees*, 12, 57–65, 1997.
20. Wolff, R.L., Pedrono, F., and Marpeau, A.M., Fatty acid composition of edible pine seeds with emphasis on North American and Mexican pines of the *Cembroides* subsection, *OCL-Ol. Corps Gras Li.,* 6, 107–110, 1999.
21. Nasri, N., Khaldi, A., Fady, B., and Triki, S., Fatty acids from seeds of *Pinus pinea* L.: Composition and population profiling, *Phytochemistry*, 66, 1729–1735, 2005.
22. Lichtenstein, A.H. and Deckelbaum, R.J., Stanol/sterol ester-containing foods and blood cholesterol levels: A Statement for Healthcare Professionals from the Nutrition Committee of the Council on Nutrition, Physical Activity, and Metabolism of the American Heart Association, *Circulation*, 103, 1177–1179, 2001.
23. Ostlund, R.E., Racette, S.B., Okeke, A., and Stenson, W.F., Phytosterols that are naturally present in commercial corn oil significantly reduce cholesterol absorption in humans, *Am. J. Clin. Nutr.*, 75, 1000–1004, 2002.
24. Ellegard, L.H., Andersson, S.W., Normen, L., and Andersson, H.A., Dietary plant sterols and cholesterol metabolism, *Nutr. Rev.*, 65, 39–45, 2007.
25. Andersson, S.W., Skinner, J., Ellegard, L., Welch, A.A., Bingham, S., Mulligan, A., Andersson, H., and Khaw, K.T., Intake of dietary plant sterols is inversely related to serum cholesterol concentration in men and women in the EPIC Norfolk population: A cross-sectional study, *Eur. J. Clin. Nutr.*, 58, 1378–1385, 2004.
26. Segura, R., Javierre, C., Lizarraga, M.A., and Ros, E., Other relevant components of nuts: Phytosterols, folate and minerals, *Br. J. Nutr.*, 96(Suppl. 2), 36S–44S, 2006.

27. Nasri, N., Fady, B., and Triki, S., Quantification of sterols and aliphatic alcohols in Mediterranean stone pine (*Pinus pinea* L.) populations, *J. Agric. Food Chem.*, 55, 2251–2255, 2007.
28. Asset, G., Bauge, E., Wolff, R.L., Fruchart, J.C., and Dallongeville, J., *Pinus pinaster* oil affects lipoprotein metabolism in apolipoprotein E-deficient mice, *J. Nutr.*, 129, 1972–1978, 1999.
29. Asset, G., Staels, B., Wolff, R.L., Bauge, E., Madj, Z., Fruchart, J.C., and Dallongeville, J., Effects of *Pinus pinaster* and *Pinus koraiensis* seed oil supplementation on lipoprotein metabolism in the rat, *Lipids*, 34, 39–44, 1999.
30. Asset, G., Bauge, E., Wolff, R.L., Fruchart, J.C., and Dallongeville, J., Comparison of maritime pine oil and fish oil effects on plasma lipoproteins in apolipoprotein E-deficient mice, *Prostag. Leukotr. Ess.*, 62, 307–310, 2000.
31. Asset, G., Leroy, A., Bauge, E., Wolff, R.L., Fruchart, J.C., and Dallongeville, J., Effects of dietary maritime pine (*Pinus pinaster*)-seed oil on high-density lipoprotein levels and in vitro cholesterol efflux in mice expressing human apolipoprotein A-I, *Br. J. Nutr.*, 84, 353–360, 2000.
32. Asset, G., Bauge, E., Wolff, R.L., Fruchart, J.C., and Dallongeville, J., Effects of dietary maritime pine seed oil on lipoprotein metabolism and atherosclerosis development in mice expressing human apolipoprotein B, *Eur. J. Nutr.*, 40, 268–274, 2001.
33. Lee, J.W., Lee, K.W., Lee, S.W., Kim, I.H., and Rhee, C., Selective increase in pinolenic acid (all-*cis*-5,9,12–18:3) in Korean pine nut oil by crystallization and its effects on LDL-receptor activity, *Lipids*, 39, 383–387, 2004.
34. Scientific Background Information on PinnoThin™ Summary, Lipid Nutrition, B.V., Wormerveer, The Netherlands, 2007.
35. Einerhand, A.W., Pasman, W., Rubingh, C., van der Berg, R., O'Shea, M., Gambelli, L., and Hendriks, H., Korean pine nut fatty acids affect appetite sensations, plasma CCK and GLP1 in overweight subjects, *FASEB J*, 20, A829, 2006.
36. Scott, C., Pasman, W., Heimerikx, J., Rubingh, C., van den Berg, R., O'Shea, M., Gambelli, L., Hendriks, H., Mennen, L., and Einerhand, A., PinnoThin™ suppresses appetite in overweight women, *Appetite*, 49, 330, 2007.
37. Heimerikx, J., Schmid, U., Einerhand, A., Stam, W., and Gambelli, L., Use of Pinolenic Acid, U.S. Patent No. 20060172023, 2006.
38. Sugano, M., Ikeda, I., Wakamatsu, K., and Oka, T., Influence of Korean pine (*Pinus koraiensis*)-seed oil containing *cis*-5,*cis*-9,*cis*-12-octodecatrienoic acid on polyunsaturated fatty acid metabolism, eicosanoid production and blood pressure of rats, *Br. J. Nutr.*, 72, 775–783, 1994.
39. Lee, K.W., Yu, K.W., Kim, K.M., Suh, H.J., Lee, S.W., and Rhee, C., Effects of pine (*Pinus koraiensis*)-seed oil supplementation on serum lipid composition in rats and immune responses in mice, *Food Sci. Biotechnol.*, 13, 358–361, 2004.
40. Bakhtin, I.V., Budaeva, V.V., Vereshchagin, A.L., Egorova, E.I., Zhukova, E.I., and Saratikov, A.S., Efficiency of Siberian pine oil in complex treating of people ill with benign hypertension [abstract], *Vopr. Pitan.*, 75, 51–53, 2006.
41. Matsuo, N., Osada, K., Kodama, T., Lim, B.O., Nakao, A., Yamada, K., and Sugano, M., Effects of γ-linolenic acid and its positional isomer pinolenic acid on immune parameters of Brown-Norway rats, *Prostag. Leukotr. Ess.*, 55, 223–229, 1996.
42. Pasquier, E., Ratnayake, W.M.N., and Wolff, R.L., Effects of $\Delta 5$ polyunsaturated fatty acids of maritime pine (*Pinus pinaster*) seed oil on the fatty acid profile of the developing brain of rats, *Lipids*, 36, 567–574, 2001.
43. Khamidullina, E.A., Purevdash, M., Ushakov, I.V., Neretina, O.V., and Medvedeva, S.A., Phenolic compounds from *Pinus sibirica* shells, *Chem. Nat. Compd.*, 41, 101–102, 2005.
44. Minami, T., Wada, S., Tokuda, H., Tanabe, G., Muraoka, O., and Tanaka, R., Potential antitumor-promoting diterpenes from the cones of *Pinus luchuensis*, *J. Nat. Prod.*, 65, 1921–1923, 2002.
45. Osguven, F. and Vursavus, K., Some physical, mechanical and aerodynamic properties of pine (*Pinus pinea*) nuts, *J. Food Eng.*, 68, 191–196, 2005.
46. Wu, X., Beecher, G.R., Holden, J.M., Haytowitz, D.B., Gebhardt, S.E., and Prior, R.L., Lipophilic and hydrophilic antioxidant capacities of common foods in the United States, *J. Agric. Food Chem.*, 52, 4026–4037, 2004.

18 Phytochemicals and Health Aspects of Pistachio (*Pistacia vera* L.)

Navindra P. Seeram, Yanjun Zhang, Susan Bowerman, and David Heber

CONTENTS

18.1 INTRODUCTION

Edible tree nuts are valued globally for their sensory, nutritional, and health attributes and are of tremendous economic importance. Aside from being high in unsaturated fatty acids and low in saturated fatty acids (SFA), tree nuts are rich sources of proteins, dietary fibers, vitamins, minerals, and antioxidant phytochemicals. In fact, a growing body of studies suggests that consumption of nuts might confer beneficial effects on coronary heart disease (CHD) [1,2]. Although the major research interest on health benefits of nuts has focused on their natural content of monounsaturated fatty acids (MUFA), emerging evidence indicates that phytochemicals present in their skins (peels), including polyphenols, may also play an important role. Polyphenols have antioxidant properties, and their presence in outer layers (skins, peels, and hulls) of fruits, vegetables, and tree nuts may offer protection against oxidative stress when they are consumed.

Pistachio (*Pistacia vera* L.) nuts are widely consumed and are of significant economic importance. The top major worldwide producer of pistachios is Iran, followed by the United States. The U.S. pistachio industry, which is located almost exclusively in California, has experienced phenomenal growth in the past 30 years, growing in production from nearly zero in 1976 to $333 million in 2002. Pistachios are unique among tree nuts in that their endocarp (shell) splits naturally prior to maturity. This allows pistachios to be marketed largely in-shell for fresh consumption, because their kernels can be easily extracted without mechanical cracking.

This chapter reviews the health benefits of pistachio in relation to the phytochemicals found in pistachio kernels and peels. Phytochemicals previously identified from pistachios include phytosterols, lutein, resveratrol, and anthocyanins. Pistachios are unique among tree nuts both in having anthocyanins in their peels and significant amounts of lutein both in the peel and the kernel of the nut.

18.2 BACKGROUND OF PISTACHIO

Among the 11 species in the genus *Pistacia* (family Anacardiaceae), only *Pistacia vera* L. produces edible nuts (known as "the pistachio"). Pistachios are native to western Asia and were distributed in the Middle East, Mediterranean, and in Europe by traders, and there is evidence that pistachios were consumed as early as 7000 BC Pistachios grew wild in the high desert regions during Biblical times. A rare delicacy, pistachios were a favorite of the Queen of Sheba, who demanded all her land's production for herself and her court. The royal nut was imported by American traders in the 1880s, primarily for U.S. citizens of Middle Eastern origin. Some 50 years later, pistachios became a popular snack food, introduced in vending machines. These imported nuts were dyed red to draw attention to them, and to cover stains from antiquated harvesting techniques. Pistachio nuts are considered among one of the prime edible nuts. Both "in-shell" and "shelled" pistachios are marketed extensively and the main producers worldwide are Iran (43%), the United States (26.5%, Sacramento and San Joaquin Valleys in California), Turkey (19.5%), Syria (8.5%), Greece (1.5%), and Italy (1%). Lesser quantities are produced in Lebanon, Tunisia, and Australia [3,4]. It should be noted that world production of pistachios has increased >300% since 1980, with nearly every country experiencing large increases.

Pistachios grow on trees in grape-like clusters and are encased in an outer skin or hull. When pistachios ripen, the hull turns rosy and the shell within splits naturally, indicating they are ready for harvest. The pistachio fruit itself consists of a single seed (kernel), encased by a thin soft and edible seed coat (testa), enclosed by a hard inedible shell (endocarp), which is further surrounded by the fleshy hull (mesocarp and epicarp), which is also inedible. However, pistachios have to be processed within 24 h after harvest to avoid hull-trapped moisture, which causes staining of the pistachio shell [5,6]. An important aspect of quality of the in-shell product is a shell that is free of staining, which is not only for cosmetic reasons, but also an indicator of developmental, pathogenic, and insect infestation problems prior to harvest [7,8]. Pistachio industries invest millions of dollars in equipment to process their pistachios quickly to avoid staining and enable sale of the nuts in their natural color shells. Therefore, bleaching methods to whiten pistachio shells are sometimes used, although this practice has been made illegal in some countries. Bleaching may adversely affect the levels of potentially beneficial phytochemicals contained in pistachios and may leave residues of various unknown bleaching agents within the nut, which may negatively impact human health [9]. Therefore, pistachios obtained from different regions of the world were monitored by our laboratory for evidence of bleaching with a variety of bleaching agents. There has been no evidence of bleaching in California nuts, but there have been some found in imported nuts (unpublished results).

18.3 PISTACHIO NUTRIENTS

In common with other tree nuts, pistachios are rich in nutrient content. Each 1-ounce (~28.3 g) serving of pistachio nuts (49 kernels) provides 165 calories, 310 mg of potassium, 3 g of dietary fiber, and 20% of the recommended dietary allowances (RDA) for vitamin B_6, copper, and manganese. Table 18.1 shows the nutritional breakdown of a 1-ounce serving of pistachio nuts.

18.4 PHYTOCHEMICAL CONSTITUENTS OF PISTACHIO

Phytochemicals (plant chemicals) are defined as "nonnutrient" or secondary metabolites produced by plants. Over the past few years, knowledge of the composition and identities of phytochemicals

TABLE 18.1
Pistachio Nutrients and Percentage of RDA

Compound	Serving Size (1 ounce or ~49 kernels)
Proximate composition	
Fat	13 g
SFA	1.5 g
MUFA	7 g
PUFA	4 g
Carbohydrate, by difference	9 g
Dietary fiber	3 g
Protein	6 g
Energy	165 kcal
Mineral	*% of RDA*
Calcium	4
Copper	20
Iron	6
Magnesium	10
Manganese	15
Phosphorus	15
Potassium	9
Sodium	8
Zinc	4
Vitamin	*% of RDA*
Vitamin A	4
Vitamin C	2
Vitamin E	6
Thiamin	15
Vitamin B_6	25
Folate	4

Source: From U.S. Department of Agriculture (USDA), National Nutrient Database for Standard Reference, Release 19, National Technical Information Service, USDA, Springfield, VA, 2006.

Note: Percentage of recommended dietary allowances (RDA) is based on a 2000 kcal diet; SFA, saturated fatty acids; MUFA, monounsaturated fatty acids; PUFA, polyunsaturated fatty acids.

found in pistachio nut and skin has rapidly expanded with the advent of highly sensitive analytical methods, allowing researchers to establish phytochemical profiles or "chemical fingerprints" of these compounds [9–14]. Phytochemicals identified from pistachio nut proper and its skins include essential oils, flavonoids, phytosterols and phytostanols, carotenoids (including lutein), resveratrol, polyphenols, organic acids, and fatty acids [9–14]. Some of these phytochemicals are summarized in Tables 18.2 through 18.4.

The skins of pistachio contain high levels of a diverse range of phytochemicals found predominantly as polyphenolic compounds (aromatic rings bearing hydroxyl group). The structural diversity of pistachio skin phenolics can be observed by the varying types and oxidation levels of

TABLE 18.2
Essential Oils from Pistachio Hull

Compound	%	Compound	%
Tricyclene	0.39	2,6-Dimethyl-1,3 (*E*), 5(*Z*), 7-octatetraene	0.26
α-Pinene	54.4		
α-Fenchene	0.34	Menthone	0.01
Camphene	3.20	Longipinene	0.07
β-Pinene	0.77	(*Z*)-Tagetone	0.08
Sabinene	tr[a]	Pentadecane	0.11
2-Carene	0.41	Camphor	0.09
1-Methyl-1H-pyrrol	0.14	(*Z*)-2-Nonenal	0.01
δ-3-Carene	3.97	Linalool	0.03
Myrcene + α-phellandrene	1.63	Bornyl acetate	1.17
p-Mentha-1(7), 8-diene	tr	Camphene hydrate[b]	0.08
α-Terpinene	0.66	Terpinen-4-ol	0.11
Limonene	6.62	Hexadecane	0.04
β-Phellandrene	0.21	Myrtenal	0.04
cis-*p*-Mentha-6,8-dien-2-yl acetate	tr	Citronellyl acetate	tr
		Neral	0.48
Pentyl furan	0.02	α-Terpineol + Borneol	1.20
(*Z*)-β-Ocimene	0.18	Heptadecane	0.02
α-Terpinene	0.52	Piperitone	0.20
(*E*)-β-Ocimene	0.43	Naphthalene	0.11
p-Cymene	0.36	δ-Cadinene	0.02
Terpinolene	18.91	*p*-Mentha-1,3-dien-7-al	tr
Fenchone	0.02	*p*-Cymen-8-ol	0.17
Nonanal	0.10	(*Z*)-Geranyl acetone	0.02
Tetradecane	tr	Piperitenone	0.01
Perillene	tr	Thymol	0.04
2,5-Dimethyl styrene[b]	1.13	Carvacrol	0.01

Source: From Kusmenoglu, S., Baser, K.H.C., and Ozek, T., *J. Essent. Oil Res.*, 7, 441, 1995. With permission.

[a] tr, trace = <0.01%.

[b] Tentative identification from gas chromatography/mass spectrometry (GC/MS) alone.

their heterolcyclic ring, their substitution patterns of hydroxylation (bearing an –OH group), the existence of stereoisomers, their glycosylation by various sugars and by conjugation with themselves to form polymers among others. We have found significant levels of these polymeric compounds (as proanthocyanidins), but their degree of polymerization has not yet been characterized (unpublished results).

The main class of pistachio phenolics includes the most abundant group, the flavonoids. Flavonoids have a basic skeletal structure of C6-C3-C6, and, based on their degree of oxidation and substitution in the 3-position, can be further subdivided into flavonols, flavanols (catechin, epicatechin, etc.), flavones, and flavanones (quercetin, kaempferol, luteolin, apigenin, etc.), isoflavones (genistein, daidzein, etc.), and anthocyanidins (cyanidin, pelargonidin, delphinidin, etc.). Anthocyanidins are water-soluble pigments, which occur naturally as glycosides (known as anthocyanins), and are responsible for the attractive red, blue, and purple colors of various fruits and many colorful vegetables.

TABLE 18.3
Phytochemical Constituents Identified from *Pistacia vera* L. Nuts

Name	Formula	MW	Plant Part	Ref.
Flavonoid				
Cyanidin-3-*O*-β-D-galactoside	$C_{21}H_{20}O_{11}$	448	Skin, kernel	[9,10]
Cyanidin-3-*O*-β-D-glucoside	$C_{21}H_{20}O_{11}$	448	Skin, kernel	[9,10]
Rutin	$C_{27}H_{30}O_{16}$	610	Skin, kernel	[9]
Luteolin	$C_{15}H_{10}O_6$	286	Skin, kernel	[9]
Apigenin	$C_{15}H_{10}O_5$	270	Skin, kernel	[9]
Naringenin	$C_{15}H_{12}O_5$	272	Skin, kernel	[9]
Quercetin	$C_{15}H_{10}O_7$	302	Skin, kernel	[9]
Eriodictyol	$C_{15}H_{10}O_6$	288	Skin, kernel	[9]
Daidzein	$C_{15}H_{10}O_4$	254	Skin, kernel	[9,25]
Genistein	$C_{15}H_{10}O_5$	270	Kernel	[25]
Phytosterol and phytostanol				
Campesterol	$C_{28}H_{48}O$	400	Kernel	[11]
Campestanol	$C_{28}H_{50}O$	402	Kernel	[11]
β-Sitosterol	$C_{29}H_{50}O$	414	Kernel	[11]
β-Sitostanol	$C_{29}H_{52}O$	416	Kernel	[11]
Δ^5-Avenasterol	$C_{29}H_{48}O$	412	Kernel	[11]
Δ^7-Avenasterol	$C_{29}H_{48}O$	412	Kernel	[29]
Stigmasterol	$C_{29}H_{48}O$	412	Kernel	[11]
Organic acid				
Propane-1,3-dioic acid	$C_3H_4O_4$	104	Kernel	[18]
(*E*)-2-Butene-1,4, dioic acid	$C_4H_4O_4$	116	Kernel	[18]
Butane-1,4-dioic acid	$C_4H_6O_4$	118	Kernel	[18]
2-Methylene-butane-1,4-dioic acid	$C_5H_6O_4$	130	Kernel	[18]
2-Hydroxy-butane-1,4-dioic acid	$C_4H_6O_5$	134	Kernel	[18]
Hexane-1,6-dioic acid	$C_6H_{10}O_4$	146	Kernel	[18]
Nonane-1,9-dioic acid	$C_9H_{16}O_4$	188	Kernel	[18]
Decane-1,10-dioic acid	$C_{10}H_{18}O_4$	202	Kernel	[18]
Dodecane-1,12-dioic acid	$C_{12}H_{22}O_4$	230	Kernel	[18]
Heptadecane-1,17-dioic acid	$C_{17}H_{32}O_4$	300	Kernel	[18]
Shikimic acid	$C_7H_{10}O_5$	174	Kernel	[18]

The anthocyanins present in the pistachio are found as glycosides of cyanidin and are cyanidin-3-galactoside (major; 696 μg/g) and cyanidin-3-glucoside (minor; 209 μg/g) (Figure 18.1) [9,10]. To the best of our knowledge, pistachios are the only tree nuts known to contain anthocyanins that impart the red-purple color to their skins. Other flavonoids that were recently identified by our laboratory in pistachio skins include quercetin (14.9 μg/g), luteolin (10.0 μg/g), eriodictyol (10.2 μg/g), rutin (1.6 μg/g), naringenin (1.2 μg/g), and apigenin (0.2 μg/g) [9] (Figure 18.1).

The edible uses of pistachio also include its oil, which is obtained from the seed but not produced commercially because of the high price of the seed. The seed yields up to 40% of this nondrying oil, which has a pleasant mild flavor and contains a number of terpenoids [15] (Table 18.2). The chemical structure of phytosterols and phytostanols found in pistachio nuts are shown in Figure 18.2.

TABLE 18.4
Fatty Acids Identified from *Pistacia vera* L. Nuts

Fatty Acid	Formula	MW	Plant Part	Ref.
14:0	$C_{14}H_{28}O_2$	228	Kernel	[16,17]
15:0	$C_{15}H_{30}O_2$	242	Kernel	[16,17]
16:0	$C_{16}H_{32}O_2$	256	Kernel	[16,17]
16:1ω7	$C_{16}H_{30}O_2$	254	Kernel	[16,17]
16:1ω9	$C_{16}H_{30}O_2$	254	Kernel	[16,17]
17:0	$C_{17}H_{34}O_2$	270	Kernel	[16,17]
18:0	$C_{18}H_{36}O_2$	284	Kernel	[16]
18:1ω9	$C_{18}H_{34}O_2$	282	Kernel	[16,17]
18:1ω11	$C_{18}H_{34}O_2$	282	Kernel	[16,17]
18:2ω6	$C_{18}H_{32}O_2$	280	Kernel	[16,17]
18:3ω3	$C_{18}H_{30}O_2$	278	Kernel	[16,17]
20:0	$C_{20}H_{40}O_2$	312	Kernel	[16]
20:1ω9	$C_{20}H_{38}O_2$	310	Kernel	[16]
22:0	$C_{22}H_{44}O_2$	340	Kernel	[16,17]

18.5 HEALTH BENEFITS OF PISTACHIOS

Epidemiological evidence has demonstrated that regular consumption of nuts might confer a beneficial effect on cardiovascular disease (CVD) risk [1,2]. Although the major research interests on health benefits of nuts has focused on their natural content of MUFA, emerging evidence indicates that various phytochemicals present in their skins may also play an important role [1]. In fact, almond and walnut skin phenolics have been shown to inhibit oxidation of low-density lipoprotein (LDL) cholesterol, a key step in atherogenesis [19,20].

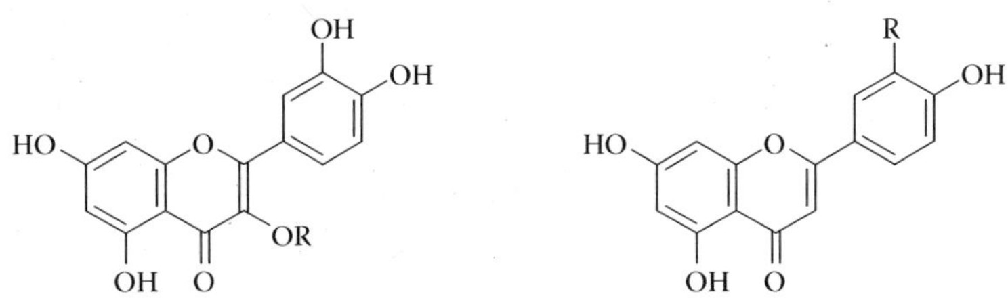

R = H, quercetin
R = rutinoside, rutin

R = OH, luteolin
R = H, apigenin

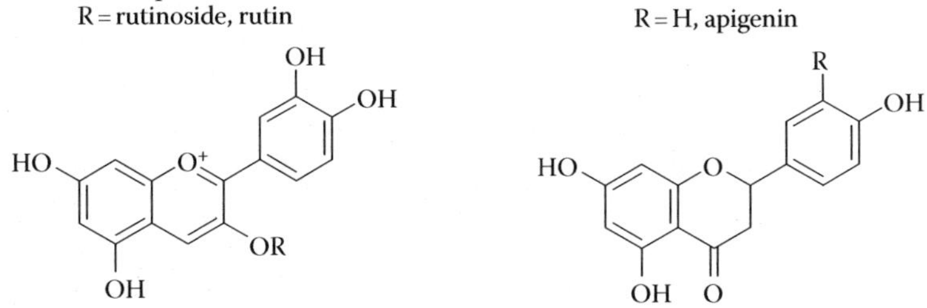

R = glucoside, cyanidin-3-glucoside
R = galactoside, cyanidin-3-galactoside

R = OH, eriodictyol
R = H, naringenin

FIGURE 18.1 Chemical structure of various flavonoids detected in pistachios.

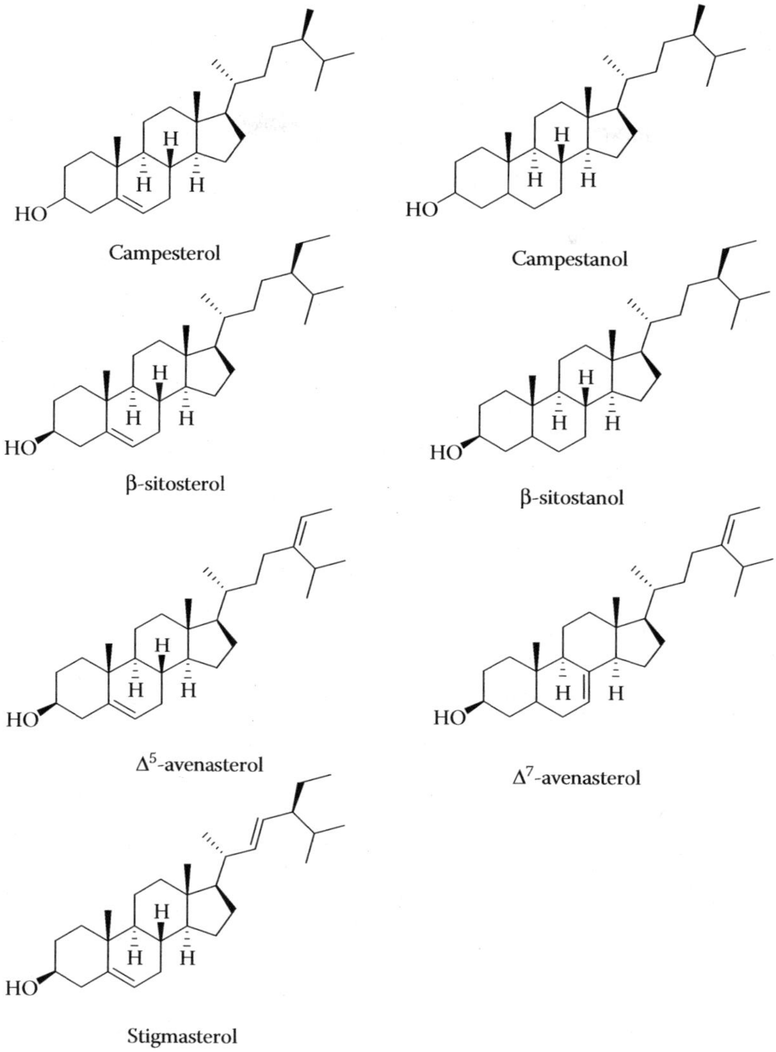

FIGURE 18.2 Chemical structure of various phytosterols and phytostanols from pistachios.

18.5.1 Pistachios and Heart Health

There is a growing body of evidence suggesting that nut consumption may be heart-healthy. In addition to their MUFA, fiber and phytochemical antioxidants, pistachios are rich sources of phytosterols, which promote heart health by inhibiting the absorption of cholesterol from the intestine through direct competition with uptake mechanisms.

Many studies have shown that a diet that incorporates tree nuts, such as pistachios, is associated with a reduced risk of CHD. In the Adventist Health Study, a prospective cohort investigation of 31,208 non-Hispanic white California Seventh-Day Adventists, extensive dietary information was obtained at baseline, along with the values of traditional coronary risk factors. These were then related to the risk of definite fatal CHD or definite nonfatal myocardial infarction (MI). Individuals who consumed nuts more than four times per week experienced substantially fewer definite fatal CHD events and definite nonfatal MI in comparison with those who consumed nuts less than once per week [21].

Edwards et al. [22] conducted a controlled, randomized crossover trial in 10 subjects with mild hypercholesterolemia in which subjects served as their own controls. For 3 weeks, subjects either consumed their regular diet or substituted 20% of their daily calorie intake with pistachio nuts; after the first 3-week period, those on the pistachio diet crossed over to consume their regular diet, and those on the regular diet crossed over into the pistachio diet. At the end of the study, subjects experienced significant decreases in total cholesterol and total cholesterol/high-density lipoprotein (HDL) cholesterol ratio, with significant increases in HDL cholesterol.

In another study of pistachio consumption among 44 men and women, subjects were randomized to either a regular diet or a pistachio diet in which pistachio nuts provided 20% of energy intake for a 3-week period. After 3 weeks on the pistachio diet, significant decreases were seen in total cholesterol, total cholesterol/HDL cholesterol ratio, and LDL/HDL ratio, and patients also experienced a decrease in plasma malondialdehyde, an important indicator of lipid peroxidation [23].

Sheridan et al. [28] recently reported on randomized crossover study with 15 subjects (11 men and 4 women) with moderate hypercholesterolemia (serum cholesterol >210 mg/dL). Subjects consumed 15% of their daily caloric intake in the form of pistachio nuts (about 2–3 ounces per day) for 4 weeks. The authors reported statistically significant reductions in triacylglycerols/HDL cholesterol and LDL/HDL ratios and statistically significant increases in HDL cholesterol [28].

18.5.2 Pistachios and Body Weight

Nuts are generally regarded as calorie-dense foods, yet several studies have shown that regular nut consumption among free-living individuals does not lead to significant weight gain. Data from the 1994–1996 Continuing Survey of Food Intakes by Individuals conducted by the USDA [8] have been used to compare body mass index (BMI) and total energy intake of nut eaters with that of nonnut eaters. Dietary intake data from a nationally representative sample were collected on two nonconsecutive 24 h recalls. Those reporting consumption of tree nuts, peanuts, or seeds on any of the 2 days were included in the "nut eater" group. The results of this survey indicated that "nut eaters" had a lower BMI compared with "nonnut eaters," even though their energy intake was higher.

BMI is an indirect measure of body fatness, which is the most useful in population-based studies, and the results of the available studies suggest an inverse relationship or no relationship between nut consumption and BMI in the U.S. population. This observation was consistent in each of the large cohort studies from which these data were reported. In the Adventist Health Study, Fraser et al. [21] reported a statistically significant negative association between consumption of nuts and BMI in a cohort of 31,200 California subjects, showing that those who ate nuts more frequently were leaner than the infrequent nut eaters. Hu and Stampfer [24] also reported a negative association between nut consumption and BMI among 86,000 females in the Nurses Health Study. There was no apparent association between BMI and nut consumption in the Physicians' Health Study [10]. BMIs by quartile of nut consumption (ranging from never to 2 times/week) were 24.9, 24.9, 25.0, and 24.7 among this cohort of 21,500 males.

The study by Edwards et al. [22], mentioned above, was designed to determine the effects of substituting 20% daily calories from chips and snack foods with pistachio nuts on lipid profiles in hypercholesterolemic subjects. It was noted, however, that in addition to improvements in total cholesterol and HDL cholesterol concentrations, no changes in body weight were seen over the course of the 3-week study.

In designing weight management programs, techniques for modifying behavior with regard to portion control and extending the time required to consume meals and snacks are often useful. It is noteworthy that pistachio nuts are most often sold in the shell, which means that each nut has to be extracted from the shell before eating. The deliberate process of opening each shell could certainly slow down consumption and serve to modify nut consumption behavior, which more typically might involve mindless snacking on handfuls of shelled nuts.

18.5.3 Incorporating Pistachios into Diet

According to qualified health claim approved by the Food and Drug Administration (FDA) in 2003 for nuts, "Scientific evidence suggests but does not prove that eating 1.5 ounces per day of most nuts, as part of a diet low in saturated fat and cholesterol, may reduce the risk of heart disease." The nuts eligible for this claim include pistachio nuts. The nuts to which the health claim could be applied were restricted by the FDA to those nuts that were specifically included in the health claim petition, which do not exceed 4 g saturated fat per 50 g of nuts. Pistachios clearly fit within this overall guideline.

The practical aspects of incorporating pistachios into the diet are simple, but some attention should be given to portion size. One or two servings per day would be a reasonable guideline for most consumers. Pistachios make colorful additions to green salads, cooked vegetables, and fruit salads as well as to cold and hot cereals, muffins, trail mix, rice dishes, pilafs, baklava, and pasta. They can also be used ground as a crispy coating for fish filets or poultry breast.

18.6 CONCLUSION

Pistachios are popularly consumed and rapidly gaining recognition for their potential impact on human health and disease. The pistachio nuts not only contain substantial levels of a diverse range of phytochemicals such as carotenoids (lutein), phytosterols, and phenolic compounds (flavonoids and resveratrol). The presence of anthocyanins imparting the red pigmentation in the skin of the pistachio is a unique feature among tree nuts and therefore the pistachio may be regarded as the "berry of tree nuts." The presence of high levels of several potentially preventive phytochemicals along with other healthful characteristics, such as a high proportion of MUFA, high nutrient density of vitamins and minerals, and relatively low calorie density, make the pistachio an attractive and healthy food source. In addition, pistachio consumption, like many other tree nuts, has not been associated with weight gain although portion control remains an important principle when incorporating pistachios into a healthy diet. Pistachios are useful taste enhancers that can be added to many foods to add nutrition, color, and flavor. Future research to evaluate the potential impact of pistachio consumption on human health should be actively pursued and should encompass carefully designed human studies.

REFERENCES

1. Kris-Etherton, P.M., Zhao, G., Binkoski, A.E., Coval, S.M., and Etherton, T.D., The effects of nuts on coronary heart disease risk, *Nutr. Rev.*, 59, 103–111, 2001.
2. Hu, F.B. and Willett, W.C., Optimal diets for prevention of coronary heart disease, *JAMA*, 288, 2569–2578, 2002.
3. California Pistachio Commission, Nutrition in a Nutshell, Published online at: http://www.pistachios.org/Data/nut_factsheet.pdf (accessed March 28, 2007).
4. International Nut Council (INC), The Cracker (No. 1, April), International Tree Nut Foundation, Reus, Spain, 1995.
5. Crane, J.C., Pistachio production problems, *Fruit Var. J.,* 38, 74, 198, 1984.
6. Ferguson, L., Kader, A., and Thompson, J., Harvesting, transporting, processing and grading, in *Pistachio Production*, Ferguson, L., Ed., Center for Fruit and Nut Crop Research and Information, University of California, Davis, CA, 1995, pp. 110–114.
7. Labavitch, J.M., Heintz, C.M., Rae, H.L., and Kader, A.A., Physiological and compositional changes associated with maturation of 'Kerman' pistachio nuts, *J. Am. Soc. Hort. Sci.*, 107, 688–692, 1982.
8. Cotton, P.A., Subar, A.F., Friday, J.E., and Cook, A., Dietary sources of nutrients among US adults, 1994 to 1996, *J. Am. Diet. Assoc.*, 104, 921–930, 2004.
9. Seeram, N.P., Zhang, Y., Henning, S.M., Lee, R., Niu, Y., Lin, G., and Heber, D., Pistachio skin phenolics are destroyed by bleaching resulting in reduced antioxidative capacities, *J. Agric. Food Chem.*, 54, 7036–7040, 2006.
10. Albert, C.M., Gaziano, J.M., Willett, W.C., and Manson, J.E., Nut consumption and decreased risk of sudden cardiac death in the Physicians' Health Study, *Arch. Intern. Med.*, 162, 1382–1387, 2002.

11. Phillips, K.M., Ruggio, D.M., and Ashraf-Khorassani, M., Phytosterol composition of nuts and seeds commonly consumed in the United States, *J. Agric. Food Chem.*, 53, 9436–9445, 2005.
12. Wu, X. and Prior, R.L., Identification and characterization of anthocyanins by high-performance liquid chromatography–electrospray ionization–tandem mass spectrometry in common foods in the United States: Vegetables, nuts, and grains, *J. Agric. Food Chem.*, 53, 3101–3113, 2005.
13. Tokusoglu, O., Unal, M.K., and Yemis, F., Determination of the phytoalexin resveratrol (3,5,4′-trihydroxystilbene) in peanuts and pistachios by high-performance liquid chromatographic diode array (HPLC-DAD) and gas chromatography–mass spectrometry (GC-MS), *J. Agric. Food Chem.*, 53, 5003–5009, 2005.
14. Kornsteiner, M., Wagner, K-.H., and Elmadfa, I., Tocopherols and total phenolics in 10 different nut types, *Food Chem.*, 98, 381–387, 2006.
15. Facciola. S., *Cornucopia—A Source Book of Edible Plants*, Kampong Publications, Vista, CA, 1990.
16. Satil, F., Azcan, N., and Baser, K.H.C., Fatty acid composition of pistachio nuts in Turkey, *Chem. Nat. Compd.*, 39, 322–324, 2003.
17. Yildiz, M., Turcan, G.S., and Ozdemir, M., Oil composition of pistachio nuts (*Pistacia vera*) from Turkey, *Fett-Lipid*, 100, 84–86, 1998.
18. Dembitsky, V.M., Goldshlag, P., and Srebnik, M., Occurrence of dicarboxylic (dioic) acids in some Mediterranean nuts, *Food Chem.*, 76, 469–473, 2002.
19. Chen, C.Y., Milbury, P.E., Lapsley, K., and Blumberg, J.B., Flavonoids from almond skins are bioavailable and act synergistically with vitamins C and E to enhance hamster and human LDL resistance to oxidation, *J. Nutr.*, 135, 1366–1373, 2005.
20. Anderson, K.J., Teuber, S.S., Gobeille, A., Cremin, P., Waterhouse, A.L., and Steinberg, F.M., Walnut polyphenolics inhibit in vitro human plasma and LDL oxidation, *J. Nutr.*, 131, 2837–2842, 2001.
21. Fraser, G.E., Sabaté, J., Beeson, W.L., and Strahan, T.M., A possible protective effect of nut consumption on risk of coronary heart disease: The Adventist Health Study, *Arch. Intern. Med.*, 152, 1416–1424, 1992.
22. Edwards, K., Kwaw, I., Matud, J., and Kurtz, I., Effect of pistachio nuts on serum lipid levels in patients with moderate hypercholesterolemia, *J. Am. Coll. Nutr.*, 18, 229–232, 1999.
23. Kocyigit, A., Koylu, A.A., and Keles, H., Effects of pistachio nuts consumption on plasma lipid profile and oxidative status in healthy volunteers, *Nutr. Metab. Cardiovasc. Dis.*, 16, 202–209, 2006.
24. Hu, F.B. and Stampfer, M.J., Nut consumption and risk of coronary heart disease: A review of epidemiologic evidence, *Curr. Ather. Rep.*, 1, 204–209, 1999.
25. Gentile, C., Tesoriere, L., Butera, D., Fazzari, M., Monastero, M., Allegra, M., and Livrea, M.A., Antioxidant activity of Sicilian pistachio (*Pistacia vera* L. var. Bronte) nut extract and its bioactive components, *J. Agric. Food Chem.*, 55, 643–648, 2007.
26. Kusmenoglu, S., Baser, K.H.C., and Ozek, T., Constituents of the essential oil from the hulls of *Pistacia vera* L., *J. Essent. Oil Res.*, 7, 441–442, 1995.
27. U.S. Department of Agriculture (USDA), National Nutrient Database for Standard Reference, Release 19, National Technical Information Service, USDA, Springfield, VA, 2006.
28. Sheridan, M.J., Cooper, J.N., Erario, M., and Cheifetz, C.E., Pistachio nut consumption and serum lipid levels, *J. Am. Coll. Nutr.*, 26, 141–148, 2007.
29. Arena, E., Campisi, S., Fallico, B., and Maccarone, E., Distribution of fatty acids and phytosterols as a criterion to discriminate geopraphic origin of pistachio seeds, *Food Chem.*, 104, 403–408, 2007.

19 Walnut Polyphenols: Structures and Functions

Toshiyuki Fukuda

CONTENTS

19.1 INTRODUCTION

Walnuts (*Juglans regia* L.) have long served as a highly nutritious food in many parts of the world. They are used as folk medicine to treat coughs, stomach pain, and cancer in Europe [1]. In Asia, walnuts provide nutritional fortification as well as relief from fluid retention, constipation, and coughs [2]. Walnuts are rich in lipids containing high proportions of unsaturated fatty acids, such as α-linolenic acid (18:3ω3) and linoleic acid (18:2ω6) [3]. A daily intake of walnuts has been reported to significantly reduce low-density lipoprotein (LDL) cholesterol concentration and to help prevent coronary heart disease (CHD) [4–7]. Although these effects are thought to be mediated by fatty acid composition of walnuts, other constituents such as polyphenols that exhibit antioxidative effects toward LDL cholesterol may be equally important [8]. The polyphenol components of walnuts were investigated by our research group in order to chemically characterize walnut constituents that are beneficial to human health. In this chapter, the isolation and characterization of 37 compounds, including four new hydrolyzable tannins and two dicarboxylic acid derivatives, are reported. The antioxidant activities of polyphenols, mainly hydrolyzable tannins, were investigated. An *in vivo* antioxidant activity in a walnut polyphenol-rich fraction (WPF) and the inhibitory effects of walnut polyphenols on α-glucosidase and α-amylase, as well as their *in vivo* hypoglycemic effects, were also examined.

19.2 ANTIOXIDANT ACTIVITY OF NUTS

It is well known that nuts are rich in lipids. Interestingly, tree nuts such as walnuts, pecans, Brazil nuts, and pine nuts have relatively long shelf life, even though they contain substantial amounts of polyunsaturated fatty acids (PUFA) [3], which are very susceptible to oxidation. Some nuts contain substantial amounts of vitamin E, which seems to act as a protector of fats and oils from oxidation [3]; however, walnuts have a lower level of vitamin E than other nuts [3], such as almonds, hazelnuts, and pine nuts. This suggests the presence of some other antioxidant substance that protects the fats and oils in walnuts from oxidation. We first compared the antioxidant effect of water, *n*-butanol, and ethyl acetate extracts prepared from 10 types of commercially available nuts using superoxide dismutase (SOD)-like activity as an index of antioxidant activity. The SOD-like activity was estimated by suppression of the superoxide anion radical ($O_2 \bullet^-$) generated in a xanthine–xanthine oxidase (XOD) system [10] in the presence of test samples. The amount of $O_2 \bullet^-$ was quantified using a nitrite kit [11] and the results are given in Table 19.1. The *n*-butanol and ethyl acetate extracts of pecans, all walnut extracts, and water extract of cashews showed strong SOD-like activity (>40%) at a final concentration of 50 μg/mL. Ethyl acetate extracts of almonds, peanuts, and macadamias also showed activity higher than 30% (unpublished data).

19.3 EXTRACTION AND CHARACTERIZATION OF WALNUT COMPONENTS

Crushed walnuts were extracted with 70% aqueous ethanol at room temperature for 24 h, and the concentrated filtrate was partitioned into *n*-hexane-, ethyl acetate-, and *n*-butanol-soluble portions. Each extract was separated by successive column chromatography over DIAION HP-20 (Mitsubishi Kasei Co., Tokyo, Japan), TOYOPEARL HW-40F (Tosoh Co., Tokyo, Japan), and MCI GEL CHP-20P (Mitsubishi Kasei Co., Tokyo, Japan) columns and preparative high-performance liquid chromatography (HPLC) to give 37 compounds (Figure 19.1). The compounds identified were as follows: casuarictin (**1**) [12], casuariin (**2**) [12], casuarinin (**3**) [12], 1,2-di-*O*-galloyl-4,6-*O*-(*S*)-HHDP-β-D-glucopyranose (**4**) [13], euprostin A (**5**) [14], gemin D (**6**) [15], 2,3-*O*-(*S*)-HHDP-D-glucopyranose (**10**) [16], isostrictinin (**11**) [17], pedunculagin (**12**) [12], 1,2,3,4,6-penta-*O*-galloyl-β-D-glucopyranose (**13**) [18], platycaryanin A (**14**) [19], praecoxin A (**15**) [20], pterocarinin A (**16**) [21], rugosin C (**17**) [22], rugosin C methyl ester (**18**), stachyuranin B (**19**) [23], stenophyllanin A (**20**) [24], strictinin (**21**) [12], tellimagrandin I (**22**) [25], tellimagrandin II (**23**) [25], 1,2,3,6-tetra-*O*-galloyl-β-D-glucopyranose (**24**) [18], 1,2,4,6-tetra-*O*-galloyl-β-D-glucopyranose (**25**) [18],

TABLE 19.1
SOD-Like Activity of Various Nut Extracts

	SOD-Like Activity (%)		
Nut	**Water Extract**	***n*-Butanol Extract**	**Ethyl Acetate Extract**
Almond	1.9	4.2	31.1
Hazelnut	9.0	4.3	21.7
Pine nut	22.2	−3.8	14.8
Peanut	3.1	−21.3	37.6
Pecan	20.7	53.1	54.5
Pistachio	8.1	7.3	31.1
Walnut	42.2	49.8	50.8
Cashew	41.8	3.2	12.8
Macadamia	7.4	2.0	31.8
Coconut	−0.7	5.8	7.4

Note: The final concentration of each extract was 50 μg/mL; SOD, superoxide dismutase.

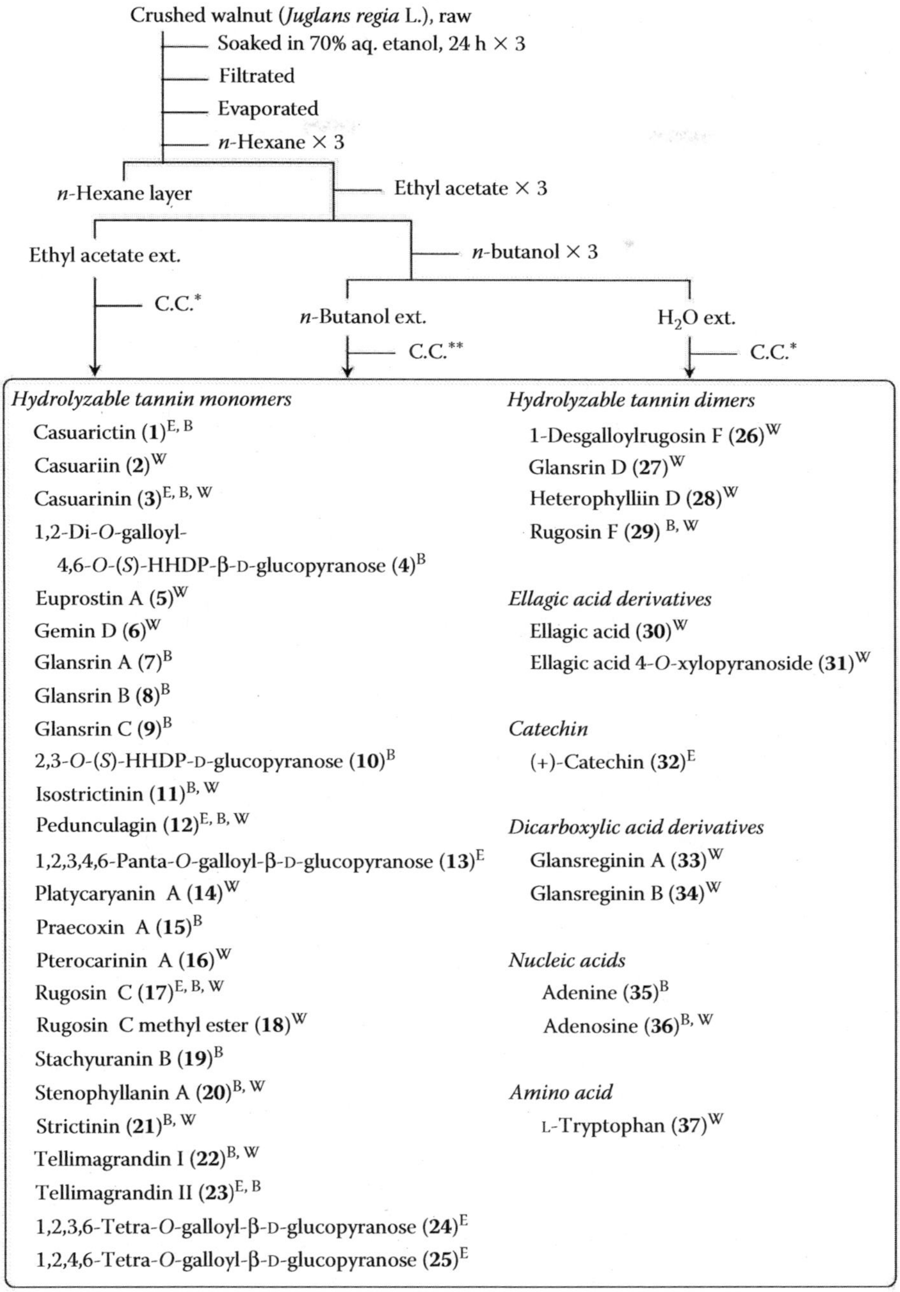

FIGURE 19.1 Isolation procedure of walnut C.C.*: column chromatography over DIAION HP-20, TOYOPEARL HW-40, MCI GEL CHP-20P, and/or YMC-GEL ODS-AQ 120-S50 columns; C.C.**: DIAION HP-20, TOYOPEARL HW-40, MCI GEL CHP-20P, and/or Develosil RPAQUEOUS ([E] from ethyl acetate extract; [B] from *n*-butanol extract; and [W] from water extract).

1-desgalloylrugosin F (**26**) [26], heterophylliin D (**28**) [27], rugosin F (**29**) [28], ellagic acid (**30**), ellagic acid 4-*O*-xylopyranoside (**31**) [29], (+)-catechin (**32**), adenine (**35**), adenosine (**36**), and L-tryptophan (**37**) by spectral analyses including nuclear magnetic resonance (NMR), mass spectrometry (MS), and circular dichroism (CD) spectra, and by degradation reactions as well as direct comparisons with authentic samples (Figures 19.2 through 19.5). Structures of new compounds,

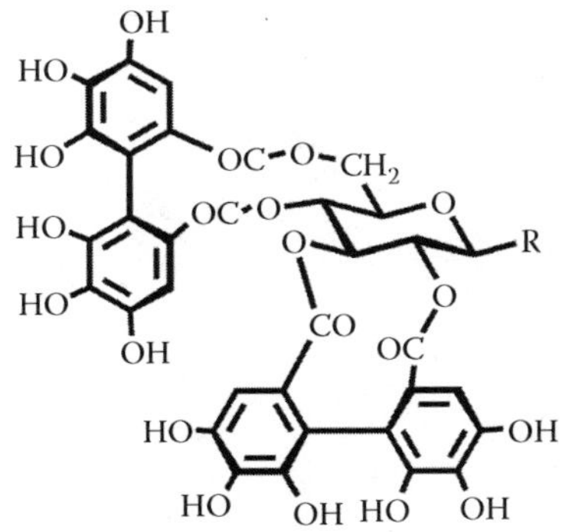

Pedunculagin (**12**): R = OH(α,β)
Casuarictin (**1**): R = OG(β)

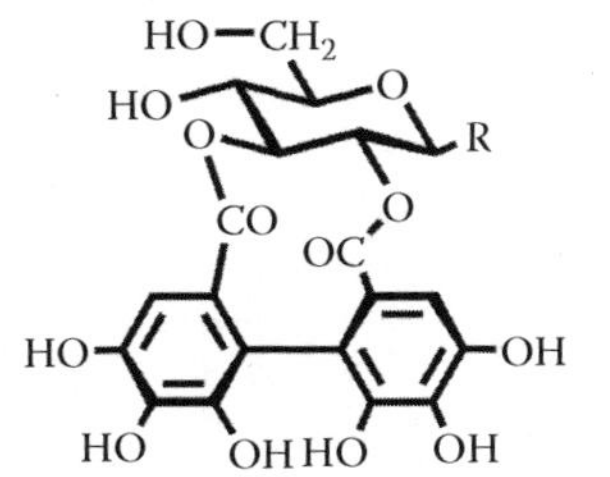

2,3-*O*-(*S*)-HHDP-D-glucopyranose (**10**): R = OH(α,β)
Isostrictinin (**11**): R = OG(β)

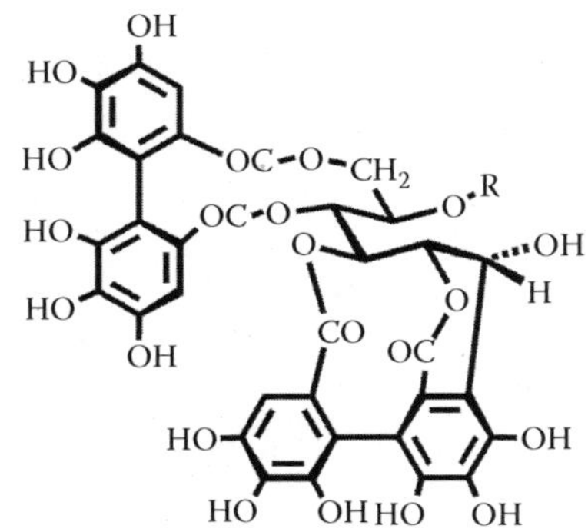

Casuariin (**2**): R = H
Casuarinin (**3**): R = G

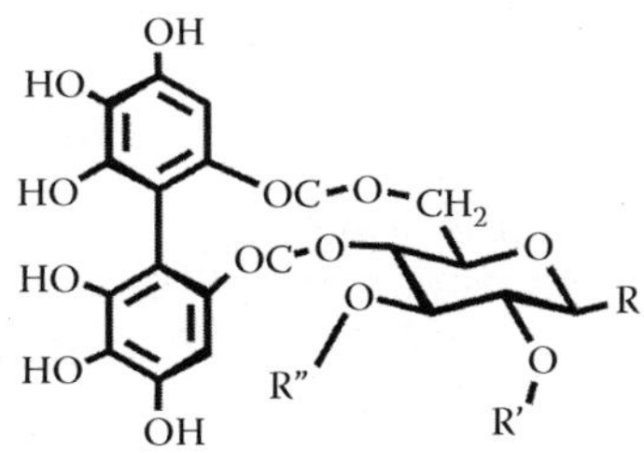

Tellimagrandin I (**22**): R = OH(α,β), R′ = R″ = G
Tellimagrandin II (**23**): R = OG(β), R′ = R″ = G
Strictinin (**21**): R = OG(β), R′ = R″ = H
Gemin D (**6**): R = OH(α,β), R′ = H, R″ = G
1,2-Di-*O*-galloyl-4,6-*O*-(*S*)-HHDP-β-D-glucopyranose (**4**): R = OG(β), R′ = G, R″ = H

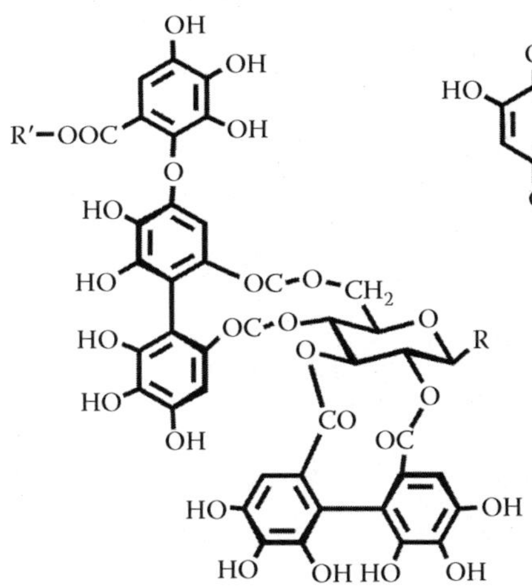

Rugosin C (**17**): R = OG(β), R′ = H
Praecoxin A (**15**): R = OH(α,β), R′ = H
Rugosin C methyl ester (**18**): R = OG(β), R′ = CH_3

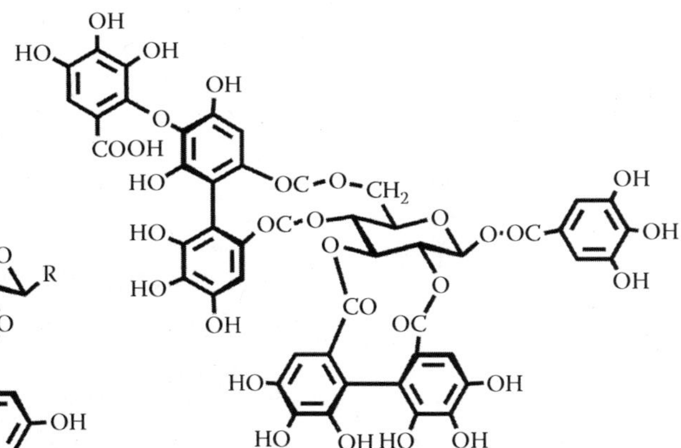

Platycaryanin A (**14**)

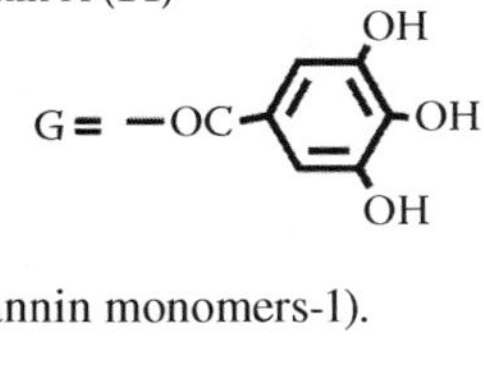

FIGURE 19.2 Chemical structures of walnut polyphenols (Hydrolyzable tannin monomers-1).

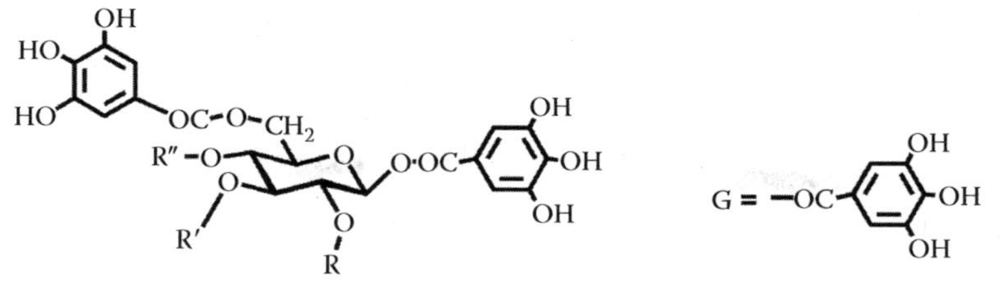

1,2,3,6-Tetra-*O*-galloyl-β-D-glucopyranose (**24**): R = R′ = G, R″ = H
1,2,4,6-Tetra-*O*-galloyl-β-D-glucopyranose (**25**): R = R″ = G, R′ = H
1,2,3,4,6-Penta-*O*-galloyl-β-D-glucopyranose (**13**): R = R′ = R″ = G

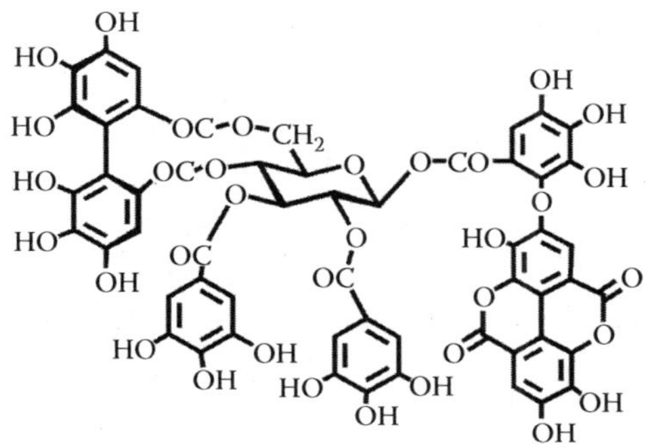

Euprostin A (**5**)

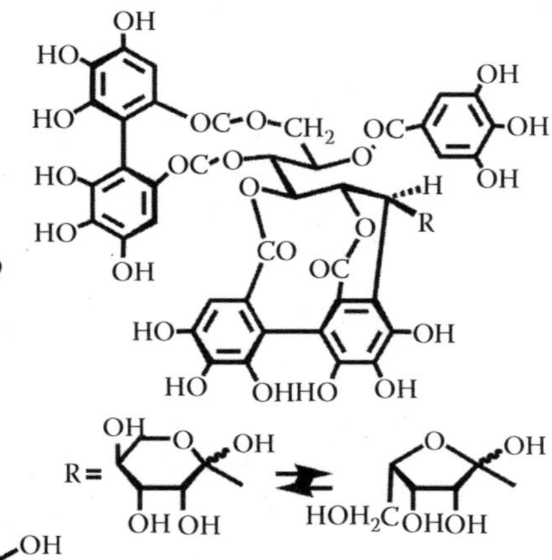

Pterocarinin A (**14**)

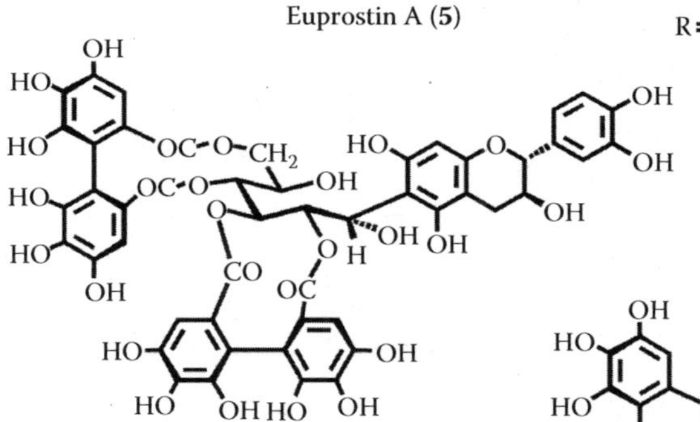

Stachyuranin B (**19**)

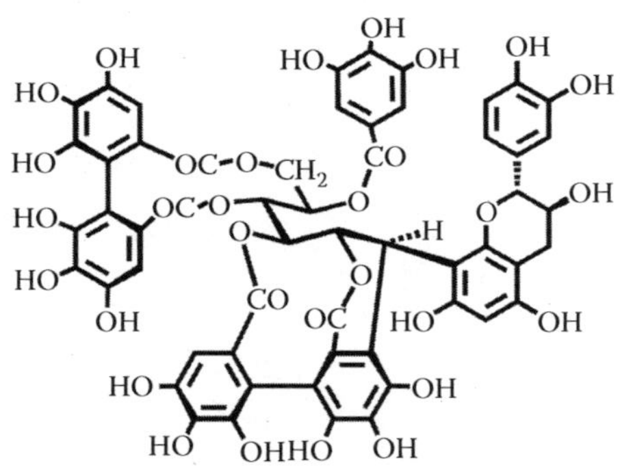

Stenophyllanin A (**20**)

FIGURE 19.3 Chemical structures of walnut polyphenols (Hydrolyzable tannin monomers-2).

glansrin A (**7**) [30], glansrin B (**8**) [30], glansrin C (**9**) [30], glansrin D (**27**) [31], glansreginin A (**33**) [31], and glansreginin B (**34**) [31] were elucidated based on spectral and chemical methods as shown in Figure 19.6

19.4 ANTIOXIDANT ACTIVITY OF POLYPHENOLS IN WALNUTS (*IN VITRO*)

The antioxidant activity of various components isolated from walnuts was evaluated using the relevant EC_{50} values from measurements of SOD-like activity and 1,1-diphenyl-2-picrylhydrazyl (DPPH)

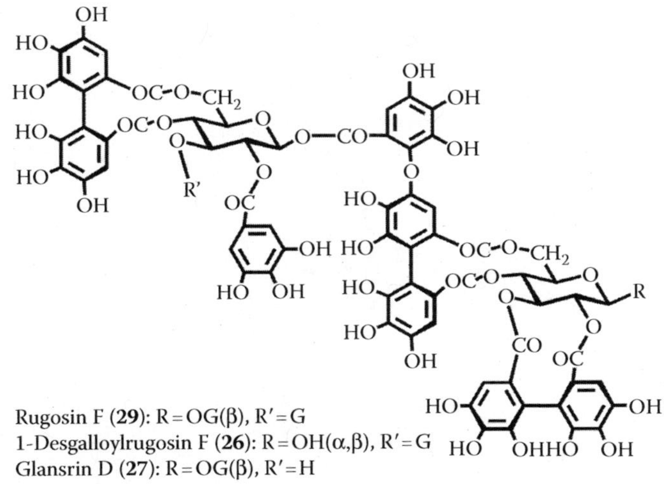

Rugosin F (**29**): R = OG(β), R′ = G
1-Desgalloylrugosin F (**26**): R = OH(α,β), R′ = G
Glansrin D (**27**): R = OG(β), R′ = H

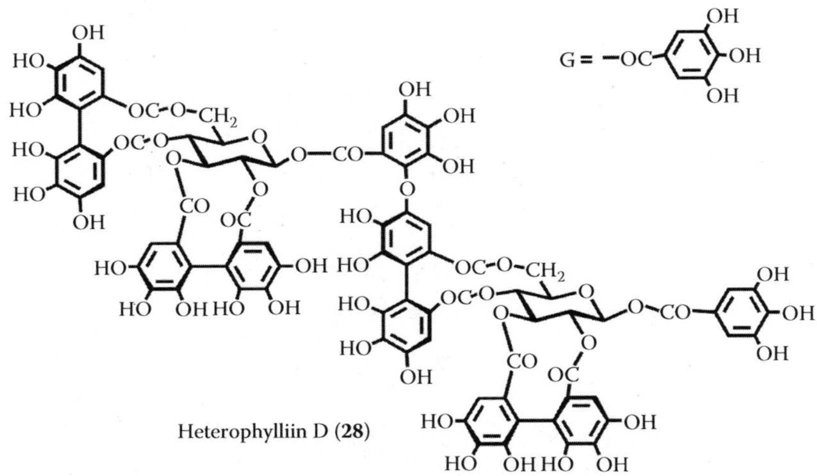

Heterophylliin D (**28**)

FIGURE 19.4 Chemical structures of walnut polyphenols (Hydrolyzable tannin dimers).

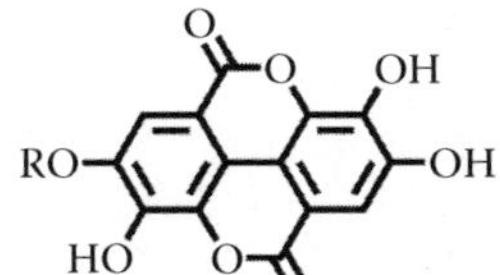

Ellagic acid (**30**): R = H
Ellagic acid 4-*O*-xylopyranoside (**31**): R = Xylose

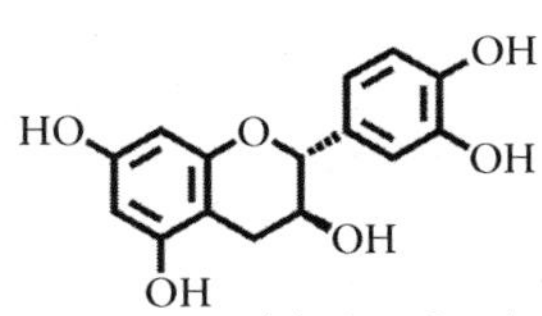

(+)-Catechin (**32**)

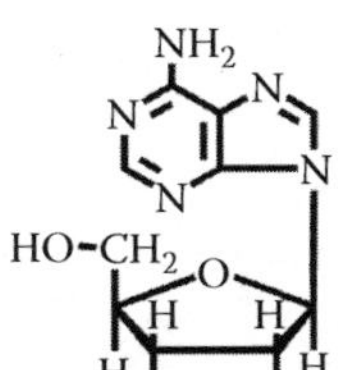

Adenosine (**36**)

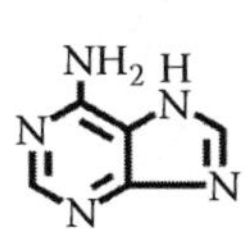

Adenine (**35**)

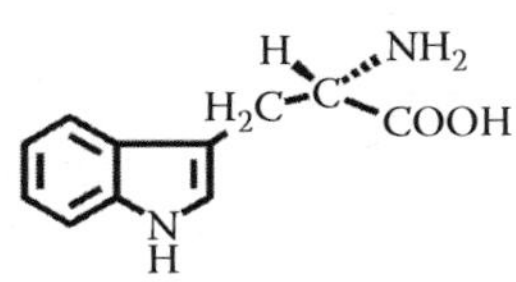

L-Tryptophan (**37**)

FIGURE 19.5 Chemical structures of low molecular weight compounds from walnuts.

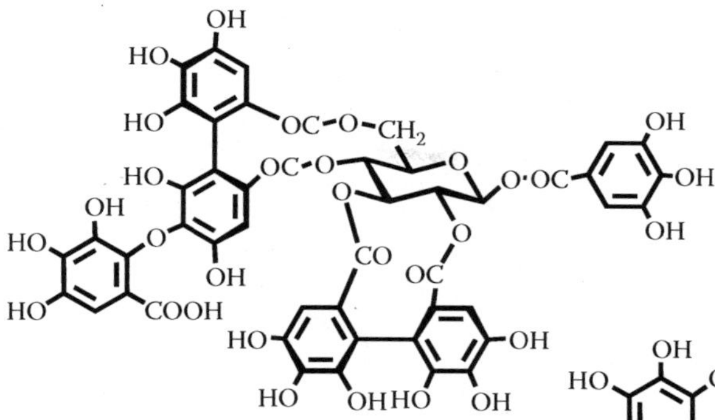

Glansrin A (**7**)

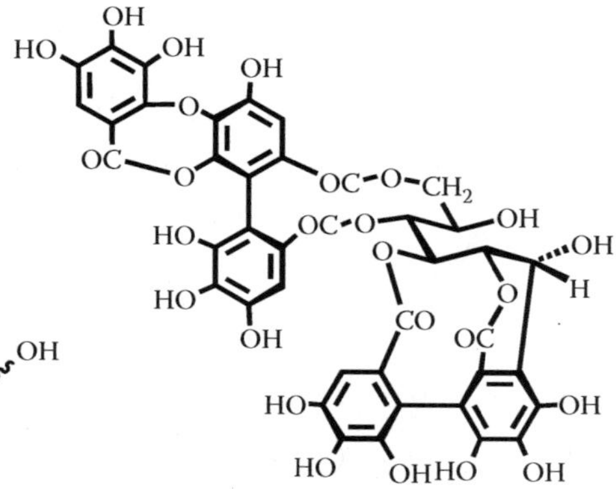

Glansrin C (**9**)

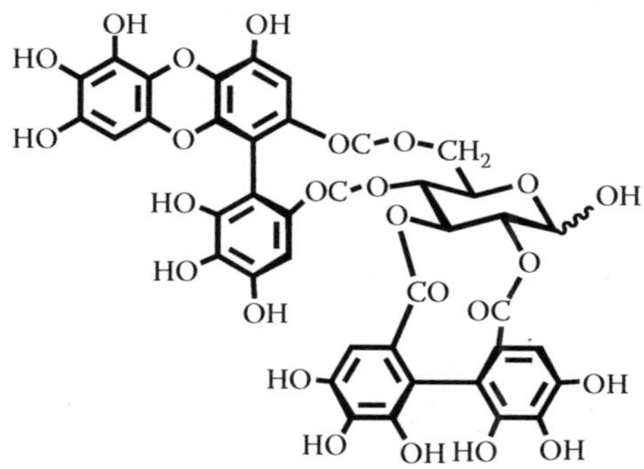

Glansrin B (**8**)

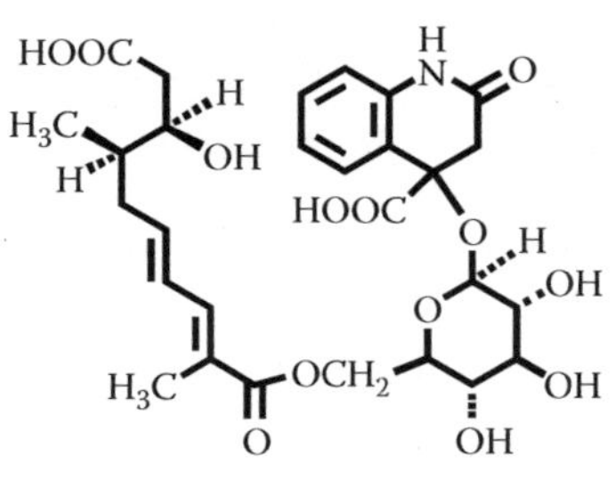

Glansreginin A (**33**)

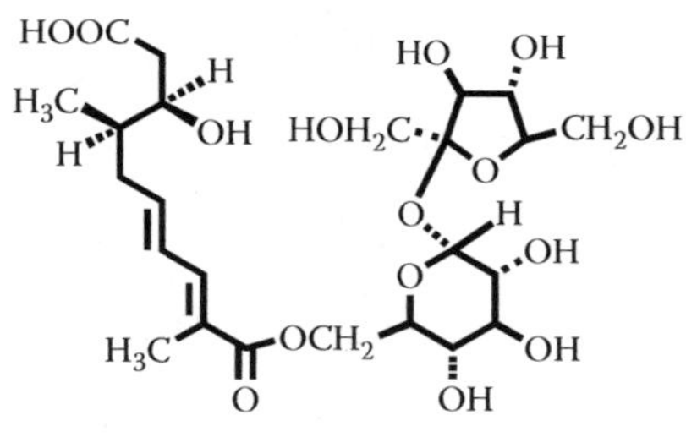

Glansreginin B (**34**)

FIGURE 19.6 Chemical structures of new compounds from walnuts.

radical-scavenging activity (Table 19.2) [30,31]. All polyphenols exhibited antioxidant activity. Gallotannins, which have several galloyl groups in the molecule, such as **13** and **24**, and ellagitannins, which have both a galloyl group and a hexahydroxyldiphenoyl (HHDP) group, such as **1**, **3**, **7**, **8**, **17**, and **22**, had the strongest activity. A complex tannin, **20**, also showed strong antioxidant activity. Ellagitannins with no galloyl group, such as **10** and **12**, had weaker effects than tannins which have galloyl groups on both SOD-like activity and DPPH radical-scavenging activity. The nonpolyphenols, such as nucleic acids and carboxylic acid derivatives, showed no antioxidant activity.

19.5 ANTIOXIDANT ACTIVITY OF WALNUT POLYPHENOLS (*IN VIVO*)

Isolation of individual compounds described above produced insufficient amounts for an *in vivo* evaluation of their antioxidant activities [32]. Instead, *in vivo* testing was carried out using a WPF,

TABLE 19.2
Antioxidant Activities of Walnut Compounds

Compound	SOD[a]-Like Activity EC_{50} (μM)	DPPH[b] Radical-Scavenging Activity EC_{50} (μM)
Hydrolyzable tannin monomers		
Casuarictin (**1**)	77.9	0.40
Casuarinin (**3**)	57.7	0.78
1,2-Di-*O*-galloyl-4,6-*O*-(*S*)-HHDP-β-D-glucopyranose (**4**)	76.3	1.27
Euprostin A (**5**)	64.6	0.81
Glansrin A (**7**)	190	0.36
Glansrin B (**8**)	41.9	0.93
Glansrin C (**9**)	21.4	0.57
2,3-*O*-(*S*)-HHDP-D-glucopyranose (**10**)	166	4.35
Isostrictinin (**11**)	47.3	1.73
Pedunculagin (**12**)	63.7	4.72
1,2,3,4,6-Penta-*O*-galloyl-β-D-glucopyranose (**13**)	50.0	0.47
Platycaryanin A (**14**)	109	0.63
Pterocarinin A (**16**)	70.2	0.44
Rugosin C (**17**)	45.3	0.34
Stenophyllanin A (**20**)	35.6	0.41
Strictinin (**21**)	48.9	2.68
Tellimagrandin I (**22**)	53.4	0.79
Tellimagrandin II (**23**)	94.8	0.44
1,2,3,6-Tetra-*O*-galloyl-β-D-glucopyranose (**24**)	76.1	0.53
1,2,4,6-Tetra-*O*-galloyl-β-D-glucopyranose (**25**)	133	0.87
Hydrolyzable tannin dimers		
1-Desgalloylrugosin F (**26**)	63.9	0.38
Glansrin D (**27**)	69.7	1.63
Heterophylliin D (**28**)	160	0.59
Rugosin F (**29**)	331	0.53
Ellagic acid derivatives		
Ellagic acid 4-*O*-xylopyranoside (**31**)	82.9	3.45
Dicarboxylic acid derivatives		
Glansreginin A (**33**)	>1000	>25
Glansreginin B (**34**)	>1000	>25
Nucleic acids		
Adenine (**35**)	>1000	>25
Adenosine (**36**)	695	>25
L-Ascorbic acid	34.6	6.25
Gallic acid	31.7	5.88

Source: From Fukuda, T., Ito, H., and Yoshida, T., *Phytochemistry*, 63, 795, 2003; Ito, H., Okuda, T., Fukuda, T., Hatano, T., and Yoshida, T., *J. Agric. Food Chem.*, 55, 672, 2007.

[a] SOD, superoxide dismutase.

[b] DPPH, 1,1-diphenyl-2-picrylhydrazyl.

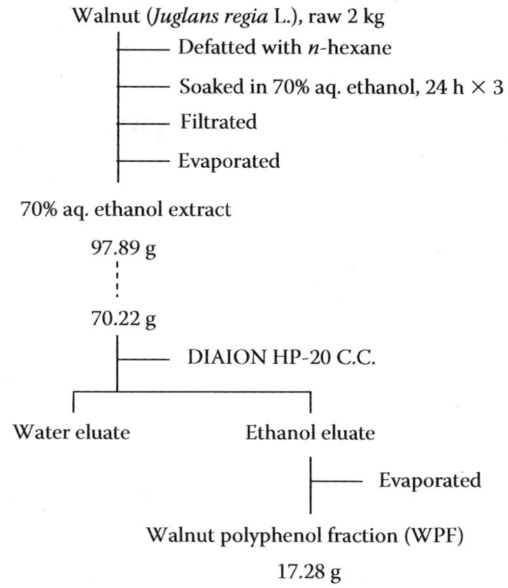

FIGURE 19.7 Preparation of WPF.

prepared as described in Figure 19.7, which was obtained in approximately 25% yield from a 70% ethanol extract. The compositions of the individual polyphenol components in WPF quantified by HPLC are shown in Table 19.3. The most abundant polyphenol was **12**, followed by **30**, **22**, and **1**. For *in vivo* evaluation, a type-2 diabetic mouse (C57BL/KsJ-db/db) was used as a hyperoxidation animal model. The experimental group of db/db mice was given WPF orally every day for 4 weeks at a dosage of 200 mg/kg. Control group mice (C57BL/KsJ-db/db) were given water instead of WPF. C57BL/KsJ-db/ + m mice were used as blank group and given water. After 4 weeks, the mice were placed in individual metabolic cages to collect their urine for 8 h, and blood was collected after overnight starvation. Enzyme-linked immunosorbent assay (ELISA) was used to measure the DNA oxidation product 8-hydroxy-2′-deoxyguanosine (8-OHdG) in the urine as a marker of *in vivo* oxidative stress [33]. As shown in Table 19.4, the level of urinary 8-OHdG modified with creatinine in

TABLE 19.3
Quantities of Polyphenols in WPF

Compound	%
Casuarictin (**1**)	4.1
Casuarinin (**3**)	1.0
Pedunculagin (**12**)	16.0
Rugosin C (**17**)	1.8
Tellimagrandin I (**22**)	6.6
Tellimagrandin II (**23**)	1.2
Ellagic acid (**30**)	15.8

Source: From Fukuda, T., Ito, H., and Yoshida, T., *BioFactors*, 21, 251, 2004. With permission.
Note: WPF, walnut polyphenol-rich fraction.

TABLE 19.4
Effect of WPF in Type-2 Diabetic Mice

				In Serum			In Urine
Group	Mouse	Number (*n*)	Body Weight (g)	Glucose (mg/dL)	Cholesterol (mg/dL)	TAG (mg/dL)	8-OHdG/CRE (ng/mg CRE)
Blank	db/ + m	5	25.62 ± 0.78**	70.0 ± 8.2**	63.5 ± 3.2**	69.3 ± 9.3**	84.0 ± 12.4**
Control	db/db	8	37.35 ± 2.44	437.4 ± 124.6	103.8 ± 19.1	177.0 ± 59.1	122.5 ± 25.5
WPF	db/db	6	36.48 ± 2.27	427.4 ± 124.0	106.7 ± 15.0	121.6 ± 37.0**	94.8 ± 24.9*

Source: From Fukuda, T., Ito, H., and Yoshida, T., *BioFactors*, 21, 251, 2004. With permission.

Note: Data are expressed as means ± standard deviation; TAG, triacylglycerols; WPF, walnut polyphenol-rich fraction. *$P < 0.05$, **$P < 0.01$, compared with the control group.

the control group of type-2 diabetic mice was significantly higher than that of the nondiabetic blank group. Conversely, compared to the control group, type-2 diabetic mice given WPF orally showed a significant reduction in 8-OHdG. The serum triacylglycerols (TAG) concentrations were also improved after WPF administration. These results show that walnut polyphenols had antioxidant activity not only *in vitro*, but *in vivo* as well.

19.6 α-GLUCOSIDASE AND α-AMYLASE INHIBITORY ACTIVITIES OF WALNUT POLYPHENOLS (*IN VITRO*)

Plant polyphenols, such as galloylglucose and green tea catechins, are known to inhibit α-amylase and/or α-glucosidase activity [34,35]. The inhibitory activity of walnut polyphenols on enzymes was investigated relative to degradation of carbohydrates. Specifically, the inhibitory effect of walnut polyphenol constituents on the α-glucosidases like sucrase and maltase was examined. Inhibition of α-glucosidase activity was measured according to a described method [9]. A crude enzyme solution was prepared using rat intestinal acetone powder. Sucrose or maltose was used as a substrate. Table 19.5 shows IC_{50} values for the inhibition of enzymatic activity by each compound. Compound **23** demonstrated stronger maltase inhibitory activity than the positive control (–)-epigallocatechin gallate (EGCG, from green tea). Similarly, inhibition of α-amylase activity by walnut polyphenols was also measured. α-Amylase type VI-B, from porcine pancreas (Sigma A-3176), was used as the enzyme and soluble starch was used as the substrate. The IC_{50} values for the inhibition of enzymatic activity by each compound are shown in Table 19.5. Compounds **1** and **23** showed 50% inhibition of α-amylase at a concentration of 3.5 and 2.0 μM, respectively, which was stronger than that of EGCG. Tannins are known to generally bind proteins to form complexes and the inhibitory effects demonstrated by walnut polyphenols were attributable to this property.

19.7 HYPOGLYCEMIC EFFECT OF WALNUT POLYPHENOL FRACTION IN NORMAL MICE

The inhibition of α-glucosidase and α-amylase activity was investigated *in vivo* using WPF. Normal mice (ddy, Clea Japan, Inc., Tokyo, Japan) were subjected to overnight fasting and blood sugar levels at the beginning were measured. The WPF was suspended in water and given orally at 500, 200, or 0 mg/kg. Thirty minutes later, a water solution of starch or sucrose was orally administered at 2 g/kg body weight. Blood sugar was measured after 30, 60, and 120 min. The results are shown

TABLE 19.5
Inhibition of Maltase, Sucrase, and α-Amylase by Walnut Polyphenols

	IC_{50} (μM)		
	α-Glucosidase		
Compound	**Maltase**	**Sucrase**	**α-Amylase**
Casuarictin (**1**)	320.3	192.2	3.5
Casuarinin (**3**)	427.1	49.1	19.2
2,3-*O*-(*S*)-HHDP-D-glucopyranose (**10**)	1389.2	1720.9	269.5
Isostrictinin (**11**)	646.2	488.6	97.7
Pedunculadin (**12**)	637.3	892.3	133.8
Rugosin C (**17**)	543.1	289.7	15.4
Strictinin (**21**)	409.8	315.2	83.5
Tellimagrandin I (**22**)	419.5	52.1	16.5
Tellimagrandin II (**23**)	458.1	26.6	2.0
EGCG	165.8	28.4	5.0

Note: EGCG, (–)-epigallocatechin gallate.

in Figure 19.8. In the sucrose-loaded mice (Figure 19.8A), the blood sugar in the WPF-administered groups decreased significantly after 30 min, then reached a maximum between 30 and 60 min, suggesting that WPF delayed sugar absorption. The starch infusion (Figure 19.8B) resulted in a significant drop in blood sugar after 30 min. These results show that walnut polyphenols acted to inhibit excessive blood sugar elevation after a meal.

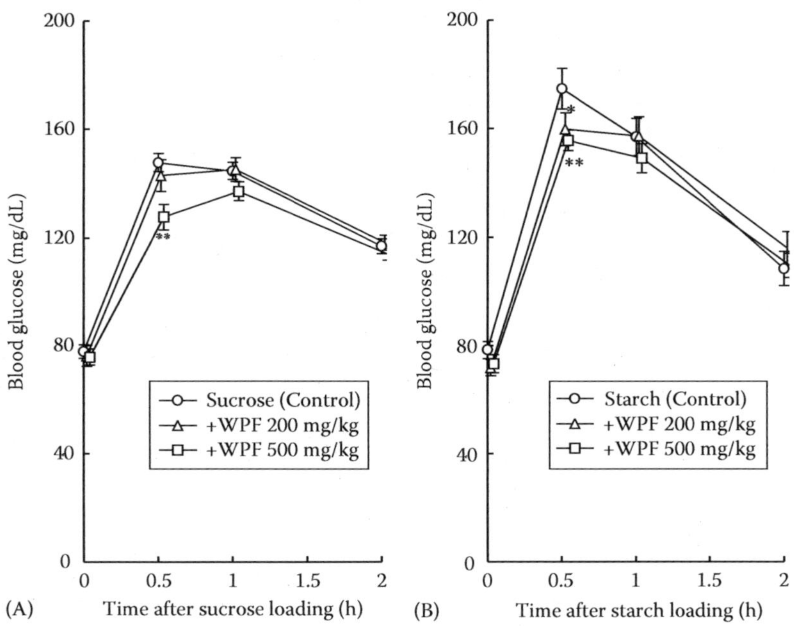

FIGURE 19.8 WPF-mediated inhibition of increases in blood glucose levels after sucrose (A) or starch (B) loading in normal mice. Results are expressed as means ± standard deviation ($n = 23$), ** $P < 0.01$, * $P < 0.05$.

19.8 QUANTITATIVE ANALYSIS OF WALNUT POLYPHENOL COMPOUNDS BY HPLC

Five commercial walnut samples were compared by HPLC analysis to measure the presence of seven major polyphenol compounds and a dicarboxylic acid derivative (**1**, **3**, **12**, **17**, **22**, **23**, **30**, and **33**). Each peak of HPLC profile was identified by comparing the retention time and ultraviolet–visible (UV) spectral pattern with those of standard compounds that were isolated from walnuts. Using calibration curves, each compound was analyzed quantitatively (Table 19.6). The results indicate virtually no difference in the content of the individual polyphenol components between raw and roasted walnuts. The results also revealed that polyphenol content in raw samples was higher in the U.S. walnuts than in Chinese walnuts. It was interesting to note that no polyphenols were detected in walnuts from which the pellicle (skin) was removed. This shows that polyphenols in walnut are present only in the pellicle portion, although a nonpolyphenol (**33**) was detected in pellicle-free walnuts.

19.9 HEALTH EFFECTS OF WALNUTS

In 2004, the U.S. Food and Drug Administration (FDA) cleared the qualified health claim, "Supportive but not conclusive research shows that eating 1.5 ounces per day of walnuts, as part of a low saturated fat and low cholesterol diet and not resulting in increased caloric intake, may reduce the risk of coronary heart disease" [36]. Several recent studies have suggested a beneficial relationship between walnuts and CHD. In this section, selected studies on the health effects of walnuts are presented.

Sabaté et al. [37] reported the effects of walnuts on serum lipid levels and blood pressure in normal men. Eighteen healthy men (mean age 30 years) were randomly assigned to a reference diet and a walnut diet (28 g of walnuts/serving, three servings/day), each to be consumed for 4 weeks in a crossover design. With the walnut diet, the mean total cholesterol concentration was 0.58 mmol/L lower than the mean concentration with the reference diet, and the LDL and high-density lipoprotein (HDL) cholesterol concentrations were 0.47 mmol/L ($P < 0.001$) and 0.06 mmol/L ($P = 0.01$) lower, respectively. The ratio of LDL cholesterol to HDL cholesterol was also lowered significantly ($P < 0.001$) by the walnut diet. Blood pressures did not differ between the two diets.

Iwamoto et al. [4] studied the effects of walnut consumption on serum lipids and blood pressure in Japanese subjects. Twenty men (mean age 23.8 years) and 20 women (mean age 23.6 years) were

TABLE 19.6
Quantities Analysis of Polyphenol Compounds in Walnuts

	Compound (%, w/w)							
Sample	**1**	**3**	**12**	**17**	**22**	**23**	**30**	**33**
Walnuts, American, raw	0.053	0.023	0.207	0.046	0.038	0.024	0.039	0.032
Walnuts, American, roasted	0.035	0.018	0.162	0.033	0.032	0.015	0.035	0.059
Walnuts, American, raw (pellicle free)	—	—	—	—	—	—	—	0.040
Walnuts, Chinese, raw	0.016	0.007	0.064	0.007	0.008	0.008	0.015	0.018
Walnuts, Japanese, raw	0.009	0.013	0.446	0.009	0.012	0.006	0.207	—

Note: HPLC conditions: **column:** Develosil RPAQUEOUS, 4.6 × 250 mm, Nomura Chemical Co., Ltd.; **mobile phase:** 10 mM H_3PO_4/10 mM KH_2PO_4/CH_3CN, A (45:45:10), B (30:30:40); **gradient:** 0–10 min, A100%, 10–40 min, A100% to B100%, 40–55 min, B100%; **flow rate:** 1 mL/min; **detection:** UV 220–400 nm, max; **column temperature:** 40°C.

randomly assigned to the average Japanese diet (reference diet) and a walnut diet (43 to 57 g of walnuts/day), each to be consumed for 4 weeks in a crossover design. Total cholesterol concentration was 0.16 mmol/L lower for men ($P = 0.05$) and 0.21 mmol/L lower for women ($P < 0.01$) when they consumed the walnut diet compared to that of the reference diet. The LDL cholesterol concentration was 0.22 mmol/L lower for women, a statistically significant effect ($P < 0.01$). The ratio of LDL cholesterol to HDL cholesterol and the apolipoprotein B concentration were lowered significantly ($P < 0.01$) by the walnut diet. Blood pressures did not differ between the two diets.

Zambón et al. [5] reported that walnuts reduced serum cholesterol levels in hypercholesterolemic men and women. Twenty-seven men (mean age 59 years) and 28 women (mean age 53 years) with polygenic hypercholesterolemia were randomly assigned to a cholesterol-lowering Mediterranean diet and a diet of similar energy and fat content in which walnuts replaced 35% of the energy obtained from monounsaturated fatty acids (MUFA), each to be consumed for 6 weeks in a crossover design. Walnuts were provided daily in amounts varying from 41 to 56 g. Forty-nine persons completed the trial. Compared with the Mediterranean diet, the walnut diet reduced mean changes of 0.28 mmol/L ($P < 0.001$) in total cholesterol concentration, 0.29 mmol/L ($P < 0.001$) in LDL cholesterol concentration, and 0.02 g/L ($P = 0.042$) in lipoprotein (a) concentration. These investigators concluded that substituting walnuts for part of the MUFA in a cholesterol-lowering Mediterranean diet further reduced total and LDL cholesterol concentrations in men and women with hypercholesterolemia.

Muñoz et al. [6] reported that a walnut-enriched diet increased the association of LDL from hypercholesterolemic men with human hepatoma HepG2 cells. Ten men with polygenic hypercholesterolemia (mean age 59 years) were randomly assigned to a cholesterol-lowering Mediterranean diet and a diet of similar energy and fat content in which walnuts replaced 35% of the energy obtained from MUFA (41 to 56 g of walnuts/day), each to be consumed for 6 weeks in a crossover design. Compared with the control diet, the walnut diet reduced serum LDL cholesterol by 6% ($P = 0.087$). In comparison with LDL obtained during the control diet, LDL obtained during the walnut diet showed a 50% increase in association rates to the LDL receptor in HepG2 cells.

Ros et al. [7] studied the effects of a walnut diet on endothelial function in hypercholesterolemic subjects. Twenty-one hypercholesterolemic men and women were randomly assigned to a cholesterol-lowering Mediterranean diet and a walnut diet. In the walnut diet, walnuts replaced 32% of the energy obtained from MUFA in the control diet (40 to 65 g of walnuts/day). Participants followed each diet for 4 weeks. Eight men and 12 women (mean age 55 years) completed the trial. Compared with the control diet, the walnut diet improved endothelium-dependent vasodilatation (from 3.6% to 5.9%, $P = 0.043$) and reduced levels of vascular cell adhesion molecule-1 (from 465 to 378 μmol/L, $P = 0.045$). The walnut diet reduced both total (4.4%, $P = 0.017$) and LDL cholesterol (6.4%, $P = 0.010$) concentrations significantly.

Tapsell et al. [38] studied the effect of a moderate-fat diet inclusive of walnuts on blood lipid profiles in patients with type 2 diabetes. Patients (men and women) were randomized into three dietary advice groups each with 30% energy as fat: low fat ($n = 21$, mean age 60.48 years), modified low fat ($n = 20$, mean age 59.30 years), and modified low fat inclusive of 30 g of walnuts per day ($n = 17$, mean age 57.71 years). Body weight, percent body fat, blood lipids, hemoglobin A1C (HbA1C), total antioxidant capacity, and erythrocyte fatty acid levels were measured at months 0, 3, and 6. The walnut group achieved a significantly greater increase in HDL cholesterol-to-total cholesterol ratio ($P = 0.049$) and HDL cholesterol concentrations ($P = 0.046$) than the other groups did. A 10% reduction in LDL cholesterol was also achieved in the walnut group.

Many human walnut clinical intervention trials have been reported in addition to the studies discussed here; they are summarized in Feldman's review [39].These clinical trials support the hypothesis that a walnut diet improves lipid metabolism and reduces the risk of CHD in normal humans, humans with polygenic hypercholesterolemia, and those with type 2 diabetes. Scientists hypothesize that the effects of diets containing walnuts are due to the amount of ω3 fatty acids, such as α-linolenic acid, and the ratio of ω6 to ω3 fatty acids in the diets.

19.10 CONCLUSION

Thirty-seven compounds were isolated from walnut extracts and their structures were characterized. All phenolic compounds had SOD-like activity and a radical-scavenging effect against DPPH. Oral administration of WPF extract caused a significant decrease in the level of urinary 8-OHdG, which is an *in vivo* marker of oxidative stress. Walnut polyphenols also had an inhibitory α-glucosidase and α-amylase activity both *in vitro* and *in vivo*. Furthermore, walnut polyphenols were detected only in the pellicle portion of the nut, suggesting that polyphenol-rich skin may protect the fatty acids from oxidation. The hydrolyzable tannins present in walnuts include those possessing antitumor [40], antivirus [41,42], and anti-Pyroli effects [43], as well as the suppression of histamine release *in vitro*[44]; therefore, data provided here add important and basic information suggesting that walnuts are beneficial to the health and may serve as functional foods for human health promotion. Nonetheless, allergic reactions to nut by certain individuals should not be forgotten.

ACKNOWLEDGMENT

The author would like to thank Prof. Takashi Yoshida (Faculty of Pharmaceutical Sciences, Matsuyama University, Ehime, Japan), Prof. Tsutomu Hatano, and Dr. Hideyuki Ito (Faculty of Pharmaceutical Sciences, Okayama University, Okayama, Japan) for their valuable suggestions and comments.

REFERENCES

1. Duke, J.A., *Handbook of Nuts*, CRC Press, London, 1989.
2. *Dictionary of Chinese Materia Mediaca Japanese Edition*, vol. 2, Shanghai Scientific & Technical Publishers, Shanghai, Shogakukan, Tokyo, 1985.
3. Kagawa, Y., *Standard Tables of Food Composition in Japan*, Kagawa Nutrition University Press, Tokyo, 2001.
4. Iwamoto, M., Sato, M., Kono, M., Hirooka, Y., Sakai, K., Takeshita, A., and Imaizumi, K., Walnuts lower serum cholesterol in Japanese men and women, *J. Nutr.*, 130, 171–176, 2000.
5. Zambón, D., Sabaté, J., Muñoz, S., Campero, B., Casals, E., Merlos, M., Laguna, J.C., and Ros, E., Substituting walnuts for monounsaturated fat improves the serum lipid profile of hypercholesterolemic men and women, *Ann. Intern. Med.*, 132, 538–546, 2000.
6. Muñoz, S., Merlos, M., Zambón, D., Rodríguez, C., Sabaté, J., Ros, E., and Laguna, J.C., Walnut-enriched diet increases the association of LDL from hypercholesterolemic men with human HepG2 cells, *J. Lipid Res.*, 42, 2069–2076, 2001.
7. Ros, E., Núñez, I., Pérez-Heras, A., Serra, M., Gilabert, R., Casals, E., and Deulofeu, R., A walnut diet improves endothelial function in hypercholesterolemic subjects, *Circulation*, 109, 1609–1614, 2004.
8. Anderson, K.J., Teuber, S.S., Gobeille, A., Cremin, P., Waterhouse, A.L., and Steinberg, F.M., Walnut polyphenolics inhibit *in vitro* human plasma and LDL oxidation, *J. Nutr.*, 131, 2837–2842, 2001.
9. Deguchi, Y., Osada, K., Uchida, K., Kimura, H., Yoshikawa, M., Kudo, T., and Yasui, H., Effects of extract of guava leaves on the development of diabetes in the db/db mouse and on the postprandial blood glucose of human subjects, *Nippon Nogeikagaku Kaishi*, 72, 923–931, 1998.
10. Hatano, T., Edamatsu, R., Hiramatsu, M., Mori, A., Fujita, Y., Yasuhara, T., Yoshida, T., and Okuda, T., Effects of the interaction of tannins with co-existing substances. VI. Effects of tannins and related polyphenols on superoxide anion radical, and on 1,1-diphenyl-2-picrylhydrazyl radical, *Chem. Pharm. Bull.*, 37, 2016–2021, 1989.
11. Oyanagi, Y., Reevaluation of assay methods and establishment of kit for superoxide dismutase activity, *Anal. Biochem.*, 142, 290–296, 1984.
12. Okuda, T., Yoshida, T., Kuwahara, M., and Yazaki, K., Structures of pedunculagin, casuarictin, strictinin, casuarinin, casuariin and stachyurin, *J. Chem. Soc. Perkin Trans.*, 1, 1765–1772, 1983.
13. Koppaka, V.R., Toxic principle of *Hippomane mancinella*. II. Structure of hipppomanin A, *Lloydia*, 40, 169–171, 1977.
14. Namba, O., Yoshida, T., Liu, Y., Chen, L., and Okuda, T., Prostratins A and B: New hydrolyzable tannin monomer and trimer from *Euphorbia prostrate* and *Loropetalum chinense*, *Planta Med.*, 56, 587, 1990.

15. Yoshida, T., Maruyama, Y., Memon, M.U., Shingu, T., and Okuda, T., Gemins D, E and F, ellagitannins from *Geum japonicum*, *Phytochemistry*, 24, 1041–1046, 1985.
16. Seikel, M.K. and Hillis, W.E., Hydrolysable tannins of *Eucalyptus delegatensis* wood, *Phytochemistry*, 9, 1115–1128, 1970.
17. Okuda, T., Yoshida, T., Hatano, T., Yazaki, K., and Ashida, M., Ellagitannins of the Casuarinaceae, Stachyuraceae and Myrtaceae, *Phytochemistry*, 21, 2871–2874, 1982.
18. Haddock, E.A., Gupta, R.K., Al-Shafi, S.M.K., Haslam, E., and Magnolato, D., The metabolism of gallic acid and hexahydroxydiphenoic acid in plants. Part 1. Introduction. Naturally occurring galloyl esters, *J. Chem. Soc. Perkin Trans.*, 1, 2515–2524, 1982.
19. Tanaka, T., Kirihara, S., Nonaka, G., and Nishioka, I., Tannins and related compounds. CXXIV. Five new ellagitannins, Platycaryanins A, B, C, and D, and Platycariin, and a new complex tannin, Strobilanin, from the fruits and bark of *Platycarya strobilacea* SIEB. *et* ZUCC., and biomimetic synthesis of *C*-glycosidic ellagitannins from glucopyranose-based ellagitannins, *Chem. Pharm. Bull.*, 41, 1708–1716, 1993.
20. Hatano, T., Yazaki, K., Okonogi, A., and Okuda, T., Tannins of Stachyurus species. II. Praecoxins A, B, C and D, four new hydrolyzable tannins from *Stachyurus praecox* leaves, *Chem. Pharm. Bull.*, 39, 1689–1693, 1991.
21. Nonaka, G., Ishimaru, K., Azuma, R., Ishimatsu, M., and Nishioka, I., Tannins and related compounds. LXXXV. Structures of novel *C*-glycosidic ellagitannins, Grandinin, and Pterocarinins A and B, *Chem. Pharm. Bull.*, 37, 2071–2077, 1989.
22. Hatano, T., Ogawa, N., Yasuhara, T., and Okuda, T., Tannins of Rosaceous plants. VIII. Hydrolyzable tannin monomers having a valoneoyl group from flower petals of *Rosa rugosa* THUNB., *Chem. Pharm. Bull.*, 38, 3308–3313, 1990.
23. Han, L., Hatano, T., Okuda, T., and Yoshida, T., Tannins of Stachyurus species. III. Stachyuranins A, B and C, three new complex tannins from *Stachyurus praecox* leaves, *Chem. Pharm. Bull.*, 43, 2109–2114, 1995.
24. Nonaka, G., Nishimura, H., and Nishioka, I., Tannins and related compounds. Part 26. Isolation and structures of Stenophyllanins A, B, and C, novel tannins from *Quercus stenophylla*, *J. Chem. Soc. Perkin Trans.*, 1, 163–172, 1985.
25. Wilkins, C.K. and Bohm, B.A., Ellagitannins from *Tellima grandiflora*, *Phytochemistry*, 15, 211–214, 1976.
26. Lee, M.-W., Tanaka, T., Nonaka, G.-I., and Nishioka, I., Hirsunin, an ellagitannin with a diarylheptanoid moiety, from *Alnus hirsute* var. *microphylla*, *Phytochemistry*, 31, 967–970, 1992.
27. Yoshida, T., Jin, Z.-X., and Okuda, T., Heterophylliins A, B, C, D, and E, ellagitannin monomers and dimers from *Corylus heterophylla* FISCH., *Chem. Pharm. Bull.*, 39, 49–54, 1991.
28. Hatano, T., Ogawa, T., Shingu, T., and Okuda, T., Tannins of Rosaceous plants. IX. Rugosins D, E, F and G, dimeric and trimeric hydrolyzable tannins with valoneoyl group(s), from flower petals of *Rosa rugosa* THUNB., *Chem. Pharm. Bull.*, 38, 3341–3346, 1990.
29. Tanaka, T., Jiang, Z.-H., and Kouno, I., Distribution of ellagic acid derivatives and a diarylheptanoid in wood of *Platycarya strobilacea*, *Phytochemistry*, 47, 851–854, 1998.
30. Fukuda, T., Ito, H., and Yoshida, T., Antioxidative polyphenols from walnuts (*Juglans regia* L.), *Phytochemistry*, 63, 795, 2003.
31. Ito, H., Okuda, T., Fukuda, T., Hatano, T., and Yoshida, T., Two novel dicarboxylic acid derivatives and a new dimeric hydrolyzable tannin from walnuts, *J. Agric. Food Chem.*, 55, 672–679, 2007.
32. Fukuda, T., Ito, H., and Yoshida, T., Effect of the walnut polyphenol fraction on oxidative stress in type 2 diabetes mice, *BioFactors*, 21, 251–253, 2004.
33. Leinonen, J., Lehtimäki, T., Toyokuni, S., Okada, K., Tanaka, T., Hiai, H., Ochi, H., Laippala, P.,Rantalaiho, V., Wirta, O., Pasternack, A., and Alho, H., New biomarker evidence of oxidative DNA damage in patients with non-insulin-dependent diabetes mellitus, *FEBS Lett.*, 417, 150–152, 1997.
34. Toda, M., Kawabata, J., and Kasai, T., Inhibitory effects of ellagi- and gallotannins on rat intestinal α-glucosidase complexes, *Biosci. Biotechnol. Biochem.*, 65, 542–547, 2001.
35. Hara, Y. and Honda, M., The inhibition of α-amylase by tea polyphenols, *Agric. Biol. Chem.*, 54, 1939–1945, 1990.
36. Food and Drug Administration (FDA), Qualified Health Claims: Letter of Enforcement Discretion—Walnuts and Coronary Heart Disease, Docket No 02P-0292, Food and Drug Administration, Washington, DC, 2004.
37. Sabaté, J., Fraser, G.E., Burke, K., Knutsen, S.F., Bennett, H., and Lindsted, K.D., Effects of walnuts on serum lipid levels and blood pressure, *N. Engl. J. Med.*, 328, 603–607, 1993.

38. Tapsell, L.C., Gillen, L.J., Patch, C.S., Batterham, M., Owen, A., Baré, M., and Kennedy, M., Including walnuts in a low-fat/modified-fat diet improves HDL cholesterol-to-total cholesterol ratios in patients with type 2 diabetes, *Diabetes Care*, 27, 2777–2783, 2004.
39. Feldman, E.B., The scientific evidence for a beneficial health relationship between walnuts and coronary heart disease, *J. Nutr.*, 132, 1062S–1101S, 2002.
40. Yoshizawa, S., Horiuchi, T., Suganuma, M., Nishiwaki, S., Yatsunami, J., Okabe, S., Okuda, T., Muto, Y., Frenkel, K., Troll, W., and Fujiki, H., *Phenolic Compounds in Food and Their Effects on Health. II. Antioxidant and Cancer Prevention*, Huang, M.-T., Ho, C.-T., and Lee, C.Y., Eds., ACS Symposium Series No. 507, American Chemical Society, Washington, DC, 1992, pp. 316–325.
41. Fukuchi, K., Sakagami, H., Okuda, T., Hatano, T., Tanuma, S., Kitajima, K., Inoue, Y., Inoue, S., Ichikawa, S., Nonoyama, M., and Konno, K., Inhibition of herpes simplex virus infection by tannins and related compounds, *Antiviral Res.*, 11, 285–297, 1989.
42. Nakashima, H., Murakami, T., Yamamoto, N., Sakagami, H., Tanuma, S., Hatano, T., Yoshida, T., and Okuda, T., Inhibition of human immunodeficiency viral replication by tannins and related compounds, *Antiviral Res.*, 18, 91–103, 1992.
43. Funatogawa, K., Hayashi, S., Shimomura, H., Yoshida, T., Hatano, T., Ito, H., and Hirai, Y., Antibacterial activity of hydrolyzable tannins derived from medicinal plants against *Helicobacter pylori*, *Microbiol. Immunol.*, 48, 251–261, 2004.
44. Kanoh, R., Hatano, T., Ito, H., Yoshida, T., and Akagi, M., Effects of tannins and related polyphenols on superoxide-induced histamine release from rat peritoneal mast cells, *Phytomedicine*, 7, 297–302, 2000.

Index

A

B

Q

R

S

T

U

V

W